AF841060

Tensor Categories

Mathematical
Surveys
and
Monographs

Volume 205

Tensor Categories

Pavel Etingof
Shlomo Gelaki
Dmitri Nikshych
Victor Ostrik

American Mathematical Society
Providence, Rhode Island

2010 *Mathematics Subject Classification.* Primary 17B37, 18D10, 19D23, 20G42.

For additional information and updates on this book, visit
www.ams.org/bookpages/surv-205

Library of Congress Cataloging-in-Publication Data

Tensor categories / Pavel Etingof, Shlomo Gelaki, Dmitri Nikshych, Victor Ostrik.
pages cm. — (Mathematical surveys and monographs ; volume 205)
Includes bibliographical references and index.
ISBN 978-1-4704-2024-6 (alk. paper)
1. Algebraic topology. 2. Tensor fields. 3. Hopf algebras. I. Etingof, P. I. (Pavel I.), 1969–
II. Gelaki, Shlomo, 1964– III. Nikshych, Dmitri, 1973– IV. Ostrik, Victor.

QA612.T46 2015
512′.57—dc23

2015006773

AMS softcover ISBN: 978-1-4704-3441-0

∞ The paper used in this book is acid-free and falls within the guidelines
established to ensure permanence and durability.
Visit the AMS home page at `http://www.ams.org/`

10 9 8 7 6 5 4 3 2 1 21 20 19 18 17 16

Dedicated to our children

Miriam and Ariela Etingof
Hadar and Klil Gelaki
Timofei, Daria, and Nadezhda Nikshych
and Tatiana, Valentina, and Yuri Ostrik

Contents

Preface

> Был один рыжий человек, у которого не было глаз и ушей. У него не было и волос, так что рыжим его называли условно. Говорить он не мог, так как у него не было рта. Носа тоже у него не было. У него не было даже рук и ног. И живота у него не было, и спины у него не было, и хребта у него не было, и никаких внутренностей у него не было. Ничего не было! Так что непонятно, о ком идет речь. Уж лучше мы о нем не будем больше говорить.
>
> Д. Хармс, Голубая тетрадь но. 10, 1937

> There was a red-haired man who had no eyes or ears. Neither did he have any hair, so he was called red-haired by convention. He couldn't speak, since he didn't have a mouth. Neither did he have a nose. He didn't even have any arms or legs. He had no stomach and he had no back and he had no spine and he had no innards whatsoever. He had nothing at all! Therefore there's no knowing whom we are even talking about. It's really better that we don't talk about him any more.
>
> D. Harms, Blue notebook # 10, 1937

Tensor categories should be thought of as counterparts of rings in the world of categories.[1] They are ubiquitous in noncommutative algebra and representation theory, and also play an important role in many other areas of mathematics, such as algebraic geometry, algebraic topology, number theory, the theory of operator algebras, mathematical physics, and theoretical computer science (quantum computation).

The definition of a monoidal category first appeared in 1963 in the work of Mac Lane [**Mac1**], and later in his classical book [**Mac2**] (first published in 1971). Mac Lane proved two important general theorems about monoidal categories – the coherence theorem and the strictness theorem, and also defined symmetric and

[1]Philosophically, the theory of tensor categories may perhaps be thought of as a theory of vector spaces or group representations without vectors, similarly to how ordinary category theory may be thought of as a theory of sets without elements. As seen from the epigraph, this idea, as well as its dismissal as "abstract nonsense" common in early years of category theory, were discussed in Russian absurdist literature several years before the foundational papers of Mac Lane and Eilenberg on category theory (1942-1945).

braided monoidal categories. Later, Saavedra-Rivano in his thesis under the direction of Grothendieck [**Sa**], motivated by the needs of algebraic geometry and number theory (more specifically, the theory of motives), developed a theory of Tannakian categories, which studies symmetric monoidal structures on abelian categories (the prototypical example being the category of representations of an algebraic group). This theory was simplified and further developed by Deligne and Milne in their classical paper [**DelM**]. Shortly afterwards, the theory of tensor categories (i.e., monoidal abelian categories) became a vibrant subject, with spectacular connections to representation theory, quantum groups, infinite dimensional Lie algebras, conformal field theory and vertex algebras, operator algebras, invariants of knots and 3-manifolds, number theory, etc., which arose from the works of Drinfeld, Moore and Seiberg, Kazhdan and Lusztig, Jones, Witten, Reshetikhin and Turaev, and many others. Initially, in many of these works tensor categories were merely a tool for solving various concrete problems, but gradually a general theory of tensor categories started to emerge, and by now there are many deep results about properties and classification of tensor categories, and the theory of tensor categories has become fairly systematic. The goal of this book is to provide an accessible introduction to this theory.

We should mention another major source of inspiration for the theory of tensor categories, which is the theory of Hopf algebras. The notion of a Hopf algebra first appeared in topology (more specifically, in the work [**Ho**] of Hopf in 1941, as an algebraic notion capturing the structure of cohomology rings of H-spaces, in particular, Lie groups). Although classical topology gives rise only to cocommutative Hopf algebras, in the 1960s operator algebraists and ring theorists (notably G. Kac) became interested in noncommutative and noncocommutative Hopf algebras and obtained the first results about them. In 1969 Sweedler wrote a textbook on this subject [**Sw**], proving the first general results about Hopf algebras. In the 1970s and 1980s a number of fundamental general results were proved about finite dimensional Hopf algebras, notably Radford's formula for the 4th power of the antipode [**Ra2**], the theorems of Larson and Radford on semisimple Hopf algebras [**LaR1, LaR2**], the freeness theorem of Nichols and Zoeller [**NicZ**], and the work of Nichols on what is now called Nichols algebras (which give rise to pointed Hopf algebras). Also, at about the same time Drinfeld developed the theory of quantum groups and the quantum double construction, and a bit later Lusztig developed the theory of quantum groups at roots of unity, which provided many interesting new examples of Hopf algebras. Around that time, it was realized that Hopf algebras could be viewed as algebraic structures arising from tensor categories with a fiber functor, i.e., tensor functor to the category of vector spaces, through the so-called reconstruction theory (which takes its origins in [**Sa**] and [**DelM**]). Since then, the theory of Hopf algebras has increasingly been becoming a part of the theory of tensor categories, and in proving some of the more recent results on Hopf algebras (such as, e.g., the classification of semisimple Hopf algebras of prime power dimension, or the classification of triangular Hopf algebras) tensor categories play a fundamental role. In fact, this is the point of view on Hopf algebras that we want to emphasize in this book: many of the important results about Hopf algebras are better understood if viewed through the prism of tensor categories. Namely, we deduce many of the most important results about Hopf algebras, especially finite dimensional and semisimple ones (such as the Fundamental Theorem on Hopf modules and bimodules, Nichols-Zoeller theorem, Larson-Radford theorems, Radford's S^4 formula, Kac-Zhu theorem, and many others) as corollaries of the general theory of tensor categories.

Let us now summarize the contents of the book, chapter by chapter.

In Chapter 1 we discuss the basics about abelian categories (mostly focusing on locally finite, or artinian, categories over a field, which is the kind of categories we will work with throughout the book). Many results here are presented without proofs, as they are well known (in fact, we view the basic theory of abelian categories as a prerequisite for reading this book). We do, however, give a more detailed discussion of the theory of locally finite categories, coalgebras, the Coend construction, and the reconstruction theory for coalgebras, for which it is harder to find an in-depth exposition in the literature. In accordance with the general philosophy of this book, we present the basic results about coalgebras (such as the Taft-Wilson theorem) as essentially categorical statements.

In Chapter 2 we develop the basic theory of monoidal categories. Here we give the detailed background on monoidal categories and functors, using a formalism that allows one not to worry much about the units, and give short proofs of the Mac Lane coherence and strictness theorems. We also develop a formalism of rigid monoidal categories, and give a number of basic examples. Finally, we briefly discuss 2-categories.

In Chapter 3 we discuss the combinatorics needed to study tensor categories, namely, the theory of $\mathbb{Z}_+$-rings (i.e., rings with a basis in which the structure constants are nonnegative integers). Such rings serve as Grothendieck rings of tensor categories, and it turns out that many properties of tensor categories actually have combinatorial origin, i.e., come from certain properties of the Grothendieck ring. In particular, this chapter contains the theory of Frobenius-Perron dimension, which plays a fundamental role in studying finite tensor categories (in particular, fusion categories), and also study $\mathbb{Z}_+$-modules over $\mathbb{Z}_+$-rings, which arise in the study of module categories. All the results in this chapter are purely combinatorial or ring-theoretical, and do not rely on category theory.

In Chapter 4 we develop the general theory of multitensor categories. Here we prove the basic results about multitensor and tensor categories, such as the exactness of tensor product and semisimplicity of the unit object, introduce a few key notions and constructions, such as pivotal and spherical structures, Frobenius-Perron dimensions of objects and categories, categorification, etc. We also provide many examples.

In Chapter 5 we consider tensor categories with a fiber functor, i.e., a tensor functor to the category of vector spaces. This leads to the notion of a Hopf algebra; we develop reconstruction theory, which establishes an equivalence between the notion of a Hopf algebra and the notion of a tensor category with a fiber functor. Then we proceed to develop the basic theory of Hopf algebras, and consider categories of modules and comodules over them. We also give a number of examples of Hopf algebras, such as Nichols Hopf algebras of dimension 2^{n+1}, Taft algebras, small quantum groups, etc., and discuss their representation theory. Then we prove a few classical theorems about Hopf algebras (such as the Cartier-Gabriel-Kostant theorem), discuss pointed Hopf algebras, quasi-Hopf algebras, and twisting.

In Chapter 6 we develop the theory of finite tensor categories, i.e., tensor categories which are equivalent, as an abelian category, to the category of representations of a finite dimensional algebra (a prototypical example being the category of representations of a finite dimensional Hopf algebra). In particular, we study the behavior of projectives in such a category, and prove a categorical version of the Nichols-Zoeller theorem (stating that a finite dimensional Hopf algebra is free over a Hopf subalgebra). We also introduce the distinguished invertible object of a finite tensor category, which is the categorical counterpart of the distinguished grouplike element (or character) of a finite dimensional Hopf algebra. Finally, we develop the theory of integrals of finite dimensional Hopf algebras.

In Chapter 7 we develop the theory of module categories over tensor and multitensor categories. Similarly to how understanding the structure of modules over a ring is necessary to understand the structure of the ring itself, the theory of module categories is now an indispensable tool in studying tensor categories. We first develop the basic theory of module categories over monoidal categories, and then pass to the abelian setting, introducing the key notion of an exact module category over a finite tensor category (somewhat analogous to the notion of a projective module over a ring). We also show that module categories arise as categories of modules over algebras in tensor categories (a categorical analog of module and comodule algebras over Hopf algebras). This makes algebras in tensor categories the main technical tool of studying module categories. We then proceed to studying the category of module functors between two module categories, and the Drinfeld center construction as an important special case of that. For Hopf algebras, this gives rise to the famous Drinfeld double construction and Yetter-Drinfeld modules. We then discuss dual categories and categorical Morita equivalence of tensor categories, prove the Fundamental Theorem for Hopf modules and bimodules over a Hopf algebra, and prove the categorical version of Radford's S^4 formula. Finally, we develop the theory of categorical dimensions of fusion categories and of Davydov-Yetter cohomology and deformations of tensor categories. At the end of the chapter, we discuss weak Hopf algebras, which are generalizations of Hopf algebras arising from semisimple module categories via reconstruction theory.

In Chapter 8 we develop the theory of braided categories, which is perhaps the most important part of the theory of tensor categories. We discuss pointed braided categories (corresponding to quadratic forms on abelian groups), quasitriangular Hopf algebras (arising from braided categories through reconstruction theory), and show that the center of a tensor category is a braided category. We develop a theory of commutative algebras in braided categories, and show that modules over such an algebra form a tensor category. We also develop the theory of factorizable, ribbon and modular categories, the S-matrix, Gauss sums, and prove the Verlinde formula and the existence of an $SL_2(\mathbb{Z})$-action. We prove the Anderson-Moore-Vafa theorem (saying that the central charge and twists of a modular category are roots of unity). Finally, we develop the theory of centralizers and projective centralizers in braided categories, de-equivariantization of braided categories, and braided G-crossed categories.

In Chapter 9 we mostly discuss results about fusion categories. This chapter is mainly concerned with applications, where we bring various tools from the previous chapters to bear on concrete problems about fusion categories. In particular, in this chapter we prove Ocneanu rigidity (the statement that fusion categories in characteristic zero have no deformations), develop the theory of dual categories, pseudo-unitary categories (showing that they have a canonical spherical structure – the categorical analog of the statement that a semisimple Hopf algebra in characteristic zero is involutive), study integral, weakly integral, group-theoretical, and weakly group-theoretical fusion categories. Next, we discuss symmetric and Tannakian fusion categories, prove Deligne's theorem on classification of such categories and discuss its nonfusion generalization stating that a symmetric tensor category of subexponential growth is the representation category of a supergroup (with a parity condition). We also give examples of symmetric categories with faster growth – Deligne's categories $\mathsf{Rep}(S_t)$, $\mathsf{Rep}(GL_t)$, $\mathsf{Rep}(O_t)$, and $\mathsf{Rep}(Sp_{2t})$. Next, we give a criterion of group-theoreticity of a fusion category, and show that any integral fusion category of prime power dimension is group theoretical. For categories of dimension p and p^2, this gives a very explicit classification (they are representation categories of an abelian group with a 3-cocycle). We introduce the notion of a solvable fusion category, and prove a categorical analog of Burnside's theorem, stating that a fusion category of dimension $p^a q^b$, where p, q are primes, is solvable. Finally, we discuss lifting theory for fusion categories (from characteristic p to characteristic zero).

Thus, Chapters 7–9 form the main part of the book.

A few disclaimers. First, we do not provide a detailed history of the subject, or a full list of references, containing all the noteworthy works; this was not our intention, and we do not even come close. Second, this book does not aim to be an exhaustive monograph on tensor categories; the subject is so vast that it would have taken a much longer text, perhaps in several volumes, to touch upon all the relevant topics. The reader will notice that we have included very little material (or none at all) on some of the key applications of tensor categories and their connections with other fields (representation theory, quantum groups, knot invariants, homotopy theory, vertex algebras, subfactors, etc.) In fact, our goal was not to give a comprehensive treatment of the entire subject, but rather to provide a background that will allow the reader to proceed to more advanced and specialized works. Also, we tried to focus on topics which are not described in detail in books or expository texts, and left aside many of those which are well addressed in the literature. Finally, as authors of any textbook, we had a preference for subjects that we understand better!

To increase the amount of material that we are able to discuss and to enable active reading, we have presented many examples and applications in the form of exercises. The more difficult exercises are provided with detailed hints, which should allow the reader to solve them without consulting other sources. At the end of Chapters 5, 7, 8, and 9 we provide a summary of some noteworthy results related to the material of the chapter that we could not discuss, and provide references. Finally, we end each chapter with references for the material discussed in the chapter.

Finally, let us discuss how this book might be used for a semester-long graduate course. Clearly, it is not feasible to go through the entire book in one semester and some omissions are necessary. The challenge is to get to Chapters 7–9 (which contain the main results) as quickly as possible. We recommend that the course start from Chapter 2, and auxiliary material from Chapters 1 and 3 be covered as needed. More specifically, we suggest the following possible course outline.

(1) Chapter 2: Sections 2.1–2.10 (monoidal categories and functors, the MacLane strictness theorem, rigidity).
(2) Chapter 4: Sections 4.1–4.9 (basic properties of tensor categories, Grothendieck rings, Frobenius-Perron dimensions).
(3) Chapter 5: Sections 5.1–5.6 (fiber functors and basic examples of Hopf algebras).
(4) Chapter 6: Sections 6.1–6.3 (properties of injective and surjective tensor functors).
(5) Chapter 7: Sections 7.1–7.12 (exact module categories, categorical Morita theory).
(6) Chapter 8: Sections 8.1–8.14 (examples from metric groups, Drinfeld centers, modular categories).
(7) Chapter 9: Sections 9.1–9.9 and 9.12 (absence of deformations, integral fusion categories, symmetric categories).

Acknowledgments. While writing this book, we were profoundly influenced by Vladimir Drinfeld, whose texts and explanations have shaped our understanding of tensor categories. We are also grateful to David Kazhdan, from whom we learned a lot about this subject. We are very grateful to Alexei Davydov, Alexander Kirillov Jr., Alex Levin, Noah Snyder, Leonid Vainerman, and Siddharth Venkatesh, who read a preliminary version of the book and made many useful comments.

The work of the first author was supported by the NSF grant DMS-1000173. The work of the second author was supported by the Israel Science Foundation (grant No. 317/09). The work of the third author was supported by the NSA grant H98230-13-1-0236. The authors were also supported by BSF grants No. 2008164 and No. 2002040.

CHAPTER 1

Abelian categories

1.1. Categorical prerequisites and notation

In this book we will assume that the reader is familiar with the basic theory of categories and functors; a discussion of this theory can be found in the classical book [**Mac2**].

The identity endofunctor of a category $\mathcal{C}$ will be denoted $\mathsf{id}_{\mathcal{C}}$. We say that a functor $F : \mathcal{C} \to \mathcal{D}$ is an *equivalence* if there is a functor $F^{-1} : \mathcal{D} \to \mathcal{C}$, called a *quasi-inverse* of F, such that $F^{-1} \circ F \cong \mathsf{id}_{\mathcal{C}}$ and $F \circ F^{-1} \cong \mathsf{id}_{\mathcal{D}}$.

A category is called *locally small* if for any objects X, Y, $\mathrm{Hom}_{\mathcal{C}}(X, Y)$ is a set, and is called *essentially small* if in addition its isomorphism classes $\mathcal{C}$ of objects form a set. In other words, an essentially small category is a category equivalent to a small category (i.e., to one where objects and morphisms form a set). All categories considered in this book will be locally small (except in the section on 2-categories), and most of them will be essentially small. So set-theoretical subtleties will not play any role in this book. [1]

For a category $\mathcal{C}$ the notation $X \in \mathcal{C}$ will mean that X is an object of $\mathcal{C}$, and the set of morphisms between $X, Y \in \mathcal{C}$ will be denoted by $\mathsf{Hom}_{\mathcal{C}}(X, Y)$. An element $\phi \in \mathsf{Hom}_{\mathcal{C}}(X, Y)$ will be usually depicted either as $\phi : X \to Y$ or as $X \xrightarrow{\phi} Y$. We denote by $\mathcal{C}^{\vee}$ the category *dual* to $\mathcal{C}$, i.e., the one obtained from $\mathcal{C}$ by reversing the direction of morphisms.

In this chapter we will briefly recall (mostly without proofs) the aspects of the theory of abelian categories which will be especially important to us in the sequel.

Unless otherwise specified, all fields are assumed to be algebraically closed.

1.2. Additive categories

DEFINITION 1.2.1. An *additive category* is a category $\mathcal{C}$ satisfying the following axioms:

[1] A well-known foundational subtlety of category theory (which, luckily, does not play an essential role in most category-theoretical considerations) is that in general, objects and morphisms of a category are defined to form a class rather than a set (this allows one to consider the category of all sets without running into Russell's paradox - the non-existence of the set of all sets). For example, the category of all sets or all vector spaces over a field is not small or even essentially small (although it is locally small). However, in this book we will usually consider categories with finiteness conditions, such as the category of finite dimensional vector spaces, so this issue is not going to be relevant most of the time. In fact, even when we work with categories that are not essentially small (such as the category of all vector spaces), we will allow ourselves to abuse terminology and speak about "the set of isomorphism classes of objects" of such a category. This does not create a confusion in our context.

(A1) Every set $\mathsf{Hom}_{\mathcal{C}}(X, Y)$ is equipped with a structure of an abelian group (written additively) such that composition of morphisms is biadditive with respect to this structure.
(A2) There exists a zero object $0 \in \mathcal{C}$ such that $\mathsf{Hom}_{\mathcal{C}}(0, 0) = 0$.
(A3) (Existence of direct sums.) For any objects $X_1, X_2 \in \mathcal{C}$ there exists an object $Y \in \mathcal{C}$ and morphisms $p_1 : Y \to X_1$, $p_2 : Y \to X_2$, $i_1 : X_1 \to Y$, $i_2 : X_2 \to Y$ such that $p_1 i_1 = \mathsf{id}_{X_1}$, $p_2 i_2 = \mathsf{id}_{X_2}$, and $i_1 p_1 + i_2 p_2 = \mathsf{id}_Y$.

In (A3), the object Y is unique up to a unique isomorphism, is denoted by $X_1 \oplus X_2$, and is called the *direct sum* of X_1 and X_2. Thus, every additive category is equipped with a bifunctor $\oplus : \mathcal{C} \times \mathcal{C} \to \mathcal{C}$.

DEFINITION 1.2.2. Let $\Bbbk$ be a field. An additive category $\mathcal{C}$ is said to be $\Bbbk$-linear (or defined over $\Bbbk$) if for any objects $X, Y \in \mathcal{C}$, $\mathsf{Hom}_{\mathcal{C}}(X, Y)$ is equipped with a structure of a vector space over $\Bbbk$, such that composition of morphisms is $\Bbbk$-linear.

DEFINITION 1.2.3. Let $F : \mathcal{C} \to \mathcal{D}$ be a functor between two additive categories. The functor F is called *additive* if the associated maps

$$\mathsf{Hom}_{\mathcal{C}}(X, Y) \to \mathsf{Hom}_{\mathcal{D}}(F(X), F(Y)), \quad X, Y \in \mathcal{C}, \tag{1.1}$$

are homorphisms of abelian groups. If $\mathcal{C}$ and $\mathcal{D}$ are $\Bbbk$-linear categories then F is called $\Bbbk$-*linear* if the homorphisms (1.1) are $\Bbbk$-linear.

PROPOSITION 1.2.4. *For any additive functor $F : \mathcal{C} \to \mathcal{D}$ there exists a natural isomorphism $F(X) \oplus F(Y) \xrightarrow{\sim} F(X \oplus Y)$.*

1.3. Definition of abelian category

Let $\mathcal{C}$ be an additive category and $f : X \to Y$ a morphism in $\mathcal{C}$. The *kernel* $\mathsf{Ker}(f)$ of f (if exists) is an object K together with a morphism $k : K \to X$ such that $fk = 0$, and if $k' : K' \to X$ is such that $fk' = 0$ then there exists a unique morphism $\ell : K' \to K$ such that $k\ell = k'$. If $\mathsf{Ker}(f)$ exists then it is unique up to a unique isomorphism.

Dually, the *cokernel* $\mathsf{Coker}(f)$ of a morphism $f : X \to Y$ in $\mathcal{C}$ (if exists) is an object C together with a morphism $c : Y \to C$ such that $cf = 0$, and if $c' : Y \to C'$ is such that $c'f = 0$ then there exists a unique morphism $\ell : C \to C'$ such that $\ell c = c'$. If $\mathsf{Coker}(f)$ exists then it is unique up to a unique isomorphism.

DEFINITION 1.3.1. An *abelian category* is an additive category $\mathcal{C}$ in which for every morphism $\varphi : X \to Y$ there exists a sequence

$$K \xrightarrow{k} X \xrightarrow{i} I \xrightarrow{j} Y \xrightarrow{c} C \tag{1.2}$$

with the following properties:

1. $ji = \varphi$,
2. $(K, k) = \mathsf{Ker}(\varphi)$, $(C, c) = \mathsf{Coker}(\varphi)$,
3. $(I, i) = \mathsf{Coker}(k)$, $(I, j) = \mathsf{Ker}(c)$.

A sequence (1.2) is called a canonical decomposition of φ. The object I is called the image of φ and is denoted by $\mathsf{Im}(\varphi)$.

In the sequel, for brevity we will write K instead of (K, k) and C instead of (C, c).

REMARK 1.3.2. In an abelian category, the canonical decomposition of a morphism is unique up to a unique isomorphism.

EXAMPLE 1.3.3. The category of abelian groups is an abelian category. The category of modules over a ring is an abelian category. The category **Vec** of vector spaces over a field $\Bbbk$ and its subcategory Vec of finite dimensional vector spaces are $\Bbbk$-linear abelian categories. More generally, the category of modules over an associative $\Bbbk$-algebra and the category of comodules over a coassociative $\Bbbk$-coalgebra (see Section 1.9 below) are $\Bbbk$-linear abelian categories.

DEFINITION 1.3.4. Let $\mathcal{C}$ be an abelian category. A morphism $f : X \to Y$ is said to be a *monomorphism* if $\mathsf{Ker}(f) = 0$. It is said to be an *epimorphism* if $\mathsf{Coker}(f) = 0$.

It is easy to see that a morphism is both a monomorphism and an epimorphism if and only if it is an isomorphism.

DEFINITION 1.3.5. A *subobject* of an object Y is an object X together with a monomorphism $i : X \to Y$. A *quotient object* of Y is an object Z with an epimorphism $p : Y \to Z$. A *subquotient object* of Y is a quotient object of a subobject of Y.

EXERCISE 1.3.6. Show that a subquotient of a subquotient of Y (in particular, a subobject of a quotient of Y) is a subquotient of Y.

For a subobject $X \subset Y$ define the quotient object $Z = Y/X$ to be the cokernel of the monomorphism $f : X \to Y$.

Let $\mathcal{C}_\alpha, \alpha \in I$, be a family of additive categories. The direct sum $\mathcal{C} = \bigoplus_{\alpha\in I} \mathcal{C}_\alpha$ is the category whose objects are sums

$$X = \bigoplus_{\alpha\in I} X_\alpha, \quad X_\alpha \in \mathcal{C}_\alpha,$$

such that almost all X_α are zero, with

$$\mathsf{Hom}_{\mathcal{C}}(X, Y) = \bigoplus_{\alpha\in I} \mathsf{Hom}_{\mathcal{C}_\alpha}(X_\alpha, Y_\alpha)$$

for $X = \bigoplus_{\alpha\in I} X_\alpha$ and $Y = \bigoplus_{\alpha\in I} Y_\alpha$. It is easy to see that $\mathcal{C}$ is an additive category. Clearly, it is abelian if all $\mathcal{C}_\alpha$ are abelian, and vice versa.

DEFINITION 1.3.7. An abelian category $\mathcal{C}$ is said to be *indecomposable* if it is not equivalent to a direct sum of two nonzero categories.

The following theorem is psychologically useful, as it allows one to think of morphisms, kernels, cokernels, subobjects, quotient objects, etc. in an abelian category in terms of usual linear algebra.

THEOREM 1.3.8 (Mitchell; [**Fr**]). *Every abelian category is equivalent, as an additive category, to a full subcategory of the category of left modules over an associative unital ring A.*

REMARK 1.3.9. If the category is $\Bbbk$-linear, the ring in Theorem 1.3.8 can be chosen to be a $\Bbbk$-algebra in such a way that the corresponding equivalence is $\Bbbk$-linear.

REMARK 1.3.10. A major drawback of Theorem 1.3.8 is that the ring A is not unique, and in many important cases there are no manageable choices of A.

1.4. Exact sequences

DEFINITION 1.4.1. A sequence of morphisms

$$\cdots \to X_{i-1} \xrightarrow{f_{i-1}} X_i \xrightarrow{f_i} X_{i+1} \to \cdots \tag{1.3}$$

in an abelian category is called *exact in degree* i if $\mathsf{Im}(f_{i-1}) = \mathsf{Ker}(f_i)$. It is called *exact* if it is exact in every degree. An exact sequence

$$0 \to X \to Y \to Z \to 0 \tag{1.4}$$

is called a *short exact sequence.*

In a short exact sequence (1.4) X is a subobject of Y and $Z \cong Y/X$ is the corresponding quotient.

DEFINITION 1.4.2. Let

$$S : 0 \to X \to Z \to Y \to 0 \qquad \text{and} \qquad S' : 0 \to X \to Z' \to Y \to 0$$

be short exact sequences. A *morphism* from S to S' is a morphism $f : Z \to Z'$ such that it restricts to the identity morphism $X \to X$, and induces the identity morphism $Y \to Y$. The set of exact sequences $0 \to X \to Z \to Y \to 0$ up to isomorphism is denoted $\mathsf{Ext}^1(Y, X)$ and is called the set of *extensions* of Y by X.

One can define an operation of addition on $\mathsf{Ext}^1(Y, X)$. Namely, let S and S' be the short exact sequences as above. Let X_{antidiag} denote the antidiagonal copy of X in $X \oplus X$ (i.e., the image of $(\mathsf{id}_X, -\mathsf{id}_X) : X \to X \oplus X$), and similarly Y_{antidiag} denote the antidiagonal copy of Y in $Y \oplus Y$. Define $S + S'$ to be the exact sequence

$$0 \to X \to Z'' \to Y \to 0, \tag{1.5}$$

where $Z'' = \tilde{Z}''/X_{\text{antidiag}}$, and $\tilde{Z}''$ is the inverse image of Y_{antidiag} in $Z \oplus Z'$, i.e., the universal object for which the following pullback diagram commutes:

$$\begin{array}{ccc} \tilde{Z}'' & \longrightarrow & Y_{\text{antidiag}} \\ \downarrow & & \downarrow \\ Z \oplus Z' & \longrightarrow & Y \oplus Y. \end{array}$$

(Alternatively, we can say that $\tilde{Z}''$ is the kernel of $\pi \circ p - \pi' \circ p' : Z \oplus Z' \to Y$, where $\pi : Z \to Y$ and $\pi' : Z' \to Y$, and p, p' are the projections of $Z \oplus Z'$ to its summands).

EXERCISE 1.4.3. (i) Show that operation (1.5) is well defined and defines a structure of an abelian group on $\mathsf{Ext}^1(Y, X)$.

(ii) Let A be an algebra over an algebraically closed field $\Bbbk$, and $\mathcal{C} = A-\mathbf{mod}$ be the category of A-modules. Show that $\mathsf{Ext}^1(Y, X) = Z(Y, X)/B(Y, X)$ with $Z(Y, X) = \mathsf{Der}(A, \mathsf{Hom}_\Bbbk(Y, X))$, the space of *derivations* of A:

$$\mathsf{Der}(A, \mathsf{Hom}_\Bbbk(Y, X)) = \{D : A \to \mathsf{Hom}_\Bbbk(Y, X) \mid D(ab) = D(a)b + aD(b),\ a, b \in A\},$$

and $B(Y, X)$ is the subspace of inner derivations, i.e., the derivations $D \in \mathsf{Der}(A, \mathsf{Hom}_\Bbbk(Y, X))$ of the form $D(a) = [f, a]$, $f \in \mathsf{Hom}_\Bbbk(Y, X)$.

1.5. Length of objects and the Jordan-Hölder theorem

Let $\mathcal{C}$ be an abelian category.

DEFINITION 1.5.1. A nonzero object X in $\mathcal{C}$ is called *simple* if 0 and X are its only subobjects. An object X in $\mathcal{C}$ is called *semisimple* if it is a direct sum of simple objects, and $\mathcal{C}$ is called *semisimple* if every object of $\mathcal{C}$ is semisimple.

LEMMA 1.5.2. *(Schur's Lemma) Let X, Y be two simple objects in $\mathcal{C}$. Then any nonzero morphism $f : X \to Y$ is an isomorphism. In particular, if X is not isomorphic to Y then $\mathsf{Hom}_{\mathcal{C}}(X, Y) = 0$, and $\mathsf{Hom}_{\mathcal{C}}(X, X)$ is a division algebra.*

Let X be an object in an abelian category $\mathcal{C}$.

DEFINITION 1.5.3. We say that X has *finite length* if there exists a filtration

$$0 = X_0 \subset X_1 \subset \cdots \subset X_{n-1} \subset X_n = X$$

such that X_i/X_{i-1} is simple for all i. Such a filtration is called a *Jordan-Hölder series* of X. We will say that this Jordan-Hölder series *contains* a simple object Y with *multiplicity* m if the number of values of i for which X_i/X_{i-1} is isomorphic to Y is m.

THEOREM 1.5.4. *(Jordan-Hölder) Suppose that X has finite length. Then any filtration of X can be extended to a Jordan-Hölder series, and any two Jordan-Hölder series of X contain any simple object with the same multiplicity, so in particular have the same length.*

DEFINITION 1.5.5. The *length* of an object X is the length of its Jordan-Hölder series (if it exists).

We will be mainly interested in abelian categories in which every object has finite length. An example of such a category is the category of finite dimensional modules over an algebra.

DEFINITION 1.5.6. An *indecomposable* object $X \in \mathcal{C}$ is an object which does not admit a non-trivial decomposition into a direct sum of its subobjects.

THEOREM 1.5.7. *(Krull-Schmidt) Any object of finite length admits a unique (up to an isomorphism) decomposition into a direct sum of indecomposable objects.*

Let $\mathcal{C}$ be an abelian $\Bbbk$-linear category where objects have finite length. If X and Y are objects in $\mathcal{C}$ such that Y is simple then we denote by $[X : Y]$ the multiplicity of Y in a Jordan-Hölder series of X. By the Jordan-Hölder Theorem 1.5.4, this quantity is well defined.

DEFINITION 1.5.8. The *Grothendieck group* $\mathsf{Gr}(\mathcal{C})$ of $\mathcal{C}$ is the free abelian group generated by isomorphism classes X_i of simple objects in $\mathcal{C}$. To every object X in $\mathcal{C}$ we can canonically associate its class $[X] \in \mathsf{Gr}(\mathcal{C})$ given by the formula

$$[X] = \sum_i [X : X_i] X_i. \tag{1.6}$$

It is obvious that if $0 \to X \to Y \to Z \to 0$ is a short exact sequence then $[Y] = [X] + [Z]$. When no confusion is possible, we will simply write X instead of $[X]$.

EXAMPLE 1.5.9. Let S be a set and let Vec_S be the abelian category of finite dimensional $\Bbbk$-vector spaces graded by S. Then $\mathsf{Gr}(\mathcal{C}) = \mathbb{Z}S$, the free abelian group on S.

EXERCISE 1.5.10. (i) Let $\mathcal{C}$ be an abelian category where objects have finite length. Show that $\mathcal{C}$ admits a unique decomposition $\mathcal{C} = \oplus_{\alpha \in I} \mathcal{C}_\alpha$ such that $\mathcal{C}_\alpha$ are indecomposable. (The categories $\mathcal{C}_\alpha$ are called the *blocks* of $\mathcal{C}$.)

(ii) Show that $\mathcal{C}$ is indecomposable if and only if any two simple objects X, Y of $\mathcal{C}$ are linked, i.e., there exists a chain $X = X_0, X_1, ..., X_n = Y$ of simple objects of $\mathcal{C}$ such that $\mathsf{Ext}^1(X_i, X_{i+1}) \neq 0$ or $\mathsf{Ext}^1(X_{i+1}, X_i) \neq 0$ for all i.

(iii) Let A be a finite dimensional algebra over an algebraically closed field $\Bbbk$, and let $\mathcal{C}$ be the category of finite dimensional A-modules. Show that blocks of $\mathcal{C}$ are labeled by characters of the center of A.

1.6. Projective and injective objects

DEFINITION 1.6.1. Let $\mathcal{C}, \mathcal{D}$ be abelian categories. An additive functor $F : \mathcal{C} \to \mathcal{D}$ is called *left* (respectively, *right*) *exact* if for any short exact sequence $0 \to X \to Y \to Z \to 0$ in $\mathcal{C}$ the sequence

$$0 \to F(X) \to F(Y) \to F(Z) \qquad \text{(respectively,} \quad F(X) \to F(Y) \to F(Z) \to 0)$$

is exact in $\mathcal{D}$. A functor is said to be *exact* if it is both left and right exact.

One also uses the terms *right exact*, *left exact*, and *exact* for contravariant functors. Namely, a contravariant functor $F : \mathcal{C} \to \mathcal{D}$ is right exact, left exact, or exact if so is the corresponding covariant functor $\mathcal{C} \to \mathcal{D}^\vee$.

EXAMPLE 1.6.2. Let X, Y be objects in an abelian category $\mathcal{C}$. The contravariant functor $\mathsf{Hom}_\mathcal{C}(-, Y)$ and the covariant functor $\mathsf{Hom}_\mathcal{C}(X, -)$ from $\mathcal{C}$ to the category of abelian groups are left exact.

EXAMPLE 1.6.3. If A is a ring and M is a right A-module, then the functor $V \mapsto M \otimes_A V$ from the category of left A-modules to the category of abelian groups is right exact.

EXERCISE 1.6.4. Show that the left adjoint to any functor between abelian categories is right exact, and the right adjoint is left exact.

DEFINITION 1.6.5. An object P in an abelian category $\mathcal{C}$ is called *projective* if the functor $\mathsf{Hom}_\mathcal{C}(P, -)$ is exact. An object I in $\mathcal{C}$ is called *injective* if the functor $\mathsf{Hom}_\mathcal{C}(-, I)$ is exact.

Let us assume that all objects of an abelian category $\mathcal{C}$ have finite length, see Definition 1.5.3.

DEFINITION 1.6.6. Let $X \in \mathcal{C}$. A *projective cover* of X is a projective object $P(X) \in \mathcal{C}$ together with an epimorphism $p : P(X) \to X$ such that if $g : P \to X$ is an epimorphism from a projective object P to X, then there exists an epimorphism $h : P \to P(X)$ such that $ph = g$.

If a projective cover $P(X)$ of X exists then it is unique up to a non-unique isomorphism.

DEFINITION 1.6.7. Let $X \in \mathcal{C}$. An *injective hull* of X is an injective object $Q(X) \in \mathcal{C}$ together with a monomorphism $i : X \to Q(X)$ such that if $g : X \to Q$ is a monomorphism from X to an injective object Q, then there exists a monomorphism $h : Q(X) \to Q$ such that $hi = g$.

If an injective hull of X exists then it is unique up to a non-unique isomorphism.

1.7. Higher Ext groups and group cohomology

Let R be a ring, and let M, N be (left) R-modules. Recall that a *projective resolution* of M is an exact sequence

$$\cdots \to P_2 \to P_1 \to P_0 \to M \to 0,$$

where P_i are projective (e.g., free) R-modules. Given such a resolution, we can define the sequence of modules and maps

$$0 \to \mathsf{Hom}_R(P_0, N) \to \mathsf{Hom}_R(P_1, N) \to \mathsf{Hom}_R(P_2, N) \to \cdots$$

which is a *complex*. Namely, denoting by d_i the map $\mathsf{Hom}(P_{i-1}, N) \to \mathsf{Hom}(P_i, N)$ for $i \geq 0$ (where $P_{-1} := 0$), we have $d_{i+1} \circ d_i = 0$. Recall also that the *cohomology* of this complex, $H^i := \mathsf{Ker}(d_{i+1})/\mathsf{Im}(d_i)$, is independent on the resolution $P^\bullet$ up to a canonical isomorphism. For $i > 0$, this cohomology is denoted by $\mathsf{Ext}^i(M, N)$ (it is easy to see that $\mathsf{Ext}^0(M, N) = \mathsf{Hom}_R(M, N)$). It is easy to show that $\mathsf{Ext}^i(M, N)$ for $i = 1$ is canonically isomorphic to $\mathsf{Ext}^1(M, N)$ defined using short exact sequences.

Let $0 \to N_1 \to N_2 \to N_3 \to 0$ be a short exact sequence of R-modules. Then it is well known that there is a long exact sequence of cohomology

$$\cdots \to \mathsf{Ext}^i(M, N_1) \to \mathsf{Ext}^i(M, N_2) \to \mathsf{Ext}^i(M, N_3) \to \mathsf{Ext}^{i+1}(M, N_1) \to \cdots$$

Let us now consider an example of this which will be especially important in this book. Let G be a group, and let A be an abelian group with an action of G (i.e., a G-module). Then the groups $\mathsf{Ext}^i_G(\mathbb{Z}, A)$ in the category of G-modules (where G acts on $\mathbb{Z}$ trivially) are called the *cohomology groups* of G with coefficients in A, and denoted $H^i(G, A)$.

The groups $H^i(G, A)$ can be defined in a much more explicit way, since there is an explicit resolution of $\mathbb{Z}$ in the category of G-modules, called the bar resolution. The terms of the bar resolution have the form $P_i := \mathbb{Z}[G^{i+1}]$, with the group action by $g(g_0, g_1, \ldots, g_i) = (gg_0, g_1, \ldots, g_i)$, and the maps $\partial_i : P_i \to P_{i-1}$ are defined by the formula

$$\begin{aligned} \partial_i(g_0, \ldots, g_i) &= (g_0g_1, g_2, \ldots, g_i) - (g_0, g_1g_2, \ldots, g_i) + \\ &\quad \cdots + (-1)^{i-1}(g_0, \ldots, g_{i-1}g_i) + (-1)^i(g_0, \ldots, g_{i-1}). \end{aligned}$$

It is easy to check that this is indeed a resolution. We have an isomorphism $\gamma_i : \mathsf{Hom}_G(P_i, A) \cong \mathsf{Fun}(G^i, A)$, given by

$$\gamma_i(h)(g_1, \ldots, g_i) := h(1, g_1, \ldots, g_i),$$

and the maps $d_i = \partial_i^* : \mathsf{Hom}_G(P_{i-1}, A) \to \mathsf{Hom}_G(P_i, A)$ upon this identification take the form

$$\begin{aligned} d_i(\mathbf{f})(g_1, \ldots, g_i) &:= g_1\mathbf{f}(g_2, \ldots, g_i) - \mathbf{f}(g_1g_2, \ldots, g_i) + \\ &\quad \cdots + (-1)^{i-1}\mathbf{f}(g_1, \ldots, g_{i-1}g_i) + (-1)^i\mathbf{f}(g_1, \ldots, g_{i-1}). \end{aligned}$$

The complex with terms $C^i = C^i(G, A) := \mathsf{Fun}(G^i, A)$ and differentials defined in this way is called *the standard complex of G with coefficients in A*. The cocycles

$\mathsf{Ker}(d_{i+1})$ and coboundaries $\mathsf{Im}(d_i)$ of this complex are denoted by $Z^i(G, A)$ and $B^i(G, A)$, respectively, and the group cohomology $H^i(G, A)$ is the cohomology of this complex.

Note that if A is a commutative ring and the G-action preserves the multiplication in A then $H^*(G, A)$ is a graded commutative ring, with multiplication induced by the Yoneda product on Ext groups.

EXAMPLE 1.7.1. For $i = 1, 2, 3, 4$, the definition of d_i looks like

$$\begin{aligned} d_1(\mathbf{f})(g) &= g\mathbf{f} - \mathbf{f}, \\ d_2(\mathbf{f})(g,h) &= g\mathbf{f}(h) - \mathbf{f}(gh) + \mathbf{f}(g), \\ d_3(\mathbf{f})(g,h,k) &= g\mathbf{f}(h,k) - \mathbf{f}(gh,k) + \mathbf{f}(g,hk) - \mathbf{f}(g,h), \\ d_4(\mathbf{f})(g,h,k,l) &= g\mathbf{f}(h,k,l) - \mathbf{f}(gh,k,l) + \mathbf{f}(g,hk,l) - \mathbf{f}(g,h,kl) + \mathbf{f}(g,h,k). \end{aligned}$$

EXAMPLE 1.7.2. (i) $H^0(G, A) = A^G$, the G-invariants in A. So if the action of G on A is trivial, $H^0(G, A) = A$.

(ii) If G acts trivially on A then $H^1(G, A) = \mathsf{Hom}(G, A)$. More generally, for any A, $H^1(G, A)$ classifies homomorphisms from G to the semidirect product $A \rtimes G$ which are right inverse to the projection $A \rtimes G \to G$, up to conjugation by elements of A.

(iii) $H^2(G, A)$ classifies abelian extensions of G by A. In particular, if G acts trivially on A, it classifies central extensions of G by A.

REMARK 1.7.3. The definition of a 1-cocycle can be generalized to the case when the group A is not necessarily abelian. In this case, writing the operation in A multiplicatively, the equation for a 1-cocycle takes the form

$$\mathbf{f}(gh) = \mathbf{f}(g) \cdot g\mathbf{f}(h).$$

Such cocycles, in general, do not form a group (only a set), but this set has an action of A via

$$(a \circ \mathbf{f})(g) := a \cdot \mathbf{f}(g) \cdot g(a)^{-1}.$$

The set of orbits of this action is called the first cohomology of G with coefficients in A and is denoted $H^1(G, A)$. Obviously, in the abelian case this coincides with the above definition. Also, it is easy to see that as in the abelian case, $H^1(G, A)$ classifies homomorphisms $G \to A \rtimes G$ which are right inverse to the standard projection, up to the action of A by conjugation.

EXAMPLE 1.7.4. Let $G = \mathbb{Z}/n\mathbb{Z}$ be a finite cyclic group. Let us compute the cohomology $H^i(G, \mathbb{Z})$. The bar resolution of $\mathbb{Z}$ is too big for a convenient computation, but luckily there is a much smaller free resolution. Namely, let $P_i = \mathbb{Z}G = \mathbb{Z}[g]/(g^n - 1)$, and let $\partial_i : P_i \to P_{i-1}$ be given by $\partial_i f = (g-1)f$ if i is odd, and $\partial_i(f) = (1 + g + \dots + g^{n-1})f$ if i is even. Using this resolution, it is easy to find that $H^{2j+1}(G, \mathbb{Z}) = 0$ and $H^{2j}(G, \mathbb{Z}) = \mathbb{Z}/n\mathbb{Z}$ for $j > 0$, while $H^0(G, \mathbb{Z}) = \mathbb{Z}$.

EXERCISE 1.7.5. Show that the graded ring $H^*(\mathbb{Z}/n\mathbb{Z}, \mathbb{Z})$ is

$$H^*(\mathbb{Z}/n\mathbb{Z}, \mathbb{Z}) = \mathbb{Z}[x]/(nx) = \mathbb{Z} \oplus x(\mathbb{Z}/n\mathbb{Z})[x],$$

where x is a generator in degree 2.

1.8. Locally finite (artinian) and finite abelian categories

Let $\Bbbk$ be a field.

DEFINITION 1.8.1. A $\Bbbk$-linear abelian category $\mathcal{C}$ is said to be *locally finite* if the following two conditions are satisfied:

(i) for any two objects X, Y in $\mathcal{C}$ the $\Bbbk$-vector space $\mathsf{Hom}_{\mathcal{C}}(X, Y)$ is finite dimensional;
(ii) every object in $\mathcal{C}$ has finite length.

In particular, the Jordan-Hölder Theorem 1.5.4 and Krull-Schmidt Theorem 1.5.7 hold in any locally finite abelian category.

REMARK 1.8.2. Locally finite abelian categories are also called *artinian* categories.

Almost all abelian categories we consider in this book are locally finite.

We will denote by $\mathcal{O}(\mathcal{C})$ the set of isomorphism classes of simple objects of a locally finite abelian category $\mathcal{C}$.

DEFINITION 1.8.3. An additive functor $F : \mathcal{C} \to \mathcal{D}$ between two locally finite abelian categories is *injective* if it is fully faithful (i.e., bijective on the sets of morphisms).[2] We say that F is *surjective* if any simple object of $\mathcal{D}$ is a subquotient of some object $F(X)$, where X is an object of $\mathcal{C}$ [3].

PROPOSITION 1.8.4. *Suppose that* $\Bbbk$ *is algebraically closed. In any locally finite category* $\mathcal{C}$ *over* $\Bbbk$ *we have* $\mathsf{Hom}_{\mathcal{C}}(X, Y) = 0$ *if* X, Y *are simple and non-isomorphic and* $\mathsf{Hom}_{\mathcal{C}}(X, X) = \Bbbk$ *for any simple object* X.

PROOF. Let $f : X \to Y$ be a morphism. By Schur's Lemma 1.5.2 either $f = 0$ or f is an isomorphism. This implies that $\mathsf{Hom}_{\mathcal{C}}(X, Y) = 0$ if X, Y are simple and non-isomorphic, and $\mathsf{Hom}(X, X)$ is a division algebra. Since $\Bbbk$ is algebraically closed, condition (i) of Definition 1.8.1 implies that $\mathsf{Hom}(X, X) = \Bbbk$ for any simple object $X \in \mathcal{C}$. □

DEFINITION 1.8.5. A $\Bbbk$-linear abelian category $\mathcal{C}$ is said to be *finite* if it is equivalent to the category A–mod of finite dimensional modules over a finite dimensional $\Bbbk$-algebra A.

Of course, such an algebra A is not canonically attached to the category $\mathcal{C}$. Instead, $\mathcal{C}$ determines the Morita equivalence class of A. For this reason, it is often better to use the following "intrinsic" definition, which is well known to be equivalent to Definition 1.8.5.

DEFINITION 1.8.6. A $\Bbbk$-linear abelian category $\mathcal{C}$ is *finite* if

(i) $\mathcal{C}$ has finite dimensional spaces of morphisms;
(ii) every object of $\mathcal{C}$ has finite length;
(iii) $\mathcal{C}$ has *enough projectives*, i.e., every simple object of $\mathcal{C}$ has a projective cover; and
(iv) there are finitely many isomorphism classes of simple objects.

---

[2]We will use the terms "injective functor" and "fully faithful functor" interchangeably.

[3]This definition does not coincide with a usual categorical definition of an *essentially surjective* functor which requires that every object of $\mathcal{D}$ be isomorphic to some $F(X)$ for an object X in $\mathcal{C}$.

Note that the first two conditions of Definition 1.8.6 are the requirement that $\mathcal{C}$ be locally finite.

To see the equivalence of Definitions 1.8.5 and 1.8.6, observe that if A is a finite dimensional algebra then $A-\mathsf{mod}$ clearly satisfies (i)-(iv), and conversely, if $\mathcal{C}$ satisfies (i)-(iv), then one can take $A = \mathsf{End}(P)^{\mathrm{op}}$, where P is a projective generator of $\mathcal{C}$ (e.g., $P = \bigoplus_{i=1}^{n} P_i$, where P_i are projective covers of all the simple objects X_i, $i = 1, \ldots, n$, of $\mathcal{C}$).

A projective generator P of $\mathcal{C}$ represents a functor $F = F_P : \mathcal{C} \to \mathsf{Vec}$ from $\mathcal{C}$ to the category of finite dimensional $\Bbbk$-vector spaces, given by the formula $F(X) = \mathsf{Hom}_{\mathcal{C}}(P, X)$. The condition that P is projective translates into the exactness property of F, and the condition that P is a generator (i.e., covers any simple object) translates into the property that F is faithful (does not kill nonzero objects or morphisms). Moreover, the algebra $A = \mathsf{End}(P)^{\mathrm{op}}$ can be alternatively defined as $\mathsf{End}(F)$, the algebra of functorial endomorphisms of F. Conversely, it is well known (and easy to show) that any exact faithful functor $F : \mathcal{C} \to \mathsf{Vec}$ is represented by a unique (up to a unique isomorphism) projective generator P.

REMARK 1.8.7. The dual category of a finite abelian category is finite. Namely, the dual to the category of finite dimensional A-modules is the category of finite dimensional A^{op}-modules, where A^{op} is the algebra A with opposite multiplication, and the duality functor between these categories is the functor of taking the dual module, $V \mapsto V^*$. Thus, in a finite abelian category, any object has both a projective cover and an injective hull.

Let A, B be finite dimensional $\Bbbk$-algebras and let $A-\mathsf{mod}, B-\mathsf{mod}$ denote the categories of finite dimensional modules over them.

DEFINITION 1.8.8. We say that an additive $\Bbbk$-linear functor

$$F : A-\mathsf{mod} \to B-\mathsf{mod}$$

is $(\otimes$-$)$*representable* if there exists a (B, A)-bimodule V such that F is naturally isomorphic to $(V \otimes_A -)$.

REMARK 1.8.9. An additive $\Bbbk$-linear functor $F : A-\mathsf{mod} \to \Bbbk-\mathsf{Vec}$ is representable if and only if it has a right adjoint.

PROPOSITION 1.8.10. *An additive $\Bbbk$-linear functor $F : A-\mathsf{mod} \to B-\mathsf{mod}$ is representable if and only if it is right exact.*

PROOF. The "only if" direction is clear, as the tensor product functor is right exact. To prove the "if" direction, let F be a right exact functor. Let $V = F(A)$. Then V is a B-module which has a commuting right action of A, i.e., is a (B, A)-bimodule. We claim that $F(X)$ may be identified with $V \otimes_A X$ for all X, naturally in X. Indeed, this is clearly true if X is free. Let $M \to N \to X \to 0$ be an exact sequence such that M, N are free. If we apply F to this sequence and use that it is right exact, $F(X)$ gets identified with the cokernel of the map $V \otimes_A M \to V \otimes_A N$, which is canonically $V \otimes_A X$. It is easy to check that this isomorphism $F(X) \to V \otimes_A X$ is independent on the choice of the sequence $M \to N \to X \to 0$, and is functorial in X, so we are done. □

COROLLARY 1.8.11. *Let $\mathcal{C}$ be a finite abelian $\Bbbk$-linear category, and let $F : \mathcal{C} \to \mathsf{Vec}$ be an additive $\Bbbk$-linear left exact functor. Then $F = \mathsf{Hom}_{\mathcal{C}}(V, -)$ for some object $V \in \mathcal{C}$.*

PROOF. Let $\mathcal{C} = A-\mathsf{mod}$ for a finite dimensional algebra A. The functor $X \mapsto F(X^*)^*$ is a right exact functor on the category $A-\mathsf{mod}^\vee$. Hence, by Proposition 1.8.10, $F(X^*)^* = X \otimes_A V$, where $V \in A-\mathsf{mod}$. Thus, $F(X^*) = (X \otimes_A V)^* = \mathrm{Hom}_A(V, X^*)$, i.e., $F(Y) = \mathsf{Hom}_{\mathcal{C}}(V, Y)$ for $Y \in \mathcal{C}$. □

REMARK 1.8.12. The Yoneda Lemma states that morphisms between functors in Corollary 1.8.11 are precisely morphisms between the representing objects in $\mathcal{C}$. That is, a morphism between functors $\mathsf{Hom}_{\mathcal{C}}(V, -)$ and $\mathsf{Hom}_{\mathcal{C}}(W, -)$ is given by the right composition with a morphism $\phi : W \to V$ in $\mathcal{C}$.

Let $\mathcal{C}$ be a finite $\Bbbk$-linear abelian category. For any Y in $\mathcal{C}$ and simple X one has

$$\dim_{\Bbbk} \mathsf{Hom}_{\mathcal{C}}(P(X), Y) = [Y : X]. \tag{1.7}$$

Let $K_0(\mathcal{C})$ denote the free abelian group generated by isomorphism classes of indecomposable projective objects of $\mathcal{C}$.

DEFINITION 1.8.13. Elements of $K_0(\mathcal{C}) \otimes_{\mathbb{Z}} \Bbbk$ will be called *virtual projective objects* of $\mathcal{C}$.

We have an obvious homomorphism

$$\gamma : K_0(\mathcal{C}) \to \mathsf{Gr}(\mathcal{C}). \tag{1.8}$$

Although the groups $K_0(\mathcal{C})$ and $\mathsf{Gr}(\mathcal{C})$ have the same rank, in general γ is neither surjective nor injective even after tensoring with $\mathbb{Q}$, see Section 6.1.

DEFINITION 1.8.14. The matrix C of γ in the natural bases (i.e., the matrix with entries $[P(X) : Y]$, where X, Y run through isomorphism classes of simple objects of $\mathcal{C}$) will be called the *Cartan matrix* of $\mathcal{C}$.

Let $F_1,\ F_2 : \mathcal{C} \to \mathsf{Vec}$ be two exact faithful functors. Define a functor

$$F_1 \otimes F_2 : \mathcal{C} \times \mathcal{C} \to \mathsf{Vec}$$

given by

$$(X,\ Y) \mapsto F_1(X) \otimes F_2(Y).$$

PROPOSITION 1.8.15. *There is a canonical algebra isomorphism*

$$\alpha_{F_1,F_2} : \mathsf{End}(F_1) \otimes \mathsf{End}(F_2) \cong \mathsf{End}(F_1 \otimes F_2)$$

given by

$$\alpha_{F_1,F_2}(\eta_1 \otimes \eta_2)|_{F_1(X)\otimes F_2(Y)} := \eta_1|_{F_1(X)} \otimes \eta_2|_{F_2(Y)},$$

where $\eta_i \in \mathsf{End}(F_i)$, $i = 1, 2$.

EXERCISE 1.8.16. Prove Proposition 1.8.15.

The following proposition is due to O. Gabber (see [**De1**], Proposition 2.14).

PROPOSITION 1.8.17. *Let $\mathcal{C}$ be a locally finite abelian category. Suppose that there exists $X \in \mathcal{C}$ such that any object of $\mathcal{C}$ is a subquotient of a direct sum of finitely many copies of X. Then $\mathcal{C}$ has a projective generator, i.e., is a finite abelian category.*

DEFINITION 1.8.18. Let $F : \mathcal{C} \to \mathcal{D}$ be an exact functor between two locally finite abelian categories. The *image* of F is the full subcategory $\mathsf{Im}F \subset \mathcal{D}$ consisting of all objects contained as subquotients in $F(X)$ for some $X \in \mathcal{C}$.

PROPOSITION 1.8.19. *If $\mathcal{C}$ in Definition 1.8.18 is finite then $\mathsf{Im}F$ is a finite abelian category.*

PROOF. First of all, it is clear that $\mathsf{Im}F$ is a locally finite abelian category. Now let P be a projective generator of $\mathcal{C}$. Then any object of $\mathsf{Im}F$ is a subquotient of $F(P)^{\oplus n}$. Thus, according to Proposition 1.8.17, the category $\mathsf{Im}F$ is finite. $\square$

1.9. Coalgebras

DEFINITION 1.9.1. A *coalgebra* (with counit) over a field $\Bbbk$ is a $\Bbbk$-vector space C together with a *comultiplicaton* (or *coproduct*) $\Delta : C \to C \otimes C$ and *counit* $\varepsilon : C \to \Bbbk$ such that Δ and ε are $\Bbbk$-linear maps and

(i) Δ is coassociative, i.e.,

$$(\Delta \otimes \mathsf{id}_C) \circ \Delta = (\mathsf{id}_C \otimes \Delta) \circ \Delta$$

as maps $C \to C^{\otimes 3}$;

(ii) one has

$$(\varepsilon \otimes \mathsf{id}_C) \circ \Delta = (\mathsf{id}_C \otimes \varepsilon) \circ \Delta = \mathsf{id}_C$$

as maps $C \to C$ (the "counit axiom").

DEFINITION 1.9.2. A *left comodule* over a coalgebra C is a vector space M together with a linear map $\pi : M \to C \otimes M$ (called the coaction map), such that for any $m \in M$, one has

$$(\Delta \otimes \mathsf{id})(\pi(m)) = (\mathsf{id} \otimes \pi)(\pi(m)), \ (\varepsilon \otimes \mathsf{id})(\pi(m)) = m.$$

Similarly, a *right comodule* over C is a vector space M together with a linear map $\pi : M \to M \otimes C$, such that for any $m \in M$, one has

$$(\pi \otimes \mathsf{id})(\pi(m)) = (\mathsf{id} \otimes \Delta)(\pi(m)), \ (\mathsf{id} \otimes \varepsilon)(\pi(m)) = m.$$

For example, C is a left and right comodule over itself with $\pi = \Delta$.

EXERCISE 1.9.3. (i) Show that if C is a coalgebra then C^* is an algebra, and if A is a finite dimensional algebra then A^* is a coalgebra in a natural way.

(ii) Show that for any coalgebra C, any (left or right) C-comodule M is a (respectively, right or left) C^*-module, and the converse is true if C is finite dimensional.

EXERCISE 1.9.4. (i) Show that any coalgebra C is a sum of finite dimensional subcoalgebras.

Hint: let $c \in C$, and let

$$(\Delta \otimes \mathsf{id}) \circ \Delta(c) = (\mathsf{id} \otimes \Delta) \circ \Delta(c) = \sum_i c_i^{(1)} \otimes c_i^{(2)} \otimes c_i^{(3)}.$$

Show that $\mathrm{span}(c_i^{(2)})$ is a subcoalgebra of C containing c.

(ii) Show that any C-comodule is a sum of finite dimensional subcomodules.

PROPOSITION 1.9.5. *Let C be a coalgebra over $\Bbbk$. Then the category of finite dimensional left (or right) C-comodules is a locally finite $\Bbbk$-linear abelian category. If C is finite dimensional, this category is finite.*

PROOF. Using Exercise 1.9.3, the category of left C-comodules is a full subcategory of the category of right C^*-modules, which is the whole category if C is finite dimensional. This implies the statement. □

EXERCISE 1.9.6. Let A, B be coalgebras, and $f : A \to B$ a homomorphism. Let $F = f_* : A-\mathsf{comod} \to B-\mathsf{comod}$ be the corresponding pushforward functor. Then F is surjective if and only if f is surjective.

DEFINITION 1.9.7. Let C be a coalgebra. A nonzero element $x \in C$ is called *grouplike* if $\Delta(x) = x \otimes x$.

This terminology will be justified in Exercises 5.2.6 and 5.3.13.

REMARK 1.9.8. There is a bijection between grouplike elements of a coalgebra C and its one-dimensional subcoalgebras, given by $x \mapsto \Bbbk x$.

EXAMPLE 1.9.9. Let X be a set. Then $\Bbbk X$, the set of formal linear combinations of elements of X, is a coalgebra, with $\Delta(x) = x \otimes x$ for $x \in X$. The grouplike elements of $\Bbbk X$ are precisely elements $x \in X$.

DEFINITION 1.9.10. Let C be a coalgebra and let g, h be grouplike elements in C. An element $x \in C$ is called *skew-primitive* (or (g, h)*-skew-primitive*) if $\Delta(x) = g \otimes x + x \otimes h$.

The subspace of (g, h)-skew-primitive elements of C will be denoted $\mathsf{Prim}_{g,h}(C)$.

REMARK 1.9.11. Let g, h be grouplike elements of a coalgebra C. A multiple of $g - h$ is always a (g, h)-skew-primitive element. Such a skew-primitive element is called *trivial*.

In fact, the notion of a skew-primitive element has a categorical meaning. Namely, we have the following proposition.

PROPOSITION 1.9.12. *Let g and h be grouplike elements of a coalgebra C. The space $\mathsf{Prim}_{g,h}(C)/\Bbbk(g-h)$ is naturally isomorphic to $\mathsf{Ext}^1(h,g)$, where g, h are regarded as 1-dimensional right C-comodules.*

PROOF. Let V be a 2-dimensional C-comodule, such that we have an exact sequence

$$0 \to g \to V \to h \to 0.$$

Then V has a basis v_0, v_1 such that

$$\pi(v_0) = v_0 \otimes g,\ \pi(v_1) = v_0 \otimes x + v_1 \otimes h.$$

The condition that this is a comodule yields that x is a skew-primitive element of type (g, h). So any extension defines a skew-primitive element, and vice versa. Also, we can change the basis by $v_0 \to v_0$, $v_1 \to v_1 + \lambda v_0$, which modifies x by adding a trivial skew-primitive element. This implies the result. □

An important class of coalgebras is the class of pointed coalgebras.

DEFINITION 1.9.13. A coalgebra C is *pointed* if any simple right C-comodule is 1-dimensional.

REMARK 1.9.14. A finite dimensional coalgebra C is pointed if and only if the algebra C^* is *basic*, i.e., the quotient $C^*/\mathrm{Rad}(C^*)$ of C^* by its radical is commutative. In this case, simple C-comodules are points of $\mathrm{Spec}(C^*/\mathrm{Rad}(C^*))$ (here Spec stands for the set of maximal ideals), which justifies the term "pointed".

In the next section we will prove the following theorem:

THEOREM 1.9.15 (Takeuchi, [**Tak2**]). *Any essentially small locally finite abelian category $\mathcal{C}$ over a field $\Bbbk$ is equivalent to the category $C-\mathsf{comod}$ for a unique pointed coalgebra C. In particular, if $\mathcal{C}$ is finite, it is equivalent to the category $A-\mathsf{mod}$ for a unique basic algebra A (namely, $A = C^*$).*

1.10. The Coend construction

Let $\mathcal{C}$ be a $\Bbbk$-linear abelian category, and $F : \mathcal{C} \to \mathsf{Vec}$ an exact, faithful functor. In this case one can define the space $\mathsf{Coend}(F)$ as follows:

$$\mathsf{Coend}(F) := (\oplus_{X\in\mathcal{C}} F(X)^* \otimes F(X))/E \tag{1.9}$$

where E is spanned by elements of the form $y_* \otimes F(f)x - F(f)^* y_* \otimes x$, $x \in F(X)$, $y_* \in F(Y)^*$, $f \in \mathsf{Hom}(X, Y)$; in other words,

$$\mathsf{Coend}(F) = \varinjlim \mathsf{End}(F(X))^*.$$

Thus we have $\mathsf{End}(F) = \varprojlim \mathsf{End}(F(X)) = \mathsf{Coend}(F)^*$, which yields a coalgebra structure on $\mathsf{Coend}(F)$. So the algebra $\mathsf{End}(F)$ (which may be infinite dimensional) carries the inverse limit topology, in which a basis of neighborhoods of zero is formed by the kernels K_X of the maps $\mathsf{End}(F) \to \mathsf{End}(F(X))$, $X \in \mathcal{C}$, and $\mathsf{Coend}(F) = \mathsf{End}(F)^\vee$, the space of continuous linear functionals on $\mathsf{End}(F)$.

The following theorem is standard (see [**Tak2**]).

THEOREM 1.10.1. *Let $\mathcal{C}$ be a $\Bbbk$-linear abelian category with an exact faithful functor $F : \mathcal{C} \to \mathsf{Vec}$. Then $\mathcal{C}$ is locally finite, and F defines an equivalence between $\mathcal{C}$ and the category of finite dimensional right comodules over $C := \mathsf{Coend}(F)$ (or, equivalently, with the category of continuous finite dimensional left $\mathsf{End}(F)$-modules).*

PROOF. We sketch the proof, leaving the details to the reader. Consider the ind-object $Q := \oplus_{X\in\mathcal{C}} F(X)^* \otimes X$. For $X, Y \in \mathcal{C}$ and $f \in \mathsf{Hom}(X, Y)$, let

$$j_f : F(Y)^* \otimes X \to F(X)^* \otimes X \oplus F(Y)^* \otimes Y \subset Q$$

be the morphism defined by the formula

$$j_f = \mathsf{id} \otimes f - F(f)^* \otimes \mathsf{id}.$$

Let I be the quotient of Q by the image of the direct sum of all j_f.

The following statements are not hard to verify:

(i) I represents the functor $F(-)^*$, i.e., $\mathsf{Hom}(X, I)$ is naturally isomorphic to $F(X)^*$; in particular, I is injective.

(ii) $F(I) = C$, and I is naturally a left C-comodule. Its comodule structure is induced by the coevaluation morphism

$$F(X)^* \otimes X \xrightarrow{\mathsf{coev}_{F(X)}} F(X)^* \otimes F(X) \otimes F(X)^* \otimes X.$$

(iii) Let us regard F as a functor $\mathcal{C} \to C - \text{comod}$. For $M \in C - \text{comod}$, let $\theta_M : M \otimes I \to M \otimes C \otimes I$ be the morphism $\pi_M \otimes \mathsf{id} - \mathsf{id} \otimes \pi_I$, and let K_M be the kernel of θ_M. Then the functor $G : C - \text{comod} \to \mathcal{C}$ given by the formula $G(M) = \mathsf{Ker}\, \theta_M$, is a quasi-inverse to F.

This completes the proof. □

EXERCISE 1.10.2. (i) Check statements (i)-(iii) in the proof of Theorem 1.10.1.

(ii) Let $\mathcal{C}$ be the category of finite dimensional representations of a finite (or, more generally, affine algebraic) group G over $\Bbbk$. Deduce directly from the definition of the object I from the proof of Theorem 1.10.1 that $I \cong \mathcal{O}(G)$, the algebra of regular functions on G, with the action of G by right multiplication (you may first consider the case when $\mathcal{C}$ is semisimple).

Proof of Theorem 1.9.15. Now let us use the Coend construction to prove Theorem 1.9.15. Consider the dual category $\mathcal{C}^\vee$ (which is also locally finite). It is known that the ind-completion of $\mathcal{C}^\vee$ is a Grothendieck category, so it has enough injectives.[4] This means that every simple object L of $\mathcal{C}^\vee$ has an injective hull Q_L in the ind-completion $\mathsf{Ind}(\mathcal{C}^\vee)$. Let $Q = \oplus_L Q_L$, where the sum is taken over isomorphism classes of simple objects of $\mathcal{C}^\vee$. Consider the (covariant) functor $F : \mathcal{C} \to \mathsf{Vec}$ given by $F(X) = \mathsf{Hom}_{\mathcal{C}^\vee}(X, Q)$. The functor F is exact and faithful, and for every simple object $L \in \mathcal{C}$, $\dim F(L) = 1$. This means that $C := \mathsf{Coend}(F)$ is a pointed coalgebra, and by Theorem 1.10.1, we have an equivalence of categories $C-\mathsf{comod} \cong \mathcal{C}$. Moreover, suppose C' is another pointed coalgebra, and $G : C'-\mathsf{comod} \to C-\mathsf{comod}$ is an equivalence. Then $Q = G(C')$ is an injective object in the ind-completion of $C-\mathsf{comod}^\vee$ such that $\dim \mathsf{Hom}(L, Q) = 1$ for any simple object L. This implies that $Q = C$, hence C' is isomorphic to C. Theorem 1.9.15 is proved.

1.11. Deligne's tensor product of locally finite abelian categories

Let $\mathcal{C}, \mathcal{D}$ be two locally finite abelian categories over a field $\Bbbk$.

DEFINITION 1.11.1. *Deligne's tensor product* $\mathcal{C} \boxtimes \mathcal{D}$ is an abelian $\Bbbk$-linear category which is universal for the functor assigning to every $\Bbbk$-linear abelian category $\mathcal{A}$ the category of right exact in both variables bilinear bifunctors $\mathcal{C} \times \mathcal{D} \to \mathcal{A}$. That is, there is a bifunctor

$$\boxtimes : \mathcal{C} \times \mathcal{D} \to \mathcal{C} \boxtimes \mathcal{D} : (X, Y) \mapsto X \boxtimes Y$$

which is right exact in both variables and is such that for any right exact in both variables bifunctor $F : \mathcal{C} \times \mathcal{D} \to \mathcal{A}$ there exists a unique right exact functor $\bar{F} : \mathcal{C} \boxtimes \mathcal{D} \to \mathcal{A}$ satisfying $\bar{F} \circ \boxtimes = F$.

PROPOSITION 1.11.2. (i) *A Deligne's tensor product $\mathcal{C} \boxtimes \mathcal{D}$ exists and is a locally finite abelian category.*

(ii) *It is unique up to a unique equivalence.*

(iii) *Let C, D be coalgebras and let $\mathcal{C} = C-\mathsf{comod}$ and $\mathcal{D} = D-\mathsf{comod}$. Then $\mathcal{C} \boxtimes \mathcal{D} = (C \otimes D)-\mathsf{comod}$.*

(iv) *The bifunctor $\boxtimes$ is exact in both variables and satisfies*

$$\mathsf{Hom}_{\mathcal{C}}(X_1, Y_1) \otimes \mathsf{Hom}_{\mathcal{D}}(X_2, Y_2) \cong \mathsf{Hom}_{\mathcal{C} \boxtimes \mathcal{D}}(X_1 \boxtimes X_2, Y_1 \boxtimes Y_2).$$

(v) *Any bilinear bifunctor $F : \mathcal{C} \times \mathcal{D} \to \mathcal{A}$ exact in each variable defines an exact functor $\bar{F} : \mathcal{C} \boxtimes \mathcal{D} \to \mathcal{A}$.*

PROOF. We only give a sketch. Part (ii) follows from the universal property in the usual way. To prove (i), take coalgebras C, D such that $\mathcal{C} = C-\mathsf{comod}$, $\mathcal{D} = D-\mathsf{comod}$ (such coalgebras exist by Theorem 1.9.15). Then one can define

[4] The notion of a Grothendieck category comes from the paper [**Gr**]; see [**KashS**] for basics of the theory of Grothendieck categories.

$\mathcal{C} \boxtimes \mathcal{D} = (C \otimes D)-\mathsf{comod}$, and it is easy to show that it satisfies the required conditions. This together with (ii) also implies (iii). Parts (iv) and (v) are routine. □

Deligne's tensor product can also be applied to functors. If $F : \mathcal{C} \to \mathcal{C}'$ and $G : \mathcal{D} \to \mathcal{D}'$ are right exact functors between locally finite abelian categories then one defines the functor $F \boxtimes G : \mathcal{C} \boxtimes \mathcal{D} \to \mathcal{C}' \boxtimes \mathcal{D}'$ using the defining universal property (see Definition 1.11.1) of $\mathcal{C} \boxtimes \mathcal{D}$. Namely, the bifunctor

$$F \times G : \mathcal{C} \times \mathcal{D} \to \mathcal{C}' \boxtimes \mathcal{D}' : (V, W) \mapsto F(V) \boxtimes G(W)$$

canonically extends to a right exact functor $F \boxtimes G : \mathcal{C} \boxtimes \mathcal{D} \to \mathcal{C}' \boxtimes \mathcal{D}'$.

1.12. The finite dual of an algebra

It follows from Exercise 1.9.3 that if A is a finite dimensional algebra, then its dual A^* is a coalgebra. However, this is no longer true if A is infinite dimensional: the dual map m^*, where $m : A \otimes A \to A$ is the multiplication of A, maps A^* to $(A \otimes A)^*$, which is bigger than $A^* \otimes A^*$, so it does not induce the structure of a coalgebra on A^*. However, it turns out that there is a natural subspace in A^* which carries the structure of a coalgebra.

Definition 1.12.1. The finite dual A^*_{fin} of A is the collection of all $f \in A^*$ that vanish on a (two-sided) ideal of finite codimension.

Note that A^*_{fin} is a subspace of A^*. Indeed, if $f,\, g \in A^*$ vanish on ideals I and J, respectively, then $f + g$ vanishes on $I \cap J$. If I and J have finite codimension, then so does $I \cap J$.

Proposition 1.12.2. *The maps $\Delta := m^*$ and $\varepsilon := u^*$, where $m : A \otimes A \to A$ and $u : \Bbbk \to A$ are the multiplication and unit of A, define a coalgebra structure on A^*_{fin}.*

Proof. Take $f \in A^*_{\mathrm{fin}}$ and let $I \subset A$ be an ideal of finite codimension such that $f \in I^\perp$. Then $\Delta(f)$ vanishes on $I \otimes A + A \otimes I$ (since $\Delta(f)(a \otimes b) = f(ab)$ for $a,\, b \in A$) and, hence, $\Delta(f) \in I^\perp \otimes I^\perp \subset A^*_{\mathrm{fin}} \otimes A^*_{\mathrm{fin}}$. The axioms of a coalgebra follow by duality. □

Remark 1.12.3. Note that if A does not have finite dimensional modules, then $A^*_{\mathrm{fin}} = 0$.

1.13. Pointed coalgebras and the coradical filtration

Let $\mathcal{C}$ be a locally finite abelian category.

Any object $X \in \mathcal{C}$ has a canonical filtration

$$0 = X_0 \subset X_1 \subset X_2 \subset \cdots \subset X_n = X \tag{1.10}$$

such that X_{i+1}/X_i is the socle (i.e., the maximal semisimple subobject) of X/X_i (in other words, X_{i+1}/X_i is the sum of all simple subobjects of X/X_i).

Definition 1.13.1. The filtration of X by X_i is called the *socle filtration* or the *coradical filtration* of X.

It is easy to show by induction that the socle filtration is a filtration of X of the smallest possible length, such that the successive quotients are semisimple. The length of the socle filtration of X is called the *Loewy length* of X, and denoted $\mathrm{Lw}(X)$. Then we have a filtration of the category $\mathcal{C}$ by Loewy length of objects: $\mathcal{C}_0 \subset \mathcal{C}_1 \subset ...$, where $\mathcal{C}_i$ denotes the full subcategory of objects of $\mathcal{C}$ of Loewy length $\leq i+1$. Clearly, the Loewy length of any subquotient of an object X does not exceed the Loewy length of X, so the categories $\mathcal{C}_i$ are closed under taking subquotients.

DEFINITION 1.13.2. The filtration of the category $\mathcal{C}$ by $\mathcal{C}_i$ is called the *socle filtration* or the *coradical filtration* of $\mathcal{C}$.

If $\mathcal{C}$ is endowed with an exact faithful functor $F : \mathcal{C} \to \mathsf{Vec}$ then we can define the coalgebra $C = \mathsf{Coend}(F)$ and its subcoalgebras $C_i = \mathsf{Coend}(F|_{\mathcal{C}_i})$, and we have $C_i \subset C_{i+1}$ and $C = \cup_i C_i$ (alternatively, we can say that C_i is spanned by matrix elements of C-comodules $F(X)$, $X \in \mathcal{C}_i$.[5] Thus we have defined an increasing filtration by subcoalgebras of any coalgebra C. This filtration is called the *coradical filtration*, and the term C_0 is called the *coradical* of C.

The "linear algebra" definition of the coradical filtration is as follows. One says that a coalgebra is *simple* if it does not have nontrivial subcoalgebras, i.e., if it is finite dimensional, and its dual is a simple (i.e., matrix) algebra. Then C_0 is the sum of all simple subcoalgebras of C. The coalgebras C_{n+1} for $n \geq 1$ are then defined inductively to be the spaces of those $x \in C$ for which

$$\Delta(x) \in C_n \otimes C + C \otimes C_0.$$

EXERCISE 1.13.3. (i) Suppose that C is a finite dimensional coalgebra, and I is the Jacobson radical of C^*. Show that $C_n^\perp = I^{n+1}$, and generalize this statement to the infinite dimensional case. This justifies the term "coradical filtration".

(ii) Show that the coproduct respects the coradical filtration, i.e., $\Delta(C_n) \subset \sum_{i=0}^n C_i \otimes C_{n-i}$.

(iii) Show that C_0 is the *direct* sum of simple subcoalgebras of C. In particular, grouplike elements of any coalgebra C are linearly independent.

Hint: Simple subcoalgebras of C correspond to finite dimensional irreducible representations of C^*.

(iv) We have defined C_i in three ways: as $\mathsf{Coend}(F|_{\mathcal{C}_i})$, as the span of matrix elements of $F(X)$, $X \in \mathcal{C}_i$, and by the "linear algebra" definition above. Show that these three definitions agree.

Let

$$\mathsf{gr}(C) := \bigoplus_{i=0}^{\infty} C_{i+1}/C_i \qquad (1.11)$$

be the associated graded coalgebra of a coalgebra C with respect to the coradical filtration. Then $\mathsf{gr}(C)$ is a $\mathbb{Z}_+$-graded coalgebra. It is easy to see from Exercise 1.13.3(i) that the coradical filtration of $\mathsf{gr}(C)$ is induced by its grading. A graded coalgebra $\bar{C}$ with this property (i.e., one isomorphic to $\mathsf{gr}(C)$ for some coalgebra C) is said to be *coradically graded*, and a coalgebra C such that $\mathsf{gr}(C) = \bar{C}$ is called a *lifting* of C.

[5]If M is a right C-comodule with coaction $\pi : M \to M \otimes C$ then a matrix element of M is an element $(f \otimes 1, \pi(m)) \in C$, where $f \in M^*$, $m \in M$.

DEFINITION 1.13.4. A coalgebra C is said to be *cosemisimple* if C is a direct sum of simple subcoalgebras.

Clearly, a coalgebra C is cosemisimple if and only if $C-\mathsf{comod}$ is a semisimple category.

PROPOSITION 1.13.5. *A locally finite abelian category $\mathcal{C}$ is semisimple if and only if $\mathcal{C}_0 = \mathcal{C}_1$. In particular, a coalgebra C is cosemisimple if and only if $C_0 = C_1$.*

PROOF. The semisimplicity of $\mathcal{C}$ means that $\mathsf{Ext}^1(X,Y) = 0$ for any simple objects X, Y of $\mathcal{C}$, which implies (by the long exact sequence of cohomology) that $\mathsf{Ext}^1(X,Y) = 0$ for all objects $X, Y \in \mathcal{C}$. □

COROLLARY 1.13.6. *(The Taft-Wilson theorem,* [**Mon**]*, Theorem 5.4.1) If C is a pointed coalgebra, then C_0 is spanned by (linearly independent) grouplike elements g, and*

$$C_1/C_0 \cong \bigoplus_{g,h} \mathsf{Prim}_{g,h}(C)/\Bbbk(g-h).$$

In particular, any non-cosemisimple pointed coalgebra contains nontrivial skew-primitive elements.

PROOF. The first statement is clear (the linear independence follows from Exercise 1.13.3(iii)). Also, it is clear that any skew-primitive element is contained in C_1. Now, if $x \in C_1$, then x is a matrix element of a C-comodule of Loewy length ≤ 2, so it is a sum of matrix elements of 2-dimensional comodules, i.e., a linear combination of grouplike and skew-primitive elements.

It remains to show that the sum $\sum_{g,h} \mathsf{Prim}_{g,h}(C)/\Bbbk(g-h) \subset C/C_0$ is direct. For this, it suffices to consider the case when C is finite dimensional. Passing to the dual algebra $A = C^*$, we see that the statement is equivalent to the claim that I/I^2 (where I is the radical of A) is isomorphic (in a natural way) to $\oplus_{g,h} \mathsf{Ext}^1(h,g)^*$.

Let p_g be a complete system of orthogonal idempotents in A/I^2, such that $h(p_g) = \delta_{hg}$. Define a pairing $I/I^2 \times \mathsf{Ext}^1(h,g) \to \Bbbk$ which sends $a \otimes \alpha$ to the upper right entry of the 2-by-2 matrix by which a acts in the extension of h by g defined by α. It is easy to see that this pairing descends to a pairing $B : p_h(I/I^2)p_g \times \mathsf{Ext}^1(h,g) \to \Bbbk$. If the extension α is nontrivial, the upper right entry cannot be zero, so B is right-non-degenerate. Similarly, if a belongs to the left kernel of B, then a acts by zero in any A-module of Loewy length 2, so $a = 0$. Thus, B is left-non-degenerate (i.e., its left kernel is zero). This implies the required isomorphism. □

EXERCISE 1.13.7. Prove the following generalization of the Taft-Wilson theorem: if C is a pointed coalgebra then for any $n \geq 1$, C_n is spanned by elements x such that $\Delta(x) = g \otimes x + x \otimes h + w$, where $w \in C_{n-1} \otimes C_{n-1}$, and g, h are grouplike elements of C.

PROPOSITION 1.13.8. *(*[**Mon**]*, Theorem 5.3.1) If C, D are coalgebras, and $f : C \to D$ is a coalgebra homomorphism such that $f|_{C_1}$ is injective, then f is injective.*

PROOF. One may assume that C and D are finite dimensional. Then the statement can be translated into the following statement about finite dimensional algebras: if A, B are finite dimensional algebras and $f : A \to B$ is an algebra homomorphism which descends to a surjective homomorphism $A \to B/\mathrm{Rad}(B)^2$, then f is surjective.

To prove this statement, let $b \in B$. Let $I = \mathrm{Rad}(B)$. We prove by induction on n that there exists $a \in A$ such that $b - f(a) \in I^n$. The base of induction is clear, so we only need to do the step of induction. So assume $b \in I^n$. We may assume that $b = b_1...b_n$, $b_i \in I$, and let $a_i \in A$ be such that $f(a_i) = b_i$ modulo I^2. Let $a = a_1...a_n$. Then $b - f(a) \in I^{n+1}$, as desired. □

1.14. Bibliographical notes

1.1-1.7. Abelian categories were introduced by Grothendieck [**Gr**] and Buchsbaum [**Bu**] (under the name of exact categories). Mitchell's Theorem 1.3.8 appeared in [**Mit**]. See the books by Freyd [**Fr**] and Kashiwara and Shapira [**KashS**] for basics of the theory of abelian categories and the books by Hilton and Stammbach [**HiS**], Gelfand and Manin [**GelfM**], and Weibel [**Weib**] for homological algebra.

1.8. For a discussion of artinian (or locally finite) categories see Gabriel's thesis [**Gabr**] (note that the terminology there is a bit different from ours). For a theory of finite abelian categories, see the book by Gabriel and Roiter [**GabR**].

1.9. For the basic theory of coalgebras see the book by Montgomery [**Mon**] and the survey paper by Chin [**Chi**].

1.10. For the Coend construction and reconstruction theory for coalgebras, see the paper by Takeuchi [**Tak2**] and the paper of Schauenburg [**Schau1**].

1.11. Construction of the tensor product of artinian categories is given by Deligne [**De1**].

1.12. The finite dual of an algebra is discussed by Cartier in [**Car2**, p. 569] (under the name "reduced dual").

1.13. Pointed coalgebras, the coradical filtration, and the Taft-Wilson theorem are discussed in [**Mon**].

CHAPTER 2

Monoidal categories

2.1. Definition of a monoidal category

A good way of thinking about category theory (which will be especially useful throughout this book) is that category theory is a refinement (or "categorification") of ordinary algebra. In other words, there exists a dictionary between these two subjects, such that usual algebraic structures are recovered from the corresponding categorical structures by passing to the set of isomorphism classes of objects.

For example, the notion of a category is a categorification of the notion of a set. Similarly, abelian categories are a categorification of abelian groups [1] (which justifies the terminology).

This dictionary goes surprisingly far, and many important constructions below will come from an attempt to enter into it a categorical "translation" of an algebraic notion.

In particular, the notion of a monoidal category is the categorification of the notion of a monoid.

Recall that a monoid may be defined as a set C with an associative multiplication operation $(x, y) \to x \cdot y$ (i.e., a semigroup), with an element 1 such that $1^2 = 1$ and the maps $x \mapsto 1 \cdot x$, $x \mapsto x \cdot 1 : C \to C$ are bijections. It is easy to show that in a semigroup, the last condition is equivalent to the usual unit axiom $1 \cdot x = x \cdot 1 = x$.[2]

As usual in category theory, to categorify the definition of a monoid, we should replace the equalities in the definition of a monoid (namely, the associativity equation $(xy)z = x(yz)$ and the equation $1^2 = 1$) by isomorphisms satisfying some consistency properties, and the word "bijection" by the word "equivalence" (of categories). This leads to the following definition.

DEFINITION 2.1.1. A *monoidal category* is a quintuple $(\mathcal{C}, \otimes, a, \mathbf{1}, \iota)$ where $\mathcal{C}$ is a category, $\otimes : \mathcal{C} \times \mathcal{C} \to \mathcal{C}$ is a bifunctor called the *tensor product* bifunctor, $a : (- \otimes -) \otimes - \xrightarrow{\sim} - \otimes (- \otimes -)$ is a natural isomorphism:

$$(2.1) \qquad a_{X,Y,Z} : (X \otimes Y) \otimes Z \xrightarrow{\sim} X \otimes (Y \otimes Z), \qquad X, Y, Z \in \mathcal{C}$$

called the *associativity constraint* (or *associativity isomorphism*), $\mathbf{1} \in \mathcal{C}$ is an object of $\mathcal{C}$, and $\iota : \mathbf{1} \otimes \mathbf{1} \xrightarrow{\sim} \mathbf{1}$ is an isomorphism, subject to the following two axioms.

---

[1] To be more precise, the set of isomorphism classes of objects in an abelian category $\mathcal{C}$ is a commutative monoid, but one usually extends it to a group by considering "virtual objects" of the form $X - Y$, $X, Y \in \mathcal{C}$.

[2] Indeed, if left and right multiplication by 1 are bijections and $1^2 = 1$, then we have $1 \cdot 1 \cdot x = 1 \cdot x$, hence $1 \cdot x = x$, and similarly $x \cdot 1 = x$.

1. The pentagon axiom. The diagram

$$(2.2)\qquad \begin{array}{ccccc} & & ((W\otimes X)\otimes Y)\otimes Z & & \\ & \swarrow^{a_{W,X,Y}\otimes \mathrm{id}_Z} & & \searrow^{a_{W\otimes X,Y,Z}} & \\ (W\otimes (X\otimes Y))\otimes Z & & & & (W\otimes X)\otimes (Y\otimes Z) \\ \downarrow{\scriptstyle a_{W,X\otimes Y,Z}} & & & & \downarrow{\scriptstyle a_{W,X,Y\otimes Z}} \\ W\otimes ((X\otimes Y)\otimes Z) & & \xrightarrow{\mathrm{id}_W\otimes a_{X,Y,Z}} & & W\otimes (X\otimes (Y\otimes Z)) \end{array}$$

is commutative for all objects W, X, Y, Z in $\mathcal{C}$.

2. The unit axiom. The functors

$$(2.3)\qquad L_{\mathbf{1}} \;:\; X\mapsto \mathbf{1}\otimes X \quad \text{and}$$

$$(2.4)\qquad R_{\mathbf{1}} \;:\; X\mapsto X\otimes \mathbf{1}$$

of left and right multiplication by $\mathbf{1}$ are autoequivalences of $\mathcal{C}$.

DEFINITION 2.1.2. The pair $(\mathbf{1},\iota)$ is called the *unit object* of $\mathcal{C}$.[3]

REMARK 2.1.3. An alternative (and, perhaps, more traditional) definition of a monoidal category is given in the next section, see Definition 2.2.8.

We see that the set of isomorphism classes of objects in a monoidal category indeed has a natural structure of a monoid, with multiplication $\otimes$ and unit $\mathbf{1}$. Thus, in the categorical-algebraic dictionary, monoidal categories indeed correspond to monoids (which explains their name).

DEFINITION 2.1.4. A *monoidal subcategory* of a monoidal category $(\mathcal{C},\otimes,a,\mathbf{1},\iota)$ is a quintuple $(\mathcal{D},\otimes,a,\mathbf{1},\iota)$, where $\mathcal{D}\subset\mathcal{C}$ is a subcategory closed under the tensor product of objects and morphisms and containing $\mathbf{1}$ and ι.

Unless otherwise specified, we will always consider full monoidal subcategories.

DEFINITION 2.1.5. Let $(\mathcal{C}, \otimes, a, \mathbf{1}, \iota)$ be a monoidal category. The monoidal category $(\mathcal{C}^{\mathrm{op}}, \otimes^{\mathrm{op}}, \mathbf{1}, a^{\mathrm{op}}, \iota)$ *opposite* to $\mathcal{C}$ is defined as follows. As a category $\mathcal{C}^{\mathrm{op}}=\mathcal{C}$, its tensor product is given by $X\otimes^{\mathrm{op}} Y := Y\otimes X$ and the associativity constraint of $\mathcal{C}^{\mathrm{op}}$ is $a^{\mathrm{op}}_{X,Y,Z} := a^{-1}_{Z,Y,X}$.

REMARK 2.1.6. The notion of the opposite monoidal category is not to be confused with the usual notion of the *dual* category, which is the category $\mathcal{C}^{\vee}$ obtained from $\mathcal{C}$ by reversing arrows (for any category $\mathcal{C}$). Note that if $\mathcal{C}$ is monoidal, so is $\mathcal{C}^{\vee}$ (in a natural way), which makes it even easier to confuse the two notions.

2.2. Basic properties of unit objects

Let $(\mathcal{C},\otimes,a,\mathbf{1},\iota)$ be a monoidal category. Define natural isomorphisms

$$(2.5)\qquad l_X:\mathbf{1}\otimes X\to X \quad\text{and}\quad r_X: X\otimes\mathbf{1}\to X$$

in such a way that $L_{\mathbf{1}}(l_X)$ and $R_{\mathbf{1}}(r_X)$ are equal, respectively, to the compositions

$$(2.6)\qquad \mathbf{1}\otimes(\mathbf{1}\otimes X) \xrightarrow{a^{-1}_{\mathbf{1},\mathbf{1},X}} (\mathbf{1}\otimes\mathbf{1})\otimes X \xrightarrow{\iota\otimes\mathrm{id}_X} \mathbf{1}\otimes X,$$

$$(2.7)\qquad (X\otimes\mathbf{1})\otimes\mathbf{1} \xrightarrow{a_{X,\mathbf{1},\mathbf{1}}} X\otimes(\mathbf{1}\otimes\mathbf{1}) \xrightarrow{\mathrm{id}_X\otimes\iota} X\otimes\mathbf{1}.$$

[3] We note that there is no condition on the isomorphism ι, so it can be chosen arbitrarily.

DEFINITION 2.2.1. Isomorphisms (2.5) are called the left and right *unit constraints* or *unit isomorphisms.*

The unit constraints provide a categorical counterpart of the unit axiom $1X = X1 = X$ of a monoid in the same sense as the associativity isomorphism provides the categorical counterpart of the associativity equation.

PROPOSITION 2.2.2. *For any object X in $\mathcal{C}$ there are equalities*

$$l_{\mathbf{1}\otimes X} = \mathsf{id}_{\mathbf{1}} \otimes l_X \quad \text{and} \quad r_{X\otimes \mathbf{1}} = r_X \otimes \mathsf{id}_{\mathbf{1}}. \tag{2.8}$$

PROOF. It follows from naturality of the left unit constraint l that the following diagram commutes

$$\begin{array}{ccc} \mathbf{1}\otimes(\mathbf{1}\otimes X) & \xrightarrow{\mathbf{1}\otimes l_X} & \mathbf{1}\otimes X \\ {\scriptstyle l_{\mathbf{1}\otimes X}}\downarrow & & \downarrow{\scriptstyle l_X} \\ \mathbf{1}\otimes X & \xrightarrow{l_X} & X. \end{array} \tag{2.9}$$

Since l_X is an isomorphism, the first identity follows. The second one follows similarly from naturality of r. □

PROPOSITION 2.2.3. *The "triangle" diagram*

$$\begin{array}{ccc} (X\otimes\mathbf{1})\otimes Y & \xrightarrow{a_{X,\mathbf{1},Y}} & X\otimes(\mathbf{1}\otimes Y) \\ & {\scriptstyle r_X\otimes \mathsf{id}_Y}\searrow \quad \swarrow{\scriptstyle \mathsf{id}_X\otimes l_Y} & \\ & X\otimes Y & \end{array} \tag{2.10}$$

is commutative for all $X, Y \in \mathcal{C}$.

PROOF. Consider the following diagram:

(2.11)

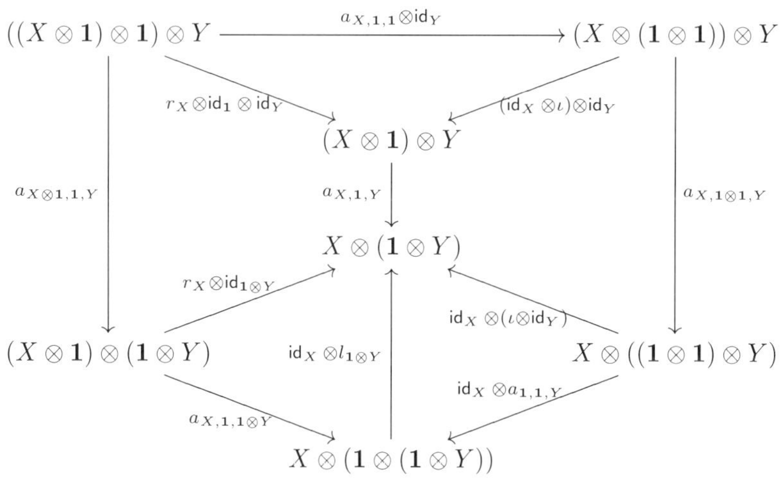

To prove the proposition, it suffices to establish the commutativity of the bottom left triangle (as any object of $\mathcal{C}$ is isomorphic to one of the form $\mathbf{1}\otimes Y$). Since the outside pentagon is commutative (by the pentagon axiom), it suffices to establish the commutativity of the other parts of the pentagon. Now, the two quadrangles are commutative due to the functoriality of the associativity isomorphisms, the commutativity of the upper triangle is the definition of r, and the commutativity of the lower right triangle holds by Proposition 2.2.2. $\square$

PROPOSITION 2.2.4. *The following diagrams commute for all objects* $X, Y \in \mathcal{C}$*:*

(2.12)

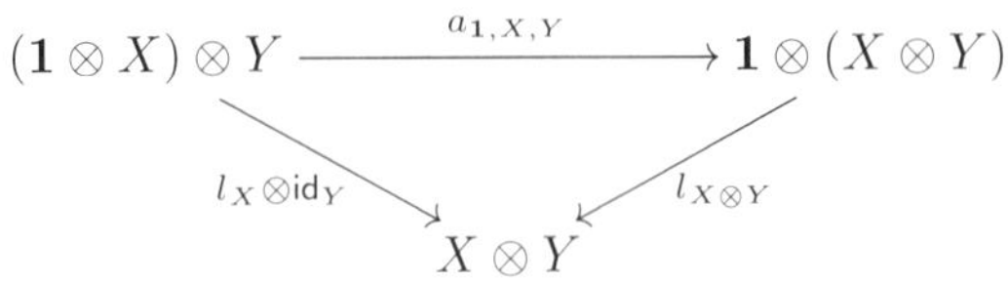

(2.13)

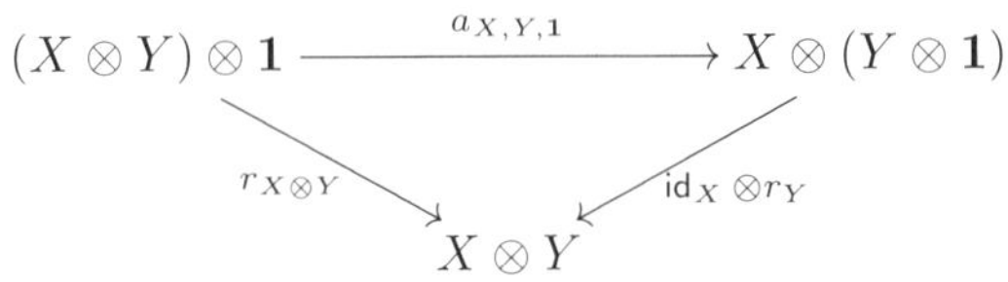

PROOF. Consider the diagram

(2.14)

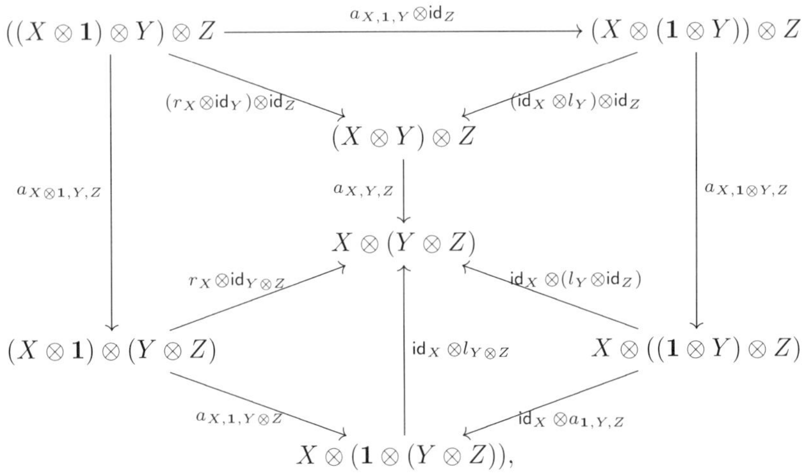

where X, Y, Z are objects in $\mathcal{C}$. The outside pentagon commutes by the pentagon axiom (2.2). The functoriality of a implies the commutativity of the two middle quadrangles. The triangle axiom (2.10) implies the commutativity of the upper triangle and the lower left triangle. Consequently, the lower right triangle commutes as well. Setting $X = \mathbf{1}$ and applying the functor $L_{\mathbf{1}}^{-1}$ to the lower right triangle, we obtain commutativity of the triangle (2.12). The commutativity of the triangle (2.13) is proved similarly. $\square$

COROLLARY 2.2.5. *In any monoidal category* $l_{\mathbf{1}} = r_{\mathbf{1}} = \iota$.

PROOF. Set $Y = Z = \mathbf{1}$ in (2.12). We have:

$$l_{\mathbf{1}} \otimes \mathrm{id}_{\mathbf{1}} = l_{\mathbf{1}\otimes\mathbf{1}} \circ a_{\mathbf{1},\mathbf{1},\mathbf{1}} = (\mathrm{id}_{\mathbf{1}} \otimes l_{\mathbf{1}}) \circ a_{\mathbf{1},\mathbf{1},\mathbf{1}}.$$

Next, setting $X = Y = \mathbf{1}$ in the triangle axiom (2.10) we obtain

$$r_{\mathbf{1}} \otimes \mathrm{id}_{\mathbf{1}} = (\mathrm{id}_{\mathbf{1}} \otimes l_{\mathbf{1}}) \circ a_{\mathbf{1},\mathbf{1},\mathbf{1}}.$$

By the definition of the unit constraint $(\mathrm{id}_{\mathbf{1}} \otimes l_{\mathbf{1}}) \circ a_{\mathbf{1},\mathbf{1},\mathbf{1}} = \iota \otimes \mathrm{id}_{\mathbf{1}}$. Hence, $r_{\mathbf{1}} \otimes \mathrm{id}_{\mathbf{1}} = l_{\mathbf{1}} \otimes \mathrm{id}_{\mathbf{1}} = \iota \otimes \mathrm{id}_{\mathbf{1}}$ and $r_{\mathbf{1}} = l_{\mathbf{1}} = \iota$ since $R_{\mathbf{1}}$ is an equivalence. $\square$

PROPOSITION 2.2.6. *The unit object in a monoidal category is unique up to a unique isomorphism.*

PROOF. Let $(\mathbf{1}, \iota)$, $(\mathbf{1}', \iota')$ be two unit objects. Let (r, l), (r', l') be the corresponding unit constraints. Then we have the isomorphism $\eta := l_{\mathbf{1}'} \circ (r'_{\mathbf{1}})^{-1} : \mathbf{1} \xrightarrow{\sim} \mathbf{1}'$.

It is easy to show using the commutativity of the above triangle diagrams that η maps ι to ι'. It remains to show that η is the only isomorphism with this property. To do so, it suffices to show that if $b : \mathbf{1} \xrightarrow{\sim} \mathbf{1}$ is an isomorphism such that the diagram

$$\begin{array}{ccc} \mathbf{1} \otimes \mathbf{1} & \xrightarrow{b \otimes b} & \mathbf{1} \otimes \mathbf{1} \\ \downarrow \iota & & \downarrow \iota \\ \mathbf{1} & \xrightarrow[b]{} & \mathbf{1} \end{array} \tag{2.15}$$

is commutative, then $b = \mathrm{id}_{\mathbf{1}}$. To see this, it suffices to note that for any morphism $c : \mathbf{1} \to \mathbf{1}$ the diagram

$$\begin{array}{ccc} \mathbf{1} \otimes \mathbf{1} & \xrightarrow{c \otimes \mathrm{id}_{\mathbf{1}}} & \mathbf{1} \otimes \mathbf{1} \\ \downarrow \iota & & \downarrow \iota \\ \mathbf{1} & \xrightarrow[c]{} & \mathbf{1} \end{array} \tag{2.16}$$

is commutative (since $\iota = r_{\mathbf{1}}$ by Corollary 2.2.5), so $b \otimes b = b \otimes \mathrm{id}_{\mathbf{1}}$ and hence $b = \mathrm{id}_{\mathbf{1}}$. □

EXERCISE 2.2.7. Verify the assertion in the proof of Proposition 2.2.6 that η maps ι to ι'.

Hint: use Propositions 2.2.3 and 2.2.4.

The results of this section show that a monoidal category can be alternatively defined as follows:

DEFINITION 2.2.8. A monoidal category is a sextuple $(\mathcal{C}, \otimes, a, \mathbf{1}, l, r)$ satisfying the pentagon axiom (2.2) and the triangle axiom (2.10).

REMARK 2.2.9. Definition 2.2.8 is perhaps more traditional than Definition 2.1.1, but Definition 2.1.1 is simpler. Besides, Proposition 2.2.6 implies that for a triple $(\mathcal{C}, \otimes, a)$ satisfying a pentagon axiom (which should perhaps be called a "*semigroup category*", as it categorifies the notion of a semigroup), being a monoidal category is a property and not a structure (similarly to how it is for semigroups and monoids).

Furthermore, one can show that the commutativity of the triangles implies that in a monoidal category one can safely identify $\mathbf{1} \otimes X$ and $X \otimes \mathbf{1}$ with X using the unit isomorphisms, and assume that the unit isomorphisms are the identities (which we will usually do from now on).[4]

In a sense, all this means that in constructions with monoidal categories, unit objects and isomorphisms always "go along for the ride", and one need not worry about them especially seriously. For this reason, below we will typically take less care dealing with them than we have done in this section.

[4] We will return to this issue later when we discuss Mac Lane's coherence theorem is Section 2.9.

PROPOSITION 2.2.10. *Let $\mathcal{C}$ be a monoidal category. Then $\mathsf{End}_{\mathcal{C}}(\mathbf{1})$ is a commutative monoid under composition. Furthermore, $f \otimes g = \iota^{-1} \circ (f \circ g) \circ \iota$ for all $f, g \in \mathsf{End}_{\mathcal{C}}(\mathbf{1})$.*

PROOF. By naturality of unit constraints of $\mathcal{C}$ we have

$$f \otimes \mathsf{id}_{\mathbf{1}} = r_{\mathbf{1}}^{-1} \circ f \circ r_{\mathbf{1}} \quad \text{and} \quad \mathsf{id}_{\mathbf{1}} \otimes g = l_{\mathbf{1}}^{-1} \circ g \circ l_{\mathbf{1}}.$$

Combining this with the identity $r_{\mathbf{1}} = l_{\mathbf{1}} = \iota$ from Corollary 2.2.5 we obtain

$$\begin{aligned} f \otimes g &= (f \otimes \mathsf{id}_{\mathbf{1}}) \circ (\mathsf{id}_{\mathbf{1}} \otimes g) = \iota^{-1} \circ (f \circ g) \circ \iota, \\ g \otimes f &= (\mathsf{id}_{\mathbf{1}} \otimes f) \circ (g \otimes \mathsf{id}_{\mathbf{1}}) = \iota^{-1} \circ (f \circ g) \circ \iota, \end{aligned}$$

whence we obtain the result. □

2.3. First examples of monoidal categories

Monoidal categories are ubiquitous. You will see one whichever way you look. Here are some examples.

EXAMPLE 2.3.1. The category **Sets** of sets is a monoidal category, where the tensor product is the Cartesian product and the unit object is a one element set; the structure morphisms a, ι, l, r are obvious. The same holds for the subcategory of finite sets, which will be denoted by Sets [5]. This example can be widely generalized: one can take the category of sets with some structure, such as groups, topological spaces, etc.

EXAMPLE 2.3.2. Any additive category (see Definition 1.2.1) is monoidal, with $\otimes$ being the direct sum functor $\oplus$, and $\mathbf{1}$ being the zero object.

The remaining examples will be especially important below. Let $\Bbbk$ be any field.

EXAMPLE 2.3.3. The category $\Bbbk-\mathbf{Vec}$ of all $\Bbbk$-vector spaces is a monoidal category, where $\otimes = \otimes_{\Bbbk}$, $\mathbf{1} = \Bbbk$, and the morphisms a, ι, l, r are the obvious ones. The same is true about the category of finite dimensional vector spaces over $\Bbbk$, denoted by $\Bbbk-\mathsf{Vec}$. We will often drop $\Bbbk$ from the notation when no confusion is possible.

More generally, if R is a commutative unital ring, then replacing $\Bbbk$ by R we can define monoidal categories $R-\mathbf{mod}$ of R-modules and $R-\mathsf{mod}$ of R-modules of finite type.

EXAMPLE 2.3.4. Let G be a group. The category $\mathbf{Rep}_{\Bbbk}(G)$ of all representations of G over $\Bbbk$ is a monoidal category, with $\otimes$ being the tensor product of representations: if for a representation V one denotes by ρ_V the corresponding map $G \to GL(V)$, then

$$\rho_{V \otimes W}(g) := \rho_V(g) \otimes \rho_W(g).$$

The unit object in this category is the trivial representation $\mathbf{1} = \Bbbk$. A similar statement holds for the category $\mathsf{Rep}_{\Bbbk}(G)$ of finite dimensional representations of G. Again, we will drop the subscript $\Bbbk$ when no confusion is possible.

[5] Here and below, the absence of a finiteness condition is indicated by the **boldface** font, while its presence is indicated by the Roman font.

EXAMPLE 2.3.5. Let G be an affine (pro)algebraic group[6] over $\Bbbk$.

The categories $\mathbf{Rep}(G)$ and $\mathsf{Rep}(G)$ of algebraic representations of G over $\Bbbk$ are monoidal categories (similarly to Example 2.3.4).

Similarly, if $\mathfrak{g}$ is a Lie algebra over $\Bbbk$, then the category of its representations $\mathbf{Rep}(\mathfrak{g})$ and the category of its finite dimensional representations $\mathsf{Rep}(\mathfrak{g})$ are monoidal categories: the tensor product is defined by

$$\rho_{V\otimes W}(a) = \rho_V(a) \otimes \mathsf{id}_W + \mathsf{id}_V \otimes \rho_W(a)$$

(where $\rho_Y : \mathfrak{g} \to \mathfrak{gl}(Y)$ is the homomorphism associated to a representation Y of $\mathfrak{g}$), and $\mathbf{1}$ is the 1-dimensional representation with the zero action of $\mathfrak{g}$.

EXAMPLE 2.3.6. Let G be a monoid (which we will usually take to be a group), and let A be an abelian group (with operation written multiplicatively). Let $\mathcal{C}_G = \mathcal{C}_G(A)$ be the category whose objects δ_g are labeled by elements of G (so there is only one object in each isomorphism class), $\mathsf{Hom}_{\mathcal{C}_G}(\delta_{g_1}, \delta_{g_2}) = \emptyset$ if $g_1 \neq g_2$, and $\mathsf{Hom}_{\mathcal{C}_G}(\delta_g, \delta_g) = A$, with the functor $\otimes$ defined by $\delta_g \otimes \delta_h = \delta_{gh}$, and the tensor product of morphisms defined by $a \otimes b = ab$. Then $\mathcal{C}_G$ is a monoidal category with the associativity isomorphism being the identity, and $\mathbf{1}$ being the unit element of G. This shows that in a monoidal category, $X \otimes Y$ need not be isomorphic to $Y \otimes X$ (indeed, it suffices to take a non-commutative monoid G).

This example has a "linear" version. Namely, let $\Bbbk$ be a field, and $\Bbbk-\mathbf{Vec}_G$ denote the category of G-graded vector spaces over $\Bbbk$, i.e., vector spaces V with a decomposition $V = \bigoplus_{g\in G} V_g$. Morphisms in this category are linear maps which preserve the grading. Define the tensor product on this category by the formula

$$(V \otimes W)_g = \bigoplus_{x,y\in G: xy=g} V_x \otimes W_y, \tag{2.17}$$

and the unit object $\mathbf{1}$ by $\mathbf{1}_1 = \Bbbk$ and $\mathbf{1}_g = 0$ for $g \neq 1$. Then, defining a, ι in an obvious way, we equip $\Bbbk-\mathbf{Vec}_G$ with the structure of a monoidal category. Similarly one defines the monoidal category $\Bbbk-\mathsf{Vec}_G$ of finite dimensional G-graded $\Bbbk$-vector spaces.

In the category $\Bbbk-\mathsf{Vec}_G$, we have pairwise non-isomorphic objects δ_g, $g \in G$, defined by the formula $(\delta_g)_x = \Bbbk$ if $x = g$ and $(\delta_g)_x = 0$ otherwise. For these objects, we have $\delta_g \otimes \delta_h \cong \delta_{gh}$. Thus the category $\mathcal{C}_G(\Bbbk^\times)$ is a non-full monoidal subcategory of $\Bbbk-\mathsf{Vec}_G$ (since the zero morphisms are missing). This subcategory can be viewed as a "basis" of $\Bbbk-\mathsf{Vec}_G$ (and $\Bbbk-\mathsf{Vec}_G$ as "the linear span" of $\mathcal{C}_G(\Bbbk^\times)$), as any object of $\Bbbk-\mathsf{Vec}_G$ is isomorphic to a direct sum of objects δ_g with non-negative integer multiplicities.

When no confusion is possible, we will denote the categories $\Bbbk-\mathsf{Vec}_G$, $\Bbbk-\mathbf{Vec}_G$ simply by Vec_G, $\mathbf{Vec}_G$.

EXERCISE 2.3.7. Let G be a group, and A an abelian group with an action $\rho : G \to \mathsf{Aut}(A)$. Define the category $\mathcal{C}_G(A, \rho)$ in the same way as $\mathcal{C}_G(A)$, except that the tensor product of morphisms is defined as follows: if $a : \delta_g \to \delta_g$ and $b : \delta_h \to \delta_h$ then $a \otimes b = ag(b)$, where $g(b) := \rho(g)b$. Show that $\mathcal{C}_G(A, \rho)$ is a monoidal category.

---

[6]Recall that an *affine algebraic group* over $\Bbbk$ is an affine algebraic variety with a group structure, such that the multiplication and inversion maps are regular, and that an *affine proalgebraic group* is an inverse limit of affine algebraic groups. A typical example of an affine proalgebraic group which is not an algebraic group is the group $G(\Bbbk[[t]])$ of formal series valued points of an affine algebraic group G defined over $\Bbbk$.

(2) There is an obvious associativity isomorphism for $\otimes$, which turns $\mathcal{T}$ into a monoidal category (with unit object being the empty tangle).

2.4. Monoidal functors and their morphisms

As we have explained, the notion of a monoidal category is a categorification of the notion of a monoid. Now we pass to categorification of morphisms between monoids, namely, to monoidal functors.

DEFINITION 2.4.1. Let $(\mathcal{C}, \otimes, \mathbf{1}, a, \iota)$ and $(\mathcal{C}', \otimes', \mathbf{1}', a', \iota')$ be two monoidal categories. A *monoidal functor* from $\mathcal{C}$ to $\mathcal{C}'$ is a pair (F, J), where $F : \mathcal{C} \to \mathcal{C}'$ is a functor, and

$$J_{X,Y} : F(X) \otimes' F(Y) \xrightarrow{\sim} F(X \otimes Y) \tag{2.22}$$

is a natural isomorphism, such that $F(\mathbf{1})$ is isomorphic to $\mathbf{1}'$ and the diagram

$$\begin{array}{ccc} (F(X) \otimes' F(Y)) \otimes' F(Z) & \xrightarrow{a'_{F(X),F(Y),F(Z)}} & F(X) \otimes' (F(Y) \otimes' F(Z)) \\ \Big\downarrow{\scriptstyle J_{X,Y} \otimes' \mathrm{id}_{F(Z)}} & & \Big\downarrow{\scriptstyle \mathrm{id}_{F(X)} \otimes' J_{Y,Z}} \\ F(X \otimes Y) \otimes' F(Z) & & F(X) \otimes' F(Y \otimes Z) \\ \Big\downarrow{\scriptstyle J_{X \otimes Y, Z}} & & \Big\downarrow{\scriptstyle J_{X, Y \otimes Z}} \\ F((X \otimes Y) \otimes Z) & \xrightarrow{F(a_{X,Y,Z})} & F(X \otimes (Y \otimes Z)) \end{array} \tag{2.23}$$

is commutative for all $X, Y, Z \in \mathcal{C}$ ("the monoidal structure axiom").

A monoidal functor F is said to be an *equivalence of monoidal categories* if it is an equivalence of ordinary categories.

REMARK 2.4.2. It is important to stress that, as seen from this definition, a monoidal functor is not just a functor between monoidal categories, but a functor with an additional structure (the isomorphism J) satisfying a certain equation (the monoidal structure axiom). As we will see in Section 2.5, this equation may have more than one solution or no solutions at all, so the same functor can be equipped with different monoidal structures or not admit any monoidal structure at all.

It turns out that if F is a monoidal functor, then there is a canonical isomorphism $\varphi : \mathbf{1}' \to F(\mathbf{1})$. This isomorphism is defined by the commutative diagram

$$\begin{array}{ccc} \mathbf{1}' \otimes' F(\mathbf{1}) & \xrightarrow{l'_{F(\mathbf{1})}} & F(\mathbf{1}) \\ \Big\downarrow{\scriptstyle \varphi \otimes' \mathrm{id}_{F(\mathbf{1})}} & & \Big\downarrow{\scriptstyle F(l_{\mathbf{1}})^{-1}} \\ F(\mathbf{1}) \otimes' F(\mathbf{1}) & \xrightarrow{J_{\mathbf{1},\mathbf{1}}} & F(\mathbf{1} \otimes \mathbf{1}) \end{array} \tag{2.24}$$

where l, r, l', r' are the unit isomorphisms for $\mathcal{C}$ and $\mathcal{C}'$ defined in (2.5).

PROPOSITION 2.4.3. *For any monoidal functor $(F, J) : \mathcal{C} \to \mathcal{C}'$, the diagrams*

$$\begin{CD} \mathbf{1}' \otimes' F(X) @>{l'_{F(X)}}>> F(X) \\ @V{\varphi \otimes' \mathsf{id}_{F(X)}}VV @VV{F(l_X)^{-1}}V \\ F(\mathbf{1}) \otimes' F(X) @>{J_{\mathbf{1},X}}>> F(\mathbf{1} \otimes X) \end{CD} \tag{2.25}$$

and

$$\begin{CD} F(X) \otimes' \mathbf{1}' @>{r'_{F(X)}}>> F(X) \\ @V{\mathsf{id}_{F(X)} \otimes' \varphi}VV @VV{F(r_X)^{-1}}V \\ F(X) \otimes' F(\mathbf{1}) @>{J_{X,\mathbf{1}}}>> F(X \otimes \mathbf{1}) \end{CD} \tag{2.26}$$

are commutative for all $X \in \mathcal{C}$.

EXERCISE 2.4.4. Prove Proposition 2.4.3.

Proposition 2.4.3 implies that a monoidal functor can be equivalently defined as follows.

DEFINITION 2.4.5. A monoidal functor $\mathcal{C} \to \mathcal{C}'$ is a triple (F, J, φ) which satisfies the monoidal structure axiom and Proposition 2.4.3.

Definition 2.4.5 is a more traditional definition of a monoidal functor.

REMARK 2.4.6. It can be seen from the above that for any monoidal functor (F, J) one can safely identify $\mathbf{1}'$ with $F(\mathbf{1})$ using the isomorphism φ, and assume that $F(\mathbf{1}) = \mathbf{1}'$ and $\varphi = \mathsf{id}_{\mathbf{1}'}$ (similarly to how we have identified $\mathbf{1} \otimes X$ and $X \otimes \mathbf{1}$ with X and assumed that $l_X = r_X = \mathsf{id}_X$). We will usually do so from now on. Proposition 2.4.3 implies that with these conventions, one has

$$J_{\mathbf{1},X} = J_{X,\mathbf{1}} = \mathsf{id}_X . \tag{2.27}$$

REMARK 2.4.7. It is clear that the composition of monoidal functors is a monoidal functor. Also, the identity functor has a natural structure of a monoidal functor.

Monoidal functors between two monoidal categories themselves form a category. Namely, one has the following notion of a morphism (or natural transformation) between two monoidal functors.

DEFINITION 2.4.8. Let $(\mathcal{C}, \otimes, \mathbf{1}, a, \iota)$ and $(\mathcal{C}', \otimes', \mathbf{1}', a', \iota')$ be two monoidal categories, and let (F^1, J^1) and (F^2, J^2) be two monoidal functors from $\mathcal{C}$ to $\mathcal{C}'$. A *morphism* (or a *natural transformation*) of monoidal functors $\eta : (F^1, J^1) \to (F^2, J^2)$ is a natural transformation $\eta : F^1 \to F^2$ such that $\eta_{\mathbf{1}}$ is an isomorphism, and the diagram

$$\begin{CD} F^1(X) \otimes' F^1(Y) @>{J^1_{X,Y}}>> F^1(X \otimes Y) \\ @V{\eta_X \otimes' \eta_Y}VV @VV{\eta_{X \otimes Y}}V \\ F^2(X) \otimes' F^2(Y) @>{J^2_{X,Y}}>> F^2(X \otimes Y) \end{CD} \tag{2.28}$$

is commutative for all $X, Y \in \mathcal{C}$.

REMARK 2.4.9. It is easy to show that if $\varphi_i : \mathbf{1}' \xrightarrow{\sim} F^i(\mathbf{1})$, $i = 1, 2$, are isomorphisms defined by (2.24) then $\eta_{\mathbf{1}} \circ \varphi_1 = \varphi_2$, so if one makes the convention that $\varphi_1 = \varphi_2 = \mathrm{id}_{\mathbf{1}'}$, one has $\eta_{\mathbf{1}} = \mathrm{id}_{\mathbf{1}'}$.

REMARK 2.4.10. It is easy to show that if $F : \mathcal{C} \to \mathcal{C}'$ is an equivalence of monoidal categories, then there exists a monoidal equivalence $F^{-1} : \mathcal{C}' \to \mathcal{C}$ such that the functors $F \circ F^{-1}$ and $F^{-1} \circ F$ are isomorphic to the identity functor as monoidal functors. Thus, for any monoidal category $\mathcal{C}$, the monoidal autoequivalences of $\mathcal{C}$ up to isomorphism form a group with respect to composition.

2.5. Examples of monoidal functors

Let us now give some examples of monoidal functors and natural transformations.

EXAMPLE 2.5.1. An important class of examples of monoidal functors is *forgetful functors* (e.g., functors of "forgetting the structure", from the categories of groups, topological spaces, etc., to the category of sets). Such functors have an obvious monoidal structure. An example especially important in this book is the forgetful functor $\mathbf{Rep}(G) \to \mathbf{Vec}$ from the representation category of a group to the category of vector spaces. More generally, if $H \subset G$ is a subgroup, then we have a forgetful (or restriction) functor $\mathbf{Rep}(G) \to \mathbf{Rep}(H)$. Still more generally, if $f : H \to G$ is a group homomorphism, then we have the pullback functor $f^* : \mathbf{Rep}(G) \to \mathbf{Rep}(H)$. All these functors are monoidal.

EXAMPLE 2.5.2. Let $f : H \to G$ be a homomorphism of groups. Then any H-graded vector space is naturally G-graded (by pushforward of grading). Thus we have a natural monoidal functor $f_* : \mathbf{Vec}_H \to \mathbf{Vec}_G$. If G is the trivial group, then f_* is just the forgetful functor $\mathbf{Vec}_H \to \mathbf{Vec}$.

EXAMPLE 2.5.3. Let $\Bbbk$ be a field, let A be a $\Bbbk$-algebra with unit, and let $\mathcal{C} = A-\mathbf{mod}$ be the category of left A-modules. Then we have a functor

$$F : M \mapsto (M \otimes_A -) : A-\mathbf{bimod} \to \mathsf{End}(\mathcal{C}). \tag{2.29}$$

This functor is naturally monoidal. A similar functor $F : A-\mathsf{bimod} \to \mathsf{End}(\mathcal{C})$ can be defined if A is a finite dimensional $\Bbbk$-algebra, and $\mathcal{C} = A-\mathsf{mod}$ is the category of finite dimensional left A-modules.

PROPOSITION 2.5.4. *The functor* (2.29) *takes values in the full monoidal subcategory* $\mathsf{End}_{re}(\mathcal{C})$ *of right exact endofunctors of* $\mathcal{C}$, *and defines an equivalence between the monoidal categories* $A-\mathsf{bimod}$ *and* $\mathsf{End}_{re}(\mathcal{C})$.

PROOF. The first statement is clear, since the tensor product functor is right exact. To prove the second statement, let us construct the quasi-inverse functor F^{-1}. Let $G \in \mathsf{End}_{re}(\mathcal{C})$. Define $F^{-1}(G)$ by the formula $F^{-1}(G) = G(A)$; this is clearly an A-bimodule, since it is a left A-module with a commuting action of $\mathrm{End}_A(A) = A^{\mathrm{op}}$ (the opposite algebra). We leave it to the reader to check that the functor F^{-1} is indeed a quasi-inverse to F (cf. Proposition 1.8.10). □

REMARK 2.5.5. A similar statement is valid without the finite dimensionality assumption, if one adds the condition that the right exact functors must commute with inductive limits.

EXAMPLE 2.5.6. Let S be a monoid, and let $\mathcal{C} = \mathsf{Vec}_S$ (see Example 2.3.6). Let us view $\mathsf{id}_\mathcal{C}$, the identity functor of $\mathcal{C}$, as a monoidal functor. It is easy to see that morphisms $\eta : \mathsf{id}_\mathcal{C} \to \mathsf{id}_\mathcal{C}$ as monoidal functors correspond to homomorphisms of monoids: $\eta : S \to \Bbbk$ (where $\Bbbk$ is equipped with the multiplication operation). In particular, $\eta(s)$ may be 0 for some s, so η does not have to be an isomorphism.

2.6. Monoidal functors between categories of graded vector spaces

Let G_1, G_2 be groups, let A be an abelian group, and let $\omega_i \in Z^3(G_i, A)$, $i = 1, 2$, be 3-cocycles (the actions of G_1, G_2 on A are assumed to be trivial). Let $\mathcal{C}_i = \mathcal{C}_{G_i}^{\omega_i}$, $i = 1, 2$, be the monoidal categories of graded vector spaces introduced in Example 2.3.8.

Any monoidal functor $F : \mathcal{C}_1 \to \mathcal{C}_2$ defines, by restriction to simple objects, a group homomorphism $f : G_1 \to G_2$. Using axiom (2.23) of a monoidal functor, we see that a monoidal structure on F is given by

$$J_{g,h} = \mu(g, h)\, \mathsf{id}_{\delta_{f(gh)}} : F(\delta_g) \otimes F(\delta_h) \xrightarrow{\sim} F(\delta_{gh}),\ g, h \in G_1, \tag{2.30}$$

where $\mu : G_1 \times G_1 \to A$ is a function such that

$$\omega_1(g, h, l)\mu(gh, l)\mu(g, h) = \mu(g, hl)\mu(h, l)\omega_2(f(g), f(h), f(l)),$$

for all $g, h, l \in G_1$. That is,

$$\omega_1 = f^*\omega_2 \cdot d_3(\mu), \tag{2.31}$$

i.e., ω_1 and $f^*\omega_2$ are cohomologous in $Z^3(G_1, A)$.

Conversely, given a group homomorphism $f : G_1 \to G_2$, any function

$$\mu : G_1 \times G_1 \to A$$

satisfying (2.31) gives rise to a monoidal functor $F : \mathcal{C}_1 \to \mathcal{C}_2$ defined by $F(\delta_g) = \delta_{f(g)}$ with the monoidal structure given by formula (2.30). This functor is an equivalence if and only if f is an isomorphism.

To summarize, monoidal functors $\mathcal{C}_{G_1}^{\omega_1} \to \mathcal{C}_{G_2}^{\omega_2}$ correspond to pairs (f, μ), where $f : G_1 \to G_2$ is a group homomorphism such that ω_1 and $f^*\omega_2$ are cohomologous, and μ is a function satisfying (2.31) (such functions are in a (non-canonical) bijection with A-valued 2-cocycles on G_1). Let $F_{f,\mu}$ denote the corresponding functor.

Let us determine natural monoidal transformations between $F_{f,\mu}$ and $F_{f',\mu'}$. Clearly, such a transformation exists if and only if $f = f'$, is always an isomorphism, and is determined by a collection of morphisms $\eta_g : \delta_{f(g)} \to \delta_{f(g)}$ (i.e., $\eta_g \in A$), satisfying the equation

$$\mu'(g, h)(\eta_g \otimes \eta_h) = \eta_{gh}\mu(g, h) \tag{2.32}$$

for all $g, h \in G_1$, i.e.,

$$\mu = \mu' \cdot d_2(\eta). \tag{2.33}$$

Conversely, every function $\eta : G_1 \to A$ satisfying (2.33) gives rise to a morphism of monoidal functors $\eta : F_{f,\mu} \to F_{f,\mu'}$ defined as above. Therefore, monoidal functors $F_{f,\mu}$ and $F_{f',\mu'}$ are isomorphic if and only if $f = f'$ and μ is cohomologous to μ'.

Thus, we have obtained the following proposition.

PROPOSITION 2.6.1. (i) *The set of isomorphisms between monoidal functors* $F_{f,\mu}, F_{f,\mu'} : \mathcal{C}_{G_1}^{\omega_1} \to \mathcal{C}_{G_2}^{\omega_2}$ *is a torsor over the group* $H^1(G_1, A) = \mathsf{Hom}(G_1, A)$.

(ii) *For a fixed homomorphism $f : G_1 \to G_2$, the set of μ parameterizing isomorphism classes of monoidal functors $F_{f,\mu}$ is a torsor over $H^2(G_1, A)$.*
(iii) *Equivalence classes of monoidal categories $\mathcal{C}_G^\omega$ are parametrized by the set $H^3(G, A)/\mathrm{Out}(G)$, where $\mathsf{Out}(G)$ denotes the group of outer automorphisms of G.* [7]

REMARK 2.6.2. The same results, including Proposition 2.6.1, are valid if we specialize to the case when $A = \Bbbk^\times$, where $\Bbbk$ is a field, replace the categories $\mathcal{C}_G^\omega$ by their "linear spans" Vec_G^ω, and require that the monoidal functors we consider are additive. To see this, it is enough to note that by definition, for any morphism η of monoidal functors, $\eta_1 \neq 0$, so equation (2.32) (with $h = g^{-1}$) implies that all η_g must be nonzero. Thus, if a morphism $\eta : F_{f,\mu} \to F_{f',\mu'}$ exists, then it is an isomorphism, and we must have $f = f'$.

REMARK 2.6.3. The above discussion implies that in the definition of the categories $\mathcal{C}_G^\omega$ and Vec_G^ω, it may be assumed without loss of generality that the cocycle ω is normalized, i.e., $\omega(g, 1, h) = 1$, and thus $l_{\delta_g} = r_{\delta_g} = \mathsf{id}_{\delta_g}$ (which is convenient in computations). Indeed, we claim that any 3-cocycle ω is cohomologous to a normalized one. To see this, it is enough to alter ω by dividing it by $d_2(\mu)$, where μ is any 2-cochain such that $\mu(g, 1) = \omega(g, 1, 1)$, and $\mu(1, h) = \omega(1, 1, h)^{-1}$.

EXAMPLE 2.6.4. Let $G = \mathbb{Z}/n\mathbb{Z}$, where $n > 1$ is an integer, and $\Bbbk = \mathbb{C}$. Consider the cohomology of $\mathbb{Z}/n\mathbb{Z}$.

Since $H^i(\mathbb{Z}/n\mathbb{Z}, \mathbb{C}) = 0$ for all $i > 0$, writing the long exact sequence of cohomology for the short exact sequence of coefficient groups

$$0 \longrightarrow \mathbb{Z} \longrightarrow \mathbb{C} \longrightarrow \mathbb{C}^\times = \mathbb{C}/\mathbb{Z} \longrightarrow 0,$$

we obtain a natural isomorphism $H^i(\mathbb{Z}/n\mathbb{Z}, \mathbb{C}^\times) \cong H^{i+1}(\mathbb{Z}/n\mathbb{Z}, \mathbb{Z})$.

As we saw in Exercise 1.7.5, the graded ring $H^*(\mathbb{Z}/n\mathbb{Z}, \mathbb{Z})$ is

$$H^*(\mathbb{Z}/n\mathbb{Z}, \mathbb{Z}) = \mathbb{Z}[x]/(nx) = \mathbb{Z} \oplus x(\mathbb{Z}/n\mathbb{Z})[x],$$

where x is a generator in degree 2. Moreover, as a module over $\mathsf{Aut}(\mathbb{Z}/n\mathbb{Z}) = (\mathbb{Z}/n\mathbb{Z})^\times$, we have $H^2(\mathbb{Z}/n\mathbb{Z}, \mathbb{Z}) \cong H^1(\mathbb{Z}/n\mathbb{Z}, \mathbb{C}^\times) = (\mathbb{Z}/n\mathbb{Z})^\vee$. Therefore, using the graded ring structure, we find that

$$H^{2m-1}(\mathbb{Z}/n\mathbb{Z}, \mathbb{C}^\times) \cong H^{2m}(\mathbb{Z}/n\mathbb{Z}, \mathbb{Z}) \cong ((\mathbb{Z}/n\mathbb{Z})^\vee)^{\otimes m}$$

as an $\mathsf{Aut}(\mathbb{Z}/n\mathbb{Z})$-module. In particular, $H^3(\mathbb{Z}/n\mathbb{Z}, \mathbb{C}^\times) = ((\mathbb{Z}/n\mathbb{Z})^\vee)^{\otimes 2}$.

Let us give an explicit formula for the 3-cocycles on $\mathbb{Z}/n\mathbb{Z}$. Modulo coboundaries, these cocycles are given by

$$\phi(i, j, k) = \varepsilon^{\frac{si(j+k-(j+k)')}{n}}, \tag{2.34}$$

where ε is a primitive nth root of unity, $s \in \mathbb{Z}/n\mathbb{Z}$, and for an integer m we denote by m' the remainder of division of m by n.

EXERCISE 2.6.5. Show that when s runs over $\mathbb{Z}/n\mathbb{Z}$, this formula defines cocycles representing all the cohomology classes in $H^3(\mathbb{Z}/n\mathbb{Z}, \mathbb{C}^\times)$.

---

[7] Recall that the group $\mathrm{Inn}(G)$ of inner automorphisms of a group G acts trivially on $H^*(G, A)$ (for any coefficient group A), and thus the action of the group $\mathsf{Aut}(G)$ on $H^*(G, A)$ factors through $\mathsf{Out}(G)$.

EXERCISE 2.6.6. Derive an explicit formula for cocycles representing cohomology classes in $H^{2m+1}(\mathbb{Z}/n\mathbb{Z},\, \mathbb{C}^\times)$ for $m \geq 1$.

Hint: use that a generator of this group has the form $x^m \otimes y$, where y is a generator of $H^1(\mathbb{Z}/n\mathbb{Z}, \mathbb{C}^\times)$.

EXERCISE 2.6.7. Describe $H^*(\mathbb{Z}/n\mathbb{Z}, \Bbbk^\times)$, where $\Bbbk$ is an algebraically closed field of any characteristic p.

Hint: Show that this cohomology coincides with the cohomology with coefficients in the group of roots of unity in $\Bbbk$; consider separately the case when p divides n and when p does not divide n.

2.7. Group actions on categories and equivariantization

Let $\mathcal{C}$ be a category. Consider the category $\mathsf{Aut}(\mathcal{C})$, whose objects are autoequivalences of $\mathcal{C}$ and whose morphisms are isomorphisms of functors. It is a monoidal subcategory of the monoidal category $\mathsf{End}(\mathcal{C})$ from Example 2.3.12.

If $\mathcal{C}$ is a monoidal category, we consider the category $\mathsf{Aut}_\otimes(\mathcal{C})$ of monoidal autoequivalences of $\mathcal{C}$.

For a group G let $\mathrm{Cat}(G)$ denote the monoidal category whose objects are elements of G, the only morphisms are the identities, and the tensor product is given by multiplication in G. In the notation of Example 2.3.6 we have $\mathrm{Cat}(G) = \mathcal{C}_G(1)$.

DEFINITION 2.7.1. Let G be a group.

(i) An *action* of G on a category $\mathcal{C}$ is a monoidal functor

$$T : \mathrm{Cat}(G) \to \mathsf{Aut}(\mathcal{C}). \tag{2.35}$$

(ii) An *action* of G on a monoidal category $\mathcal{C}$ is a monoidal functor

$$T : \mathrm{Cat}(G) \to \mathsf{Aut}_\otimes(\mathcal{C}). \tag{2.36}$$

In these situations we also say that G *acts* on $\mathcal{C}$.

Let G be a group acting on a category $\mathcal{C}$. Let $g \mapsto T_g$ denote the corresponding action (2.35). For any $g \in G$ let $T_g \in \mathsf{Aut}(\mathcal{C})$ be the corresponding functor, and for any $g, h \in G$ let $\gamma_{g,h}$ be the isomorphism $T_g \circ T_h \simeq T_{gh}$ that defines the monoidal structure on the functor $\mathrm{Cat}(G) \to \mathsf{Aut}(\mathcal{C})$.

DEFINITION 2.7.2. A *G-equivariant object* in $\mathcal{C}$ is a pair $(X,\, u)$ consisting of an object X of $\mathcal{C}$ and a family of isomorphisms $u = \{u_g : T_g(X) \xrightarrow{\sim} X \mid g \in G\}$, such that the diagram

$$\begin{array}{ccc} T_g(T_h(X)) & \xrightarrow{T_g(u_h)} & T_g(X) \\ {\scriptstyle \gamma_{g,h}(X)}\downarrow & & \downarrow{\scriptstyle u_g} \\ T_{gh}(X) & \xrightarrow{u_{gh}} & X \end{array}$$

commutes for all $g, h \in G$. One defines morphisms of equivariant objects to be morphisms in $\mathcal{C}$ commuting with u_g, $g \in G$.

The *category of G-equivariant objects of $\mathcal{C}$*, or the *G-equivariantization* of $\mathcal{C}$, will be denoted by $\mathcal{C}^G$. There is an obvious forgetful functor

$$\mathrm{Forg} : \mathcal{C}^G \to \mathcal{C}. \tag{2.37}$$

A similar definition can be made for monoidal categories $\mathcal{C}$, replacing $\mathsf{Aut}(\mathcal{C})$ with $\mathsf{Aut}_{\otimes}(\mathcal{C})$. When $\mathcal{C}$ is a monoidal category, the category $\mathcal{C}^G$ inherits a structure of a monoidal category such that the functor (2.37) is a monoidal functor.

EXERCISE 2.7.3. Show that actions of a group G on the category Vec viewed as an *abelian* category correspond to elements of $H^2(G, \Bbbk^\times)$, while any action of G on Vec viewed as a *monoidal* category is trivial.

2.8. The Mac Lane strictness theorem

As we have seen above, it is simpler to work with monoidal categories in which the associativity and unit constrains are the identity maps.

DEFINITION 2.8.1. A monoidal category $\mathcal{C}$ is *strict* if for all objects X, Y, Z in $\mathcal{C}$ one has equalities $(X \otimes Y) \otimes Z = X \otimes (Y \otimes Z)$ and $X \otimes \mathbf{1} = X = \mathbf{1} \otimes X$, and the associativity and unit constraints are the identity maps.

EXAMPLE 2.8.2. The category $\mathsf{End}(\mathcal{C})$ of endofunctors of a category $\mathcal{C}$ (see Example 2.3.12) is strict.

EXAMPLE 2.8.3. Let $\overline{\text{Sets}}$ be the category whose objects are non-negative integers, and $\mathsf{Hom}_{\overline{\mathsf{Sets}}}(m, n)$ is the set of maps from $\{0, ..., m-1\}$ to $\{0, ..., n-1\}$. Define the tensor product functor on objects by $m \otimes n = mn$, and for

$$f_1 : m_1 \to n_1 \quad \text{and} \quad f_2 : m_2 \to n_2$$

define $f_1 \otimes f_2 : m_1 m_2 \to n_1 n_2$ by

$$(f_1 \otimes f_2)(m_2 x + y) = n_2 f_1(x) + f_2(y), 0 \le x \le m_1 - 1,\ 0 \le y \le m_2 - 1.$$

Then $\overline{\text{Sets}}$ is a strict monoidal category. Moreover, we have a natural inclusion $\overline{\text{Sets}} \hookrightarrow \text{Sets}$ (recall that Sets stands for the category of finite sets), which is obviously a monoidal equivalence.

EXAMPLE 2.8.4. This is a linear version of the previous example. Let $\Bbbk$ be a field. Let $\Bbbk - \overline{\mathsf{Vec}}$ be the category whose objects are non-negative integers, and $\mathsf{Hom}_{\Bbbk - \overline{\mathsf{Vec}}}(m, n)$ is the set of matrices with m columns and n rows over $\Bbbk$ (and the composition of morphisms is the product of matrices). Define the tensor product functor on objects by $m \otimes n = mn$, and for $f_1 : m_1 \to n_1, f_2 : m_2 \to n_2$, define $f_1 \otimes f_2 : m_1 m_2 \to n_1 n_2$ to be the Kronecker product of f_1 and f_2. Then $\Bbbk - \overline{\mathsf{Vec}}$ is a strict monoidal category. Moreover, we have a natural inclusion $\Bbbk - \overline{\mathsf{Vec}} \hookrightarrow \Bbbk - \mathsf{Vec}$, which is obviously a monoidal equivalence.

Similarly, for any group G one can define a strict monoidal category $\Bbbk - \overline{\mathsf{Vec}}_G$, whose objects are $\mathbb{Z}_+$-valued functions on G with finitely many nonzero values, and which is monoidally equivalent to $\Bbbk - \mathsf{Vec}_G$. We leave this definition to the reader.

On the other hand, some of the most important monoidal categories, such as **Sets**, **Vec**, $\mathbf{Vec}_G$, Sets, Vec, Vec_G, should be regarded as non-strict (at least if one defines them in the usual way). It is even more indisputable that the categories $\mathbf{Vec}_G^\omega$, Vec_G^ω for cohomologically nontrivial ω are not strict.

However, the following remarkable theorem of Mac Lane implies that in practice, one may always assume that a monoidal category is strict.

THEOREM 2.8.5. *Any monoidal category is monoidally equivalent to a strict monoidal category.*

PROOF. We will establish an equivalence between $\mathcal{C}$ and the monoidal category of right $\mathcal{C}$-module endofunctors of $\mathcal{C}$, which we will discuss in more detail in Chapter 7. The non-categorical counterpart of this result is the fact that every monoid M is isomorphic to the monoid consisting of maps from M to itself commuting with the right multiplication.

For a monoidal category $\mathcal{C}$ let $\mathcal{C}^{\wr}$ be the monoidal category defined as follows. The objects of $\mathcal{C}^{\wr}$ are pairs $(F,\, c)$ where $F : \mathcal{C} \to \mathcal{C}$ is a functor and

$$c_{X,Y} : F(X) \otimes Y \xrightarrow{\sim} F(X \otimes Y)$$

is a natural isomorphism such that the following diagram is commutative for all objects X, Y, Z in $\mathcal{C}$:

$$\begin{array}{ccc} & (F(X) \otimes Y) \otimes Z & \\ {\scriptstyle c_{X,Y} \otimes \mathrm{id}_Z} \swarrow & & \searrow {\scriptstyle a_{F(X),Y,Z}} \\ F(X \otimes Y) \otimes Z & & F(X) \otimes (Y \otimes Z) \\ \downarrow {\scriptstyle c_{X \otimes Y, Z}} & & \downarrow {\scriptstyle c_{X, Y \otimes Z}} \\ F((X \otimes Y) \otimes Z) & \xrightarrow[F(a_{X,Y,Z})]{} & F(X \otimes (Y \otimes Z)). \end{array} \tag{2.38}$$

A morphism $\theta : (F^1,\, c^1) \to (F^2,\, c^2)$ in $\mathcal{C}^{\wr}$ is a natural transformation $\theta : F^1 \to F^2$ such that the following square commutes for all objects X, Y in $\mathcal{C}$:

$$\begin{array}{ccc} F^1(X) \otimes Y & \xrightarrow{c^1_{X,Y}} & F^1(X \otimes Y) \\ \downarrow {\scriptstyle \theta_X \otimes \mathrm{id}_Y} & & \downarrow {\scriptstyle \theta_{X \otimes Y}} \\ F^2(X) \otimes Y & \xrightarrow[c^2_{X,Y}]{} & F^2(X \otimes Y), \end{array} \tag{2.39}$$

Composition of morphisms is the vertical composition of natural transformations. The tensor product of objects is given by $(F^1,\, c^1) \otimes (F^2,\, c^2) = (F^1F^2, c)$ where c is given by a composition

$$F^1F^2(X) \otimes Y \xrightarrow{c^1_{F^2(X),Y}} F^1(F^2(X) \otimes Y) \xrightarrow{F^1(c^2_{X,Y})} F^1F^2(X \otimes Y) \tag{2.40}$$

for all $X, Y \in \mathcal{C}$, and the tensor product of morphisms is the composition of natural transformations. Thus $\mathcal{C}^{\wr}$ is a strict tensor category (the unit object is the identity functor).

Consider a functor of left multiplication $L : \mathcal{C} \to \mathcal{C}^{\wr}$ given by

$$L(X) = (X \otimes -,\, a_{X,-,-}), \quad L(f) = (f \otimes -).$$

Note that the diagram (2.38) for L is nothing but the pentagon diagram (2.2).

We will show that this functor L is a monoidal equivalence. First of all, note that any $(F,\, c)$ in $\mathcal{C}^{\wr}$ is isomorphic to $L(F(\mathbf{1}))$.

Let us now show that L is fully faithful. Let $\theta : L(X) \to L(Y)$ be a morphism in $\mathcal{C}^{\wr}$. Define $f : X \to Y$ to be the composition

$$X \xrightarrow{r_X^{-1}} X \otimes \mathbf{1} \xrightarrow{\theta_1} Y \otimes \mathbf{1} \xrightarrow{r_Y} Y, \tag{2.41}$$

where r is the unit constraint. We claim that for all Z in $\mathcal{C}$ one has $\theta_Z = f \otimes \mathsf{id}_Z$ (so that $\theta = L(f)$ and L is full). Indeed, this follows from the commutativity of the following diagram

$$
\begin{CD}
X \otimes Z @>{r_X^{-1} \otimes \mathsf{id}_Z}>> (X \otimes \mathbf{1}) \otimes Z @>{a_{X,\mathbf{1},Z}}>> X \otimes (\mathbf{1} \otimes Z) @>{X \otimes l_Z}>> X \otimes Z \\
@V{f \otimes \mathsf{id}_Z}VV @V{\theta_{\mathbf{1}} \otimes \mathsf{id}_Z}VV @V{\theta_{\mathbf{1} \otimes Z}}VV @V{\theta_Z}VV \\
Y \otimes Z @>>{r_Y^{-1} \otimes \mathsf{id}_Z}> (Y \otimes \mathbf{1}) \otimes Z @>>{a_{Y,\mathbf{1},Z}}> Y \otimes (\mathbf{1} \otimes Z) @>>{Y \otimes l_Z}> Y \otimes Z,
\end{CD}
\tag{2.42}
$$

where the rows are the identity morphisms by the triangle axiom (2.10), the left square commutes by the definition of f, the right square commutes by the naturality of θ, and the central square commutes since θ is a morphism in $\mathcal{C}^{\wr}$.

Next, if $L(f) = L(g)$ for some morphisms f, g in $\mathcal{C}$ then $f \otimes \mathsf{id}_{\mathbf{1}} = g \otimes \mathsf{id}_{\mathbf{1}}$ so that $f = g$ by the naturality of r. So L is faithful. Thus, it is an equivalence.

Finally, we define a monoidal functor structure

$$\phi : \mathbf{1}_{\mathcal{C}^{\wr}} \xrightarrow{\sim} L(\mathbf{1}_{\mathcal{C}}) \quad \text{and} \quad J_{X,Y} : L(X) \circ L(Y) \xrightarrow{\sim} L(X \otimes Y)$$

on L by $\phi = l^{-1} : (\mathsf{id}_{\mathcal{C}}, \mathsf{id}) \xrightarrow{\sim} (\mathbf{1} \otimes -,\, a_{\mathbf{1},-,-})$ and

$$
\begin{aligned}
J_{X,Y} = a_{X,Y,-}^{-1} : ((X \otimes (Y \otimes -)), (\mathsf{id}_X \otimes a_{Y,-,-} \circ a_{X,Y\otimes -,-})) \\
\xrightarrow{\sim} ((X \otimes Y) \otimes -,\, a_{X \otimes Y,-,-}).
\end{aligned}
$$

The diagram (2.39) for the latter natural isomorphism is the pentagon diagram in $\mathcal{C}$. For the functor L the hexagon diagram (2.23) in the definition of a monoidal functor also reduces to the pentagon diagram in $\mathcal{C}$. The square diagrams (2.25) and (2.26) reduce to triangles, one of which is the triangle axiom (2.10) for $\mathcal{C}$ and another is (2.13). □

REMARK 2.8.6. The nontrivial nature of Mac Lane's strictness theorem is demonstrated by the following instructive example, which shows that even though a monoidal category is always **equivalent** to a strict category, it need **not** be **isomorphic** to one. (By definition, an isomorphism of monoidal categories is a monoidal equivalence which is an isomorphism of categories).

Namely, let $\mathcal{C}$ be the category $\mathcal{C}_G^{\omega}(A)$. If ω is cohomologically nontrivial, this category is clearly not isomorphic to a strict one. However, by Mac Lane's strictness theorem, it is equivalent to a strict category $\mathcal{C}^{\wr}$.

In fact, in this example a strict category $\mathcal{C}^{\wr}$ monoidally equivalent to $\mathcal{C}$ can be constructed quite explicitly, as follows. Let $\widetilde{G}$ be another group with a surjective homomorphism $f : \widetilde{G} \to G$ such that the 3-cocycle $f^*\omega$ is cohomologically trivial. Such $\widetilde{G}$ always exists, e.g., a free group (since the cohomology of a free group in degrees higher than 1 is trivial). Let $\mathcal{C}^{\wr}$ be the category whose objects δ_g are labeled by elements of $\widetilde{G}$, $\mathsf{Hom}(\delta_g, \delta_h) = A$ if g, h have the same image in G, and $\mathsf{Hom}(\delta_g, \delta_h) = \emptyset$ otherwise. This category has an obvious tensor product, and a monoidal structure defined by the 3-cocycle $f^*\omega$. We have an obvious monoidal functor $F : \mathcal{C}^{\wr} \to \mathcal{C}$ defined by the homomorphism $f : \widetilde{G} \to G$, and it is an equivalence, even though not an isomorphism. However, since the cocycle $f^*\omega$ is cohomologically trivial, the category $\mathcal{C}^{\wr}$ is isomorphic to the same category with the trivial associativity isomorphism, which is strict.

REMARK 2.8.7. A category is called *skeletal* if it has only one object in each isomorphism class. The axiom of choice implies that any category is equivalent to a skeletal one. Also, by Mac Lane's strictness theorem, any monoidal category is monoidally equivalent to a strict one. However, Remark 2.8.6 shows that a monoidal category need not be monoidally equivalent to a category which is skeletal and strict at the same time. Indeed, as we have seen, to make a monoidal category strict, it may be necessary to add new objects to it (which are isomorphic, but not equal to already existing ones). In fact, the desire to avoid adding such objects is the reason why we sometimes use nontrivial associativity isomorphisms, even though Mac Lane's strictness theorem tells us we do not have to. This also makes precise the sense in which the categories Sets, Vec, Vec_G, are "more strict" than the category Vec_G^ω for cohomologically nontrivial ω. Namely, the first three categories are monoidally equivalent to strict skeletal categories $\overline{\text{Sets}}$, $\overline{\mathsf{Vec}}$, $\overline{\mathsf{Vec}_G}$, while the category Vec_G^ω is not monoidally equivalent to a strict skeletal category.

EXERCISE 2.8.8. Show that any monoidal category $\mathcal{C}$ is monoidally equivalent to a skeletal monoidal category $\overline{\mathcal{C}}$. Moreover, $\overline{\mathcal{C}}$ can be chosen in such a way that $l_X, r_X = \mathsf{id}_X$ for all objects $X \in \overline{\mathcal{C}}$.

Hint: without loss of generality one can assume that $\mathbf{1} \otimes X = X \otimes \mathbf{1} = X$ and $l_X, r_X = \mathsf{id}_X$ for all objects $X \in \mathcal{C}$. Now in every isomorphism class i of objects of $\mathcal{C}$ fix a representative X_i, so that $X_1 = \mathbf{1}$, and for any two classes i, j fix an isomorphism $\mu_{ij} : X_i \otimes X_j \to X_{i \cdot j}$, so that $\mu_{i1} = \mu_{1i} = \mathsf{id}_{X_i}$. Let $\overline{\mathcal{C}}$ be the full subcategory of $\mathcal{C}$ consisting of the objects X_i, with tensor product defined by $X_i \overline{\otimes} X_j = X_{i \cdot j}$, and with all the structure transported using the isomorphisms μ_{ij}. Then $\overline{\mathcal{C}}$ is the required skeletal category, monoidally equivalent to $\mathcal{C}$.

2.9. The coherence theorem

In a monoidal category, one can form n-fold tensor products of any ordered sequence of objects $X_1, ..., X_n$. Namely, such a product can be attached to any parenthesizing of the expression $X_1 \otimes ... \otimes X_n$, and such products are, in general, distinct objects of $\mathcal{C}$.

However, for $n = 3$, the associativity isomorphism gives a canonical identification of the two possible parenthesizings, $(X_1 \otimes X_2) \otimes X_3$ and $X_1 \otimes (X_2 \otimes X_3)$. An easy combinatorial argument then shows that one can identify any two parenthesized products of $X_1, ..., X_n$, $n \geq 3$, using a chain of associativity isomorphisms.

EXERCISE 2.9.1. Show that the number of ways in which an n-fold product can be parenthesized is given by the *Catalan number* $\frac{1}{n+1}\binom{2n}{n}$.

We would like to say that for this reason we can completely ignore parentheses in computations in any monoidal category, identifying all possible parenthesized products with each other. But this runs into the following problem: for $n \geq 4$ there may be two or more different chains of associativity isomorphisms connecting two different parenthesizings, and a priori it is not clear that they provide the same identification.

Luckily, for $n = 4$, this is settled by the pentagon axiom, which states exactly that the two possible identifications are the same. But what about $n > 4$?

This problem is solved by the following theorem of Mac Lane, which is the first important result in the theory of monoidal categories.

THEOREM 2.9.2. *(Coherence Theorem) Let $X_1, \ldots, X_n \in \mathcal{C}$. Let P_1, P_2 be any two parenthesized products of $X_1, \ldots, X_n$ (in this order) with arbitrary insertions of the unit object $\mathbf{1}$. Let $f, g : P_1 \to P_2$ be two isomorphisms, obtained by composing associativity and unit isomorphisms and their inverses possibly tensored with identity morphisms. Then $f = g$.*

PROOF. We derive this theorem as a corollary of the Mac Lane's strictness Theorem 2.8.5. Let $L : \mathcal{C} \to \mathcal{C}'$ be a monoidal equivalence between $\mathcal{C}$ and a strict monoidal category $\mathcal{C}'$. Consider a diagram in $\mathcal{C}$ representing f and g and apply L to it. Over each arrow of the resulting diagram representing an associativity isomorphism, let us build a rectangle as in (2.23), and do similarly for the unit morphisms. This way we obtain a prism one of whose faces consists of identity maps (associativity and unit isomorphisms in $\mathcal{C}'$) and whose sides are commutative. Hence, the other face is commutative as well, i.e., $f = g$. □

REMARK 2.9.3. As we mentioned, Theorem 2.9.2 implies that any two parenthesized products of $X_1, \ldots, X_n$ with insertions of unit objects are indeed canonically isomorphic, and thus one can safely identify all of them with each other and ignore bracketings in calculations in a monoidal category. We will do so from now on, unless confusion is possible.

2.10. Rigid monoidal categories

Let $(\mathcal{C}, \otimes, \mathbf{1}, a, \iota)$ be a monoidal category, and let X be an object of $\mathcal{C}$. In what follows, we suppress the unit constraints l and r.

DEFINITION 2.10.1. An object X^* in $\mathcal{C}$ is said to be a *left dual* of X if there exist morphisms $\mathsf{ev}_X : X^* \otimes X \to \mathbf{1}$ and $\mathsf{coev}_X : \mathbf{1} \to X \otimes X^*$, called the *evaluation* and *coevaluation*, such that the compositions

$$X \xrightarrow{\mathsf{coev}_X \otimes \mathsf{id}_X} (X \otimes X^*) \otimes X \xrightarrow{a_{X,X^*,X}} X \otimes (X^* \otimes X) \xrightarrow{\mathsf{id}_X \otimes \mathsf{ev}_X} X, \tag{2.43}$$

$$X^* \xrightarrow{\mathsf{id}_{X^*} \otimes \mathsf{coev}_X} X^* \otimes (X \otimes X^*) \xrightarrow{a^{-1}_{X^*,X,X^*}} (X^* \otimes X) \otimes X^* \xrightarrow{\mathsf{ev}_X \otimes \mathsf{id}_{X^*}} X^* \tag{2.44}$$

are the identity morphisms.

DEFINITION 2.10.2. An object ${}^*X$ in $\mathcal{C}$ is said to be a *right dual* of X if there exist morphisms $\mathsf{ev}'_X : X \otimes {}^*X \to \mathbf{1}$ and $\mathsf{coev}'_X : \mathbf{1} \to {}^*X \otimes X$ such that the compositions

$$X \xrightarrow{\mathsf{id}_X \otimes \mathsf{coev}'_X} X \otimes ({}^*X \otimes X) \xrightarrow{a^{-1}_{X,{}^*X,X}} (X \otimes {}^*X) \otimes X \xrightarrow{\mathsf{ev}'_X \otimes \mathsf{id}_X} X, \tag{2.45}$$

$${}^*X \xrightarrow{\mathsf{coev}'_X \otimes \mathsf{id}_{{}^*X}} ({}^*X \otimes X) \otimes {}^*X \xrightarrow{a_{{}^*X,X,{}^*X}} {}^*X \otimes (X \otimes {}^*X) \xrightarrow{\mathsf{id}_{{}^*X} \otimes \mathsf{ev}'_X} {}^*X \tag{2.46}$$

are the identity morphisms.

REMARK 2.10.3. It is obvious that if X^* is a left dual of an object X then X is a right dual of X^* with $\mathsf{ev}'_{X^*} = \mathsf{ev}_X$ and $\mathsf{coev}'_{X^*} = \mathsf{coev}_X$, and vice versa. Therefore, ${}^*(X^*) \cong X \cong ({}^*X)^*$ for any object X admitting left and right duals. Also, in any monoidal category, $\mathbf{1}^* = {}^*\mathbf{1} = \mathbf{1}$ with the evaluation and coevaluation morphisms ι and ι^{-1}. Also note that changing the order of tensor product switches left duals and right duals, so to any statement about right duals there corresponds a symmetric statement about left duals.

EXERCISE 2.10.4. Let $\mathcal{C}$ be a category and $\mathsf{End}(\mathcal{C})$ be the monoidal category of endofunctors of $\mathcal{C}$. Show that a left (respectively, right) dual of $F \in \mathsf{End}(\mathcal{C})$ is the same thing as a functor left (respectively, right) adjoint to F. This justifies our terminology.

PROPOSITION 2.10.5. *If $X \in \mathcal{C}$ has a left (respectively, right) dual object, then it is unique up to a unique isomorphism.*

PROOF. Let X_1^*, X_2^* be two left duals of X. Denote by e_1, c_1, e_2, c_2 the corresponding evaluation and coevaluation morphisms. Then we have a morphism $\alpha : X_1^* \to X_2^*$ defined as the composition

$$X_1^* \xrightarrow{\mathsf{id}_{X_1^*} \otimes c_2} X_1^* \otimes (X \otimes X_2^*) \xrightarrow{a^{-1}_{X_1^*, X, X_2^*}} (X_1^* \otimes X) \otimes X_2^* \xrightarrow{e_1 \otimes \mathsf{id}_{X_2^*}} X_2^*.$$

Similarly, one defines a morphism $\beta : X_2^* \to X_1^*$. We claim that $\beta \circ \alpha$ and $\alpha \circ \beta$ are the identity morphisms, so α is an isomorphism. Indeed, consider the following diagram:

$$\begin{array}{ccccc}
X_1^* & \xrightarrow{\mathsf{id} \otimes c_1} & X_1^* \otimes X \otimes X_1^* & & \\
\downarrow{\scriptstyle \mathsf{id} \otimes c_2} & & \downarrow{\scriptstyle \mathsf{id} \otimes c_2 \otimes \mathsf{id}} & \searrow{\scriptstyle \mathsf{id}} & \\
X_1^* \otimes X \otimes X_2^* & \xrightarrow[\mathsf{id} \otimes c_1]{} & X_1^* \otimes X \otimes X_2^* \otimes X \otimes X_1^* & \xrightarrow[\mathsf{id} \otimes e_2 \otimes \mathsf{id}]{} & X_1^* \otimes X \otimes X_1^* \\
\downarrow{\scriptstyle e_1 \otimes \mathsf{id}} & & \downarrow{\scriptstyle e_1 \otimes \mathsf{id}} & & \downarrow{\scriptstyle e_1 \otimes \mathsf{id}} \\
X_2^* & \xrightarrow[\mathsf{id} \otimes c_1]{} & X_2^* \otimes X \otimes X_1^* & \xrightarrow[e_2 \otimes \mathsf{id}]{} & X_1^*.
\end{array}$$

Here we suppress the associativity constraints. It is clear that the three small squares commute. The triangle in the upper right corner commutes by axiom (2.43) applied to X_2^*. Hence, the perimeter of the diagram commutes. The composition through the top row is the identity by (2.44) applied to X_1^*. The composition through the bottom row is $\beta \circ \alpha$ and so $\beta \circ \alpha = \mathsf{id}_{X_1^*}$. The proof of $\alpha \circ \beta = \mathsf{id}_{X_2^*}$ is completely similar.

Moreover, it is easy to check that $\alpha : X_1^* \to X_2^*$ is the only isomorphism which preserves the evaluation and coevaluation morphisms. This proves the proposition for left duals. The proof for right duals is similar. □

If X, Y are objects in $\mathcal{C}$ which have left duals X^*, Y^* and $f : X \to Y$ is a morphism, one defines the *left dual* $f^* : Y^* \to X^*$ of f by

$$\begin{aligned}
f^* := Y^* &\xrightarrow{\mathsf{id}_{Y^*} \otimes \mathsf{coev}_X} Y^* \otimes (X \otimes X^*) \xrightarrow{a^{-1}_{Y^*, X, X^*}} (Y^* \otimes X) \otimes X^* \\
&\xrightarrow{(\mathsf{id}_{Y^*} \otimes f) \otimes \mathsf{id}_{X^*}} (Y^* \otimes Y) \otimes X^* \xrightarrow{\mathsf{ev}_Y \otimes \mathsf{id}_{X^*}} X^*.
\end{aligned} \tag{2.47}$$

Similarly, if X, Y are objects in $\mathcal{C}$ which have right duals ${}^*X, {}^*Y$ and $f : X \to Y$ is a morphism one defines the *right dual* ${}^*f : {}^*Y \to {}^*X$ of f by

$$\begin{aligned}
{}^*f := {}^*Y &\xrightarrow{\mathsf{coev}'_X \otimes \mathsf{id}_{{}^*Y}} ({}^*X \otimes X) \otimes {}^*Y \xrightarrow{a_{{}^*X, X, {}^*Y}} {}^*X \otimes (X \otimes {}^*Y) \\
&\xrightarrow{\mathsf{id}_{{}^*X} \otimes (f \otimes \mathsf{id}_{{}^*Y})} {}^*X \otimes (Y \otimes {}^*Y) \xrightarrow{\mathsf{id}_{{}^*X} \otimes \mathsf{ev}'_Y} {}^*X.
\end{aligned} \tag{2.48}$$

EXERCISE 2.10.6. Let $\mathcal{C}, \mathcal{D}$ be monoidal categories. Suppose $F = (F, J, \varphi)$ is a monoidal functor, $F : \mathcal{C} \to \mathcal{D}$. Let X be an object in $\mathcal{C}$ with a left dual X^*. Prove that $F(X^*)$ is a left dual of $F(X)$ with the evaluation and coevaluation given by

$$\mathsf{ev}_{F(X)} \quad : \quad F(X^*) \otimes F(X) \xrightarrow{J_{X^*,X}} F(X^* \otimes X) \xrightarrow{F(\mathsf{ev}_X)} F(\mathbf{1}) \xrightarrow{\varphi^{-1}} \mathbf{1},$$

$$\mathsf{coev}_{F(X)} \quad : \quad \mathbf{1} \xrightarrow{\varphi} F(\mathbf{1}) \xrightarrow{F(\mathsf{coev}_X)} F(X \otimes X^*) \xrightarrow{J^{-1}_{X,X^*}} F(X) \otimes F(X^*).$$

State and prove a similar result for right duals.

EXERCISE 2.10.7. Let $\mathcal{C}$ be a monoidal category, let U, V, W be objects in $\mathcal{C}$, and let $f : V \to W$, $g : U \to V$ be morphisms in $\mathcal{C}$. Prove that

(a) If U, V, W have left (respectively, right) duals then $(f \circ g)^* = g^* \circ f^*$ (respectively, ${}^*(f \circ g) = {}^*g \circ {}^*f$).

(b) If U, V have left (respectively, right) duals then $U \otimes V$ has a left dual $V^* \otimes U^*$ (respectively, right dual ${}^*V \otimes {}^*U$).

PROPOSITION 2.10.8. *Let $\mathcal{C}$ be a monoidal category and let V be an object in $\mathcal{C}$.*

(a) *If V has a left dual V^* then there are natural adjunction isomorphisms*

(2.49) $$\mathsf{Hom}_{\mathcal{C}}(U \otimes V, W) \xrightarrow{\sim} \mathsf{Hom}_{\mathcal{C}}(U, W \otimes V^*),$$

(2.50) $$\mathsf{Hom}_{\mathcal{C}}(V^* \otimes U, W) \xrightarrow{\sim} \mathsf{Hom}_{\mathcal{C}}(U, V \otimes W).$$

(b) *If V has a right dual ${}^*V$ then there are natural adjunction isomorphisms*

(2.51) $$\mathsf{Hom}_{\mathcal{C}}(U \otimes {}^*V, W) \xrightarrow{\sim} \mathsf{Hom}_{\mathcal{C}}(U, W \otimes V),$$

(2.52) $$\mathsf{Hom}_{\mathcal{C}}(V \otimes U, W) \xrightarrow{\sim} \mathsf{Hom}_{\mathcal{C}}(U, {}^*V \otimes W).$$

PROOF. An isomorphism in (2.49) is given by $f \mapsto (f \otimes \mathsf{id}_{V^*}) \circ (\mathsf{id}_U \otimes \mathsf{coev}_V)$ and has the inverse $g \mapsto (\mathsf{id}_W \otimes \mathsf{ev}_V) \circ (g \otimes \mathsf{id}_V)$. Other isomorphisms are similar and are left to the reader as an exercise. □

REMARK 2.10.9. Proposition 2.10.8 says, in particular, that when a left (respectively, right) dual of V exists, then the functor $V^* \otimes -$ is the left adjoint of $V \otimes -$ (respectively, $- \otimes V^*$ is the right adjoint of $- \otimes V$).

REMARK 2.10.10. Proposition 2.10.8 provides another proof of Proposition 2.10.5. Namely, setting $U = \mathbf{1}$ and $V = X$ in (2.50), we obtain a natural isomorphism $\mathsf{Hom}_{\mathcal{C}}(X^*, W) \cong \mathsf{Hom}_{\mathcal{C}}(\mathbf{1}, X \otimes W)$ for any left dual X^* of X. Hence, if Y_1, Y_2 are two such duals then there is a natural isomorphism $\mathsf{Hom}_{\mathcal{C}}(Y_1, W) \cong \mathsf{Hom}_{\mathcal{C}}(Y_2, W)$, whence there is a canonical isomorphism $Y_1 \cong Y_2$ by the Yoneda Lemma. The proof for right duals is similar.

DEFINITION 2.10.11. An object in a monoidal category is called *rigid* if it has left and right duals. A monoidal category $\mathcal{C}$ is called *rigid* if every object of $\mathcal{C}$ is rigid.

EXAMPLE 2.10.12. The category Vec of finite dimensional $\Bbbk$-vector spaces is rigid: the right and left dual to a finite dimensional vector space V are its dual space V^*, with the evaluation map $\mathsf{ev}_V : V^* \otimes V \to \Bbbk$ being the contraction, and the coevaluation map $\mathsf{coev}_V : k \to V \otimes V^*$ being the usual embedding. On the other hand, the category **Vec** of all $\Bbbk$-vector spaces is not rigid, since for infinite dimensional spaces there is no coevaluation maps (indeed, suppose that

$c : \Bbbk \to V \otimes Y$ is a coevaluation map, and consider the subspace V' of V spanned by the first component of $c(1)$; this subspace is finite dimensional, and yet the composition $V \to V \otimes Y \otimes V \to V$, which is supposed to be the identity map, lands in V' - a contradiction).

EXAMPLE 2.10.13. Let $\Bbbk$ be a field. The category $\mathsf{Rep}(G)$ of finite dimensional representations of a group G over $\Bbbk$ is rigid: for a finite dimensional representation V, the (left or right) dual representation V^* is the usual dual space (with the evaluation and coevaluation maps as in Example 2.10.12), and with the G-action given by $\rho_{V^*}(g) = (\rho_V(g)^{-1})^*$. Similarly, the category $\mathsf{Rep}(\mathfrak{g})$ of finite dimensional representations of a Lie algebra $\mathfrak{g}$ is rigid, with $\rho_{V^*}(a) = -\rho_V(a)^*$.

EXAMPLE 2.10.14. The category Vec_G (see Example 2.3.6) is rigid if and only if the monoid G is a group; namely, $\delta_g^* = {}^*\delta_g = \delta_{g^{-1}}$ (with the obvious structure maps). More generally, for any group G and 3-cocycle $\omega \in Z^3(G, \Bbbk^\times)$, the category Vec_G^ω is rigid. Namely, assume for simplicity that the cocycle ω is normalized (as we know, we can do so without loss of generality). Then we can define duality as above, and normalize the coevaluation morphisms of δ_g to be the identities. The evaluation morphisms will then be defined by the formula $\mathsf{ev}_{\delta_g} = \omega(g, g^{-1}, g)\,\mathsf{id}_{\mathbf{1}}$.

It follows from Proposition 2.10.5 that in a monoidal category $\mathcal{C}$ with left (respectively, right) duals, one can define a contravariant left (respectively, right) *duality functor* by

$$X \mapsto X^*,\ f \mapsto f^* : \mathcal{C} \to \mathcal{C}$$

(respectively, by $X \mapsto {}^*X$, $f \mapsto {}^*f$) for every object X and morphism f in $\mathcal{C}$. By Exercise 2.10.7(ii), these are monoidal functors $\mathcal{C}^\vee \to \mathcal{C}^{\mathrm{op}}$, where the monoidal structure of the opposite category $\mathcal{C}^{\mathrm{op}}$ is given in Definition 2.1.5. Hence, the functors $X \mapsto X^{**}$, $X \mapsto {}^{**}X$ are monoidal. Also, it follows from Proposition 2.10.8(a) that the functors of left and right duality, when they are defined, are fully faithful.

Moreover, it follows from Remark 2.10.3 that in a rigid monoidal category, the functors of left and right duality are mutually quasi-inverse monoidal equivalences of categories $\mathcal{C}^\vee \xrightarrow{\sim} \mathcal{C}^{\mathrm{op}}$ (so for rigid categories, the notions of dual and opposite category are the same up to equivalence). This implies that the functors $X \mapsto X^{**}$ and $X \mapsto {}^{**}X$ are mutually quasi-inverse monoidal autoequivalences. We will see later in Example 7.19.5 that these autoequivalences may be nontrivial; in particular, it is possible that objects V^* and ${}^*V$ are not isomorphic.

EXERCISE 2.10.15. Show that if $\mathcal{C}, \mathcal{D}$ are rigid monoidal categories, $F_1, F_2 : \mathcal{C} \to \mathcal{D}$ are monoidal functors, and $\eta : F_1 \to F_2$ is a morphism of monoidal functors, then η is an isomorphism (as we have seen in Remark 2.5.6, this is false for non-rigid categories).

EXERCISE 2.10.16. Let A be an algebra. Show that $M \in A-\mathbf{bimod}$ has a left (respectively, right) dual if and only if it is finitely generated projective when considered as a left (respectively, right) A-module. Similarly, if A is commutative, $M \in A$–$\mathbf{mod}$ has left and right duals if and only if it is finitely generated projective.

2.11. Invertible objects and Gr-categories

Let $\mathcal{C}$ be a rigid monoidal category.

DEFINITION 2.11.1. An object X in $\mathcal{C}$ is *invertible* if $\mathsf{ev}_X : X^* \otimes X \to \mathbf{1}$ and $\mathsf{coev}_X : \mathbf{1} \to X \otimes X^*$ are isomorphisms.

Clearly, this notion categorifies the notion of an invertible element in a monoid.

EXAMPLE 2.11.2. Let G be a group.

(1) The objects δ_g in Vec_G^ω (see Example 2.3.8) are invertible.
(2) The invertible objects in $\mathsf{Rep}(G)$ (see Example 2.3.4) are precisely the 1-dimensional representations of G.

PROPOSITION 2.11.3. *Let X be an invertible object in $\mathcal{C}$. Then*

(i) ${}^*X \cong X^*$ *and* X^* *is invertible;*
(ii) *if Y is another invertible object then $X \otimes Y$ is invertible.*

PROOF. Dualizing coev_X and ev_X we get isomorphisms $X \otimes {}^*X \cong \mathbf{1}$ and ${}^*X \otimes X \cong \mathbf{1}$. Hence ${}^*X \cong {}^*X \otimes X \otimes X^* \cong X^*$. In any rigid category the evaluation and coevaluation morphisms for ${}^*X$ can be defined by $\mathsf{ev}_{{}^*X} := {}^*\mathsf{coev}_X$ and $\mathsf{coev}_{{}^*X} := {}^*\mathsf{ev}_X$, so ${}^*X$ is invertible. The second statement follows from the fact that $\mathsf{ev}_{X\otimes Y}$ can be defined as a composition of ev_X and ev_Y, and similarly $\mathsf{coev}_{X\otimes Y}$ can be defined as a composition of coev_Y and coev_X. □

Proposition 2.11.3 implies that invertible objects of $\mathcal{C}$ form a monoidal subcategory $\mathrm{Inv}(\mathcal{C})$ of $\mathcal{C}$.

DEFINITION 2.11.4. A *Gr-category*, or a *categorical group*, is a rigid monoidal category in which every object is invertible and all morphisms are isomorphisms.

The second condition of Definition 2.11.4 means that a Gr-category is a groupoid. In fact, it is precisely a group object in the category of groupoids.

The next theorem provides a classification of Gr-categories.

THEOREM 2.11.5. *Monoidal equivalence classes of Gr-categories are in bijection with triples (G, A, ω), where G is a group, A is a G-module, and ω is an orbit in $H^3(G, A)$ under the action of* $\mathsf{Out}(G)$.

PROOF. We may assume that a Gr-category $\mathcal{C}$ is skeletal, i.e., there is only one object in each isomorphism class, and objects form a group G. Also, by Proposition 2.2.10, $\mathsf{End}_{\mathcal{C}}(\mathbf{1})$ is an abelian group; let us denote it by A. Then for any $g \in G$ we can identify $\mathsf{End}_{\mathcal{C}}(g)$ with A, by sending $f \in \mathsf{End}_{\mathcal{C}}(g)$ to $f \otimes \mathsf{id}_{g^{-1}} \in \mathsf{End}_{\mathcal{C}}(\mathbf{1}) = A$. Then we have an action of G on A by

$$a \in \mathsf{End}_{\mathcal{C}}(\mathbf{1}) \mapsto g(a) := \mathsf{id}_g \otimes a \in \mathsf{End}_{\mathcal{C}}(g).$$

Let us now consider the associativity isomorphism. It is defined by a function $\omega : G \times G \times G \to A$. The pentagon relation gives

$$\omega(g_1g_2, g_3, g_4)\omega(g_1, g_2, g_3g_4) = \omega(g_1, g_2, g_3)\omega(g_1, g_2g_3, g_4)g_1(\omega(g_2, g_3, g_4)), \tag{2.53}$$

for all $g_1, g_2, g_3, g_4 \in G$, which means that ω is a 3-cocycle of G with coefficients in the (in general, nontrivial) G-module A. We see that any such 3-cocycle defines a rigid monoidal category, which we call $\mathcal{C}_G^\omega(A)$. The analysis of monoidal equivalences between such categories is similar to the case when A is a trivial G-module and yields that for a given group G and G-module A, equivalence classes of $\mathcal{C}_G^\omega(A)$ are parametrized by $H^3(G, A)/\mathsf{Out}(G)$. □

2.12. 2-categories

The notion of a 2-category extends the notion of a category, in the sense that in a 2-category one has in addition to objects and morphisms between them, also "morphisms between morphisms". Here is the formal definition.

DEFINITION 2.12.1. A *strict 2-category* $\underline{\mathcal{C}}$ consists of objects A, $B, \ldots$, 1-morphisms between objects $f : A \to B, \ldots$ and 2-morphisms $\alpha : f \Rightarrow g, \ldots$ between 1-morphisms $f, g : A \to B$ such that the following axioms are satisfied:

(1) The objects together with the 1-morphisms form a category $\mathcal{C}$. The composition of 1-morphisms is denoted by $\circ$.
(2) For any fixed pair of objects (A, B), the 1-morphisms from A to B together with the 2-morphisms between them form a category $\underline{\mathcal{C}}(A, B)$. The unital associative composition (along 1-morphisms) $\beta \cdot \alpha : f \Rightarrow h$ of two 2-morphisms $\alpha : f \Rightarrow g$, $\beta : g \Rightarrow h$ is called the *vertical composition*

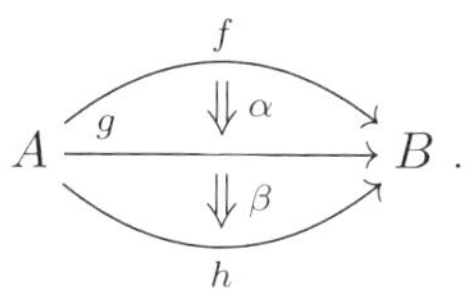

The identities are denoted by $\mathsf{id}_f : f \Rightarrow f$; 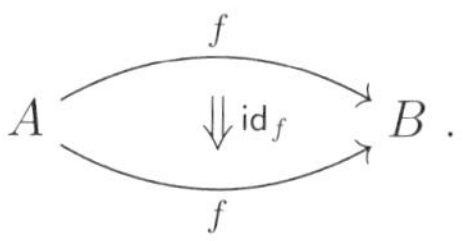

(3) There is a unital associative *horizontal composition* (along objects) $\beta \circ \alpha : h \circ f \Rightarrow i \circ g$ of 2-morphisms $\alpha : f \Rightarrow g$, $\beta : h \Rightarrow i$, where $f, g : A \to B$ and $h, i : B \to C$:

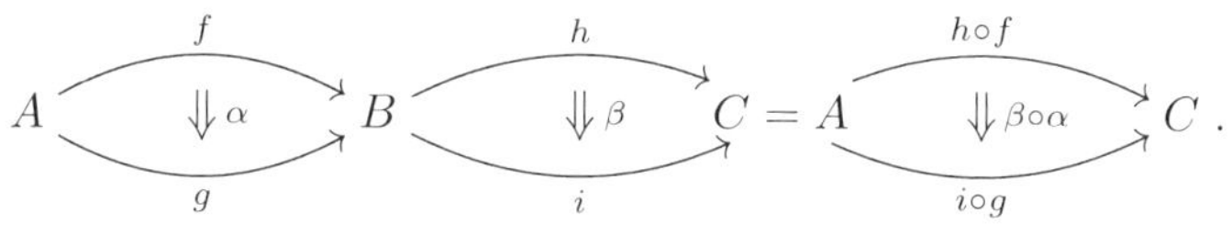

The identities are $\mathsf{id}_{\mathsf{id}_A} : \mathsf{id}_A \Rightarrow \mathsf{id}_A$; 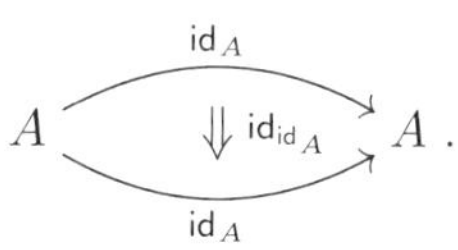

(4) For any triple of objects (A, B, C), 1-morphisms $f, g, h : A \to B$ and $i, j, k : B \to C$, and 2-morphisms $\alpha : f \Rightarrow g$, $\beta : g \Rightarrow h$, $\gamma : i \Rightarrow j$, $\delta : j \Rightarrow k$, we have $(\delta \circ \beta) \cdot (\gamma \circ \alpha) = (\delta \cdot \gamma) \circ (\beta \cdot \alpha)$ ("interchange law").
(5) The horizontal composition preserves vertical units, i.e., for any objects A, B, C, and 1-morphisms $f : A \to B$, $i : B \to C$, we have $\mathsf{id}_i \circ \mathsf{id}_f = \mathsf{id}_{i \circ f}$.

DEFINITION 2.12.2. A *2-category* $\underline{\mathcal{C}}$ consists of the same data as a strict 2-category $\underline{\mathcal{C}}$, except that the composition of 1-morphisms is required to be unital associative only up to *associativity and unital constraints*. Namely, there exist natural families of invertible 2-morphisms

$$\alpha_{f,g,h} : h \circ (g \circ f) \Rightarrow (h \circ g) \circ f,\ \lambda_f : f \circ \mathsf{id}_A \Rightarrow f,\ \rho_f : \mathsf{id}_B \circ f \Rightarrow f,$$

satisfying the *pentagon axiom*

$$\alpha_{f,g,i\circ h} \cdot \alpha_{g\circ f,h,i} = (\mathsf{id}_i \circ \alpha_{h,g,f}) \cdot \alpha_{f,h\circ g,i} \cdot (\alpha_{g,h,i} \circ \mathsf{id}_f)$$

and the *triangle axiom*

$$(\mathsf{id}_{\mathsf{id}_A} \circ \lambda_f) \cdot \alpha_{f,\mathsf{id}_B,g} = \rho_r \circ \mathsf{id}_f .$$

REMARK 2.12.3. The objects, 1-morphisms and 2-morphisms are called in some texts, 0-cells, 1-cells and 2-cells, respectively.

EXAMPLE 2.12.4. (*The 2-category of categories*) The objects are categories, the 1-morphisms are functors between categories, and the 2-morphisms are natural transformations between functors. The horizontal composition of natural transformations is the so called *Godement product*.

EXAMPLE 2.12.5. (*Rings and bimodules*) Rings are the objects, bimodules are the 1-morphisms, and homomorphisms between bimodules are the 2-morphisms.

EXAMPLE 2.12.6. Recall that the notion of a monoid is a special case of the notion of a category; namely, a monoid is the same thing as a category with one object (the morphisms of this category are the elements of the corresponding monoid). Similarly, the notion of a monoidal category is a special case of the notion of a 2-category [**Mac2**]: a monoidal category is the same thing as a 2-category with one object. Namely, the 1-morphisms and 2-morphisms of such a 2-category are the objects and morphisms of the corresponding monoidal category, and composition of 1-morphisms is the tensor product functor.

Below we will see other examples of 2-categories: a multitensor category (see Remark 4.3.7) and the 2-category of module categories over a multitensor category (see Remark 7.12.15).

2.13. Bibliographical notes

2.1.-2.2. Monoidal categories were introduced by Bénabou [**Ben1**] as "categories with multiplication". The pentagon and triangle axioms (2.2) and (2.10) were introduced by Mac Lane [**Mac1**]. Definition 2.1.1 (with unit being an idempotent that can be cancelled) and Proposition 2.2.10 appeared in the paper [**Sa**] by Saavedra Rivano. Note that Proposition 2.2.10 is a categorical version of the famous Eckmann-Hilton argument. Proposition 2.2.4 is due to Kelly [**Ke**].

2.3. Categories of tangles considered in Example 2.3.14 were introduced by Turaev [**Tu2**] and Yetter [**Ye2**]. These categories are important for applications to low-dimensional topology. A detailed description of such categories can be found in books by Kassel, Turaev, and Bakalov-Kirillov [**Kas, Tu4, BakK**].

2.4-2.6. For a discussion of monoidal functors, see e.g. the papers by Joyal and Street [**JoyS5**] and Müger [**Mu6**].

2.7. For a discussion of equivariantization, see e.g. [**DrGNO2**].

2.8-2.9. The proof of the Mac Lane strictness Theorem 2.8.5 presented here is given by Joyal and Street [**JoyS5**]. The coherence Theorem 2.9.2 is due to Mac Lane [**Mac1**]. Remark 2.8.7 is due to Kuperberg [**Ku**].

2.10. The notions of duality and of a rigid monoidal category appeared independently in many classical works, in particular in the papers by Saavedra Rivano [**Sa**] and Kelly [**Ke**]. A convenient way to do computations involving duality, e.g., in the proof of Proposition 2.10.8, is via the graphical calculus (see, e.g., Kassel's book [**Kas**, Chapter XIV]).

2.11. Definition 2.11.4 of a Gr-category goes back to Sính [**Sin**]. Gr-categories are also known as "2-groups". A theorem of Verdier establishes an equivalence between the category of Gr-categories (with monoidal functors as morphisms) and the category of *crossed modules*, see the paper by Barrett and Mackaay [**BarM**] for definition and discussion.

2.12. Good references for the theory of 2-categories are the books of Mac Lane [**Mac2**] and Lenster [**Le**].

CHAPTER 3

$\mathbb{Z}_+$-rings

3.1. Definition of a $\mathbb{Z}_+$-ring

Let $\mathbb{Z}_+$ denote the semi-ring of non-negative integers.

DEFINITION 3.1.1. Let A be a ring which is free as a $\mathbb{Z}$-module.

(i) A $\mathbb{Z}_+$*-basis* of A is a basis $B = \{b_i\}_{i\in I}$ such that $b_i b_j = \sum_{k\in I} c_{ij}^k b_k$, where $c_{ij}^k \in \mathbb{Z}_+$.

(ii) A $\mathbb{Z}_+$*-ring* is a ring with a fixed $\mathbb{Z}_+$-basis and with identity 1 which is a non-negative linear combination of the basis elements.

(iii) A *unital* $\mathbb{Z}_+$*-ring* is a $\mathbb{Z}_+$-ring such that 1 is a basis element.

REMARK 3.1.2. Every $\mathbb{Z}_+$-ring is assumed to have an identity, but according to our terminology a $\mathbb{Z}_+$-ring is not necessarily unital.

Let A be a $\mathbb{Z}_+$-ring, and let I_0 be the set of $i \in I$ such that b_i occurs in the decomposition of 1. Let $\tau : A \to \mathbb{Z}$ denote the group homomorphism defined by

$$\tau(b_i) = \begin{cases} 1 & \text{if} \quad i \in I_0 \\ 0 & \text{if} \quad i \notin I_0. \end{cases} \tag{3.1}$$

DEFINITION 3.1.3. A $\mathbb{Z}_+$-ring A with basis $\{b_i\}_{i\in I}$ is called a *based ring* if there exists an involution $i \mapsto i^*$ of I such that the induced map

$$a = \sum_{i\in I} a_i b_i \mapsto a^* = \sum_{i\in I} a_i b_{i^*},\ a_i \in \mathbb{Z},$$

is an anti-involution of the ring A, and

$$\tau(b_i b_j) = \begin{cases} 1 & \text{if} \quad i = j^* \\ 0 & \text{if} \quad i \neq j^*. \end{cases} \tag{3.2}$$

PROPOSITION 3.1.4. *In any based ring one has* $1 = \sum_{i\in I_0} b_i$.

PROOF. We have $1 = \sum_{i\in I_0} a_i b_i$, $a_i > 0$. Hence $1 = \sum_{i\in I_0} a_i b_i^*$. Thus, for $j \in I_0$

$$a_j = \tau(b_j(\sum_{i\in I_0} a_i b_i^*)) = \tau(b_j \cdot 1) = \tau(b_j) = 1,$$

as desired. □

EXERCISE 3.1.5. (i) Show that in a $\mathbb{Z}_+$-ring, $i, j \in I_0$, $i \neq j$ implies that $b_i^2 = b_i$, $b_i b_j = 0$, and in a based ring $i^* = i$ for $i \in I_0$.

(ii) Show that for a given $\mathbb{Z}_+$-ring A, being a (unital) based ring is a *property*, not an additional structure.

PROPOSITION 3.1.6. *In any based ring, the number* $c_{ij}^{k^*}$ *is invariant under cyclic permutations of* i, j, k.

PROOF. It is easy to see that $c_{ij}^{k^*} = \tau(b_i b_j b_k)$, and $\tau(xy) = \tau(yx)$. Thus, $c_{ij}^{k^*}$ is cyclically symmetric. □

DEFINITION 3.1.7. A *multifusion ring* is a based ring of finite rank. A *fusion ring* is a unital based ring of finite rank.

PROPOSITION 3.1.8. *Let A be a multifusion ring with basis $\{b_i\}$. Then for any $x \in A$, the element $Z(x) := \sum_i b_i x b_i^*$ is central in A.*

PROOF. It suffices to show that in $A \otimes A$,

$$\sum_{i\in I} b_k b_i \otimes b_i^* = \sum_{i\in I} b_i \otimes b_i^* b_k, \qquad \text{for all } k \in I; \tag{3.3}$$

then

$$b_k Z(x) = \sum_{i\in I} b_k b_i x b_i^* = \sum_{i\in I} b_i x b_i^* b_k = Z(x) b_k.$$

We compute

$$\begin{aligned}\sum_i b_k b_i \otimes b_i^* &= \sum_{r,i} c_{ki}^r b_r \otimes b_i^* \\ &= \sum_{r,i} c_{r^*k}^{i^*} b_r \otimes b_i^* \\ &= \sum_r b_r \otimes b_r^* b_k,\end{aligned}$$

where the equality $c_{ki}^r = c_{r^*k}^{i^*}$ is established in Proposition 3.1.6. This implies the proposition. □

EXAMPLE 3.1.9. Here are some examples of $\mathbb{Z}_+$-rings.

(i) The ring of matrices $\mathsf{Mat}_n(\mathbb{Z})$ is a multifusion ring, with the basis consisting of the elementary matrices E_{ij} (i.e., I is the set of pairs (i, j), $i, j = 1, ..., n$), and $*$ being the transposition. The set I_0 is the set of diagonal pairs (i, i). This ring is not a fusion ring unless $n = 1$ (since $|I_0| = n$).

(ii) If G is a group, then the group ring $\mathbb{Z}G$ is a unital based ring, with the basis of group elements, and $g^* = g^{-1}$ for $g \in G$. This is a fusion ring if G is finite.

(iii) The center $C(\mathbb{Z}G)$ for a finite group G is a unital $\mathbb{Z}_+$-ring, with basis consisting of the sums of group elements over conjugacy classes C of G, $b_C := \sum_{g\in C} g$. This ring admits an involution induced by $g^* = g^{-1}$, but is not a based (or fusion) ring for a nonabelian G, since $\tau(b_C b_{C^{-1}}) = |C|$, which may be > 1.

(iv) The ring R_G of complex representations of a finite group G, with basis of irreducible representations, is a commutative fusion ring (the involution $*$ is the operation of taking the dual representation). For example, if G is abelian, then $R_G = \mathbb{Z}G^\vee$, where $G^\vee$ is the character group of G. If $G = S_3$, the symmetric group on 3 letters, then R_G has basis 1, χ, V with relations

$$\chi V = V\chi = V,\ \chi^2 = 1,\ V^2 = 1 + \chi + V$$

and the involution $*$ being the identity map.

(v) The ring of real representations of a finite group G is a unital $\mathbb{Z}_+$-ring with involution, but it is not a fusion ring, since for a real irreducible representation V, $V \otimes V^*$ may contain more than one copy of the trivial representation (e.g., take V to be the defining 2-dimensional representation of the group $\mathbb{Z}/3\mathbb{Z}$ of rotations of an equilateral triangle[1]). The same happens over any field of characteristic zero which is not algebraically closed. Also, the representation ring over an algebraically closed field of positive characteristic is not, in general, a fusion ring, for essentially the same reason. For example, if V is the 2-dimensional irreducible representation of S_3 in characteristic 2 then $V \otimes V^*$ contains in its composition series two copies of the trivial representation (since in characteristic 2, the trivial representation is indistinguishable from the sign representation).

(vi) Let G be a compact Lie group or a reductive complex algebraic group. Then the ring R_G of representations of G is a unital based ring (which is not a fusion ring if G is infinite). For example, if $G = SU(2)$ or $G = SL(2, \mathbb{C})$, then R_G has $* = \mathrm{id}$ and basis V_i for $i \in \mathbb{Z}_+$, with

$$V_i V_j = \sum_{l=0}^{\min(i,j)} V_{i+j-2l}. \tag{3.4}$$

Formula (3.4) is called the *Clebsch-Gordan rule*.

(vii) The Yang-Lee fusion ring has two basis elements 1 and X with the multiplication rule $X^2 = 1 + X$ and the identity involution. It appears as a subring of the Verlinde fusion ring, see Example 4.10.6.

(viii) The *Ising fusion ring* has 3 basis elements 1, χ, V with the multiplication rules

$$\chi^2 = 1, \quad \chi V = V\chi = V, \quad V^2 = 1 + \chi$$

and the identity involution. It arises in the Ising model of statistical mechanics. It is a special case of the Tambara-Yamagami fusion ring, see Example 4.10.5.

3.2. The Frobenius-Perron theorem

The following classical theorem from linear algebra [**Gant**, XIII.2] plays a crucial role in the theory of tensor categories.

THEOREM 3.2.1. *(Frobenius-Perron) Let B be a square matrix with non-negative real entries.*

(1) *B has a non-negative real eigenvalue. The largest non-negative real eigenvalue $\lambda(B)$ of B dominates the absolute values of all other eigenvalues μ of B: $|\mu| \leq \lambda(B)$ (in other words, the spectral radius of B is an eigenvalue). Moreover, there is an eigenvector of B with non-negative entries and eigenvalue $\lambda(B)$.*

(2) *If B has strictly positive entries then $\lambda(B)$ is a simple positive eigenvalue, and the corresponding eigenvector can be normalized to have strictly positive entries. Moreover, $|\mu| < \lambda(B)$ for any other eigenvalue μ of B.*

(3) *If a matrix B with non-negative entries has an eigenvector $\mathbf{v}$ with strictly positive entries, then the corresponding eigenvalue is $\lambda(B)$.*

---

[1]However, this ring, as well as the ring $C(\mathbb{Z}G)$ for a finite G, are weak fusion rings, see Section 3.8

PROOF. Let B be an n-by-n matrix with non-negative entries. Let us first show that B has a non-negative real eigenvalue and that there is an eigenvector with non-negative real entries corresponding to this eigenvalue.

If B has an eigenvector $\mathbf{v}$ with non-negative entries and eigenvalue 0, then there is nothing to prove. Otherwise, let Σ be the set of column vectors $\mathbf{x} \in \mathbb{R}^n$ with non-negative entries x_i, $i = 1, \ldots, n$, and $s(\mathbf{x}) := \sum_{i=1}^n x_i$ equal to 1 (this is a simplex). Define a continuous map

$$f_B : \Sigma \to \Sigma, \quad \mathbf{x} \mapsto \frac{B\mathbf{x}}{s(B\mathbf{x})}.$$

This is well defined because $s(B\mathbf{x}) > 0$ for every $\mathbf{x} \in \Sigma$ (as we have assumed that B does not have a null vector with non-negative entries). By the Brouwer fixed point theorem, this map has a fixed point. For any fixed point $\mathbf{e}$ we have $B\mathbf{e} = \nu\mathbf{e}$, where $\nu > 0$, as desired.

Now let $\lambda = \lambda(B)$ be the maximal non-negative eigenvalue of B for which there is an eigenvector with non-negative entries. Let us denote this eigenvector by $\mathbf{f} = (f_1, \ldots, f_n)$.

Now let us prove (2). Assume that B has strictly positive entries. Then $B\mathbf{f} = \lambda\mathbf{f}$ has strictly positive entries, so $\mathbf{f}$ must have strictly positive entries as well, and $\lambda > 0$. If $\mathbf{d} = (d_1, \ldots, d_n)$ is another real eigenvector of B with eigenvalue λ, let z be the smallest of the numbers d_i/f_i, $i = 1, \ldots, n$. Then the vector $\mathbf{v} = \mathbf{d} - z\mathbf{f}$ satisfies $B\mathbf{v} = \lambda\mathbf{v}$, has non-negative entries, and at least one of its entries is equal to zero. Hence $\mathbf{v} = 0$ and λ is a simple eigenvalue.

Now let $\mathbf{y} = (y_1, \ldots, y_n) \in \mathbb{C}^n$ be a row vector. Define the norm $|\mathbf{y}| := \sum |y_j| f_j$. Then

$$|\mathbf{y}B| = \sum_j |\sum_i y_i b_{ij}| f_j \le \sum_{i,j} |y_i| b_{ij} f_j = \lambda |\mathbf{y}|,$$

and the equality holds if and only if all the complex numbers y_i which are nonzero have the same argument. So if $\mathbf{y}B = \mu\mathbf{y}$, then $|\mu| \le \lambda$, and if $|\mu| = \lambda$ then all y_i which are nonzero have the same argument, so we can renormalize $\mathbf{y}$ to have non-negative entries. This implies that $\mu = \lambda$. Thus, (2) is proved (since eigenvalues of B coincide with eigenvalues of B^T).

Now let us return to the general case (B has non-negative entries), and finish the proof of (1) (i.e., prove that the spectral radius $\rho(B)$ of B equals $\lambda(B)$). To do so, let B_N be a sequence of matrices with strictly positive entries that converges to B (for instance, $B_N = B + \frac{1}{N} I$, where I is the matrix consisting of ones). Since the spectral radius is a continuous function of the matrix, we see that $\rho(B_N) \to \rho(B)$ as $N \to \infty$. Also, $\rho(B_N) = \lambda(B_N)$ for all N by (2). Hence, $\rho(B)$ is an eigenvalue of B, and there is an eigenvector of B with non-negative entries for this eigenvalue. This implies that $\rho(B) = \lambda(B)$, hence (1).

Assume that B has a row eigenvector $\mathbf{y}$ with strictly positive entries and eigenvalue μ. Then

$$\mu\mathbf{y}\mathbf{f} = \mathbf{y}B\mathbf{f} = \lambda\mathbf{y}\mathbf{f},$$

which implies $\mu = \lambda$, as $\mathbf{y}\mathbf{f} \neq 0$. This implies (3) (for the matrix B^T), since by (1), $\lambda(B) = \lambda(B^T)$. □

3.3. The Frobenius-Perron dimensions

Let A be a $\mathbb{Z}_+$-ring with $\mathbb{Z}_+$-basis I.

DEFINITION 3.3.1. We will say that A is *transitive* if for any $X, Z \in I$ there exist $Y_1, Y_2 \in I$ such that XY_1 and Y_2X contain Z with a nonzero coefficient.

EXERCISE 3.3.2. Show that any unital based ring is transitive. (Take Y_1 to be a suitable summand of X^*Z and Y_2 to be a suitable summand of ZX^*).

Let A be a transitive unital $\mathbb{Z}_+$-ring of finite rank. Define a group homomorphism $\mathsf{FPdim} : A \to \mathbb{C}$ as follows. For $X \in I$, let $\mathsf{FPdim}(X)$ be the maximal non-negative eigenvalue of the matrix of left multiplication by X. It exists by the Frobenius-Perron theorem, since this matrix has non-negative entries. Let us extend FPdim from the basis I to A by additivity.

DEFINITION 3.3.3. The function FPdim is called the *Frobenius-Perron dimension.*

PROPOSITION 3.3.4. *Let $X \in I$.*

(1) *The number $\alpha = \mathsf{FPdim}(X)$ is an algebraic integer, and for any algebraic conjugate α' of α we have $\alpha \geq |\alpha'|$.*
(2) $\mathsf{FPdim}(X) \geq 1$.

PROOF. (1) Note that α is an eigenvalue of the integer matrix N_X of left multiplication by X, hence α is an algebraic integer. The number α' is a root of the characteristic polynomial of N_X, so it is also an eigenvalue of N_X. Thus by the Frobenius-Perron theorem $\alpha \geq |\alpha'|$.

(2) Let r be the number of algebraic conjugates of α. Then $\alpha^r \geq |N(\alpha)|$ where $N(\alpha)$ is the norm of α. This implies the statement since $|N(\alpha)| \geq 1$. □

REMARK 3.3.5. Thus, the Frobenius-Perron dimensions of all elements of A are algebraic integers.

PROPOSITION 3.3.6. (1) *The function $\mathsf{FPdim} : A \to \mathbb{C}$ is a ring homomorphism.*

(2) *There exists a unique, up to scaling, nonzero element $R \in A_{\mathbb{C}} := A \otimes_{\mathbb{Z}} \mathbb{C}$ such that $XR = \mathsf{FPdim}(X)R$ for all $X \in A$, and it satisfies the equality $RY = \mathsf{FPdim}(Y)R$ for all $Y \in A$. After an appropriate normalization this element has positive coefficients, and thus $\mathsf{FPdim}(R) > 0$.*
(3) *FPdim is the unique character of A which takes non-negative values on I, and these values are actually strictly positive.*
(4) *If $X \in A$ has non-negative coefficients with respect to the basis of A, then $\mathsf{FPdim}(X)$ is the largest non-negative eigenvalue $\lambda(N_X)$ of the matrix N_X of multiplication by X.*

REMARK 3.3.7. Note that in Proposition 3.3.6, the condition that A is unital cannot be removed. E.g., $A = \mathrm{Mat}_n(\mathbb{Z})$ with basis of elementary matrices $\{E_{ij}\}$ does not admit any homomorphisms to $\mathbb{C}$ for $n \geq 2$.

PROOF. Consider the matrix M of right multiplication by $\sum_{X \in I} X$ in A in the basis I. By transitivity, this matrix has strictly positive entries, so by Theorem 3.2.1, part (2), it has a unique, up to scaling, eigenvector $R \in A_{\mathbb{C}}$ with eigenvalue $\lambda(M)$ (the maximal positive eigenvalue of M). Furthermore, this eigenvector can be normalized to have strictly positive entries.

Since R is unique, it satisfies the equation $XR = d(X)R$ for some function $d : A \to \mathbb{C}$. Indeed, XR is also an eigenvector of M with eigenvalue $\lambda(M)$, so it must be proportional to R. Furthermore, it is clear that d is a character of A. Since R has positive entries, $d(X) = \mathsf{FPdim}(X)$ for $X \in I$. This implies (1). We also see that $\mathsf{FPdim}(X) > 0$ for $X \in I$ (as R has strictly positive coefficients), and hence $\mathsf{FPdim}(R) > 0$.

Now, by transitivity, R is the unique, up to scaling, solution of the system of linear equations $XR = \mathsf{FPdim}(X)R$ (as the matrix N of left multiplication by $\sum_{X\in I} X$ also has positive entries). Hence, for any $Y \in A$, $RY = d'(Y)R$ for some number $d'(Y)$ (as both RY and R are solutions of the above system of equations). Moreover, it is clear that the assignment $Y \to d'(Y)$ is a character. Applying FPdim to both sides of the equation $RY = d'(Y)R$ and using that $\mathsf{FPdim}(R) > 0$, we find $d' = \mathsf{FPdim}$, proving (2).

If χ is another character of A taking nonnegative values on I, then the vector with entries $\chi(Y)$, $Y \in I$ is an eigenvector of the matrix N of left multiplication by the element $\sum_{X\in I} X$. Because of transitivity of A the matrix N has positive entries. By the Frobenius-Perron theorem there exists a positive number λ such that $\chi(Y) = \lambda\,\mathsf{FPdim}(Y)$. Since χ is a character, $\lambda = 1$, which completes the proof.

Finally, part (4) follows from part (2) and the Frobenius-Perron theorem (part (3)). □

DEFINITION 3.3.8. An element $R \in A \otimes_{\mathbb{Z}} \mathbb{R}$ as in Proposition 3.3.6 will be called a *regular element* of A.

The terminology of Definition 3.3.8 will be justified in Example 4.5.5.

PROPOSITION 3.3.9. *Let A be as above and $* : I \to I$ be a bijection which extends to an anti-automorphism of A. Then* FPdim *is invariant under* $*$.

PROOF. Let $X \in I$. Then the matrix of right multiplication by X^* is the transpose of the matrix of left multiplication by X modified by the permutation $*$. Thus the required statement follows from Proposition 3.3.6(2). □

COROLLARY 3.3.10. *Let A be a fusion ring and let X be a basis element of A. If* $\mathsf{FPdim}(X) = 1$ *then* $XX^* = X^*X = 1$.

PROOF. It is sufficient to show that $XX^* = \mathbf{1}$. This follows from the facts that $\mathbf{1}$ is contained in the decomposition of XX^* and

$$\mathsf{FPdim}(XX^*) = \mathsf{FPdim}(X)\,\mathsf{FPdim}(X^*) = 1. \qquad \square$$

Note that an element R in Proposition 3.3.6 is only defined up to normalization (i.e., up to multiplication by a positive real number). However, if A is a fusion ring, there is a certain specific normalization which is preferable.

PROPOSITION 3.3.11. *If A is a fusion ring then the element*

$$R = \sum_{Y\in I} \mathsf{FPdim}(Y)Y$$

is a regular element.

PROOF. Using Proposition 3.1.6 and Proposition 3.3.9, we have

$$\begin{aligned} XR &= \sum_Y \mathsf{FPdim}(Y)XY = \sum_{Y,Z} \mathsf{FPdim}(Y)c_{XY}^Z Z = \sum_{Y,Z} \mathsf{FPdim}(Y)c_{Z^*X}^{Y^*} Z \\ &= \sum_{Y,Z} \mathsf{FPdim}(Y)c_{X^*Z}^{Y} Z = \sum_Z \mathsf{FPdim}(X^*Z)Z \\ &= \mathsf{FPdim}(X^*)(\sum_Z \mathsf{FPdim}(Z)Z) = \mathsf{FPdim}(X)R. \end{aligned}$$

□

DEFINITION 3.3.12. The element R normalized as in Proposition 3.3.11 (i.e., so that the coefficient of $1 \in I$ is 1) is called *the canonical regular element* of A. The number $\mathsf{FPdim}(R) = \sum_{X \in I} \mathsf{FPdim}(X)^2$ is called the *Frobenius-Perron dimension* of A and will be denoted by $\mathsf{FPdim}(A)$.

Note that $\mathsf{FPdim}(A)$ is an algebraic integer. Moreover, $\mathsf{FPdim}(A) - 1$ is totally[2] non-negative since $\mathsf{FPdim}(A) - 1 = \mathsf{FPdim}(\sum_{X \in I, X \neq 1} XX^*)$, and the matrix of multiplication by $\sum_{X \in I, X \neq 1} XX^*$ is symmetric and non-negative definite, so all its eigenvalues are non-negative.

Now let A_1, A_2 be transitive unital $\mathbb{Z}_+$-rings of finite rank with $\mathbb{Z}_+$-bases I_1, I_2.

PROPOSITION 3.3.13. *Let $f : A_1 \to A_2$ be a unital homomorphism, whose matrix in the bases I_1, I_2 has non-negative entries. Then*

(1) *f preserves Frobenius-Perron dimensions.*
(2) *Suppose that for any Y in I_2 there exists $X \in I_1$ such that the coefficient of Y in $f(X)$ is nonzero. If R is a regular element of A_1 then $f(R)$ is a regular element of A_2.*
(3) *If in the situation of (2), A_1, A_2 are fusion rings and R_1, R_2 are their canonical regular elements, then $f(R_1) = \frac{\mathsf{FPdim}(A_1)}{\mathsf{FPdim}(A_2)} R_2$. In particular, $\frac{\mathsf{FPdim}(A_1)}{\mathsf{FPdim}(A_2)}$ is an algebraic integer.*

PROOF. (1) The function $X \mapsto \mathsf{FPdim}(f(X))$ is a character of A_1 with non-negative values on the basis. By Proposition 3.3.6(3), $\mathsf{FPdim}(f(X)) = \mathsf{FPdim}(X)$ for all X in I.

(2) By part (1) we have

$$f(\sum_{X \in I_1} X)f(R) = \mathsf{FPdim}(f(\sum_{X \in I_1} X))f(R). \tag{3.5}$$

But $f(\sum_{X \in I_1} X)$ has strictly positive coefficients in I_2, hence $f(R)$ is a regular element of A_2.

(3) The equality $f(R_1) = \frac{\mathsf{FPdim}(A_1)}{\mathsf{FPdim}(A_2)} R_2$ follows from (1), (2) by applying FPdim to both sides. By taking the coefficient of 1, we see that $\frac{\mathsf{FPdim}(A_1)}{\mathsf{FPdim}(A_2)}$ is an algebraic integer. □

We observed in Remark 3.3.5 that the values of Frobenius-Perron dimensions in a $\mathbb{Z}_+$-ring are algebraic integers. Below we show that for fusion rings there are further strong restrictions. We will need a classical result of Kronecker.

[2]We say that an algebraic number has some property totally if it does so under any embedding of algebraic numbers into complex numbers. E.g., we can talk about totally real numbers, totally non-negative numbers, etc.

LEMMA 3.3.14. *Let $q \in \mathbb{C}^\times$ be an algebraic integer such that all its algebraic conjugates have absolute values ≤ 1. Then q is a root of unity.*

PROOF. Let ℓ denote the degree of the minimal polynomial $p \in \mathbb{Z}[t]$ of q. Let $q = q_1, \ldots, q_\ell$ be the roots of p. For all $k = 1, 2, \ldots$ let

$$p_k(t) = \Pi_{j=1}^{\ell}(t - q_j^k).$$

The coefficients of p_k are symmetric polynomials on $q_1, \ldots, q_\ell$ with integer coefficients, hence by the Fundamental Theorem on symmetric functions, they are polynomials with integer coefficients of the coefficients of $p = p_1$. Thus, $p_k \in \mathbb{Z}[t]$ for each $k = 1, 2, \ldots$.

Observe that the set of integral monic polynomials of degree ℓ having all their roots inside the unit disk is finite (indeed, the absolute values of coefficients of such polynomials are bounded by $\max_{n=1,\ldots,\ell}\binom{\ell}{n}$). Thus, for each i, the sequence q_i^k, $k \geq 1$, takes finitely many values. This implies that each q_i is a root of unity. $\square$

Recall that for a square matrix B with non-negative real entries its largest non-negative real eigenvalue is denoted by $\lambda(B)$.

PROPOSITION 3.3.15. *(Kronecker) Let B be a matrix with non-negative integer entries, such that $\lambda(BB^T) = \lambda(B)^2$. If $\lambda(B) < 2$ then $\lambda(B) = 2\cos(\pi/n)$ for some integer $n \geq 2$.*

PROOF. Let $\lambda(B) = q + q^{-1}$. Since the ring of algebraic integers is integrally closed, q is an algebraic integer. Also, $|q| = 1$ since $\lambda(B)$ is real. Moreover, all conjugates of $\lambda(B)^2$ are non-negative (since they are eigenvalues of the matrix BB^T, which is symmetric and non-negative definite), so all conjugates of $\lambda(B)$ are real. Thus, if q_* is a conjugate of q then $q_* + q_*^{-1}$ is real with absolute value < 2 (by the Frobenius-Perron theorem), so $|q_*| = 1$. By Lemma 3.3.14, q is a root of unity: $q = e^{2\pi ik/m}$, where k and m are coprime. By the Frobenius-Perron theorem, $k = \pm 1$, and m is even (indeed, if $m = 2p + 1$ is odd then $|q^p + q^{-p}| > |q + q^{-1}|$). So $q = e^{\pi i/n}$ for some integer $n \geq 2$, and we are done. $\square$

COROLLARY 3.3.16. *Let A be a fusion ring, and $X \in A$ a basis element. Then if $\mathsf{FPdim}(X) < 2$ then $\mathsf{FPdim}(X) = 2\cos(\pi/n)$, for some integer $n \geq 3$.*

PROOF. This follows from Proposition 3.3.15, since $\mathsf{FPdim}(XX^*) = \mathsf{FPdim}(X)^2$. $\square$

3.4. $\mathbb{Z}_+$-modules

DEFINITION 3.4.1. Let A be a $\mathbb{Z}_+$-ring with basis $\{b_i\}$. A $\mathbb{Z}_+$-*module* over A is an A-module M with a fixed $\mathbb{Z}$-basis $\{m_l\}_{l\in L}$ such that all the structure constants a_{il}^k (defined by the equality $b_i m_l = \sum_k a_{il}^k m_k$) are non-negative integers.

The direct sum of $\mathbb{Z}_+$-modules is also a $\mathbb{Z}_+$-module whose basis is the union of the bases of the summands. We say that a $\mathbb{Z}_+$-module is *indecomposable* if it is not isomorphic to a nontrivial direct sum of $\mathbb{Z}_+$-modules.

DEFINITION 3.4.2. A $\mathbb{Z}_+$-module M over a $\mathbb{Z}_+$-ring A is called *irreducible* if it has no proper $\mathbb{Z}_+$-submodules (in other words, the $\mathbb{Z}$-span of any proper subset of the basis of M is not an A-submodule).

EXERCISE 3.4.3. (i) Prove that a $\mathbb{Z}_+$-module over a based ring is irreducible if and only if it is indecomposable.

(ii) Give an example of a $\mathbb{Z}_+$-module over a $\mathbb{Z}_+$-ring which is not irreducible but is indecomposable.

PROPOSITION 3.4.4. *Let M be an irreducible $\mathbb{Z}_+$-module over a fusion ring A with a $\mathbb{Z}_+$-basis $\{m_l\}_{l\in L}$. For each $X \in A$ let $[X]|_M$ denote the matrix of the operator of multiplication by X on M. There exists a unique up to a scalar common eigenvector $\mathbf{m}$ of all matrices $[X]|_M$, $X \in A$, which has strictly positive entries. The corresponding eigenvalue of $[X]|_M$ is $\mathsf{FPdim}(X)$.*

PROOF. Consider the following element of A:

$$Z = \sum_{i,j} b_i b_j b_i^*.$$

It is easy to check that Z has strictly positive coefficients in the basis $\{b_i\}$. Also, by Proposition 3.1.8, Z belongs to the center of A.

Define $\mathbf{m}$ to be an eigenvector of $[Z]|_M$ with positive entries. This eigenvector exists and is unique up to a positive scalar by Theorem 3.2.1(2). Since Z is in the center of A, we conclude that $\mathbf{m}$ is a common eigenvector of the matrices $[X]|_M$. The assertion about the corresponding eigenvalue of $[X]|_M$ follows from Proposition 3.3.6(3). □

DEFINITION 3.4.5. The vector $\mathbf{m}$ from Proposition 3.4.4 is called a *regular element* of M. The coefficients of $\mathbf{m}$ in the basis $\{m_i\}$ are called *Frobenius-Perron dimensions* of basis elements of M.

Unlike Frobenius-Perron dimensions of elements of A that are unique by Proposition 3.3.6, the Frobenius-Perron dimensions of elements of M are determined only up to a common positive factor.

PROPOSITION 3.4.6. *Let A be a based ring of finite rank over $\mathbb{Z}$. Then there exists only finitely many irreducible $\mathbb{Z}_+$-modules over A.*

PROOF. First of all, it is clear that an irreducible $\mathbb{Z}_+$-module M over A is of finite rank over $\mathbb{Z}$. Let $\{m_l\}_{l\in L}$ be the basis of M. Let us consider the element $b := \sum_{i\in I} b_i$ of A. Let $b^2 = \sum_i n_i b_i$ and let $N = \max_{i\in I} n_i$ (N exists since I is finite). For any $l \in L$ let $bm_l = \sum_{k\in L} d_{lk} m_k$ and let $d_l := \sum_{k\in L} d_{lk} > 0$. Let $l_0 \in I$ be such that $d := d_{l_0}$ equals $\min_{l\in L} d_l$. Let $b^2 m_{l_0} = \sum_{l\in L} c_l m_l$. Calculating $b^2 m_{l_0}$ in two ways — as $(b^2)m_{l_0}$ and as $b(bm_{l_0})$, and computing the sum of the coefficients, we have:

$$Nd \geq \sum_l c_l \geq d^2$$

and consequently $d \leq N$. So there are only finitely many possibilities for $|L|$, values of c_i and consequently for expansions $b_i m_l$ (since each m_l appears in bm_{l_0}). The proposition is proved. □

EXERCISE 3.4.7. (a) Classify irreducible $\mathbb{Z}_+$-modules over $\mathbb{Z}G$ (Answer: such modules are in bijection with subgroups of G up to conjugacy).

(b) Classify irreducible $\mathbb{Z}_+$-modules over the ring of representations of S_3, see Example 3.1.9(v). Consider all the cases: $\mathrm{char}(\Bbbk) \neq 2, 3$, $\mathrm{char}(\Bbbk) = 2$, $\mathrm{char}(\Bbbk) = 3$.

(c) Classify irreducible $\mathbb{Z}_+$-modules over the Yang-Lee and Ising based rings from Example 3.1.9(vii) and (viii).

3.5. Graded based rings

Let A be a unital based ring with a $\mathbb{Z}_+$-basis $B = \{b_i\}_{i \in I}$.

DEFINITION 3.5.1. Let G be a group. A grading of A by G is a partition

$$B = \sqcup_{g \in G} B_g$$

of the basis of A into disjoint subsets such that for all g, $h \in G$ and for all basis elements $b_i \in B_g$ and $b_j \in B_h$ the product $b_i b_j$ is a $\mathbb{Z}_+$-linear combination of elements of B_{gh}.

Note that if $b_i \in B_g$ then $b_i^* \in B_{g^{-1}}$. We will call B_g the *component* of B corresponding to $g \in G$, and B_1 (where 1 denotes the identity element of G) the *trivial component* of B. Note that the trivial component includes all basis elements contained in the decomposition of $b_i b_{i^*}$, $i \in I$. In particular, the trivial component contains b_i for all $i \in I_0$.

If A is G-graded, we have a decomposition

$$A = \bigoplus_{g \in G} A_g,$$

where A_g is the $\mathbb{Z}$-span of $B_g, g \in G$. We will say that A_g is the *component* of A corresponding to g. The subring A_1 will be called the *trivial component* of A.

The grading is called *faithful* if all the components $B_g, g \in G$, are nonempty.

Let A be a fusion ring faithfully graded by a finite group G. Let $R \in A \otimes_{\mathbb{Z}} \mathbb{C}$ be the regular element of A (see Definition 3.3.11) and let

$$R = \sum_{g \in G} R_g$$

be its decomposition with respect to the grading. For all $g \in G$ set

$$\mathsf{FPdim}(A_g) = \mathsf{FPdim}(R_g)$$

and call it the *Frobenius-Perron dimension* of the component A_g of A.

THEOREM 3.5.2. *Let A be a fusion ring faithfully graded by a finite group G. The Frobenius-Perron dimensions of the components A_g are equal for all $g \in G$, so* $\mathsf{FPdim}(A) = |G| \, \mathsf{FPdim}(A_1)$. *In particular, $|G|$ divides* $\mathsf{FPdim}(A)$.

PROOF. We have $RR_h = \mathsf{FPdim}(R_h)R = \mathsf{FPdim}(A_h)R$. Taking the gh-component of this equation, we find

$$R_g R_h = \mathsf{FPdim}(A_h) R_{gh}.$$

Similarly, $R_g R = \mathsf{FPdim}(A_g)R$, which yields $R_g R_h = \mathsf{FPdim}(A_g) R_{gh}$. But $R_{gh} \neq 0$, so we get $\mathsf{FPdim}(A_g) = \mathsf{FPdim}(A_h)$, i.e., $\mathsf{FPdim}(A_g) = \mathsf{FPdim}(A)/|G|$. □

PROPOSITION 3.5.3. *Let A be a fusion ring with a fusion subring A_0 (i.e., a subring spanned by a *-stable subset of the basis) such that* $\mathsf{FPdim}(A) = 2 \, \mathsf{FPdim}(A_0)$. *Then A has a faithful grading by $\mathbb{Z}/2\mathbb{Z}$ with the trivial component A_0.*

PROOF. Write A as a direct sum $A = A_0 \oplus A_1$, where A_1 is a $\mathbb{Z}_+$-module over A_0 (such a decomposition exists by Exercise 3.4.3(i)). Let R and R_0 be the regular elements of A and A_0 and let R_1 be the regular element of A_1 such that $R = R_0 + R_1$. Note that $R_0R_1 = R_1R_0 = \mathsf{FPdim}(A_0)R_1$.

We have $R^2 = \mathsf{FPdim}(A)R$. On the other hand,

$$R^2 = R_0^2 + R_0R_1 + R_1R_0 + R_1^2 = \mathsf{FPdim}(A_0)R_0 + 2\,\mathsf{FPdim}(A_0)R_1 + R_1^2.$$

Comparing the two expressions and using that $\mathsf{FPdim}(A) = 2\,\mathsf{FPdim}(A_0)$, we obtain $R_1^2 = \mathsf{FPdim}(A_0)R_0$. This means that the product of any pair of simple objects in A_1 belongs to A_0, i.e., that A is $\mathbb{Z}/2\mathbb{Z}$-graded. □

REMARK 3.5.4. Taking in Proposition 3.5.3 $A = \mathbb{Z}G$, where G is a finite group, one recovers a classical result in group theory: a subgroup of index 2 is normal.

DEFINITION 3.5.5. A fusion ring A is said to be *weakly integral* if $\mathsf{FPdim}(A) \in \mathbb{Z}$. It is called *integral* if the Frobenius-Perron dimensions of all elements of A are integers.

Let us analyze the structure of weakly integral fusion rings.

LEMMA 3.5.6. *Let A be a fusion ring. Let X_1, X_2, Y be non-negative integral combinations of basis elements in A such that $Y = X_1 + X_2$. If $\mathsf{FPdim}(Y)$ is an integer then so are $\mathsf{FPdim}(X_1)$ and $\mathsf{FPdim}(X_2)$*

PROOF. The Frobenius-Perron eigenvalue of a non-negative matrix has strictly largest real part among its algebraic conjugates. Since $\mathsf{FPdim}(X_1)+\mathsf{FPdim}(X_2)$ is an integer, the algebraic numbers $\mathsf{FPdim}(X_1)$ and $\mathsf{FPdim}(X_2)$ have both strictly largest and strictly smallest real parts among their conjugates, hence they are integers. □

PROPOSITION 3.5.7. *Let A be a weakly integral fusion ring. Then there is an elementary abelian 2-group E, a set of distinct square free positive integers n_x, $x \in E$, with $n_0 = 1$, and a faithful grading $A = \oplus_{x\in E} A(n_x)$ such that $\mathsf{FPdim}(X) \in \mathbb{Z}\sqrt{n_x}$ for each $X \in A(n_x)$. Moreover, the map $x \mapsto n_x$ is an inclusion of E into $\mathbb{Q}^\times/(\mathbb{Q}^\times)^2$.*

PROOF. First, let us show that every basis element of A has dimension $\sqrt{N}$ for some $N \in \mathbb{Z}$. Indeed, since $\mathsf{FPdim}(\sum_i b_ib_{i^*}) = \mathsf{FPdim}(A)$ is an integer, we conclude from Lemma 3.5.6 that each $\mathsf{FPdim}(b_ib_{i^*})$ is an integer, whence $\mathsf{FPdim}(b_i)$ is a square root of an integer.

Let $A(1) \subset A$ be the based subring of A generated by all basis elements of integer dimension. Observe that for each square free $n \in \mathbb{Z}$ the basis elements of A whose dimension is in $\mathbb{Z}\sqrt{n}$ generate an $A(1)$-sub-bimodule $A(n)$ of A. Let

$$E = \{n \text{ is square free } \mid A(n) \neq 0\}.$$

It is clear that for $X \in A(n)$ and $Y \in A(m)$ their product XY is in $A((nm)')$ where l' denotes the square free part of l. This defines a commutative group operation on E and a grading of A by E. Since the order of every $e \in E$ is at most two, E is an elementary abelian 2-group. □

COROLLARY 3.5.8. *Let A be a fusion ring of odd Frobenius-Perron dimension. Then A is integral.*

PROOF. This follows from the fact that $\mathsf{FPdim}(A) = |E|\,\mathsf{FPdim}(A(1))$, where $A(1)$ is integral, and $|E| = 2^N$, for some positive integer N. □

3.7. Complexified $\mathbb{Z}_+$-rings and $*$-algebras

DEFINITION 3.7.1. A *$*$-algebra* over $\mathbb{C}$ is a $\mathbb{C}$-algebra A with a map $* : A \to A$ which is anti-linear (i.e., such that $(\lambda a)^* = \bar{\lambda} a^*$ for $\lambda \in \mathbb{C}$) and such that $(ab)^* = b^* a^*$ and $a^{**} = a$ for all $a, b \in A$. A *positive trace* on A is a linear functional $l : A \to \mathbb{C}$ such that $l(ab) = l(ba)$, $l(a^*) = \overline{l(a)}$, and $l(aa^*) > 0$ for $a \neq 0$.

EXAMPLE 3.7.2. Let $A = \mathsf{Mat}_n(\mathbb{C})$. For $a \in \mathsf{Mat}_n(\mathbb{C})$ let a^* be the usual Hermitian conjugate matrix. Define $l : A \to \mathbb{C}$ by $l(a) := p\,\mathsf{Tr}(a)$, $p > 0$. Then A is a $*$-algebra and l is a positive trace.

Let $A_1, \dots, A_r$ be $*$-algebras and $A = A_1 \oplus \cdots \oplus A_r$. Let $l_i : A_i \to \mathbb{C}$ be a positive trace, $i = 1, \dots, r$. Then the linear functional $l : A \to \mathbb{C}$ defined by $l(a_1, \dots, a_r) = l_1(a_1) + \cdots + l_r(a_r)$ is a positive trace on A. In this situation we write $(A, l) = \bigoplus_{i=1}^r (A_i, l_i)$. The following proposition is classical.

PROPOSITION 3.7.3. *Any finite dimensional $*$-algebra with a positive trace is semisimple and isomorphic to $\bigoplus_{i=1}^r (A_i, l_i)$, where each (A_i, l_i) is as in Example 3.7.2.*

PROOF. Let $\mathsf{End}\, A$ denote the algebra of linear operators $A \to A$. Equip $\mathsf{End}\, A$ with the $*$-algebra structure corresponding to the Hermitian scalar product $(a, b) = l(ab^*)$ on A. The left action of A on itself defines an injective $*$-homomorphism $A \hookrightarrow \mathsf{End}\, A$. Thus A identifies with a $*$-subalgebra of the algebra of operators in a finite dimensional Hilbert space. Such an algebra is clearly semisimple since every submodule N of any A-module M has the orthogonal complement $N^\perp$ such that $M = N \oplus N^\perp$. So the $*$-algebra A is isomorphic to a product of matrix algebras $A_1, \dots, A_r$ equipped with their standard anti-involutions. For each $i \in \{1, \dots, r\}$ the restriction $l|_{A_i}$ is proportional to the usual matrix trace, and the coefficient is positive because l is. □

COROLLARY 3.7.4. *Let A be a finite dimensional $*$-algebra with a positive trace l. If $e \in A$, $e^2 = e$, $e \neq 0$ then $l(e) > 0$.* □

REMARK 3.7.5. Corollary 3.7.4 is not an immediate consequence of the inequality $l(ee^*) > 0$ because we do not assume that $e^* = e$.

LEMMA 3.7.6. *Let A be a finite dimensional $*$-algebra with a positive trace l. If $a \in A$ and $l(a^n) = 0$ for all $n \in \mathbb{N}$ then a is nilpotent.*

PROOF. If a is not nilpotent then there exists a nonzero idempotent e of the form $\sum_{i=1}^{N} \lambda_i a^i$ (this can be seen using the Jordan normal form theorem). Then $l(e) = 0$, which contradicts Corollary 3.7.4. □

COROLLARY 3.7.7. ([**Lus2**, 1.2(a)]) *For any multifusion ring A the algebra $A \otimes_{\mathbb{Z}} \mathbb{C}$ is semisimple.*

PROOF. It is clear that $A \otimes_{\mathbb{Z}} \mathbb{C}$ is a finite dimensional $*$-algebra with positive trace being the anti-linear extension of the involution τ defined in (3.1). So the statement follows from Proposition 3.7.3. □

3.8. Weak based rings

One can define the notion of a *weak (unital) based ring*, which generalizes the notion of a (unital) based ring. Namely, the definition of a weak based ring is obtained from the definition of a based ring by relaxing the condition that $\tau(b_i b_j) = \delta_{ij^*}$: instead one requires that $\tau(b_i b_j) = d_i \delta_{ij^*}$, where d_i are positive integers. A weak based ring of finite rank is called a *weak multifusion ring*, and it is called a *weak fusion ring* if it is unital (see [**EtG10**]).

A basic example of a weak unital based ring is the center ZG of the group algebra $\mathbb{Z}G$ of a finite group G, with basis $b_i = \sum_{g \in C_i} g$ labeled by conjugacy classes C_i of G. In this case, $d_i = |C_i|$. Another example is the convolution algebra $\mathsf{Fun}(H \backslash G / H, \mathbb{Z})$ of integer-valued functions on G which are two-sided invariant under a subgroup $H \subset G$, with basis $b_i = \chi_{S_i}$ of characteristic functions of double cosets S_i of H in G; in this case $d_i = |S_i|/|H|$.

Weak based rings enjoy similar properties to usual based rings; in particular, in the case of finite rank (weak multifusion and weak fusion rings), there is a theory of Frobenius-Perron dimensions. However, a detailed discussion of weak based rings is outside of the scope of this book.

EXERCISE 3.8.1. Generalize the results of this chapter on based rings to the case of weak based rings.

3.9. Bibliographical notes

3.1. The terminology on $\mathbb{Z}_+$-rings is borrowed from the paper by Lusztig [**Lus2**] as well as [**EtKh, Os1**]. Such rings (with a different terminology) were considered by Davydov in [**Da2, Da3**]. For basic facts about based and fusion rings see [**Os1**]. Proposition 3.1.8 is borrowed from [**ENO2**] (see Theorem 8.2).

3.2. The Frobenius-Perron theorem was proved by Perron in 1907 and Frobenius in 1912. This theorem has many applications in various fields of pure and applied mathematics. For more details on this theorem, see [**Gant**, Chapter XIII].

3.3. The notion of the Frobenius-Perron dimension is motivated by the notion of the index of a subfactor (see the paper by Jones, [**Jon1**]). Frobenius-Perron dimensions for commutative fusion rings were defined and used in the book by Fröhlich and Kerler, [**FroK**]. A theory of Frobenius-Perron dimensions for general fusion rings and categories was developed in [**ENO2**], and in the not necessarily semisimple case in [**E, EtO1**]. The notion of a transitive $\mathbb{Z}_+$-ring is defined in [**E**, Section 3]. Proposition 3.3.4 is contained in [**ENO2**]. For Proposition 3.3.6 see [**ENO2**, Section 8], and [**E**, Section 3]. The regular element of a fusion ring and its Frobenius-Perron dimension are defined in [**ENO2**, 8.2] (for $\mathbb{Z}_+$-rings arising from finite tensor categories, this is done in [**E**, Section 3]). Proposition 3.3.13 is modeled after Proposition 8.8 in [**ENO2**]. Corollary 3.3.16 was pointed out to us by Müger (see [**ENO2**, Remark 8.4]). It is closely related to the Jones Index Theorem in the theory of subfactors ([**Jon1**]); namely, the index of a subfactor in [**Jon1**], when it is < 4, is the square of the Frobenius-Perron dimension of an object of a certain fusion category.

3.4. The terminology on $\mathbb{Z}_+$-modules is borrowed from [**Lus2, EtKh, Os1**]. Proposition 3.4.4 is taken from [**ENO2**, Section 8] (see Proposition 8.5). Proposition 3.4.6 is from the paper by Gannon [**Gann2**] (see also [**Os1**, Proposition 2.1]).

3.5. The theory of graded based rings is motivated by the theory of graded fusion categories, discussed briefly in [**ENO2**, 5.9], and then developed further in [**GelaN, GeNN, DrGNO2, ENO4**]. For basic material on graded based rings, see [**GelaN**]. Proposition 3.5.7 is taken from [**GelaN**, Theorem 3.10]. Corollary 3.5.8 is also from [**GelaN**].

3.6. The notion of the adjoint subring of a fusion ring is motivated by the notion of the adjoint subcategory of a fusion category, introduced in [**ENO2**, 8.5], and developed in [**GelaN**]. The material on the adjoint subring, the universal grading, and nilpotent based rings is taken from [**GelaN**].

3.7. The material on complexified Grothendieck rings is taken from [**DrGNO2**, Appendix F].

3.8. The material on weak based rings is taken from [**EtG10**, 4.1].

CHAPTER 4

Tensor categories

4.1. Tensor and multitensor categories

Let $\Bbbk$ be an algebraically closed field.

DEFINITION 4.1.1. Let $\mathcal{C}$ be a locally finite $\Bbbk$-linear abelian rigid monoidal category. We will call $\mathcal{C}$ a *multitensor category* over $\Bbbk$ if the bifunctor $\otimes : \mathcal{C} \times \mathcal{C} \to \mathcal{C}$ is bilinear on morphisms. We will say that $\mathcal{C}$ is *indecomposable* if $\mathcal{C}$ is not equivalent to a direct sum of nonzero multitensor categories. If in addition $\mathsf{End}_{\mathcal{C}}(\mathbf{1}) \cong \Bbbk$ then we will call $\mathcal{C}$ a *tensor category*.

A *multifusion category* is a finite semisimple multitensor category. A *fusion category* is a multifusion category with $\mathsf{End}_{\mathcal{C}}(\mathbf{1}) \cong \Bbbk$, i.e., a finite semisimple tensor category.

EXAMPLE 4.1.2. The following are examples of tensor categories (cf. Section 2.3): the category Vec of finite dimensional $\Bbbk$-vector spaces (a fusion category), the category $\mathsf{Rep}(G)$ of finite dimensional $\Bbbk$-representations of a group G (or algebraic representations of an affine algebraic group G), the category $\mathsf{Rep}(\mathfrak{g})$ of finite dimensional representations of a Lie algebra $\mathfrak{g}$, and the category Vec_G^{ω} of G-graded finite dimensional $\Bbbk$-vector spaces with associativity defined by a 3-cocycle ω. If G is a finite group then Vec_G^{ω} is a fusion category, and $\mathsf{Rep}(G)$ is a fusion category if $\mathrm{char}(\Bbbk) = 0$, or $\mathrm{char}(\Bbbk)$ is coprime to $|G|$.

EXAMPLE 4.1.3. Let A be a finite dimensional semisimple algebra over $\Bbbk$. Let $\mathcal{C} = A-\mathsf{bimod}$ be the category of finite dimensional A-bimodules with bimodule tensor product over A:

$$(M, N) \mapsto M \otimes_A N.$$

Then $\mathcal{C}$ is a multifusion category with the unit object $\mathbf{1} = A$, the left dual defined by $M \mapsto \mathsf{Hom}({}_AM, {}_AA)$, and the right dual defined by $M \mapsto \mathsf{Hom}(M_A, A_A)$[1]. The category $\mathcal{C}$ is fusion if and only if A is simple, in which case it is equivalent to Vec. More generally, if A has n matrix blocks, the category $\mathcal{C}$ can be alternatively described as the category whose objects are n-by-n matrices of vector spaces, $V = (V_{ij})$, and the tensor product is matrix multiplication:

$$(V \otimes W)_{il} = \bigoplus_{j=1}^{n} V_{ij} \otimes W_{jl}.$$

This category will be denoted by $\mathsf{Mat}_n(\mathsf{Vec})$.

[1]Note that if A is a finite dimensional non-semisimple algebra then the category of finite dimensional A-bimodules is not rigid, since the duality functors defined as above do not satisfy the rigidity axioms (cf. Exercise 2.10.16).

In a similar way, one can define the multitensor category $\mathsf{Mat}_n(\mathcal{C})$ of n-by-n matrices of objects of a given multitensor category $\mathcal{C}$. If $\mathcal{C}$ is a multifusion category, so is $\mathsf{Mat}_n(\mathcal{C})$.

4.2. Exactness of the tensor product

PROPOSITION 4.2.1. *Let $\mathcal{C}$ be a multitensor category. Then the bifunctor $\otimes$: $\mathcal{C} \times \mathcal{C} \to \mathcal{C}$ is exact in both factors (i.e.,* biexact*).*

PROOF. By Proposition 2.10.8, the functors $(V \otimes -)$ and $(- \otimes V)$ have left and right adjoint functors (the functors of tensoring with the corresponding duals). The statement follows since any functor between abelian categories that has left and right adjoint functors is exact (Exercise 1.6.4). □

REMARK 4.2.2. The proof of Proposition 4.2.1 shows that the biadditivity of the functor $\otimes$ holds automatically in any rigid monoidal abelian category. However, this is not the case for bilinearity of $\otimes$, and thus the condition of bilinearity of tensor product in the definition of a multitensor category is not redundant.

This may be illustrated by the following example. Let $\mathcal{C}$ be the category of finite dimensional $\mathbb{C}$-bimodules in which the left and right actions of $\mathbb{R}$ coincide. This category is $\mathbb{C}$-linear abelian; namely, it is semisimple with two simple objects $\mathbb{C}_+ = \mathbf{1}$ and $\mathbb{C}_-$, both equal to $\mathbb{C}$ as a real vector space, with bimodule structures $(a,b)z = azb$ and $(a,b)z = az\bar{b}$, respectively. It is also rigid monoidal, with $\otimes$ being the tensor product of bimodules. But the tensor product functor is not $\mathbb{C}$-bilinear on morphisms (it is only $\mathbb{R}$-bilinear).

DEFINITION 4.2.3. A *multiring category* over $\Bbbk$ is a locally finite $\Bbbk$-linear abelian monoidal category $\mathcal{C}$ with bilinear and biexact tensor product. If in addition $\mathsf{End}_{\mathcal{C}}(\mathbf{1}) = \Bbbk$, we will call $\mathcal{C}$ a *ring category.*

EXAMPLE 4.2.4. The category Vec_G of finite dimensional vector spaces graded by a monoid G is a ring category, with tensor product given by the formula (2.17).

Thus, the difference between Definition 4.1.1 of a (multi)tensor category and Definition 4.2.3 is that for the latter we do not require existence of duals, but instead require biexactness of the tensor product. Note that Proposition 4.2.1 implies that any multitensor category is a multiring category, and any tensor category is a ring category.

DEFINITION 4.2.5. Let $\mathcal{C}$, $\mathcal{D}$ be multiring categories over $\Bbbk$, and let $F : \mathcal{C} \to \mathcal{D}$ be an exact and faithful $\Bbbk$-linear functor.

(i) F is said to be a *quasi-tensor functor* if it is equipped with a functorial isomorphism $J : F(-) \otimes F(-) \to F(- \otimes -)$, and $F(\mathbf{1}) = \mathbf{1}$.
(ii) A quasi-tensor functor (F, J) is said to be a *tensor functor* if it is a monoidal functor in the sense of Definition 2.4.1, i.e., satisfies diagram (2.23).

REMARK 4.2.6. The requirement that a (quasi)tensor functor be exact and faithful is imposed only inside this book; in other sources, this term may be used in more general settings, when these conditions do not necessarily hold.

EXAMPLE 4.2.7. The functors of Examples 2.5.1, 2.5.2, and Section 2.6 (for the categories Vec_G^{ω}) are tensor functors. The identity functor $\mathsf{Vec}_G^{\omega_1} \to \mathsf{Vec}_G^{\omega_2}$ for non-cohomologous 3-cocycles ω_1, ω_2 does not admit a structure of a tensor functor, but it can be made quasi-tensor by any choice of J.

PROPOSITION 4.2.8. *For any pair of morphisms f_1, f_2 in a multiring category $\mathcal{C}$ one has* $\mathrm{Im}(f_1 \otimes f_2) = \mathsf{Im}(f_1) \otimes \mathsf{Im}(f_2)$.

PROOF. Let I_1, I_2 be the images of f_1, f_2. Then the morphisms $f_i : X_i \to Y_i$, $i = 1, 2$, have decompositions $X_i \to I_i \to Y_i$, where the sequences

$$X_i \to I_i \to 0, \qquad 0 \to I_i \to Y_i, \quad i = 1, 2,$$

are exact. Tensoring the sequence $X_1 \to I_1 \to 0$ with I_2, and using Proposition 4.2.1 we get the exact sequence

$$X_1 \otimes I_2 \to I_1 \otimes I_2 \to 0$$

Tensoring X_1 with the sequence $X_2 \to I_2 \to 0$, we get the exact sequence

$$X_1 \otimes X_2 \to X_1 \otimes I_2 \to 0.$$

Combining these, we get an exact sequence $X_1 \otimes X_2 \to I_1 \otimes I_2 \to 0$. Arguing similarly, we see that the sequence $0 \to I_1 \otimes I_2 \to Y_1 \otimes Y_2$ is exact. This implies the statement. □

PROPOSITION 4.2.9. *If $\mathcal{C}$ is a multiring category with left duals, then the left dualization functor is exact. The same applies to right duals.*

PROOF. Let $0 \to X \to Y \to Z \to 0$ be an exact sequence. We need to show that the sequence $0 \to Z^* \to Y^* \to X^* \to 0$ is exact. Let T be any object of $\mathcal{C}$, and consider the sequence

$$0 \to \mathsf{Hom}_{\mathcal{C}}(T, Z^*) \to \mathsf{Hom}_{\mathcal{C}}(T, Y^*) \to \mathsf{Hom}_{\mathcal{C}}(T, X^*).$$

By Proposition 2.10.8, it can be written as

$$0 \to \mathsf{Hom}_{\mathcal{C}}(T \otimes Z, \mathbf{1}) \to \mathsf{Hom}_{\mathcal{C}}(T \otimes Y, \mathbf{1}) \to \mathsf{Hom}_{\mathcal{C}}(T \otimes X, \mathbf{1}).$$

Using the definition of a multiring category, we see that the last sequence is exact. This implies that the sequence $0 \to Z^* \to Y^* \to X^*$ is exact.

Similarly, consider the sequence

$$0 \to \mathsf{Hom}_{\mathcal{C}}(X^*, T) \to \mathsf{Hom}_{\mathcal{C}}(Y^*, T) \to \mathsf{Hom}_{\mathcal{C}}(Z^*, T).$$

By Proposition 2.10.8, it can be written as

$$0 \to \mathsf{Hom}_{\mathcal{C}}(\mathbf{1}, X \otimes T) \to \mathsf{Hom}_{\mathcal{C}}(\mathbf{1}, Y \otimes T) \to \mathsf{Hom}_{\mathcal{C}}(\mathbf{1}, Z \otimes T),$$

which is exact since $0 \to X \otimes T \to Y \otimes T \to Z \otimes T$ is exact by the definition of a multiring category. This implies that the sequence $Z^* \to Y^* \to X^* \to 0$ is exact. □

PROPOSITION 4.2.10. *A finite ring category $\mathcal{C}$ with left duals is a tensor category (i.e., it also has right duals).*

PROOF. We know that the left duality functor $X \mapsto X^*$ is fully faithful (see Proposition 2.10.8), exact (see Proposition 4.2.9), and monoidal (see Exercise 2.10.7 (b)). In particular, this functor respects the Frobenius-Perron dimension by Proposition 3.3.13 (1). We claim that this functor sends simple objects to simple ones. Let $d_1 < d_2 < \ldots < d_r$ be the list of possible Frobenius-Perron dimensions of simple objects in $\mathcal{C}$. Let $\mathcal{O}_i(\mathcal{C}) \subset \mathcal{O}(\mathcal{C})$ be the isomorphism classes of simple objects L with $\mathsf{FPdim}(L) = d_i$, $i = 1, \ldots, r$.

LEMMA 4.2.11. *Let $L \in \mathcal{O}_i(\mathcal{C})$. Then L^* is simple and there exists $L' \in \mathcal{O}_i(\mathcal{C})$ such that $L \simeq (L')^*$.*

PROOF. We prove the lemma by induction in i. The base ($i = 1$) is obvious. If L^* is not simple then there exists a nonzero map $K \to L^*$, where K is simple with $\mathsf{FPdim}(K) < \mathsf{FPdim}(L^*) = d_i$. By the induction assumption $K \simeq (K')^*$, where K' is simple with $\mathsf{FPdim}(K') < d_i$. Since the duality functor is full, we get a nonzero map $L \to K'$, which is a contradiction. Thus we have a map $L \mapsto L^*$ from $\mathcal{O}_i(\mathcal{C})$ to itself. Again by the fullness of duality we see that this map is injective. Since the set $\mathcal{O}_i(\mathcal{C})$ is finite, it is also surjective, which proves the lemma. □

By Lemma 4.2.11 the functor $X \mapsto X^{**}$ induces a permutation of $\mathcal{O}(\mathcal{C})$. Thus there exists a finite power $F = (**)^n$ of the double duality functor such that $F(L) \simeq L$ for any $L \in \mathcal{O}(\mathcal{C})$. Also, the functor F is exact and fully faithful. Let P be the projective cover of $L \in \mathcal{O}(\mathcal{C})$. Since the functor F is full we see that the object $F(P)$ has a unique simple quotient $F(L) \simeq L$. It follows that we have a surjection $P \to F(P)$. Since the length of $F(P)$ equals the length of P, this must be an isomorphism $P \simeq F(P)$. Thus F is surjective on isomorphism classes of projective objects. Since any object of $\mathcal{C}$ is isomorphic to a cokernel of a map of projective objects and the functor F is fully faithful, we get that F is surjective on isomorphism classes of objects and hence is an equivalence. Thus the duality functor $X \mapsto X^*$ is surjective on isomorphism classes of objects. The proposition is proved. □

PROPOSITION 4.2.12. *Let P be a projective object in a multiring category $\mathcal{C}$. If $X \in \mathcal{C}$ has a left dual, then the object $P \otimes X$ is projective. Similarly, if $X \in \mathcal{C}$ has a right dual, then the object $X \otimes P$ is projective.*

PROOF. In the first case by Proposition 2.10.8 we have

$$\mathsf{Hom}_{\mathcal{C}}(P \otimes X, Y) = \mathsf{Hom}_{\mathcal{C}}(P, Y \otimes X^*),$$

which is an exact functor of Y, since the functors $(- \otimes X^*)$ and $\mathsf{Hom}_{\mathcal{C}}(P, -)$ are exact. So $P \otimes X$ is projective. The second statement is similar. □

COROLLARY 4.2.13. *If $\mathcal{C}$ is a multiring category with left duals (e.g., a multitensor category), then $\mathbf{1} \in \mathcal{C}$ is a projective object if and only if $\mathcal{C}$ is semisimple.*

PROOF. If $\mathbf{1}$ is projective then by Proposition 4.2.12 $X \cong \mathbf{1} \otimes X$ is projective for any $X \in \mathcal{C}$. This implies that $\mathcal{C}$ is semisimple. The converse is obvious. □

REMARK 4.2.14. Corollary 4.2.13 can be regarded as a categorical analog of Maschke's theorem for group algebras saying that for a finite group G the category $\mathsf{Rep}(G)$ of finite dimensional representations of G over $\Bbbk$ is semisimple if and only if $|G| \neq 0$ in $\Bbbk$. The latter condition is equivalent to the trivial representation $\Bbbk$ being projective, as the map $\Bbbk G \to \Bbbk : g \mapsto 1$ is the right inverse to the map $\Bbbk \to \Bbbk G : 1 \mapsto \frac{1}{|G|} \sum_{g \in G} g$.

There is a generalization of this theorem for finite dimensional Hopf algebras, see [**Mon**]. It says that the value of the counit on a nonzero integral is nonzero. This result (well known in the theory of Hopf algebras) is a special case of Corollary 4.2.13.

4.3. Semisimplicity of the unit object

THEOREM 4.3.1. *Let $\mathcal{C}$ be a multiring category. Then $\mathsf{End}_{\mathcal{C}}(\mathbf{1})$ is a semisimple algebra, so it is isomorphic to a direct sum of finitely many copies of $\Bbbk$.*

PROOF. By Proposition 2.2.10, $\mathsf{End}(\mathbf{1})$ is a commutative algebra, so it is sufficient to show that for any $a \in \mathsf{End}(\mathbf{1})$ such that $a^2 = 0$ we have $a = 0$. Let $J = \mathsf{Im}(a)$. Then by Proposition 4.2.8 $J \otimes J = \mathsf{Im}(a \otimes a) = \mathsf{Im}(a^2 \otimes 1) = 0$.

Now let $K = \mathsf{Ker}(a)$. Then by Proposition 4.2.8, $K \otimes J$ is the image of $1 \otimes a$ on $K \otimes \mathbf{1}$. But since $K \otimes \mathbf{1}$ is a subobject of $\mathbf{1} \otimes \mathbf{1}$, this is the same as the image of $a \otimes 1$ on $K \otimes \mathbf{1}$, which is zero. So $K \otimes J = 0$.

Now tensoring the exact sequence $0 \to K \to \mathbf{1} \to J \to 0$ with J, and applying the definition of a multiring category, we get that $J = 0$, so $a = 0$. □

Let $\{p_i\}_{i \in I}$ be the primitive idempotents of the algebra $\mathsf{End}(\mathbf{1})$. Let $\mathbf{1}_i$ be the image of p_i. Then we have $\mathbf{1} = \bigoplus_{i \in I} \mathbf{1}_i$.

COROLLARY 4.3.2. *In any multiring category $\mathcal{C}$ the unit object $\mathbf{1}$ is isomorphic to a direct sum of pairwise non-isomorphic indecomposable objects.*

EXERCISE 4.3.3. Prove that $\mathbf{1}_i \otimes \mathbf{1}_j = 0$ for $i \neq j$. Prove that there are canonical isomorphisms $\mathbf{1}_i \otimes \mathbf{1}_i \cong \mathbf{1}_i$, and that each $\mathbf{1}_i$ has both left and right duals such that $\mathbf{1}_i^* \cong {}^*\mathbf{1}_i \cong \mathbf{1}_i$.

REMARK 4.3.4. Let $\mathcal{C}$ be a multiring category and let $\mathcal{C}_{ij} := \mathbf{1}_i \otimes \mathcal{C} \otimes \mathbf{1}_j$.

(1) We have a decomposition

$$\mathcal{C} = \oplus_{i,j \in I} \mathcal{C}_{ij}. \tag{4.1}$$

Thus every indecomposable object of $\mathcal{C}$ belongs to some $\mathcal{C}_{ij}$.

(2) The tensor product maps $\mathcal{C}_{ij} \times \mathcal{C}_{kl}$ to $\mathcal{C}_{il}$, and it is zero unless $j = k$.

(3) The categories $\mathcal{C}_{ii}$ are ring categories with unit objects $\mathbf{1}_i$ (which are tensor categories if $\mathcal{C}$ is rigid).

(4) If $X \in \mathcal{C}_{ij}$ has a left or right dual, then this dual belongs to $\mathcal{C}_{ji}$.

DEFINITION 4.3.5. The subcategories $\mathcal{C}_{ij}$ will be called the *component* subcategories of $\mathcal{C}$.

EXERCISE 4.3.6. Prove the statements of Remark 4.3.4.

Remark 4.3.4 motivates the terms "multiring category" and "multitensor category", as such a category gives us multiple ring categories, respectively tensor categories $\mathcal{C}_{ii}$.

REMARK 4.3.7. Any multitensor category $\mathcal{C} = \oplus_{i,j \in I} \mathcal{C}_{ij}$ may be viewed as a 2-category in the following way: the objects are the elements of I, the 1-morphisms from j to i are the categories $\mathcal{C}_{ij}$, and the 2-morphisms are the morphisms in $\mathcal{C}$.

THEOREM 4.3.8. (i) *In a ring category with left duals, the unit object $\mathbf{1}$ is simple.*

(ii) *In a multiring category with left duals, the unit object $\mathbf{1}$ is semisimple, and is a direct sum of pairwise non-isomorphic simple objects $\mathbf{1}_i$.*

PROOF. Clearly, (i) implies (ii) (by applying (i) to the component categories $\mathcal{C}_{ii}$). So it is enough to prove (i).

Let X be a simple subobject of $\mathbf{1}$ (it exists, since $\mathbf{1}$ has finite length). Let

$$0 \longrightarrow X \longrightarrow \mathbf{1} \longrightarrow Y \longrightarrow 0 \tag{4.2}$$

be the corresponding exact sequence. By Proposition 4.2.9, the left dualization functor is exact, so we get an exact sequence

$$0 \longrightarrow Y^* \longrightarrow \mathbf{1} \longrightarrow X^* \longrightarrow 0. \tag{4.3}$$

Tensoring this sequence with X on the left, we obtain

$$0 \longrightarrow X \otimes Y^* \longrightarrow X \longrightarrow X \otimes X^* \longrightarrow 0. \tag{4.4}$$

Since X is simple and $X \otimes X^* \neq 0$ (because the coevaluation morphism is nonzero) we obtain that $X \otimes X^* \cong X$. So we have a surjective composition morphism $\mathbf{1} \to X \otimes X^* \to X$. From this and (4.2) we have a nonzero composition morphism $\mathbf{1} \twoheadrightarrow X \hookrightarrow \mathbf{1}$. Since $\mathsf{End}_{\mathcal{C}}(\mathbf{1}) = \Bbbk$, this morphism is a nonzero scalar, whence $X = \mathbf{1}$. □

COROLLARY 4.3.9. *In a ring category with left duals, the evaluation morphisms are monomorphisms and the coevaluation morphisms are epimorphisms.*

REMARK 4.3.10. If $\mathcal{C}$ is a ring category with left duals and $F : \mathcal{C} \to \mathcal{D}$ is an exact $\Bbbk$-linear monoidal functor into a multiring category $\mathcal{D}$, then F is faithful, i.e., a tensor functor. Indeed, let $0 \neq X \in \mathcal{C}$. Since F is exact and $\mathrm{coev}_X : \mathbf{1} \to X \otimes X^*$ is a monomorphism, we have a monomorphism $\mathbf{1} \to F(X) \otimes F(X^*)$, which implies that $F(X) \neq 0$. Now, let f be a morphism in $\mathcal{C}$ such that $F(f) = 0$. Then again by the exactness of F, we have $F(\mathrm{Im}(f)) = 0$. Thus $\mathrm{Im}(f) = 0$ and hence $f = 0$.

EXERCISE 4.3.11. Let $\mathcal{C}$ be a multiring category with left duals and let $X \in \mathcal{C}_{ij}$ and $Y \in \mathcal{C}_{jk}$ be nonzero objects.

(1) Show that $X \otimes Y \neq 0$.
(2) Deduce that $\mathrm{length}(X \otimes Y) \geq \mathrm{length}(X)\mathrm{length}(Y)$.
(3) Show that if $\mathcal{C}$ is a ring category with left duals then an invertible object in $\mathcal{C}$ is simple.
(4) Let X be an object in a multiring category with left duals such that $X \otimes X^* \cong \mathbf{1}$. Show that X is invertible.

EXAMPLE 4.3.12. An example of a ring category where the unit object is not semisimple is the category $\mathcal{C}$ of finite dimensional representations of the quiver of type A_2. Such representations are triples (V, W, A), where V, W are finite dimensional vector spaces, and $A : V \to W$ is a linear operator. The tensor product on such triples is defined by the formula

$$(V, W, A) \otimes (V', W', A') = (V \otimes V', W \otimes W', A \otimes A'),$$

with obvious associativity isomorphisms, and the unit object $(\Bbbk, \Bbbk, \mathsf{id})$. Of course, this category has neither left nor right duals.

4.4. Absence of self-extensions of the unit object

Assume that $\Bbbk$ has characteristic 0.

THEOREM 4.4.1. *Let $\mathcal{C}$ be a finite ring category over $\Bbbk$ with simple object $\mathbf{1}$. Then* $\mathsf{Ext}^1(\mathbf{1}, \mathbf{1}) = 0$.

PROOF. Assume the contrary, and suppose that V is a nontrivial extension of $\mathbf{1}$ by itself. Let P be the projective cover of $\mathbf{1}$. Then $\mathsf{Hom}(P, V)$ is a 2-dimensional space, with a filtration induced by the filtration on V, and both quotients naturally isomorphic to $E := \mathsf{Hom}(P, \mathbf{1})$. Let v_0, v_1 be a basis of $\mathsf{Hom}(P, V)$ compatible to the filtration, i.e., v_0 spans the 1-dimensional subspace defined by the filtration. Let $A = \mathsf{End}(P)$ (this is a finite dimensional algebra). Let $\varepsilon : A \to \Bbbk$ be the character defined by the (right) action of A on E. Then the matrix of $a \in A$ in the basis v_0, v_1 has the form

$$[a]_1 = \begin{pmatrix} \varepsilon(a) & \chi_1(a) \\ 0 & \varepsilon(a) \end{pmatrix} \tag{4.5}$$

where $\chi_1 \in A^*$ is nonzero. Since $a \to [a]_1$ is a homomorphism, χ_1 is a derivation: $\chi_1(xy) = \chi_1(x)\varepsilon(y) + \varepsilon(x)\chi_1(y)$.

Now consider the representation $V \otimes V$. Using the exactness of the tensor product, we see that the space $\mathsf{Hom}(P, V \otimes V)$ is 4-dimensional, and has a 3-step filtration, with successive quotients $E, E \oplus E, E$, and basis $v_{00}; v_{01}, v_{10}; v_{11}$, consistent with this filtration. The matrix of $a \in \mathsf{End}(P)$ in this basis is

$$[a]_2 = \begin{pmatrix} \varepsilon(a) & \chi_1(a) & \chi_1(a) & \chi_2(a) \\ 0 & \varepsilon(a) & 0 & \chi_1(a) \\ 0 & 0 & \varepsilon(a) & \chi_1(a) \\ 0 & 0 & 0 & \varepsilon(a) \end{pmatrix}. \tag{4.6}$$

Since $a \to [a]_2$ is a homomorphism, we find

$$\chi_2(ab) = \varepsilon(a)\chi_2(b) + \chi_2(a)\varepsilon(b) + 2\chi_1(a)\chi_1(b), \quad a, b \in A.$$

We can now proceed further (i.e., consider $V \otimes V \otimes V$ etc.) and define for every positive n, a linear function $\chi_n \in A^*$ which satisfies the equation

$$\chi_n(ab) = \sum_{j=0}^{n} \binom{n}{j} \chi_j(a)\chi_{n-j}(b),$$

where $\chi_0 = \varepsilon$. Thus for any $s \in \Bbbk$, we can define $\phi_s : A \to \Bbbk((t))$ by $\phi_s(a) = \sum_{m \geq 0} \chi_m(a) s^m t^m / m!$ (this is where we use that $\Bbbk$ has characteristic 0) and we find that ϕ_s is a family of pairwise distinct homomorphisms. This is a contradiction, as A is a finite dimensional algebra and can have only finitely many 1-dimensional representations over any extension field. We are done. □

COROLLARY 4.4.2. *If a finite ring category $\mathcal{C}$ over a field of characteristic zero has a unique simple object* $\mathbf{1}$, *then $\mathcal{C}$ is equivalent to the category* Vec.

REMARK 4.4.3. The conclusion of Corollary 4.4.2 does not hold when the ground field has a positive characteristic. We give a counterexample: the category $\mathsf{Rep}(\mathbb{Z}/p\mathbb{Z})$ over a field of characteristic p.

4.5. Grothendieck ring and Frobenius-Perron dimension

In Section 1.5 we introduced the notion of the Grothendieck group $\mathsf{Gr}(\mathcal{C})$ of a locally finite abelian category $\mathcal{C}$.

Now let $\mathcal{C}$ be a multiring category over $\Bbbk$. Let X_i, $i \in I$, be representatives of the isomorphism classes of simple objects in $\mathcal{C}$. The tensor product on $\mathcal{C}$ induces a natural multiplication on $\mathsf{Gr}(\mathcal{C})$ defined by the formula

$$X_i X_j := [X_i \otimes X_j] = \sum_{k \in I} [X_i \otimes X_j : X_k] X_k, \qquad i, j \in I. \tag{4.7}$$

Here and below, if X, Y are objects of $\mathcal{C}$ and Y is simple, recall that we denote $[X : Y]$ the multiplicity of Y in the Jordan-Hölder series of X.

LEMMA 4.5.1. *The above multiplication on* $\mathsf{Gr}(\mathcal{C})$ *is associative.*

PROOF. Since the tensor product functor is exact,

$$[(X_i \otimes X_j) \otimes X_p : X_l] = \sum_k [X_i \otimes X_j : X_k][X_k \otimes X_p : X_l], \ \forall i, j, p, l.$$

On the other hand,

$$[X_i \otimes (X_j \otimes X_p) : X_l] = \sum_k [X_j \otimes X_p : X_k][X_i \otimes X_k : X_l].$$

Therefore, the associativity of the multiplication follows from the isomorphism $(X_i \otimes X_j) \otimes X_p \cong X_i \otimes (X_j \otimes X_p)$. □

Thus, $\mathsf{Gr}(\mathcal{C})$ is a $\mathbb{Z}_+$-ring with the unit $[\mathbf{1}]$.

DEFINITION 4.5.2. $\mathsf{Gr}(\mathcal{C})$ is called the *Grothendieck ring* of $\mathcal{C}$.

REMARK 4.5.3. Formula (4.7) is called the *fusion rule* (or fusion rules) of $\mathcal{C}$ (a terminology coming from physics). We will see below that fusion rules do not determine the tensor category $\mathcal{C}$ (even when $\mathcal{C}$ is semisimple).

PROPOSITION 4.5.4. *If* $\mathcal{C}$ *is a ring category with left duals then* $\mathsf{Gr}(\mathcal{C})$ *is a transitive unital* $\mathbb{Z}_+$*-ring in the sense of Definition* 3.3.1.

PROOF. Recall from Theorem 4.3.8 that the unit object $\mathbf{1}$ in $\mathcal{C}$ is simple. So $\mathsf{Gr}(\mathcal{C})$ is unital. This implies that for any simple objects X, Z of $\mathcal{C}$, the object $X \otimes X^* \otimes Z$ contains Z as a composition factor (as $X \otimes X^*$ contains $\mathbf{1}$ as a composition factor), so one can find a simple object Y_1 occurring in $X^* \otimes Z$ such that Z occurs in $X \otimes Y_1$. Similarly, the object $Z \otimes X^* \otimes X$ contains Z as a composition factor, so one can find a simple object Y_2 occurring in $Z \otimes X^*$ such that Z occurs in $Y_2 \otimes X$. Thus $\mathsf{Gr}(\mathcal{C})$ is transitive. □

In particular, Proposition 4.5.4 shows that for a ring category $\mathcal{C}$ with left duals, we can talk about Frobenius-Perron dimensions of objects and regular objects, see Section 3.3.

EXAMPLE 4.5.5. Let $\mathcal{C} = \mathsf{Rep}(G)$ be the category of finite dimensional representations of a finite group G, and let A be its Grothendieck ring. Let $\Bbbk G$ denote the regular representation of G. Then by Proposition 2.10.8, for any $X, Y \in \mathcal{C}$

$$\dim_\Bbbk \mathsf{Hom}(X \otimes \Bbbk G, Y) = \dim_\Bbbk \mathsf{Hom}(\Bbbk G, {}^*X \otimes Y) = \dim_\Bbbk(X) \dim_\Bbbk(Y).$$

Hence, $X \otimes \Bbbk G = \dim(X) \Bbbk G$, so $\mathsf{FPdim}(X) = \dim_\Bbbk(X)$ for all X, and a regular element of A is $R = \Bbbk G$ up to scaling.

REMARK 4.5.6. Let $\mathcal{C}$ and $\mathcal{D}$ be multiring categories and let $F : \mathcal{C} \to \mathcal{D}$ be a quasi-tensor functor. Then F defines a homomorphism of unital $\mathbb{Z}_+$-rings $[F] : \mathsf{Gr}(\mathcal{C}) \to \mathsf{Gr}(\mathcal{D})$.

PROPOSITION 4.5.7. *Let $\mathcal{C}$ and $\mathcal{D}$ be tensor categories with finitely many classes of simple objects. If $F : \mathcal{C} \to \mathcal{D}$ is a quasi-tensor functor, then $\mathsf{FPdim}_{\mathcal{D}}(F(X)) = \mathsf{FPdim}_{\mathcal{C}}(X)$ for any X in $\mathcal{C}$.*

PROOF. This follows from Proposition 3.3.13. □

REMARK 4.5.8. Proposition 4.5.7 fails in the situation when $\mathcal{C}$ has infinitely many classes of simple objects, see Remark 5.12.10.

Thus, we see that (multi)ring categories categorify rings (which justifies the terminology), while quasi-tensor (in particular, tensor) functors between them categorify unital ring homomorphisms. Note that Remark 4.3.4 may be regarded as a categorical analog of the Peirce decomposition in classical algebra.

EXERCISE 4.5.9. Show that an object X of a finite tensor category $\mathcal{C}$ is invertible if and only if $\mathsf{FPdim}(X) = 1$.

4.6. Deligne's tensor product of tensor categories

Let $\mathcal{C}, \mathcal{D}$ be multiring categories. Recall the definition of Deligne's tensor product of locally finite abelian categories from Section 1.11.

PROPOSITION 4.6.1. *Deligne's tensor product $\mathcal{C} \boxtimes \mathcal{D}$ is a multiring category, which is a ring category, a multitensor category and a tensor category respectively if so are $\mathcal{C}$ and $\mathcal{D}$.*

PROOF. If $\mathcal{C}$ is a multiring category then the tensor product functor $\otimes : \mathcal{C} \times \mathcal{C} \to \mathcal{C}$ extends to an exact functor $T_{\mathcal{C}} : \mathcal{C} \boxtimes \mathcal{C} \to \mathcal{C}$, which is equipped with an isomorphism of functors $T_{\mathcal{C}} \circ (\mathsf{id} \boxtimes T_{\mathcal{C}}) \cong T_{\mathcal{C}} \circ (T_{\mathcal{C}} \boxtimes \mathsf{id})$ (the associativity isomorphism) which satisfies the pentagon relation (where we regard Deligne's tensor product as a strictly associative operation). Thus, in the situation of the proposition, we can define an exact functor

$$(T_{\mathcal{C}} \boxtimes T_{\mathcal{D}}) \circ (23) : \mathcal{C} \boxtimes \mathcal{D} \boxtimes \mathcal{C} \boxtimes \mathcal{D} \to \mathcal{C} \boxtimes \mathcal{D},$$

where (23) stands for the permutation of components 2 and 3. Now, for $X, Y \in \mathcal{C} \boxtimes \mathcal{D}$, set $X \otimes Y := ((T_{\mathcal{C}} \boxtimes T_{\mathcal{D}}) \circ (23))(X \boxtimes Y)$. The associativity isomorphism for $\otimes$ comes from the associativity isomorphisms for $T_{\mathcal{C}}$ and $T_{\mathcal{D}}$, and it satisfies the pentagon relation because so do the associativity isomorphisms for $T_{\mathcal{C}}$ and $T_{\mathcal{D}}$. The existence of the unit object and the unit axiom are obvious. This implies that $\mathcal{C} \boxtimes \mathcal{D}$ is a multiring category, which is a ring category if (and only if) so are $\mathcal{C}$ and $\mathcal{D}$. If the categories $\mathcal{C}$ and $\mathcal{D}$ admit duals, then so does $\mathcal{C} \boxtimes \mathcal{D}$ (namely, its duality functor is simply Deligne's tensor product of the duality functors for $\mathcal{C}$ and $\mathcal{D}$). The proposition is proved. □

COROLLARY 4.6.2. *Deligne's tensor product of (multi)tensor categories is a (multi)tensor category. Deligne's tensor product of (multi)fusion categories is a (multi)fusion category.*

4.7. Quantum traces, pivotal and spherical categories

DEFINITION 4.7.1. Let $\mathcal{C}$ be a rigid monoidal category, V be an object in $\mathcal{C}$, and $a \in \mathsf{Hom}_{\mathcal{C}}(V, V^{**})$. Define its *left categorical (or quantum) trace*

$$\mathsf{Tr}^L(a) : \mathbf{1} \xrightarrow{\mathsf{coev}_V} V \otimes V^* \xrightarrow{a \otimes \mathsf{id}_{V^*}} V^{**} \otimes V^* \xrightarrow{\mathsf{ev}_{V^*}} \mathbf{1}. \tag{4.8}$$

Similarly, if $a \in \mathsf{Hom}(V, {}^{**}V)$ then we can define the *right* quantum trace

$$\mathsf{Tr}^R(a) : \mathbf{1} \xrightarrow{\mathsf{coev}_{*V}} {}^*V \otimes V \xrightarrow{\mathsf{id}_{*V} \otimes a} {}^*V \otimes {}^{**}V \xrightarrow{\mathsf{ev}_{**V}} \mathbf{1}. \tag{4.9}$$

REMARK 4.7.2. By definition, $\mathsf{Tr}^L(a)$, $\mathsf{Tr}^R(a) \in \mathsf{End}_{\mathcal{C}}(\mathbf{1})$. When $\mathcal{C}$ is a tensor category over $\Bbbk$, $\mathsf{Tr}^L(a)$ and $\mathsf{Tr}^R(a)$ can be regarded as elements of $\Bbbk$.

The following proposition shows that usual linear algebra formulas hold for the quantum trace.

PROPOSITION 4.7.3. *If $a \in \mathsf{Hom}_{\mathcal{C}}(V, V^{**})$, $b \in \mathsf{Hom}_{\mathcal{C}}(W, W^{**})$ then*

(1) $\mathsf{Tr}^L(a) = \mathsf{Tr}^R(a^*)$;
(2) $\mathsf{Tr}^L(a \oplus b) = \mathsf{Tr}^L(a) + \mathsf{Tr}^L(b)$ *(in additive categories);*
(3) $\mathsf{Tr}^L(a \otimes b) = \mathsf{Tr}^L(a)\,\mathsf{Tr}^L(b)$;
(4) *If $c \in \mathsf{Hom}(V, V)$ then* $\mathsf{Tr}^L(ac) = \mathsf{Tr}^L(c^{**}a)$, $\mathsf{Tr}^R(ac) = \mathsf{Tr}^R({}^{**}ca)$.

Equalities similar to (2) and (3) also hold for right quantum traces.

EXERCISE 4.7.4. Prove Proposition 4.7.3.

When $\mathcal{C}$ is a multitensor category, it is useful to generalize Proposition 4.7.3(2) as follows.

PROPOSITION 4.7.5. *If $a \in \mathsf{Hom}_{\mathcal{C}}(V, V^{**})$ and $W \subset V$ such that $a(W) \subset W^{**}$ then $\mathsf{Tr}^L_V(a) = \mathsf{Tr}^L(a|_W) + \mathsf{Tr}^L(a|_{V/W})$. That is, Tr^L is additive on exact sequences. The same statement holds for right quantum traces.*

EXERCISE 4.7.6. Prove Proposition 4.7.5.
Hint: use the 3-step filtration on $V^* \otimes V$ induced by the exact sequence $0 \to W \to V \to V/W \to 0$.

DEFINITION 4.7.7. Let $\mathcal{C}$ be a rigid monoidal category. A *pivotal structure* on $\mathcal{C}$ is an isomorphism of monoidal functors $a_X : X \xrightarrow{\sim} X^{**}$.

That is, a pivotal structure is a collection of isomorphisms $a_X : X \xrightarrow{\sim} X^{**}$ natural in X and satisfying $a_{X \otimes Y} = a_X \otimes a_Y$ for all objects X, Y in $\mathcal{C}$.

DEFINITION 4.7.8. A rigid monoidal category $\mathcal{C}$ equipped with a pivotal structure is said to be *pivotal*.

EXERCISE 4.7.9. Show that if a is a pivotal structure on $\mathcal{C}$ then $a_{V^*} = (a_V)^{*-1}$. Hence, $a_{V^{**}} = a_V^{**}$.

EXAMPLE 4.7.10. The categories Vec, Vec_G, $\mathsf{Rep}(G)$ have a canonical pivotal structure coming from the natural isomorphisms of vector spaces $V \xrightarrow{\sim} V^{**}$ for objects V.

Let a be a pivotal structure on a rigid monoidal category $\mathcal{C}$.

DEFINITION 4.7.11. The *dimension* Tr^L of an object X with respect to a is

$$\dim_a(X) = \mathsf{Tr}(a_X) \in \mathsf{End}_{\mathcal{C}}(\mathbf{1}).$$

Thus, in a tensor category over $\Bbbk$, dimensions are elements of $\Bbbk$. Also, it follows from Exercise 4.7.9 that $\dim_a(V) = \dim_a(V^{**})$.

PROPOSITION 4.7.12. *If $\mathcal{C}$ is a tensor category, then the function $[X] \mapsto \dim_a(X)$ is a character of the Grothendieck ring $\mathsf{Gr}(\mathcal{C})$.*

PROOF. Proposition 4.7.5 implies that $\dim_a$ is additive on exact sequences, which means that it gives rise to a well-defined linear map from $\mathsf{Gr}(\mathcal{C})$ to $\Bbbk$. The fact that this map is a character follows from the obvious fact that $\dim_a(\mathbf{1}) = 1$ and Proposition 4.7.3(3). □

COROLLARY 4.7.13. *Dimensions of objects in a pivotal finite tensor category are algebraic integers in* $\Bbbk$ [2].

PROOF. This follows from the fact that a character of any ring that is finitely generated as a $\mathbb{Z}$-module takes values in algebraic integers. □

DEFINITION 4.7.14. A pivotal structure a on a tensor category $\mathcal{C}$ is *spherical* if $\dim_a(V) = \dim_a(V^*)$ for any object V in $\mathcal{C}$. A tensor category is *spherical* if it is equipped with a spherical structure.

Since $\dim_a$ is additive on exact sequences, it suffices to require the property $\dim_a(V) = \dim_a(V^*)$ only for simple objects V.

THEOREM 4.7.15. *Let $\mathcal{C}$ be a spherical category and V be an object of $\mathcal{C}$. Then for any $x \in \mathsf{Hom}_{\mathcal{C}}(V, V)$ one has*

$$\mathsf{Tr}^L(a_V x) = \mathsf{Tr}^R(x a_V^{-1}).$$

PROOF. We first note that $\mathsf{Tr}^R(a_X^{-1}) = \dim_a(X^*)$ for any object X by Proposition 4.7.3(1) and Exercise 4.7.9(1). Now let us prove the proposition in the special case when V is semisimple. Thus $V = \bigoplus_i Y_i \otimes V_i$, where V_i are vector spaces and Y_i are simple objects. Then $x = \bigoplus_i x_i \otimes \mathsf{id}_{V_i}$ with $x_i \in \mathsf{End}_{\Bbbk}(Y_i)$ and $a_V = \bigoplus \mathsf{id}_{Y_i} \otimes a_{V_i}$ (by naturality of a). Hence

$$\begin{aligned} \mathsf{Tr}^L(a_V x) &= \sum_i \mathsf{Tr}(x_i) \dim(V_i), \\ \mathsf{Tr}^R(x a_V^{-1}) &= \sum_i \mathsf{Tr}(x_i) \dim(V_i^*). \end{aligned}$$

This implies the result for a semisimple V.

Consider now the general case. Then V has the socle filtration (1.10)

$$0 = V_0 \subset V_1 \subset V_2 \subset \cdots \subset V_n = V \tag{4.10}$$

such that V_{i+1}/V_i is a maximal semisimple subobject in V/V_i. This filtration is preserved by x and by a (i.e., $a : V_i \to V_i^{**}$). Since traces are additive on exact sequences by Proposition 4.7.5, this implies that the general case of the required statement follows from the semisimple case. □

EXERCISE 4.7.16. Let $\mathsf{Aut}_{\otimes}(\mathsf{id}_{\mathcal{C}})$ be the group of natural monoidal automorphisms of the identity endofunctor of $\mathcal{C}$. Show that the set of isomorphism classes of pivotal structures on $\mathcal{C}$ is a torsor over $\mathsf{Aut}_{\otimes}(\mathsf{id}_{\mathcal{C}})$, and the set of isomorphism classes of spherical structures on $\mathcal{C}$ is a torsor over the subgroup $\mathsf{Aut}_{\otimes}(\mathsf{id}_{\mathcal{C}})_2$ in $\mathsf{Aut}_{\otimes}(\mathsf{id}_{\mathcal{C}})$ of elements which have order ≤ 2 on simple objects.

---

[2] If $\Bbbk$ has positive characteristic, by an algebraic integer in $\Bbbk$ we mean an element of a finite subfield of $\Bbbk$.

4.8. Semisimple multitensor categories

Let $\Bbbk$ be an algebraically closed field. In this section we will more closely consider semisimple multitensor categories over $\Bbbk$. They have some important additional properties compared to the general case.

PROPOSITION 4.8.1. *Let $\mathcal{C}$ be a semisimple multitensor category over $\Bbbk$ and let V be an object in $\mathcal{C}$. Then ${}^*V \cong V^*$. Hence, $V \cong V^{**}$.*

PROOF. We may assume that V is simple.

We claim that the unique simple object X such that $\mathsf{Hom}_{\mathcal{C}}(\mathbf{1}, V \otimes X) \neq 0$ is V^*. Indeed, $\mathsf{Hom}_{\mathcal{C}}(\mathbf{1}, V \otimes X) \cong \mathsf{Hom}_{\mathcal{C}}({}^*X, V)$ which is nonzero if and only if ${}^*X \cong V$, i.e., $X \cong V^*$. Similarly, the unique simple object X such that $\mathsf{Hom}_{\mathcal{C}}(V \otimes X, \mathbf{1}) \neq 0$ is ${}^*V$. But since $\mathcal{C}$ is semisimple, $\dim_{\Bbbk} \mathsf{Hom}_{\mathcal{C}}(\mathbf{1}, V \otimes X) = \dim_{\Bbbk} \mathsf{Hom}_{\mathcal{C}}(V \otimes X, \mathbf{1})$, which implies the result. □

REMARK 4.8.2. The result of Proposition 4.8.1 is false without the semisimplicity assumption, see Example 7.19.5.

Proposition 4.8.1 gives rise to the following question.

QUESTION 4.8.3. Does every semisimple tensor category admit a pivotal structure? A spherical structure?

This is the case for all known examples. The general answer is unknown to us at the moment of writing (even for fusion categories over ground fields of characteristic zero).

PROPOSITION 4.8.4. *Let $\mathcal{C}$ be a semisimple tensor category over $\Bbbk$, let V be a simple object in $\mathcal{C}$ and $a : V \xrightarrow{\sim} V^{**}$ be an isomorphism. Then $\mathsf{Tr}^R(a) \neq 0$ and $\mathsf{Tr}^L(a) \neq 0$.*

PROOF. The trace of a is the composition $\mathbf{1} \to V \otimes V^* \to \mathbf{1}$, where both morphisms are nonzero. If this composition is zero then there is a nonzero morphism $(V \otimes V^*)/\mathbf{1} \to \mathbf{1}$, which means that $[V \otimes V^* : 1] \geq 2$. Since $\mathcal{C}$ is semisimple, this implies that $\dim_{\Bbbk} \mathsf{Hom}(1, V \otimes V^*)$ is at least 2. Hence, $\dim_{\Bbbk} \mathsf{Hom}(V, V) \geq 2$ which contradicts the simplicity of V. □

REMARK 4.8.5. The above result is false for non-semisimple categories over fields of positive characteristic. For example, let $\mathcal{C} = \mathsf{Rep}_{\Bbbk}(GL_p(\mathbb{F}_p))$, the representation category of the group $GL_p(\mathbb{F}_p)$ over a field $\Bbbk$ of characteristic p. Let V be the p-dimensional vector representation of $GL_p(\mathbb{F}_p)$ (which is clearly irreducible). Let $a : V \to V^{**}$ be the identity map. Then $\mathsf{Tr}(a) = \dim_{\Bbbk}(V) = p = 0$ in $\Bbbk$.

A counterexample in characteristic 0 is the vector representation of the Lie superalgebra $\mathfrak{gl}(n|n)$, or the Steinberg representation (irreducible of dimension p) of the small quantum group $u_q(\mathfrak{sl}_2)$, where q is a root of unity of order p (see Chapter 5).

4.9. Grothendieck rings of semisimple tensor categories

PROPOSITION 4.9.1. *If $\mathcal{C}$ is a semisimple multitensor category then $\mathsf{Gr}(\mathcal{C})$ is a based ring. If $\mathcal{C}$ is a semisimple tensor category then $\mathsf{Gr}(\mathcal{C})$ is a unital based ring. If $\mathcal{C}$ is a (multi)fusion category, then $\mathsf{Gr}(\mathcal{C})$ is a (multi)fusion ring.*

PROOF. The $\mathbb{Z}_+$-basis in $\mathsf{Gr}(\mathcal{C})$ consists of the isomorphism classes of simple objects of $\mathcal{C}$. The set I_0 consists of the classes of simple subobjects of $\mathbf{1}$. The involution $*$ is the duality map (by Proposition 4.8.1, it does not matter whether to use left or right duality). This implies the first two statements. The last statement is clear. □

REMARK 4.9.2. The conclusion of Proposition 4.9.1 fails for non-semisimple tensor categories. This happens because axiom (3.2) in the definition of a based ring fails, i.e., one can have $[X \otimes X^* : \mathbf{1}] > 1$ for a simple object X. E.g., we saw in Example 3.1.9(v) that this is the case when X is the 2-dimensional irreducible representation of the group S_3 over a field of characteristic 2.

EXAMPLE 4.9.3. Let $\mathcal{C}$ be the category of finite dimensional representations of the Lie algebra $\mathfrak{sl}_2(\mathbb{C})$. Then the simple objects of this category are irreducible representations V_m of dimension $m+1$ for $m = 0, 1, 2, \ldots$; $V_0 = \mathbf{1}$. The Grothendieck ring of $\mathcal{C}$ is determined by the Clebsch-Gordan rule, cf. (3.4):

$$V_i \otimes V_j = \bigoplus_{l=0}^{\min(i,j)} V_{i+j-2l}. \tag{4.11}$$

The duality map on this ring is the identity.

Let $\mathcal{C}$ be a semisimple multitensor category with simple objects $\{X_i\}_{i\in I}$. Let I_0 be the subset of I such that $\mathbf{1} = \oplus_{i\in I_0} X_i$. Let $H_{ij}^l := \mathsf{Hom}(X_l, X_i \otimes X_j)$ (if $X_p \in \mathcal{C}_{ij}$ with $p \in I$ and $i, j \in I_0$, we will identify the spaces H_{pj}^p and H_{ip}^p with $\Bbbk$ using the left and right unit morphisms).

We have $X_i \otimes X_j = \bigoplus_l H_{ij}^l \otimes X_l$. Hence,

$$\begin{aligned} (X_{i_1} \otimes X_{i_2}) \otimes X_{i_3} &\cong \bigoplus_{i_4} \bigoplus_{j} H_{i_1 i_2}^j \otimes H_{j i_3}^{i_4} \otimes X_{i_4} \\ X_{i_1} \otimes (X_{i_2} \otimes X_{i_3}) &\cong \bigoplus_{i_4} \bigoplus_{l} H_{i_1 l}^{i_4} \otimes H_{i_2 i_3}^{l} \otimes X_{i_4}. \end{aligned}$$

Thus the associativity constraint reduces to a collection of linear isomorphisms

$$\Phi_{i_1 i_2 i_3}^{i_4} : \bigoplus_j H_{i_1 i_2}^j \otimes H_{j i_3}^{i_4} \cong \bigoplus_l H_{i_1 l}^{i_4} \otimes H_{i_2 i_3}^l. \tag{4.12}$$

The matrix blocks of these isomorphisms,

$$(\Phi_{i_1 i_2 i_3}^{i_4})_{jl} : H_{i_1 i_2}^j \otimes H_{j i_3}^{i_4} \to H_{i_1 l}^{i_4} \otimes H_{i_2 i_3}^l \tag{4.13}$$

are called *6j-symbols* because they depend on six indices.

EXAMPLE 4.9.4. Let $\mathcal{C}$ be the category of finite dimensional representations of the Lie algebra $\mathfrak{sl}_2(\mathbb{C})$. Then the spaces H_{ij}^l are 0- or 1-dimensional. In fact, it is obvious from the Clebsch-Gordan rule that the map $(\Phi_{i_1 i_2 i_3}^{i_4})_{jl}$ is a map between nonzero (i.e., 1-dimensional) spaces if and only if the numbers i_1, i_2, i_3, i_4, j, l are edge lengths of a tetrahedron with faces corresponding to the four H-spaces ($i_1 i_2 j$, $j i_3 i_4$,$i_1 l i_4$, $i_2 i_3 l$, such that the perimeter of every face is even (this tetrahedron is allowed to be in Euclidean 3-space, the Euclidean plane, or hyperbolic 3-space, so the only conditions are the triangle inequalities on the faces). In this case, the $6j$-symbol can be regarded as a number, provided we choose a basis vector in every nonzero H_{ij}^l. Under an appropriate normalization of basis vectors these numbers are the *Racah coefficients* or *classical 6j-symbols*.

EXERCISE 4.9.5. Write down explicitly the relation on $6j$-symbols coming from the pentagon identity. If $\mathcal{C} = \mathsf{Rep}(\mathfrak{sl}_2(\mathbb{C}))$ this relation is called the *Elliott-Biedenharn relation*.

EXERCISE 4.9.6. More generally, if $\mathcal{C} = \mathsf{Rep}(U_q(\mathfrak{sl}_2))$, where q is not a root of unity (see Section 5.6 below), then the numbers $(\Phi^{i_4}_{i_1 i_2 i_3})_{jl}$ are called the q-*Racah coefficients* or the *quantum* $6j$*-symbols*.

Further, the evaluation and coevaluation maps define elements

$$\alpha_{ij} \in (H^j_{ii^*})^* \quad \text{and} \quad \beta_{ij} \in H^j_{ii^*}, \qquad j \in I_0. \tag{4.14}$$

Now the axioms of a rigid monoidal category, i.e., the triangle and pentagon identities and the rigidity axioms translate into non-linear algebraic equations with respect to the $6j$-symbols $(\Phi^{i_4}_{i_1 i_2 i_3})_{jl}$ and vectors α_{ij}, β_{ij}.

4.10. Categorification of based rings

Proposition 4.9.1 gives rise to the following general question of *categorification* of based rings, which is one of the main problems in the structure theory of tensor categories.

DEFINITION 4.10.1. Given a based ring R, its *categorification* over $\Bbbk$ is a multifusion category over $\Bbbk$ together with an isomorphism of based rings $R \xrightarrow{\sim} \mathsf{Gr}(\mathcal{C})$.

REMARK 4.10.2. We use the word "categorification" in a very narrow sense suitable for our book, and there are many other meanings of this word in the literature.

How can one find all possible categorifications (up to equivalence of multitensor categories) of a given based ring R?

It is clear from the discussion of $6j$-symbols in Section 4.9 that this problem is equivalent to finding all solutions of the system of algebraic equations coming from the axioms of the rigid monoidal category modulo the group of automorphisms of the spaces H^k_{ij} ("gauge transformations"). In general, this problem is very difficult because the system of equations involved is nonlinear, contains many unknowns and is usually over-determined. In particular, it is not clear a priori whether for a given R this system has at least one solution, and if it does, how many solutions are there, e.g., whether the set of these solutions is finite. It is therefore amazing that the theory of tensor categories allows one to solve the categorification problem in a number of nontrivial cases.

Here is the simplest result in this direction.

Let $\mathbb{Z}G$ be the unital based group ring of a group G, with basis $\{g \mid g \in G\}$ and involution $g^* = g^{-1}$, see Example 3.1.9(ii).

PROPOSITION 4.10.3. *The categorifications of* $\mathbb{Z}G$ *are* Vec^ω_G, *and they are parametrized by* $H^3(G, \Bbbk^\times)$.

PROOF. This follows from the results of Section 2.6. □

REMARK 4.10.4. The tensor equivalence classes of categories Vec^ω_G are parametrized by $H^3(G, \Bbbk^\times)/\mathsf{Out}(G)$, where $\mathsf{Out}(G)$ is the group of outer automorphisms of G.

EXAMPLE 4.10.5. (Tambara-Yamagami fusion rings) Let G be a finite group, and TY_G be an extension of the unital based ring $\mathbb{Z}G$:

$$TY_G := \mathbb{Z}G \oplus \mathbb{Z}X,$$

where X is a new basis vector with $gX = Xg = X$, $X^2 = \sum_{g \in G} g$. This is a fusion ring with $X^* = X$. It is easy to see that $\mathsf{FPdim}(g) = 1$, $\mathsf{FPdim}(X) = |G|^{1/2}$. It is possible to show that these rings are categorifiable if and only if G is abelian, and such categorifications are parametrized by symmetric isomorphisms $G \cong G^\vee$ and a choice of sign.

EXAMPLE 4.10.6. (Verlinde rings for $\mathfrak{sl}_2$). Let k be a non-negative integer. Define a unital $\mathbb{Z}_+$-ring $\mathrm{Ver}_k = \mathrm{Ver}_k(\mathfrak{sl}_2)$ with basis V_i, $i = 0, ..., k$ ($V_0 = 1$), with duality given by $V_i^* = V_i$ and multiplication given by the truncated Clebsch-Gordan rule:

$$V_i V_j = \sum_{l=\max(i+j-k,\,0)}^{\min(i,\,j)} V_{i+j-2l}. \tag{4.15}$$

It other words, one computes the product by the usual Clebsch-Gordan rule, and then deletes the terms that are not defined (V_i with $i > k$) and also their mirror images with respect to the point $k + 1$. We will later see that this ring admits categorifications coming from quantum groups at roots of unity.

Note that $\mathrm{Ver}_0 = \mathbb{Z}$, $\mathrm{Ver}_1 = \mathbb{Z}[\mathbb{Z}_2]$, $\mathrm{Ver}_2 = TY_{\mathbb{Z}_2}$. The latter is the Ising fusion ring from Example 3.1.9(viii).

EXERCISE 4.10.7. Show that $\mathsf{FPdim}(V_j) = [j+1]_q := \frac{q^{j+1} - q^{-j-1}}{q - q^{-1}}$, where $q = e^{\frac{\pi i}{k+2}}$.

Note that the Verlinde ring has a subring Ver_k^0 spanned by V_j with even j. If $k = 3$, this ring has basis $1, X = V_2$ with $X^2 = X + 1, X^* = X$. This is the Yang-Lee fusion ring from Example 3.1.9(vii). In the Yang-Lee ring, $\mathsf{FPdim}(X)$ is the golden ratio $\frac{1+\sqrt{5}}{2}$.

Note that one can define the generalized Yang-Lee fusion rings YL_n $n \in \mathbb{Z}_+$, with basis $1, X$, multiplication $X^2 = 1 + nX$ and duality $X^* = X$. It is, however, shown in [**Os2**] that these rings are not categorifiable over $\mathbb{C}$ when $n > 1$.

REMARK 4.10.8. It is tempting to say that any $\mathbb{Z}_+$-ring R has a canonical categorification over any field $\Bbbk$: one can take the skeletal semisimple category $\mathcal{C} = \mathcal{C}_R$ over $\Bbbk$ whose Grothendieck group is R, define the tensor product functor on $\mathcal{C}$ according to the multiplication in R, and then "define" the associativity isomorphism to be the identity (which appears to make sense because the category is skeletal, and therefore by the associativity of R one has $(X \otimes Y) \otimes Z = X \otimes (Y \otimes Z)$). However, a more careful consideration shows that this approach does not actually work. Namely, such "associativity isomorphism" fails to be functorial with respect to morphisms; in other words, if $g : Y \to Y$ is a morphism, then $(\mathsf{id}_X \otimes g) \otimes \mathsf{id}_Z$ is not always equal to $\mathsf{id}_X \otimes (g \otimes \mathsf{id}_Z)$.

To demonstrate this explicitly, denote the simple objects of the category $\mathcal{C}$ by X_i, $i = 1, ..., r$, and let $X_i \otimes X_j = \oplus_l N_{ij}^l X_l$. Take $X = X_i$, $Y = mX_j$, and

$Z = X_l$; then g is an m-by-m matrix over $\Bbbk$. The algebra $\mathsf{End}((X \otimes Y) \otimes Z) = \mathsf{End}(X \otimes (Y \otimes Z))$ is equal to $\oplus_s \mathsf{Mat}_{mn_s}(\Bbbk)$, where

$$n_s = \sum_p N_{ij}^p N_{pl}^s = \sum_q N_{iq}^s N_{jl}^q,$$

and in this algebra we have

$$\begin{aligned}(\mathsf{id}_X \otimes g) \otimes \mathsf{id}_Z &= \bigoplus_{p=1}^{r} \mathsf{id}_{N_{ij}^p} \otimes g \otimes \mathsf{id}_{N_{pl}^s}, \\ \mathsf{id}_X \otimes (g \otimes \mathsf{id}_Z) &= \bigoplus_{q=1}^{r} \mathsf{id}_{N_{iq}^s} \otimes g \otimes \mathsf{id}_{N_{jl}^q}.\end{aligned}$$

We see that these two matrices are, in general, different, which shows that the identity "associativity isomorphism" is not natural.

4.11. Tensor subcategories

DEFINITION 4.11.1. A *(multi)tensor subcategory* in a (multi)tensor category $\mathcal{D}$ is a full subcategory $\mathcal{C} \subset \mathcal{D}$ which is closed under taking subquotients, tensor products, and duality (in particular, contains the unit object of $\mathcal{C}$).

EXERCISE 4.11.2. (i) Let G be a group and let $\mathcal{C} = \mathsf{Vec}_G^\omega$ be the category of G-graded vector spaces with the associativity constraint defined by a 3-cocycle $\omega \in Z^3(G, \Bbbk^\times)$. Show that tensor subcategories of $\mathcal{C}$ are in bijection with subgroups of G. Namely, the subcategory corresponding to a subgroup $H \subset G$ consists of vector spaces supported on H.

(ii) Let G be a finite group and suppose that $\Bbbk$ is algebraically closed of characteristic 0. Show that tensor subcategories of $\mathsf{Rep}(G)$ are in bijection with normal subgroups $N \subset G$. Namely, the subcategory corresponding to a normal subgroup $N \subset G$ consists of representations of G trivial on N and, hence, is equivalent to $\mathsf{Rep}(G/N)$.

Let $\mathcal{C}$ be a fusion category and $A = \mathsf{Gr}(\mathcal{C}) \otimes_{\mathbb{Z}} \mathbb{C}$. We extend the involution $X \mapsto X^*$ to A by anti-linearity. Let $l : A \to \mathbb{C}$ be the linear functional such that $l([X]) = \dim_{\mathbb{C}} \mathsf{Hom}_{\mathcal{C}}(\mathbf{1}, X)$ for any $X \in \mathcal{C}$. Then $(A, *)$ is a $*$-algebra and $l : A \to \mathbb{C}$ is a positive trace.

LEMMA 4.11.3. *Let $\mathcal{C}$ be a fusion category. For any simple object X in $\mathcal{C}$ there exists a positive integer N such that $\mathsf{Hom}_{\mathcal{C}}(\mathbf{1}, X^{\otimes N}) \neq 0$.*

PROOF. It is obvious that $[X] \in \mathsf{Gr}(\mathcal{C}) \subset \mathsf{Gr}(\mathcal{C}) \otimes_{\mathbb{Z}} \mathbb{C}$ is not nilpotent. So the result follows from Lemma 3.7.6. □

COROLLARY 4.11.4. *Let $\mathcal{D} \neq 0$ be a full abelian subcategory of a fusion category $\mathcal{C}$ which is closed under subquotients and tensor products. Then $\mathcal{D}$ contains the unit object $\mathbf{1}$ and is rigid, i.e., is a tensor subcategory of $\mathcal{C}$.*

PROOF. Let X be a simple object in $\mathcal{D}$. Then $X^{\otimes n} \in \mathcal{D}$ for all $n \geq 1$. For N as in Lemma 4.11.3 we get $\mathsf{Hom}(\mathbf{1}, X^{\otimes N}) \neq 0$, hence $\mathsf{Hom}(X^*, X^{\otimes N-1}) \neq 0$. So $\mathbf{1}, X^* \in \mathcal{D}$, as desired. □

4.12. Chevalley property of tensor categories

THEOREM 4.12.1. *(Chevalley) Let $\Bbbk$ be a field of characteristic zero. Then the tensor product of two simple finite dimensional representations of any group or Lie algebra over $\Bbbk$ is semisimple.*

PROOF. Let V be a finite dimensional vector space over a field $\Bbbk$ (of any characteristic), and $G \subset GL(V)$ be a Zariski closed subgroup.

LEMMA 4.12.2. *If V is a completely reducible representation of G, then G is reductive.*

PROOF. Let V be a nonzero rational representation of an affine algebraic group G. Let U be the unipotent radical of G. Let $V^U \subset V$ be the subspace of invariants. Since U is a normal subgroup, V^U is a subrepresentation. Since U is unipotent, $V^U \neq 0$. So if V is irreducible, then $V^U = V$, i.e., U acts trivially. Thus, U acts trivially on any completely reducible representation of G. So if V is completely reducible and $G \subset GL(V)$, then $U = 1$ and hence G is reductive. □

Now let G be any group, and V, W be two finite dimensional irreducible representations of G. Let G_V, G_W be the Zariski closures of the images of G in $GL(V)$ and $GL(W)$, respectively. Then by Lemma 4.12.2, G_V, G_W are reductive. Let G_{VW} be the Zariski closure of the image of G in $G_V \times G_W$. Let U be the unipotent radical of G_{VW}. Let $p_V : G_{VW} \to G_V$, $p_W : G_{VW} \to G_W$ be the projections. Since p_V is surjective, $p_V(U)$ is a normal unipotent subgroup of G_V, so $p_V(U) = 1$. Similarly, $p_W(U) = 1$. So $U = 1$, and G_{VW} is reductive.

Let G'_{VW} be the closure of the image of G in $GL(V \otimes W)$. Then G'_{VW} is a quotient of G_{VW}, so it is also reductive. Since $\mathrm{char}(\Bbbk) = 0$, this implies that the representation $V \otimes W$ is completely reducible as a representation of G'_{VW}, hence of G.

This proves Chevalley's theorem for groups. The proof for Lie algebras is similar. □

DEFINITION 4.12.3. A tensor category $\mathcal{C}$ is said to have the *Chevalley property* if the category $\mathcal{C}_0$ of semisimple objects of $\mathcal{C}$ is a tensor subcategory.

Thus, Chevalley's theorem says that the category of finite dimensional representations of any group or Lie algebra over a field of characteristic zero has the Chevalley property.

PROPOSITION 4.12.4. *A tensor category in which every simple object is invertible has the Chevalley property.*

PROOF. Obvious. □

REMARK 4.12.5. Tensor categories in which all simple objects are invertible are called pointed, see Section 5.11.

Recall from Section 1.13 that for any object X in a locally finite abelian category we denote by $\mathrm{Lw}(X)$ the Loewy length of X.

Let $\mathcal{C}_0 \subset \mathcal{C}_1 \subset \cdots$ be the socle filtration of $\mathcal{C}$, see Definition 1.13.2.

PROPOSITION 4.12.6. *In a tensor category with the Chevalley property,*

$$\mathrm{Lw}(X \otimes Y) \leq \mathrm{Lw}(X) + \mathrm{Lw}(Y) - 1. \tag{4.16}$$

Thus $\mathcal{C}_i \otimes \mathcal{C}_j \subset \mathcal{C}_{i+j}$.

PROOF. Let $X(i)$, $0 \le i \le m$, $Y(j)$, $0 \le j \le n$, be the successive quotients of the socle filtrations of X, Y. Then $Z := X \otimes Y$ has a filtration with successive quotients $Z(r) = \oplus_{i+j=r} X(i) \otimes Y(j)$, $0 \le r \le m+n$. Because of the Chevalley property, these quotients are semisimple. This implies the statement. $\square$

REMARK 4.12.7. It is clear that the converse to Proposition 4.12.6 holds as well: equation (4.16) (for simple X and Y) implies the Chevalley property.

EXERCISE 4.12.8. Let $\Bbbk$ be a field of characteristic p, and let G be a finite group. Show that the category $\mathsf{Rep}_{\Bbbk}(G)$ has the Chevalley property if and only if G has a normal p-Sylow subgroup.

4.13. Groupoids

The most basic examples of multitensor categories arise from groupoids. Recall that a *groupoid* is a small category where all morphisms are isomorphisms. Thus, a groupoid $\mathcal{G}$ entails a set X of objects of $\mathcal{G}$ and a set G of morphisms of $\mathcal{G}$, the source and target maps $s, t : G \to X$, the composition map $\mu : G \times_X G \to G$ (where the fibered product is defined using s in the first factor and t in the second factor), the unit morphism map $u : X \to G$, and the inversion map $i : G \to G$ satisfying certain natural axioms.

Here are some examples of groupoids.

(1) Any group G is a groupoid $\mathcal{G}$ with a single object whose set of morphisms to itself is G.
(2) Let X be a set and let $G = X \times X$. Then the *product groupoid* $\mathcal{G}(X) := (X, G)$ is a groupoid in which s is the first projection, t is the second projection, u is the diagonal map, and i is the permutation of factors. In this groupoid for any $x, y \in X$ there is a unique morphism from x to y.
(3) A more interesting example is the *transformation groupoid* $T(G, X)$ arising from the action of a group G on a set X. The set of objects of $T(G, X)$ is X, and arrows correspond to triples (g, x, y) where $y = gx$ with an obvious composition law. In other words, the set of morphisms is $G \times X$ and $s(g, x) = x$, $t(g, x) = gx$, $u(x) = (1, x)$, $i(g, x) = (g^{-1}, gx)$.

Let $\mathcal{G} = (X, G, \mu, s, t, u, i)$ be a groupoid whose set of objects X is finite and let $\mathcal{C}(\mathcal{G})$ be the category of finite dimensional vector spaces graded by the set G of morphisms of $\mathcal{G}$, i.e., vector spaces of the form $V = \oplus_{g \in G} V_g$. Introduce a tensor product on $\mathcal{C}(\mathcal{G})$ by the formula

$$(V \otimes W)_g = \bigoplus_{(g_1, g_2): g_1 g_2 = g} V_{g_1} \otimes W_{g_2}. \tag{4.17}$$

Then $\mathcal{C}(\mathcal{G})$ is a multitensor category. The unit object is $\mathbf{1} = \oplus_{x \in X} \mathbf{1}_x$, where $\mathbf{1}_x$ is a 1-dimensional vector space which sits in degree id_x in G. The left and right duals are defined by $(V^*)_g = ({}^*V)_g = V_{g^{-1}}$.

We invite the reader to check that the component subcategories $\mathcal{C}(\mathcal{G})_{xy}$ are the categories of vector spaces graded by $\mathsf{Mor}(y, x)$.

We see that $\mathcal{C}(\mathcal{G})$ is a tensor category if and only if $\mathcal{G}$ is a group, which is the case of Vec_G already considered in Example 2.3.6. Note also that if $X = \{1, ..., n\}$ then $\mathcal{C}(\mathcal{G}(X))$ is naturally equivalent to $\mathsf{Mat}_n(\mathsf{Vec})$, see Example 4.1.3.

EXERCISE 4.13.1. Let C_i be isomorphism classes of objects in a finite groupoid $\mathcal{G}$, $n_i = |C_i|$, $x_i \in C_i$ be representatives of C_i, and $G_i = \mathsf{Aut}(x_i)$ be the corresponding automorphism groups. Show that $\mathcal{C}(\mathcal{G})$ is (non-canonically) monoidally equivalent to $\bigoplus_i \mathsf{Mat}_{n_i}(\mathsf{Vec}_{G_i})$.

REMARK 4.13.2. The condition of local finiteness in Definition 4.1.1 is not superfluous: there exists a rigid monoidal $\Bbbk$-linear abelian category with bilinear tensor product which contains objects of infinite length. An example of such a category is the category $\mathcal{C}$ of Jacobi matrices of finite dimensional vector spaces. Namely, the objects of $\mathcal{C}$ are semi-infinite matrices $V = \{V_{ij}\}_{i,j\in\mathbb{Z}_+}$ of finite dimensional vector spaces V_{ij} with finitely many nonzero diagonals, and morphisms are matrices of linear maps. The tensor product in this category is defined by the formula

$$(V \otimes W)_{il} = \sum_j V_{ij} \otimes W_{jl}, \tag{4.18}$$

and the unit object $\mathbf{1}$ is defined by the condition $\mathbf{1}_{ij} = \Bbbk^{\delta_{ij}}$. The left and right duality functors coincide and are given by the formula

$$(V^*)_{ij} = (V_{ji})^*. \tag{4.19}$$

The evaluation map is the direct sum of the canonical maps $V_{ij}^* \otimes V_{ij} \to \mathbf{1}_{jj}$, and the coevaluation map is a direct sum of the canonical maps $\mathbf{1}_{ii} \to V_{ij} \otimes V_{ij}^*$.

Note that the category $\mathcal{C}$ is a subcategory of the category $\tilde{\mathcal{C}}$ of $\mathcal{G}(\mathbb{Z}_+)$-graded vector spaces with finite dimensional homogeneous components. Note also that the category $\tilde{\mathcal{C}}$ is not closed under the tensor product defined by (4.18) but the category $\mathcal{C}$ is.

EXERCISE 4.13.3.
(1) Show that if X is a finite set then the group of invertible objects of the category $\mathcal{C}(\mathcal{G}(X))$ is isomorphic to $\mathsf{Aut}(X)$ (the group of permutations of X).
(2) Let $\mathcal{C}$ be the category of Jacobi matrices of vector spaces from Example 4.13.2. Show that the statement of Exercise 4.3.11(d) fails for $\mathcal{C}$. Thus the finite length condition is important in Exercise 4.3.11.

4.14. The adjoint subcategory and universal grading

In this section we extend the definitions and results of Section 3.6 to tensor categories.

DEFINITION 4.14.1. A *Serre subcategory* of an abelian category $\mathcal{C}$ is a full subcategory $\mathcal{S}$ such that for all short exact sequences

$$0 \to X \to Y \to Z \to 0$$

in $\mathcal{C}$, the object Y belongs to $\mathcal{S}$ if and only if both X and Z do.

It follows that a Serre subcategory contains an object X if and only if it contains all Jordan-Hölder composition factors of X. Thus, Serre subcategories of $\mathcal{C}$ are in bijection with subsets of the set of isomorphism classes of simple objects of $\mathcal{C}$.

A *grading* of a tensor category $\mathcal{C}$ by a group G is a decomposition

$$\mathcal{C} = \bigoplus_{g\in G} \mathcal{C}_g,$$

where $\mathcal{C}_g \subset \mathcal{C}$, $g \in G$, are abelian subcategories, such that the tensor product $\otimes$ maps $\mathcal{C}_g \times \mathcal{C}_h$ to $\mathcal{C}_{gh}$. The subcategory $\mathcal{C}_1$ is a tensor subcategory; it is called the *trivial component* of the grading. In this case we also say that $\mathcal{C}$ is an *extension* of $\mathcal{C}_1$.

A grading is said to be *trivial* if $\mathcal{C}_1 = \mathcal{C}$. It is said to be *faithful* if $\mathcal{C}_g \neq 0$ for all $g \in G$. For any two faithful gradings of $\mathcal{C}$ there is a common refinement, so there exists a *universal grading* of $\mathcal{C}$.

Clearly, gradings of $\mathcal{C}$ are in bijection with gradings of its Grothendieck ring $\mathsf{Gr}(\mathcal{C})$.

The following definition is a direct analog of Definition 3.6.5.

DEFINITION 4.14.2. The group corresponding to the universal grading of $\mathcal{C}$ will be called the *universal grading group and denoted by* $U_{\mathcal{C}}$.

PROPOSITION 4.14.3. *The group* $\mathsf{Aut}_{\otimes}(\mathsf{id}_{\mathcal{C}})$ *of tensor automorphisms of the identity functor* $\mathsf{id}_{\mathcal{C}}$ *is canonically isomorphic to* $\mathsf{Hom}(U_{\mathcal{C}}, \Bbbk^\times)$.

PROOF. The homomorphism $f_\psi \in \mathsf{Hom}(U_{\mathcal{C}}, \Bbbk^\times)$ corresponding to $\psi \in \mathsf{Aut}_{\otimes}(\mathsf{id}_{\mathcal{C}})$ is defined as follows: assigning to a simple $X \in \mathcal{C}$ the element $\psi_X \in \mathsf{Aut}(X) = \Bbbk^\times$, one gets a grading of $\mathcal{C}$ by $\Bbbk^\times$, which defines a homomorphism $f_\psi : U_{\mathcal{C}} \to \Bbbk^\times$ by the universality of $U_{\mathcal{C}}$. It is easy to check that the assignment $\psi \mapsto f_\psi$ is a group isomorphism. □

REMARK 4.14.4. The universal grading of $\mathcal{C}$ is faithful.

Recall that the notion of the adjoint subring of a based ring was introduced in Section 3.6.

Let $\mathcal{C}$ be a semisimple tensor category and let $\mathsf{Gr}(\mathcal{C})$ be its Grothendieck based ring. Let $\mathsf{Gr}(\mathcal{C})_{ad}$ be the adjoint based subring of $\mathsf{Gr}(\mathcal{C})$, see Definition 3.6.1.

DEFINITION 4.14.5. The *adjoint subcategory* of $\mathcal{C}$ is the tensor Serre subcategory $\mathcal{C}_{ad} \subset \mathcal{C}$ generated by objects from $\mathsf{Gr}(\mathcal{C})_{ad}$. Equivalently, $\mathcal{C}_{ad}$ is the smallest tensor Serre subcategory of $\mathcal{C}$ containing all objects $X \otimes X^*$, $X \in \mathcal{O}(\mathcal{C})$.

It is clear from the results of Section 3.6 that $\mathcal{C}_{ad}$ is the trivial component of the universal grading of $\mathcal{C}$.

EXAMPLE 4.14.6. Let G be a finite group and let $\mathcal{C} = \mathsf{Rep}(G)$ be the category of finite dimensional representations of G. Then $\mathcal{C}_{ad} = \mathsf{Rep}(G/Z(G))$, where $Z(G)$ is the center of G. Similarly, if G is a complex reductive algebraic group and $\mathcal{C} = \mathsf{Rep}(G)$, then $\mathcal{C}_{ad} = \mathsf{Rep}(G_{ad})$, where $G_{ad} = G/Z(G)$ is the adjoint group of G. This justifies our terminology.

PROPOSITION 4.14.7. *There is a one-to-one correspondence between Serre tensor subcategories* $\mathcal{D} \subset \mathcal{C}$ *containing* $\mathcal{C}_{ad}$ *and subgroups* $G \subset U_{\mathcal{C}}$, *namely*

$$\mathcal{D} \mapsto G_{\mathcal{D}} := \{g \in U_{\mathcal{C}} \mid \mathcal{D} \cap \mathcal{C}_g \neq 0\}, \quad G \mapsto \mathcal{D}_G := \bigoplus_{g \in G} \mathcal{C}_g.$$

PROOF. This follows from the fact that $\mathcal{C} = \oplus_{g \in U_{\mathcal{C}}} \mathcal{C}_g$, where $\mathcal{C}_1 = \mathcal{C}_{ad}$, and $\mathrm{Gr}(\mathcal{C}_g)$ are indecomposable $\mathrm{Gr}(\mathcal{C}_1)$-modules. □

Let $\mathcal{D} \subset \mathcal{C}$ be a tensor subcategory. Among all gradings of $\mathcal{C}$ trivial on $\mathcal{D}$ there is a universal one, namely, the grading by $U_{\mathcal{C}}/G$, where $G \subset U_{\mathcal{C}}$ is the normal subgroup generated by the image of $U_{\mathcal{D}}$ under the natural homomorphism $U_{\mathcal{D}} \to U_{\mathcal{C}}$.

COROLLARY 4.14.8. *There is a one-to-one correspondence between equivalence classes of faithful gradings of $\mathcal{C}$ and Serre tensor subcategories $\mathcal{D} \subset \mathcal{C}$ containing $\mathcal{C}_{ad}$ such that the subgroup $G_{\mathcal{D}} \subset U_{\mathcal{C}}$ is normal. Namely, one associates to $\mathcal{D}$ the universal grading of $\mathcal{C}$ trivial on $\mathcal{D}$; one associates to a grading its trivial component.*

PROOF. For every normal subgroup $G \subset U_{\mathcal{C}}$ the homomorphism $U_{\mathcal{C}} \to U_{\mathcal{C}}/G$ gives a faithful grading of $\mathcal{C}$ by $U_{\mathcal{C}}/G$. By the universality of the grading $\mathcal{O}(\mathcal{C}) \to U_{\mathcal{C}}$, one thus gets a one-to-one correspondence between normal subgroups of $U_{\mathcal{C}}$ and equivalence classes of faithful gradings of $\mathcal{C}$. Now use Proposition 4.14.7. □

REMARK 4.14.9. If the Grothendieck ring of $\mathcal{C}$ is commutative (e.g., if $\mathcal{C}$ is braided, see Section 8 for definition) then $U_{\mathcal{C}}$ is abelian. So in this case the normality condition in Corollary 4.14.8 holds automatically.

Recall that the notion of commutator of a subring of a based ring was introduced in Definition 3.6.9.

DEFINITION 4.14.10. Let $\mathcal{C}$ be a tensor category. Suppose that the Grothendieck ring of $\mathcal{C}$ is commutative. Let $\mathcal{K}$ be a Serre tensor subcategory of $\mathcal{C}$. The *commutator* of $\mathcal{K}$ is the Serre tensor subcategory $\mathcal{K}^{co} \subset \mathcal{C}$ generated by all simple objects $X \in \mathcal{C}$ such that $X \otimes X^* \in \mathcal{K}$.

Equivalently, $\mathcal{K}^{co}$ is the biggest Serre tensor subcategory $\mathcal{R} \subset \mathcal{C}$ such that $\mathcal{R}_{ad} \subset \mathcal{K}$. We have $(\mathcal{K}^{co})_{ad} \subset \mathcal{K} \subset (\mathcal{K}_{ad})^{co}$.

REMARK 4.14.11. We have $\mathsf{Gr}(\mathcal{K}^{co}) = \mathsf{Gr}(\mathcal{K})^{co}$.

Let

$$\mathcal{C} = \bigoplus_{g \in G} \mathcal{C}_g$$

be a tensor category faithfully graded by a group G. By Theorem 3.5.2 the Frobenius-Perron dimensions of components $\mathcal{C}_g$ are equal for all $g \in G$.

For a tensor category $\mathcal{C}$ we define $\mathcal{C}^{(0)} = \mathcal{C}$, $\mathcal{C}^{(1)} = \mathcal{C}_{ad}$, and $\mathcal{C}^{(n)} = (\mathcal{C}^{(n-1)})_{ad}$ for every integer $n \geq 1$.

DEFINITION 4.14.12. The non-increasing sequence of tensor subcategories of $\mathcal{C}$

$$\mathcal{C} = \mathcal{C}^{(0)} \supseteq \mathcal{C}^{(1)} \supseteq \cdots \supseteq \mathcal{C}^{(n)} \supseteq \cdots \tag{4.20}$$

will be called the *upper central series* of $\mathcal{C}$.

EXAMPLE 4.14.13. For every group H let $Z(H)$ denote its center. Let G be a finite group and $\mathcal{C} = \mathsf{Rep}(G)$. Let

$$\{1\} = C^0(G) \subseteq C^1(G) \subseteq \cdots \subseteq C^n(G) \subseteq \cdots$$

be the upper central series of G; i.e., $C^0(G) := \{1\}, C^1(G) := Z(G)$ and for $n \geq 1$ the subgroup $C^n(G)$ is defined by $C^n(G)/C^{n-1}(G) = Z(G/C^{n-1}(G))$. Then $\mathcal{C}^{(n)} = \mathsf{Rep}(G/C^n(G))$, so that our definition of the upper central series agrees with the classical one.

DEFINITION 4.14.14. A tensor category $\mathcal{C}$ is *nilpotent* if its upper central series converges to Vec; i.e., $\mathcal{C}^{(n)} = \mathsf{Vec}$ for some n. The smallest such n is called the *nilpotency class* of $\mathcal{C}$.

REMARK 4.14.15. A tensor category $\mathcal{C}$ is nilpotent if and only if its Grothendieck ring $\mathsf{Gr}(\mathcal{C})$ is nilpotent in the sense of Definition 3.6.8.

REMARK 4.14.16. A tensor category $\mathcal{C}$ is nilpotent if and only if there is a sequence of Serre tensor subcategories

$$\mathcal{C}_0 = \mathsf{Vec} \subsetneq \mathcal{C}_1 \subsetneq \cdots \subsetneq \mathcal{C}_N = \mathcal{C}$$

such that for each $i = 1, \dots, N$ the category $\mathcal{C}_i$ has faithful grading by a group G_i with the trivial component $\mathcal{C}_{i-1}$. We say that $\mathcal{C}$ is *cyclically nilpotent* if each G_i is a cyclic group.

REMARK 4.14.17. Let G be a finite group and $\mathcal{C} = \mathsf{Rep}(G)$. Then $\mathcal{C}$ is nilpotent if and only if G is nilpotent.

EXERCISE 4.14.18. Let $\mathcal{C}$ be a nilpotent tensor category.

(i) Show that every Serre tensor subcategory $\mathcal{E} \subset \mathcal{C}$ is nilpotent.
(ii) Show that if $F : \mathcal{C} \to \mathcal{D}$ is a surjective tensor functor, then $\mathcal{D}$ is nilpotent.

4.15. Equivariantization of tensor categories

Recall that in Section 2.7 we introduced the notions of a group action and equivariantization in the setting of monoidal categories.

For a multitensor category $\mathcal{C}$ we will denote by $\mathsf{Aut}_\otimes(\mathcal{C})$ the monoidal category of tensor autoequivalences of $\mathcal{C}$.

As in Section 2.7 for a group G we denote by $\mathrm{Cat}(G)$ the monoidal category whose objects are elements of G, the only morphisms are the identities, and the tensor product is given by multiplication in G.

DEFINITION 4.15.1. An *action* of a group G on a multitensor category $\mathcal{C}$ is a monoidal functor

$$\mathrm{Cat}(G) \to \mathsf{Aut}_\otimes(\mathcal{C}). \tag{4.21}$$

The category $\mathcal{C}^G$ of G-equivariant objects in $\mathcal{C}$ (see Definition 2.7.2) is a multitensor category. It is tensor if and only if $\mathcal{C}$ is tensor. The forgetful functor $\mathcal{C}^G \to \mathcal{C}$ is a tensor functor. If $\Bbbk$ is an algebraically closed field of characteristic 0 and $\mathcal{C}$ is a fusion category then so is $\mathcal{C}^G$.

EXAMPLE 4.15.2. Let $\mathcal{C} = \mathsf{Vec}$ be the trivial fusion category. Every group G has a unique action on Vec, namely the trivial action. We claim that $\mathsf{Vec}^G \cong \mathsf{Rep}(G)$, the category of representations of G. Indeed, a G-equivariant object in Vec is a vector space V together with a collection of automorphisms $u_g : V \simeq V$, $g \in G$, satisfying compatibility conditions (2.7.2). These conditions say precisely that the assignment $g \mapsto u_g$ is a representation of G.

EXERCISE 4.15.3. This is a generalization of Example 4.15.2. Let N be a normal subgroup of a group G. Show that the quotient group G/N acts on $\mathsf{Rep}(N)$ and $\mathsf{Rep}(N)^{G/N} \cong \mathsf{Rep}(G)$.

EXAMPLE 4.15.4. Let G be a finite group. It acts on the category Vec_G of G-graded vector spaces by conjugation. The resulting equivariantization Vec_G^G is the Drinfeld center of Vec_G, see Definition 7.13.1.

DEFINITION 4.15.5. Let G be a finite group acting on a tensor category $\mathcal{C}$. A *crossed product* category $\mathcal{C} \rtimes G$ is defined as follows. We set $\mathcal{C} \rtimes G = \mathcal{C} \boxtimes \mathsf{Vec}_G$ as an abelian category (here $\boxtimes$ denotes Deligne's tensor product of abelian categories) and define a tensor product by

$$(X \boxtimes g) \otimes (Y \boxtimes h) := (X \otimes T_g(Y)) \boxtimes gh, \qquad X, Y \in \mathcal{C}, \quad g, h \in G. \tag{4.22}$$

The unit object is $\mathbf{1} \boxtimes e$ and the associativity and unit constraints come from those of $\mathcal{C}$.

Note that $\mathcal{C} \rtimes G$ is a G-graded tensor category,

$$\mathcal{C} \rtimes G = \bigoplus_{g \in G} (\mathcal{C} \rtimes G)_g, \qquad \text{where } (\mathcal{C} \rtimes G)_g = \mathcal{C} \otimes (\mathbf{1} \boxtimes g).$$

In particular, $\mathcal{C} \rtimes G$ contains $\mathcal{C} = \mathcal{C} \otimes (\mathbf{1} \boxtimes e)$ as a tensor subcategory.

EXAMPLE 4.15.6. For the trivial action of G on Vec we have $\mathsf{Vec} \rtimes G = \mathsf{Vec}_G$.

REMARK 4.15.7. We will see later that the equivariantization category $\mathcal{C}^G$ and the crossed product $\mathcal{C} \rtimes G$ are related to each other by a certain form of duality, see Example 7.12.19.

REMARK 4.15.8. Let X be a simple object of $\mathcal{C}$ and $G_X \subset G$ the stabilizer of its isomorphism class. We have a central extension $1 \to \Bbbk^\times \to \tilde{G}_X \to G_X \to 1$, where $\tilde{G}_X$ is the set of pairs consisting of an element $h \in G_X$ and an isomorphism $T_h(X) \xrightarrow{\sim} X$. (Here $h \mapsto T_h$ denotes the action.) The set of isomorphism classes of simple objects of $\mathcal{C}^G$ whose image under the forgetful functor $\mathcal{C}^G \to \mathcal{C}$ contains X is in natural bijection with the set $\mathrm{Irr}^1(\tilde{G}_X)$ of irreducible finite dimensional representations of $\tilde{G}_X$ such that each $\lambda \in \Bbbk^\times \subset \tilde{G}_X$ acts as multiplication by λ.

PROPOSITION 4.15.9. *If $\mathcal{C}$ is a tensor category and Y is the object of $\mathcal{C}^G$ corresponding to $V \in Irr^1(\tilde{G}_X)$ as above then*

$$\mathsf{FPdim}(Y) = \dim_{\Bbbk}(V) \cdot [G : G_X] \cdot \mathsf{FPdim}(X). \tag{4.23}$$

PROOF. Let Forg : $\mathcal{C}^G \to \mathcal{C}$ denote the canonical forgetful tensor functor. The stated bijection is defined by restricting the equivariant structure of $Y \in \mathcal{C}_G$ such that $X \subset \mathrm{Forg}(Y)$ to $V := \mathsf{Hom}_{\mathcal{C}}(X, \mathrm{Forg}(Y))$. Equation (4.23) follows from the formula $\mathsf{FPdim}(Y) = \mathsf{FPdim}(\mathrm{Forg}(Y))$, which is a consequence of Proposition 4.5.7. □

4.16. Multitensor categories over arbitrary fields

We have developed the theory of multitensor categories over an algebraically closed field, but the definition of a (finite) (multi)tensor and (multi)fusion category actually makes sense over any field $\Bbbk$. For instance, if G is a finite group, then the category of finite dimensional representations of G over $\Bbbk$ is a tensor category over $\Bbbk$.

Moreover, if $\Bbbk \subset \Bbbk'$ is a field extension, and $\mathcal{C}$ is a locally finite abelian category over $\Bbbk$, then we can naturally define the extension of scalars – the locally finite abelian category $\mathcal{C}' := \mathcal{C} \otimes_{\Bbbk} \Bbbk'$ over $\Bbbk'$ (which is (multi)tensor if so was $\mathcal{C}$), such that every object X of $\mathcal{C}$ gives rise to an object $X' := X \otimes_{\Bbbk} \Bbbk'$ of $\mathcal{C}'$. This procedure, however, is less innocent than one might think: for instance, for a simple object X of $\mathcal{C}$, the object X' may not be simple (e.g., take X to be the 2-dimensional rotation representation of $\mathbb{Z}/3\mathbb{Z}$ over $\Bbbk = \mathbb{R}$, and $\Bbbk' = \mathbb{C}$).

One defines $\mathcal{C}$ to be *split* if X' is simple for any simple X (for any choice of $\Bbbk'$). This is equivalent to saying that $\mathsf{End}(X) \cong \Bbbk$ for any simple X. For example, the category of representations of a finite group G over $\Bbbk$ is split if and only if all irreducible G-modules over $\overline{\Bbbk}$ are defined over $\Bbbk$.

If $\mathcal{C}$ is a (multi)fusion category over any field $\Bbbk$ then the Grothendieck ring of $\mathcal{C}$ is a weak (multi)fusion ring (see Section 3.8); namely, if b_i is the class of a simple object X_i then numbers $d_i = \tau(b_i b_i^*)$ equal $\dim D_i$, where $D_i := \mathrm{End}(X_i)$. This gives rise to the problem of categorification of a given weak fusion ring R over a given field $\Bbbk$. This problem is quite nontrivial, and even for simplest weak fusion rings leads to the study of torsion in the Brauer group of $\Bbbk$. It is discussed in the paper [**EtG10**]. For example, it is shown in [**EtG10**] that the fusion ring with basis 1 and b with $b^2 = n \cdot 1$, $n \geq 1$, is categorifiable over a field of characteristic zero if and only if $n = 4^m$ for a non-negative integer m, but over a global field and over $\mathbb{R}$, only $n = 1$ and $n = 4$ are possible.

Finally, note that a split (multi)fusion category can be defined over any commutative ring $\Bbbk$ (not necessarily a field). We leave it to the reader to give a precise definition, and just note that the monoidal structure of such a category is determined by a collection of isomorphisms (4.12) for $H_{ij}^m = \Bbbk^{N_{ij}^m}$, $N_{ij}^m \geq 0$, which satisfy the pentagon relation, and the evaluation and coevaluation morphisms are defined similarly.

4.17. Bibliographical notes

4.1. Tensor categories with a commutativity isomorphism were first considered by Saavedra Rivano in [**Sa**] and later by Deligne and Milne in [**DelM**]. General tensor categories (in particular, fusion categories) became a popular subject in the 1980s, with the advent of quantum groups, quantum knot invariants, and conformal field theory. Some basic references on this subject are the books by Bakalov and Kirillov [**BakK**], Kassel [**Kas**], Turaev [**Tu4**] and the paper by Moore and Seiberg [**MooS1**]. The term "multifusion category" appears in [**ENO2**] and "multitensor category" in [**EtO1**], although such categories (at least in the commutative case) were considered much earlier, in [**DelM**].

4.2. Proposition 4.2.1 is Proposition 1.16 in [**DelM**] (see also [**BakK**], Proposition 2.1.8). Proposition 4.2.12 is a slightly more general version of Kazhdan and Lusztig [**KazL5**], Corollary 2 (p.441), see also [**EtO1**], Proposition 2.1.

4.3. Theorem 4.3.8 is due to Deligne and Milne, [**DelM**], Proposition 1.17. The decomposition of a multitensor category into component categories is considered in [**ENO2, EtO1**].

4.4. The results on absence of self-extensions are taken from [**EtO1**] (see Theorem 2.17).

4.5. The material on Grothendieck rings and Frobenius-Perron dimensions (in particular, Proposition 4.5.7) is taken from [**EtO1**], see also [**E**].

4.6. This section is just the explanation that Deligne's tensor product (defined in [**De1**]) of (multi)tensor categories is itself a (multi)tensor category.

4.7. The notions of pivotal and spherical categories and quantum traces are discussed in Barrett and Westbury [**BarW**]. In the setting of ribbon categories, these notions are considered earlier by Reshetikhin and Turaev in [**RT1, RT2**]. See also the paper by Müger [**Mu2**] for a review of these notions.

4.8. Proposition 4.8.1 is taken from [**Os1**]. Proposition 4.8.4 is taken from [**BakK**].

4.9. For classical and quantum $6j$-symbols see the books by Carter, Flath, and Saito [**CarFS**] and Biedenharn and Louck [**BiL**].

4.10. The notion of categorification was first suggested by Crane and Frenkel in [**CraF**]. Tambara-Yamagami fusion rings and their categorifications are considered in [**TaY**]. On Verlinde rings and their categorifications see [**BakK**].

4.11. The results of this section are taken from [**DrGNO2**, Appendix F].

4.12. Chevalley's theorem is proved in [**Che**], p.88. The notion of Chevalley property for tensor categories was introduced in [**AEG**].

4.13. The book by Renault [**Ren**] is a good reference about groupoids. See also the reviews by Brown [**Bro**] and Weinstein [**Wein**].

4.14. The notion of the adjoint subcategory of a fusion category was introduced in [**ENO2**], 8.5, and developed in [**GelaN**]. The material on the adjoint subcategory and the universal grading is taken from [**GelaN**].

4.15. Our exposition of the theory of equivariantization of tensor categories follows [**DrGNO2**], Section 4 (which develops the ideas of Bruguiéres [**Bru**] and Múger [**Mu1**]).

4.16. Our discussion of multitensor categories over arbitrary fields follows [**EtG10**]. An earlier reference is Morrison and Snyder [**MorS**], 2.1.

CHAPTER 5

Representation categories of Hopf algebras

5.1. Fiber functors

Let $\mathcal{C}$ be a ring category over a field $\Bbbk$.

DEFINITION 5.1.1. A *quasi-fiber functor* on $\mathcal{C}$ is an exact faithful functor

$$F : \mathcal{C} \to \mathsf{Vec}$$

from $\mathcal{C}$ to the category of finite dimensional $\Bbbk$-vector spaces, such that $F(\mathbf{1}) = \Bbbk$, equipped with a natural isomorphism

$$J : F(X) \otimes F(Y) \to F(X \otimes Y), \qquad X, Y \in \mathcal{C}.$$

If in addition J is a tensor structure (i.e., satisfies axioms in Definition 2.4.1), one says that F is a *fiber functor*.

In other words, a (quasi-)fiber functor is a (quasi-)tensor functor to the category of vector spaces.

Let G be a group.

EXAMPLE 5.1.2. The forgetful functors $\mathsf{Vec}_G \to \mathsf{Vec}$ and $\mathsf{Rep}(G) \to \mathsf{Vec}$ are fiber functors.

EXAMPLE 5.1.3. Let $\omega \in Z^3(G, \Bbbk^\times)$ be a cohomologically non-trivial 3-cocycle. The forgetful functor $\mathsf{Vec}_G^\omega \to \mathsf{Vec}$ is quasi-fiber, for any choice of the isomorphism J. It follows from the results of Section 2.6 that Vec_G^ω does not admit a fiber functor.

EXAMPLE 5.1.4. The functor $\mathrm{Loc}(X) \to \mathsf{Vec}$ on the category of local systems of finite dimensional $\Bbbk$-vector spaces on a connected topological space X which attaches to a local system E its fiber E_x at a point $x \in X$ is a fiber functor, which justifies the terminology. (Note that if X is Hausdorff, then this functor can be identified with the above mentioned forgetful functor $\mathsf{Rep}(\pi_1(X, x)) \to \mathsf{Vec}$).

5.2. Bialgebras

Let $\mathcal{C}$ be a finite ring category and let $F : \mathcal{C} \to \mathsf{Vec}$ be a fiber functor with the tensor functor structure $J_{XY} : F(X) \otimes F(Y) \to F(X \otimes Y)$, $X, Y \in \mathcal{C}$. Consider the algebra $H := \mathsf{End}(F)$. This algebra has two additional structures: a comultiplication (or coproduct) $\Delta : H \to H \otimes H$ and a counit $\varepsilon : H \to \Bbbk$ defined as follows. Let

$$\alpha : \mathsf{End}(F) \otimes \mathsf{End}(F) \xrightarrow{\sim} \mathsf{End}(F \boxtimes F),$$

where $F \boxtimes F : \mathcal{C} \boxtimes \mathcal{C} \to \mathsf{Vec}$ is Deligne's tensor product of functors, see Section 1.11, denote the natural isomorphism. The comultiplication is defined by the formula

$$\Delta(a) = \alpha^{-1}(\widetilde{\Delta}(a)),$$

where $\widetilde{\Delta}(a) \in \mathsf{End}(F \boxtimes F)$ is given by

$$\widetilde{\Delta}(a)_{X,Y} = J_{X,Y}^{-1} a_{X \otimes Y} J_{X,Y},$$

and the counit is defined by the formula

$$\varepsilon(a) = a_{\mathbf{1}} \in \Bbbk.$$

THEOREM 5.2.1. *Let $H = \mathsf{End}(F)$ be as above.*

(i) *The algebra H is a coalgebra with comultiplication Δ and counit ε.*
(ii) *The maps Δ and ε are unital algebra homomorphisms.*

PROOF. The coassociativity of Δ follows from axiom (2.23) of a monoidal functor. The counit axiom follows from (2.25) and (2.26). Finally, observe that for all $\eta, \nu \in \mathsf{End}(F)$ the images under α of both $\Delta(\eta)\Delta(\nu)$ and $\Delta(\eta\nu)$ have components $J_{X,Y}^{-1}(\eta\nu)_{X \otimes Y} J_{X,Y}$; hence, Δ is an algebra homomorphism (which is obviously unital). The fact that ε is a unital algebra homomorphism is clear. □

DEFINITION 5.2.2. An algebra H equipped with a comultiplication Δ and a counit ε satisfying properties (i),(ii) of Theorem 5.2.1 is called a *bialgebra*. A homomorphism of bialgebras is a unital algebra homomorphism between bialgebras which preserves the coproduct and counit.

Thus, Theorem 5.2.1 claims that the algebra $H = \mathsf{End}(F)$ has a natural structure of a bialgebra.

Now let H be any bialgebra (not necessarily finite dimensional). Then the category $\mathbf{Rep}(H)$ of representations (i.e., left modules) of H and its subcategory $\mathsf{Rep}(H)$ of finite dimensional representations of H are naturally monoidal categories (and the same applies to right modules). Indeed, one can define the tensor product of two H-modules X, Y to be the usual tensor product of vector spaces $X \otimes Y$, with the action of H defined by the formula

$$\rho_{X \otimes Y}(a) = (\rho_X \otimes \rho_Y)(\Delta(a)), \ a \in H$$

(where $\rho_X : H \to \mathsf{End}(X), \rho_Y : H \to \mathsf{End}(Y)$), the associativity isomorphism to be the obvious one, and the unit object to be the 1-dimensional space $\Bbbk$ with the action of H given by the counit, $a \to \varepsilon(a)$. Moreover, the forgetful functor $\text{Forget} : \mathsf{Rep}(H) \to \mathsf{Vec}$ is a fiber functor.

Thus, one has the following theorem.

THEOREM 5.2.3. *The assignments*

$$(\mathcal{C}, F) \mapsto H = \mathsf{End}(F), \quad H \mapsto (\mathsf{Rep}(H), \text{Forget}) \tag{5.1}$$

are mutually inverse bijections between (1) finite ring categories $\mathcal{C}$ with a fiber functor $F : \mathcal{C} \to \mathsf{Vec}$, up to tensor equivalence and isomorphism of tensor functors and (2) isomorphism classes of finite dimensional bialgebras H over $\Bbbk$.

PROOF. Straightforward from the above. □

Theorem 5.2.3 is called *the reconstruction theorem for finite dimensional bialgebras* (as it reconstructs the bialgebra H from the category of its modules using a fiber functor).

REMARK 5.2.4. Below we will prove a number of versions of this theorem, which apply to ring categories with additional structures and/or properties. All of these theorems admit a stronger "categorical" version. Namely, to give a stronger version of Theorem 5.2.3, let **Fbialg** be the category of finite dimensional bialgebras over $\Bbbk$ (with morphisms being isomorphisms), and let **FRCF** be the category of finite ring categories with a fiber functor, with morphisms being isomorphism classes of equivalences compatible with fiber functors. Then the stronger form of Theorem 5.2.3 says that the functor $\mathsf{Rep} : \mathbf{Fbialg} \to \mathbf{FRCF}$, $H \mapsto \mathsf{Rep}(H)$, is an equivalence. This follows from our arguments in the same way as Theorem 5.2.3.

We will not formulate such categorical versions for the other reconstruction theorems below, and leave this to the reader.

EXERCISE 5.2.5. Show that the axioms of a bialgebra are self-dual in the following sense: if H is a finite dimensional bialgebra with multiplication $\mu : H\otimes H \to H$, unit $i : \Bbbk \to H$, comultiplication $\Delta : H \to H \otimes H$ and counit $\varepsilon : H \to \Bbbk$, then H^* is also a bialgebra, with the multiplication Δ^*, unit ε^*, comultiplication μ^*, and counit i^*.

EXERCISE 5.2.6.
(i) Let G be a finite monoid, and $\mathcal{C} = \mathsf{Vec}_G$. Let $F : \mathcal{C} \to \mathsf{Vec}$ be the forgetful functor. Show that $H = \mathsf{End}(F)$ is the bialgebra $\mathsf{Fun}(G, \Bbbk)$ of $\Bbbk$-valued functions on G, with comultiplication
$$\Delta(f)(x, y) = f(xy), \ \ x, y \in G,$$
where we identify $H \otimes H$ with $\mathsf{Fun}(G \times G, \Bbbk)$, and counit $\varepsilon(f) = f(1)$.
(ii) Show that $\mathsf{Fun}(G, \Bbbk)^* = \Bbbk G$, the monoid algebra of G (with basis G and product $x \cdot y = xy$), with coproduct $\Delta(x) = x \otimes x$, and counit $\varepsilon(x) = 1$, $x \in G$. Note that the bialgebra $\Bbbk G$ may be defined for any G (not necessarily finite).

EXERCISE 5.2.7. Let H be a $\Bbbk$-algebra, $\mathcal{C} = \mathbf{Rep}(H)$ be the category of H-modules, and $F : \mathcal{C} \to \mathbf{Vec}$ be the forgetful functor (we do not assume finite dimensionality). Assume that $\mathcal{C}$ is a $\Bbbk$-linear abelian monoidal category with a bilinear and biexact tensor product, and that F is a tensor functor. Show that this endows H with the structure of a bialgebra, such that F defines a monoidal equivalence $\mathcal{C} \to \mathbf{Rep}(H)$.

Note that not only modules, but also comodules over a bialgebra H form a monoidal category. Indeed, for a finite dimensional bialgebra, this is clear, as right (respectively, left) modules over H is the same thing as left (respectively, right) comodules over H^*. In general, if X, Y are, say, right H-comodules, then the right comodule $X \otimes Y$ is the usual tensor product of X, Y with the coaction map defined as follows: if $x \in X, y \in Y$, $\pi_X(x) = \sum x_i \otimes a_i$, $\pi_Y(y) = \sum y_j \otimes b_j$, then
$$\pi_{X\otimes Y}(x \otimes y) = \sum x_i \otimes y_j \otimes a_i b_j.$$
For a bialgebra H, the monoidal category of right H-comodules will be denoted by $H - \mathbf{comod}$, and the subcategory of finite dimensional comodules by $H - \mathrm{comod}$.

5.3. Hopf algebras

In the next few sections, we will review some of the most important basic results about bialgebras and Hopf algebras.

Let us now consider the additional structure on the bialgebra $H = \mathsf{End}(F)$ from the previous section in the case when the category $\mathcal{C}$ has left duals. In this case, one can define a linear map $S : H \to H$ by the formula

$$S(a)_X = a^*_{X^*},$$

where we use the natural identification of $F(X)^*$ with $F(X^*)$ (see Exercise 2.10.6).

PROPOSITION 5.3.1. *("The antipode axiom.") Let $\mu : H \otimes H \to H$ and $i : \Bbbk \to H$ be the multiplication and the unit maps of H. Then*

$$\mu \circ (\mathsf{id} \otimes S) \circ \Delta = i \circ \varepsilon = \mu \circ (S \otimes \mathsf{id}) \circ \Delta \tag{5.2}$$

as maps $H \to H$.

PROOF. For any $b \in \mathsf{End}(F \otimes F)$ the linear map $\mu \circ (\mathsf{id} \otimes S)(\alpha^{-1}(b))_X$, $X \in \mathcal{C}$ is given by

(5.3)

$$F(X) \xrightarrow{\mathsf{coev}_{F(X)}} F(X) \otimes F(X)^* \otimes F(X) \xrightarrow{b_{X,X^*}} F(X) \otimes F(X)^* \otimes F(X) \xrightarrow{\mathsf{ev}_{F(X)}} F(X),$$

where we suppress the identity isomorphisms, the associativity constraint, and the isomorphism $F(X)^* \cong F(X^*)$. Indeed, it suffices to check (5.3) for $b = \eta \otimes \nu$, where $\eta, \nu \in H$, which is straightforward.

Now the first equality of the proposition follows from the commutativity of the diagram

$$\begin{array}{ccc}
F(X) & \xrightarrow{\mathsf{coev}_{F(X)}} & F(X) \otimes F(X)^* \otimes F(X) \\
\downarrow{\scriptstyle \mathsf{id}} & & \downarrow{\scriptstyle J_{X,X^*}} \\
F(X) & \xrightarrow{F(\mathsf{coev}_X)} & F(X \otimes X^*) \otimes F(X) \\
\downarrow{\scriptstyle \eta_1} & & \downarrow{\scriptstyle \eta_{X \otimes X^*}} \\
F(X) & \xrightarrow{F(\mathsf{coev}_X)} & F(X \otimes X^*) \otimes F(X) \\
\downarrow{\scriptstyle \mathsf{id}} & & \downarrow{\scriptstyle J^{-1}_{X,X^*}} \\
F(X) & \xleftarrow{\mathsf{ev}_{F(X)}} & F(X) \otimes F(X)^* \otimes F(X),
\end{array} \tag{5.4}$$

for any $\eta \in \mathsf{End}(F)$.

Namely, the commutativity of the upper and the lower squares follows from the fact that upon identification of $F(X)^*$ with $F(X^*)$, the morphisms $\mathsf{ev}_{F(X)}$ and $\mathsf{coev}_{F(X)}$ are given by the diagrams of Exercise 2.10.6. The middle square commutes by the naturality of η. The composition of left vertical arrows gives $\varepsilon(\eta)\,\mathsf{id}_{F(X)}$, while the composition of the top, right, and bottom arrows gives $\mu \circ (\mathsf{id} \otimes S) \circ \Delta(\eta)$.

The second equality is proved similarly. □

DEFINITION 5.3.2. An *antipode* on a bialgebra H is a linear map $S : H \to H$ which satisfies the equalities of Proposition 5.3.1.

EXERCISE 5.3.3. Show that the antipode axiom is self-dual in the following sense: if H is a finite dimensional bialgebra with antipode S_H, then the dual bialgebra H^* from Example 5.2.5 also admits an antipode $S_{H^*} = S_H^*$.

EXERCISE 5.3.4. Let H be a bialgebra (not necessarily finite dimensional). Show that the finite dual H^*_{fin} (see Definition 1.12.1) is a bialgebra.

The following is a "linear algebra" analog of the fact that the left dual, when it exists, is unique up to a unique isomorphism.

PROPOSITION 5.3.5. *An antipode on a bialgebra H is unique if exists.*

PROOF. The proof essentially repeats the proof of uniqueness of left dual, cf. Proposition 2.10.5. Let S, S' be two antipodes for H. Then

$$\begin{aligned} S &= \mu\circ(S\otimes[\mu\circ(\mathrm{id}\otimes S')\circ\Delta])\circ\Delta \\ &= \mu\circ(\mathrm{id}\otimes\mu)\circ(S\otimes\mathrm{id}\otimes S')\circ(\mathrm{id}\otimes\Delta)\circ\Delta \\ &= \mu\circ(\mu\otimes\mathrm{id})\circ(S\otimes\mathrm{id}\otimes S')\circ(\Delta\otimes\mathrm{id})\circ\Delta \\ &= \mu\circ([\mu\circ(S\otimes\mathrm{id})\circ\Delta]\otimes S')\circ\Delta = S', \end{aligned}$$

where we used the antipode properties (5.2) of S and S', associativity of μ, and coassociativity of Δ. □

PROPOSITION 5.3.6. *If S is an antipode on a bialgebra H then S is an antihomomorphism of algebras with unit and of coalgebras with counit.*

PROOF. Let

$$(\Delta\otimes\mathrm{id})\circ\Delta(a)=(\mathrm{id}\otimes\Delta)\circ\Delta(a)=\sum_i a_i^1\otimes a_i^2\otimes a_i^3,$$

$$(\Delta\otimes\mathrm{id})\circ\Delta(b)=(\mathrm{id}\otimes\Delta)\circ\Delta(b)=\sum_j b_j^1\otimes b_j^2\otimes b_j^3.$$

Then using the definition of the antipode, we have

$$S(ab)=\sum_i S(a_i^1b)a_i^2S(a_i^3)=\sum_{i,j}S(a_i^1b_j^1)a_i^2b_j^2S(b_j^3)S(a_i^3)=S(b)S(a).$$

Thus S is an antihomomorphism of algebras (which is obviously unital). The fact that it is an antihomomorphism of coalgebras then follows using the self-duality of the axioms (see Exercises 5.2.5, 5.3.3), or can be shown independently by a similar argument. □

COROLLARY 5.3.7. (i) *If H is a bialgebra with an antipode S, then the category $\mathcal{C}=\mathsf{Rep}(H)$ has left duals. Namely, for any object X, the left dual X^* is the usual dual space of X, with action of H given by*

$$\rho_{X^*}(a)=\rho_X(S(a))^*, a\in H,$$

and the usual evaluation and coevaluation morphisms of the category Vec.

(ii) *If in addition S is invertible, then $\mathcal{C}$ also admits right duals, i.e., is rigid (in other words, $\mathcal{C}$ is a tensor category). Namely, for any object X, the right dual ${}^*X$ is the usual dual space of X, with action of H given by*

$$\rho_{{}^*X}(a)=\rho_X(S^{-1}(a))^*,$$

and the usual evaluation and coevaluation morphisms of the category Vec.

PROOF. Part (i) follows from the antipode axiom and Proposition 5.3.6. Part (ii) follows from part (i) and the fact that the operation of taking the left dual is inverse to the operation of taking the right dual. □

REMARK 5.3.8. A similar statement holds for finite dimensional comodules. Namely, if X is a finite dimensional right comodule over a bialgebra H with an antipode, then the left dual is the dual vector space X^* with the coaction defined by

$$(\pi_{X^*}(f), x \otimes \phi) := ((\mathrm{id} \otimes S)(\pi_X(x)), f \otimes \phi),$$

$x \in X, f \in X^*, \phi \in H^*$. If S is invertible, then the right dual ${}^*X$ is defined by the same formula with S replaced by S^{-1}.

REMARK 5.3.9. The fact that S is an antihomomorphism of coalgebras is the "linear algebra" version of the categorical fact that dualization changes the order of tensor product (Proposition 2.10.7(ii)).

DEFINITION 5.3.10. A bialgebra equipped with an invertible antipode S is called a *Hopf algebra*.

REMARK 5.3.11. We note that many authors use the term "Hopf algebra" for any bialgebra with an antipode (not necessarily invertible).

Thus, Corollary 5.3.7 states that if H is a Hopf algebra then $\mathrm{Rep}(H)$ is a tensor category. So, we get the following *reconstruction theorem for finite dimensional Hopf algebras.*

THEOREM 5.3.12. *The assignments*

$$(\mathcal{C}, F) \mapsto H = \mathrm{End}(F), \quad H \mapsto (\mathrm{Rep}(H), \text{Forget}) \tag{5.5}$$

are mutually inverse bijections between (1) equivalence classes of finite tensor categories $\mathcal{C}$ with a fiber functor F, up to tensor equivalence and isomorphism of tensor functors, and (2) isomorphism classes of finite dimensional Hopf algebras over $\Bbbk$.

PROOF. Straightforward from the above. □

EXERCISE 5.3.13. The bialgebra of functions $\mathrm{Fun}(G, \Bbbk)$ on a finite monoid G from Exercise 5.2.6(i) is a Hopf algebra if and only if G is a group. In this case, the antipode is given by the formula $S(f)(x) = f(x^{-1})$, $x \in G$.

More generally, if G is an affine algebraic group over $\Bbbk$, or still more generally an affine group scheme, then the algebra $\mathcal{O}(G)$ of regular functions on G is a Hopf algebra, with the comultiplication, counit, and antipode defined as in the finite case.

Similarly, $\Bbbk G$ is a Hopf algebra if and only if G is a group, with $S(x) = x^{-1}$, $x \in G$.

EXERCISE 5.3.14. Show that if g is a grouplike element of a Hopf algebra H (see Definition 1.9.7), then g is invertible, with $g^{-1} = S(g)$. Also, show that the product of two grouplike elements is grouplike. In particular, grouplike elements of any Hopf algebra H form a group, denoted $\mathbf{G}(H)$. Show that $\mathbf{G}(H)$ can also be defined as the group of isomorphism classes of 1-dimensional H-comodules under tensor multiplication.

PROPOSITION 5.3.15. *If H is a finite dimensional bialgebra with an antipode S, then S is invertible, so H is a Hopf algebra.*

PROOF. Let H_n be the image of S^n. Since S is an antihomomorphism of algebras and coalgebras, H_n is a sub-bialgebra of H. Let m be the smallest n such that $H_n = H_{n+1}$ (it exists because H is finite dimensional). We need to show that $m = 0$. If not, we can assume that $m = 1$ by replacing H with H_{m-1}.

We have a map $S' : H_1 \to H_1$ inverse to $S|_{H_1}$. For $a \in H$, let the triple coproduct of a be

$$(\Delta \otimes \mathsf{id})(\Delta(a)) = \sum_i a_i^1 \otimes a_i^2 \otimes a_i^3.$$

Consider the element

$$b = \sum_i S'(S(a_i^1))S(a_i^2)a_i^3.$$

On the one hand, collapsing the last two factors using the antipode axiom, we have $b = S'(S(a))$. On the other hand, writing b as

$$b = \sum_i S'(S(a_i^1))S(S'(S(a_i^2)))a_i^3$$

and collapsing the first two factors using the antipode axiom, we get $b = a$. Thus $a = S'(S(a))$ and thus $a \in H_1$, so $H = H_1$, a contradiction. □

REMARK 5.3.16. Proposition 5.3.15 is a special case of Proposition 4.2.10.

EXERCISE 5.3.17. Let $(H, \mu, i, \Delta, \varepsilon, S)$ be a Hopf algebra with the multiplication μ, the unit i, the comultiplication Δ, the counit ε, and the antipode S. Let μ^{op} and Δ^{op} be obtained from μ and Δ by permutation of components.

(i) Show that

$$H_{\mathrm{op}} := (H, \mu^{\mathrm{op}}, i, \Delta, \varepsilon, S^{-1}), \tag{5.6}$$

$$H^{\mathrm{cop}} := (H, \mu, i, \Delta^{\mathrm{op}}, \varepsilon, S^{-1}), \tag{5.7}$$

$$H_{\mathrm{op}}^{\mathrm{cop}} := (H, \mu^{\mathrm{op}}, i, \Delta^{\mathrm{op}}, \varepsilon, S) \tag{5.8}$$

are Hopf algebras. Show that H is isomorphic to $H_{\mathrm{op}}^{\mathrm{cop}}$, and H_{op} to H^{cop}. Deduce that $\mathsf{Rep}(H_{\mathrm{op}}) \cong \mathsf{Rep}(H)^{\mathrm{op}}$, where $\mathsf{Rep}(H)^{\mathrm{op}}$ is the category opposite to $\mathsf{Rep}(H)$, see Definition 2.1.5.

(ii) Suppose that a bialgebra H is a commutative ($\mu = \mu^{\mathrm{op}}$) or cocommutative ($\Delta = \Delta^{\mathrm{op}}$). Let S be an antipode on H. Show that $S^2 = \mathsf{id}_H$.

DEFINITION 5.3.18. The Hopf algebras (or bialgebras) in (5.6) and (5.7) will be called the opposite and co-opposite Hopf algebras (or bialgebras) of H.

EXERCISE 5.3.19. Show that if A, B are bialgebras, bialgebras with antipode, or Hopf algebras, then so is the tensor product $A \otimes B$.

5.4. Reconstruction theory in the infinite setting

In this section we would like to generalize the reconstruction theory to the situation when the category $\mathcal{C}$ is not assumed to be finite.

Let $\mathcal{C}$ be a ring category over $\Bbbk$. Then the coalgebra $\mathsf{Coend}(F)$ (defined in Section 1.10) also carries a multiplication and unit, dual to the comultiplication and counit of $\mathsf{End}(F)$. More precisely, since $\mathsf{End}(F)$ may now be infinite dimensional, the algebra $\mathsf{End}(F \otimes F)$ is in general isomorphic not to the usual tensor product $\mathsf{End}(F) \otimes \mathsf{End}(F)$, but rather to its completion $\mathsf{End}(F)\widehat{\otimes}\,\mathsf{End}(F)$ with respect to the inverse limit topology. Thus, the comultiplication of $\mathsf{End}(F)$ is a continuous linear map $\Delta : \mathsf{End}(F) \to \mathsf{End}(F)\widehat{\otimes}\,\mathsf{End}(F)$. The dual Δ^* of this map defines a multiplication on $\mathsf{Coend}(F)$.

If $\mathcal{C}$ has left duals, the bialgebra $\mathsf{Coend}(F)$ acquires an antipode, defined in the same way as in the finite dimensional case. This antipode is invertible if there are also right duals (i.e., if $\mathcal{C}$ is rigid). Thus, Theorem 1.10.1 implies the following "infinite" extensions of the reconstruction theorems.

THEOREM 5.4.1. *The assignments*

$$(\mathcal{C}, F) \mapsto H = \mathsf{Coend}(F), \quad H \mapsto (H - \text{Comod}, \text{Forget}) \tag{5.9}$$

are mutually inverse bijections between the following pairs of sets:

(1) *ring categories $\mathcal{C}$ over $\Bbbk$ with a fiber functor F, up to tensor equivalence and isomorphism of tensor functors, and bialgebras over $\Bbbk$, up to isomorphism;*
(2) *ring categories $\mathcal{C}$ over $\Bbbk$ with left duals with a fiber functor F, up to tensor equivalence and isomorphism of tensor functors, and bialgebras over $\Bbbk$ with an antipode, up to isomorphism;*
(3) *tensor categories $\mathcal{C}$ over $\Bbbk$ with a fiber functor F, up to tensor equivalence and isomorphism of tensor functors, and Hopf algebras over $\Bbbk$, up to isomorphism.*

REMARK 5.4.2. This theorem allows one to give a categorical proof of Proposition 5.3.5, deducing it from the fact that a left dual, when it exists, is unique up to a unique isomorphism.

REMARK 5.4.3. Corollary 5.3.15 is not true, in general, in the infinite dimensional case: there exist bialgebras H with a non-invertible antipode S, see [**Tak1**]. Therefore, there exist ring categories with simple object $\mathbf{1}$ and left duals that do not have right duals, i.e., are not tensor categories (namely, H – comod).

Reconstruction theory was first developed by T. Tannaka and M. Krein ([**Tan**], [**Kr**]) for compact topological groups, to reconstruct such a group from the category of its representations; it was a nonabelian analog of Pontryagin's duality and is called the *Tannaka-Krein duality*. Later this duality was adapted by P. Cartier and Grothendieck's school to algebraic groups, which led to creation of the theory of Tannakian categories (see [**Car1**] and [**Sa, DelM**]). The following example summarizes some of the main results of this theory, and explains how to use this theory to define various completions of an abstract group.

EXAMPLE 5.4.4. (i) Let $\mathcal{C}$ be the category of algebraic representations of an affine algebraic (or, more generally, proalgebraic) group G over $\Bbbk$. Let $F : \mathcal{C} \to \mathsf{Vec}$ be the forgetful functor. Then it is easy to check that $\mathsf{Coend}(F) = \mathcal{O}(G)$, the Hopf algebra of regular functions on G. More generally, the same holds for affine group schemes.

(ii) Let $\mathcal{C}$ be the category of finite dimensional representations over $\Bbbk$ of any (discrete) group G, and let the functor F be as in (i). It is easy to check that $\mathsf{Coend}(F)$ is a commutative Hopf algebra. So $\mathsf{Coend}(F) = \mathcal{O}(\widehat{G}_{\text{alg}})$, where $\widehat{G}_{\text{alg}} = \operatorname{Spec} \mathsf{Coend}(F)$ is a certain proalgebraic group canonically attached to G (its closed points are characters of the algebra $\mathsf{Coend}(F)$). The group $\widehat{G}_{\text{alg}}$ is called *the proalgebraic completion of* G.

(iii) Let G be as in (ii), $\text{char}(\Bbbk) = 0$, and $\mathcal{C}$ be the category of semisimple finite dimensional representations of G. By Chevalley's theorem, $\mathcal{C}$ is a tensor category. In this case, $\operatorname{Spec}\mathsf{Coend}(F) = \widehat{G}_{\mathrm{red}}$, the *proreductive completion* of G.

(iv) Let $\mathrm{char}(\Bbbk) = 0$, and $\mathcal{C}$ be the category of triangular representations of G, i.e., finite dimensional representations whose composition factors are 1-dimensional. Then $\operatorname{Spec}\mathsf{Coend}(F) = \widehat{G}_{\mathrm{solv}}$, called the *prosolvable completion* of G.

(v) Let $\mathcal{C}$ be the category of unipotent representations of G, i.e., finite dimensional representations whose composition factors are trivial. Then $\operatorname{Spec}\mathsf{Coend}(F) = \widehat{G}_{\mathrm{unip}}$, called the *prounipotent completion* of G.

(vi) Let $\mathcal{C}$ be the category of finite dimensional representations of G which factor through a finite group. Then $\operatorname{Spec}\mathsf{Coend}(F) = \widehat{G}_{\mathrm{fin}}$, called the *profinite completion* of G.

Note that all the completions $\widehat{G}_*$ defined above are equipped with a canonical homomorphism $\xi : G \to \widehat{G}_*$, since every $g \in G$ acts as an automorphism of F. Moreover, the group $\widehat{G}_*$ is an inverse limit of algebraic groups, so it carries an inverse limit topology (where the topology on honest algebraic groups is discrete), and the image of ξ is dense in $\widehat{G}_*$ in this topology.

We note, however, that the map ξ need not be injective. For example, if G has no nontrivial finite dimensional representations (e.g., G is a finitely generated infinite simple group[1]) then $\mathcal{C} = \mathsf{Vec}$, and $\widehat{G}_*$ is the trivial group.

(vii) Let $\mathfrak{g}$ be a Lie algebra over a field of characteristic zero (not necessarily finite dimensional). Similarly to (ii)-(vi), one can define the completions $\widehat{\mathfrak{g}}_{\mathrm{alg}}, \widehat{\mathfrak{g}}_{\mathrm{red}}, \widehat{\mathfrak{g}}_{\mathrm{tr}}, \widehat{\mathfrak{g}}_{\mathrm{unip}}$, which are proalgebraic groups defined using the categories of all, semisimple, triangular, and unipotent finite dimensional representations of $\mathfrak{g}$.

5.5. More examples of Hopf algebras

Let us give a few more examples of Hopf algebras. As we have seen, to define a Hopf algebra, it suffices to give an associative unital algebra H, and define a coproduct on generators of H (this determines a Hopf algebra structure on H uniquely if it exists). This is what we'll do in the examples below.

EXAMPLE 5.5.1. (Enveloping algebras) Let $\mathfrak{g}$ be a Lie algebra, and let $H = U(\mathfrak{g})$ be the universal enveloping algebra of $\mathfrak{g}$. Define the coproduct on H by setting $\Delta(x) = x \otimes 1 + 1 \otimes x$ for all $x \in \mathfrak{g}$. It is easy to show that this extends to the whole H, and that H equipped with this Δ is a Hopf algebra. Moreover, it is easy to see that the tensor category $\mathsf{Rep}(H)$ is equivalent to the tensor category $\mathsf{Rep}(\mathfrak{g})$.

This example motivates the following definition.

DEFINITION 5.5.2. An element x of a bialgebra H is called *primitive* if $\Delta(x) = x \otimes 1 + 1 \otimes x$. The space of primitive elements of H is denoted $\mathsf{Prim}(H)$.

In the notation of Section 1.9 we have $\mathsf{Prim}(H) = \mathsf{Prim}_{1,1}(H)$.

[1] A finitely generated infinite simple group G cannot have nontrivial finite dimensional representations over any field. Indeed, otherwise G would be a linear group, and a finitely generated linear group is known to be residually finite.

EXERCISE 5.5.3.
(i) Show that $\mathrm{Prim}(H)$ is a Lie algebra under the commutator.
(ii) Show that if x is a primitive element then $\varepsilon(x) = 0$, and in presence of an antipode $S(x) = -x$.

Let V be a vector space, and SV be the symmetric algebra of V. Then SV is a Hopf algebra (namely, it is the universal enveloping algebra of the abelian Lie algebra V).

PROPOSITION 5.5.4. *Let* $\Bbbk$ *have characteristic zero.*
(i) *One has* $\mathrm{Prim}(SV) = V$.
(ii) *Let* $\mathfrak{g}$ *be a Lie algebra over* $\Bbbk$. *Then* $\mathrm{Prim}(U(\mathfrak{g})) = \mathfrak{g}$.

PROOF. (i) It suffices to assume that V is finite dimensional. Then an element $P \in SV$ is a polynomial on V^*, and $\Delta(P)$ is the polynomial $P(x+y)$ on $V^* \times V^*$. Thus, P is primitive if and only if $P(x+y) = P(x) + P(y)$. If this equation holds, then by setting $x = y = 0$ we see that $P(0) = 0$; moreover replacing y with ty and tending t to zero, we get that $\partial_y P$ is independent of x. This implies that P is a homogeneous linear function, i.e., an element of V.

(ii) Consider the symmetrization map $s : S\mathfrak{g} \to U(\mathfrak{g})$ defined by the formula

$$x_1 \otimes \ldots \otimes x_n \mapsto \frac{1}{n!} \sum_{\sigma \in S_n} x_{\sigma(1)} \ldots x_{\sigma(n)}.$$

It is easy to see that this map is an isomorphism of coalgebras. Thus, it suffices to prove the statement when $\mathfrak{g}$ is commutative, which is part (i). □

REMARK 5.5.5. Note that this proposition fails in characteristic p. Indeed, the element x^p is primitive in SV, where V is 1-dimensional and spanned by x. The proof fails because in characteristic p, derivative being constant does not imply that a polynomial is linear.

EXAMPLE 5.5.6. (Taft algebras) Let q be a primitive n-th root of unity in $\Bbbk$ ($n \geq 2$). Let H be the algebra (of dimension n^2) generated over $\Bbbk$ by g and x satisfying the following relations: $g^n = 1$, $x^n = 0$ and $gxg^{-1} = qx$. Define the coproduct on H by $\Delta(g) = g \otimes g$, $\Delta(x) = x \otimes g + 1 \otimes x$. It is easy to show that this extends to a Hopf algebra structure on H. This Hopf algebra H is called the *Taft algebra*. For $n = 2$, one obtains the Sweedler Hopf algebra of dimension 4. Note that H is neither commutative nor cocommutative, and $S^2 \neq 1$ on H (as $S^2(x) = qx$).

Note that the skew-primitive element x in Example 5.5.6 is nontrivial, i.e., it is not a difference of two grouplike elements, cf. Remark 1.9.11.

EXERCISE 5.5.7. Let x be a (g, h)-skew-primitive element in a Hopf algebra H (see Definition 1.9.10).
(i) Show that $\varepsilon(x) = 0$, $S(x) = -g^{-1}xh^{-1}$.
(ii) Show that if $a, b \in H$ are grouplike elements, then axb is an (agb, ahb)-skew-primitive element.

EXAMPLE 5.5.8. (Nichols Hopf algebras of dimension 2^{n+1}) Let $H = \mathbb{C}[\mathbb{Z}/2\mathbb{Z}] \ltimes \wedge(x_1, \ldots, x_n)$, where the generator g of $\mathbb{Z}/2\mathbb{Z}$ acts on x_i by $gx_ig^{-1} = -x_i$. Define the coproduct on H by making g grouplike, and setting $\Delta(x_i) := x_i \otimes g + 1 \otimes x_i$

(so x_i are skew-primitive elements). Then H is a Hopf algebra of dimension 2^{n+1}. For $n = 1$, H is the Sweedler Hopf algebra from the previous example.

EXERCISE 5.5.9. Show that the Hopf algebras of Examples 5.5.1, 5.5.6, and 5.5.8 are well defined.

EXERCISE 5.5.10. (Semidirect product Hopf algebras) Let H be a Hopf algebra, and G a group of automorphisms of H. Let A be the semidirect product $H \rtimes \Bbbk G$.

(i) Show that A admits a unique structure of a Hopf algebra in which $\Bbbk G$ and H are Hopf subalgebras.
(ii) Show that the action of G on H induces an action of G on $\mathsf{Rep}(H)$ such that $\mathsf{Rep}(A) = \mathsf{Rep}(H)^G$, see Section 4.15.

5.6. The quantum group $\mathbf{U_q}(\mathfrak{sl}_2)$

Let us consider the Lie algebra $\mathfrak{sl}_2$. Recall that there is a basis $h, e, f \in \mathfrak{sl}_2$ such that

$$[h, e] = 2e, \quad [h, f] = -2f, \quad [e, f] = h. \tag{5.10}$$

This motivates the following definition.

DEFINITION 5.6.1. Let $q \in \Bbbk^\times$, $q \neq \pm 1$. The *quantum group* $U_q(\mathfrak{sl}_2)$ is generated by elements E, F and an invertible element K with defining relations

$$KEK^{-1} = q^2E, \quad KFK^{-1} = q^{-2}F, \quad [E, F] = \frac{K - K^{-1}}{q - q^{-1}}. \tag{5.11}$$

THEOREM 5.6.2. *There exists a unique Hopf algebra structure on $U_q(\mathfrak{sl}_2)$, given by*

$$\begin{aligned} \Delta(K) &= K \otimes K, \\ \Delta(E) &= E \otimes K + 1 \otimes E, \\ \Delta(F) &= F \otimes 1 + K^{-1} \otimes F, \end{aligned}$$

with

$$\varepsilon(K) = 1,\ \varepsilon(E) = \varepsilon(F) = 0,\ S(K) = K^{-1},\ S(E) = -EK^{-1},\ S(F) = -KF.$$

Thus, K is a grouplike element and E, F are skew-primitive elements.

EXERCISE 5.6.3. Prove Theorem 5.6.2.

There are also several versions of quantum $\mathfrak{sl}_2$ with the number q being replaced by a variable v. Namely, first of all, one can use the above definition (with q replaced by v) to define a Hopf algebra $U_v^{\mathrm{rat}}(\mathfrak{sl}_2)$ over the field $\Bbbk(v)$ of rational functions of v. Secondly, we have two important $\Bbbk[v, v^{-1}]$-lattices (or orders) in this Hopf algebra – the De Concini-Kac quantum group $U_v^{DK}(\mathfrak{sl}_2)$ and the Lusztig quantum group $U_v^L(\mathfrak{sl}_2)$.

Namely, $U_v^{DK}(\mathfrak{sl}_2)$ is the $\Bbbk[v, v^{-1}]$-subalgebra of $U_v^{\mathrm{rat}}(\mathfrak{sl}_2)$ generated by $K^{\pm 1}, E, F$.

To define $U_v^L(\mathfrak{sl}_2)$, we need the following definition.

DEFINITION 5.6.4. The *q-analog* of n is

$$[n]_q = \frac{q^n - q^{-n}}{q - q^{-1}}.$$

The *q-analog of the factorial* is

$$[n]_q! = \prod_{l=1}^{n} [l]_q = \frac{(q-q^{-1})\cdots(q^n-q^{-n})}{(q-q^{-1})^n}.$$

Now, $U_v^L(\mathfrak{sl}_2)$ is the $\Bbbk[v,v^{-1}]$-subalgebra of $U_v^{\mathrm{rat}}(\mathfrak{sl}_2)$ generated by $K^{\pm 1}$ and the divided powers $E^{(n)} := \frac{E^n}{[n]_v!}$ and $F^{(n)} = \frac{F^n}{[n]_v!}$, $n \geq 1$.

It is easy to check that both $U_v^{DK}(\mathfrak{sl}_2)$ and $U_v^L(\mathfrak{sl}_2)$ are closed with respect to the coproduct, counit, and antipode, i.e., they are Hopf algebras over $\Bbbk[v,v^{-1}]$. Moreover, we have a tautological inclusion $U_v^{DK}(\mathfrak{sl}_2) \to U_v^L(\mathfrak{sl}_2)$. Note also that for any $q \in \Bbbk^\times$, $U_q(\mathfrak{sl}_2) = U_v^{DK}(\mathfrak{sl}_2)/(v-q)$.

REMARK 5.6.5. Heuristically, $K = q^h$, and thus

$$\lim_{q\to 1} \frac{K-K^{-1}}{q-q^{-1}} = h.$$

So in the limit $q \to 1$, the relations of $U_q(\mathfrak{sl}_2)$ degenerate into the relations of $U(\mathfrak{sl}_2)$, and thus $U_q(\mathfrak{sl}_2)$ should be viewed as a Hopf algebra deformation of the enveloping algebra $U(\mathfrak{sl}_2)$. In fact, one can make this heuristic idea into a precise statement, see e.g. [**Kas**].

If q is a root of unity, then one can define three interesting versions of quantum $\mathfrak{sl}_2$ with parameter q. Assume for simplicity that the order of q is an odd number $\ell \geq 3$. Then, first of all, we have the De Concini-Kac quantum group $U_q(\mathfrak{sl}_2) = U_q^{DK}(\mathfrak{sl}_2)$, already defined above. Secondly, we have the Lusztig quantum group with divided powers, $U_q^L(\mathfrak{sl}_2) = U_v^L(\mathfrak{sl}_2)/(v-q)$. And finally, we can define a finite dimensional version of quantum $\mathfrak{sl}_2$. Namely, let $u_q(\mathfrak{sl}_2)$ be the quotient of $U_q(\mathfrak{sl}_2)$ by the additional relations

$$E^\ell = F^\ell = K^\ell - 1 = 0.$$

Then it is easy to show that $u_q(\mathfrak{sl}_2)$ is a Hopf algebra (with the coproduct inherited from $U_q(\mathfrak{sl}_2)$). This Hopf algebra is called *the small quantum group* attached to $\mathfrak{sl}_2$. We have $\dim_\Bbbk(u_q(\mathfrak{sl}_2)) = \ell^3$.

We have a chain of Hopf algebra maps $U_q^{DK}(\mathfrak{sl}_2) \to u_q(\mathfrak{sl}_2) \to U_q^L(\mathfrak{sl}_2)$, whose composition is the reduction to $v = q$ of the inclusion $U_v^{DK}(\mathfrak{sl}_2) \to U_v^L(\mathfrak{sl}_2)$. But this reduction itself is not an inclusion, since it factors through a finite dimensional Hopf algebra.

Also, one may consider the quotient H of $U_q^L(\mathfrak{sl}_2)$ by the relations $E = F = 0, K = 1$. One may show that H is isomorphic to the ordinary enveloping algebra $U(\mathfrak{sl}_2)$, generated by $e = E^{(\ell)}$ and $f = F^{(\ell)}$. Thus, we have a surjective Hopf algebra homomorphism $F_q : U_q^L(\mathfrak{sl}_2) \to U(\mathfrak{sl}_2)$, called the *quantum Frobenius map*.

The above constructions make sense if $\Bbbk$ is any commutative ring. In particular, we can take $\Bbbk = \mathbb{Z}$. This allows one to define the above quantum groups over the Laurent polynomial ring $\mathbb{Z}[v,v^{-1}]$ and the cyclotomic ring $\mathbb{Z}[q]$. The latter allows us to reduce to positive characteristic. Namely, assume that ℓ is a prime. Then there is a natural homomorphism $\mathbb{Z}[q] \to \mathbb{F}_\ell$, sending q to 1. So we can tensor the small quantum group with $\mathbb{F}_\ell$ using this homomorphism. Then we get:

$$u_q^L(\mathfrak{sl}_2) \otimes_{\mathbb{Z}[q]} \mathbb{F}_\ell = U^{\mathrm{res}}(\mathfrak{sl}_2)$$

where the superscripts res stands for the restricted enveloping algebra of $\mathfrak{sl}_2$ over $\mathbb{F}_\ell$ (i.e., the quotient of the usual enveloping algebra by the ideal generated by $E^\ell, F^\ell, H^\ell - H$).

Similarly, reductions modulo ℓ of the quantum groups $U_q^{DK}(\mathfrak{sl}_2)$ and $U_q^L(\mathfrak{sl}_2)$ in an appropriate sense give the ordinary enveloping algebra of $\mathfrak{sl}_2$, respectively the enveloping algebra of $\mathfrak{sl}_2$ with divided powers, and the quantum Frobenius map reduces to the ordinary Frobenius map on the enveloping algebra with divided powers (dual to the Frobenius map on the corresponding group SL_2), see Remark 5.8.7.

5.7. The quantum group $\mathbf{U_q}(\mathfrak{g})$

The example of the previous section can be generalized to the case of any finite dimensional simple Lie algebra. Namely, let $\mathfrak{g}$ be a simple Lie algebra of rank r, and let $A = (a_{ij})$, $i, j = 1, ..., r$, be its Cartan matrix. Recall that there exist unique relatively prime positive integers $d_i, i = 1, \dots r$ such that $d_i a_{ij} = d_j a_{ji}$. Let $q \in \mathbb{k}^\times, q_i := q^{d_i}$, and suppose that $q_i \neq \pm 1$.

DEFINITION 5.7.1. The *quantum group* $U_q(\mathfrak{g})$ is generated by elements E_i, F_i and invertible elements K_i, $i = 1, ..., r$, with defining relations

$$K_iK_j = K_jK_i, \quad K_iE_jK_i^{-1} = q_i^{a_{ij}}E_j, \quad K_iF_jK_i^{-1} = q_i^{-a_{ij}}F_j, \tag{5.12}$$

$$[E_i, F_j] = \delta_{ij}\frac{K_i - K_i}{q_i - q_i^{-1}}, \tag{5.13}$$

and the *q-Serre relations*:

$$\sum_{l=0}^{1-a_{ij}} \frac{(-1)^l}{[l]_{q_i}![1 - a_{ij} - l]_{q_i}!} E_i^{1-a_{ij}-l}E_jE_i^l = 0, i \neq j \tag{5.14}$$

and

$$\sum_{l=0}^{1-a_{ij}} \frac{(-1)^l}{[l]_{q_i}![1 - a_{ij} - l]_{q_i}!} F_i^{1-a_{ij}-l}F_jF_i^l = 0, i \neq j. \tag{5.15}$$

More generally, the same definition can be made for any symmetrizable Kac-Moody algebra $\mathfrak{g}$.

THEOREM 5.7.2. *(see e.g. [**ChP**]) There exists a unique Hopf algebra structure on $U_q(\mathfrak{g})$, given by*

- $\Delta(K_i) = K_i \otimes K_i$;
- $\Delta(E_i) = E_i \otimes K_i + 1 \otimes E_i$;
- $\Delta(F_i) = F_i \otimes 1 + K_i^{-1} \otimes F_i$.
- $\varepsilon(K_i) = 1, \ \varepsilon(E_i) = \varepsilon(F_i) = 0$,
- $S(K_i) = K_i^{-1}, \ S(E_i) = -E_iK_i^{-1}, \ S(F_i) = -K_iF_i$.

REMARK 5.7.3. Similarly to the case of $\mathfrak{sl}_2$, in the limit $q \to 1$, these relations degenerate into the relations for $U(\mathfrak{g})$, so $U_q(\mathfrak{g})$ should be viewed as a Hopf algebra deformation of the enveloping algebra $U(\mathfrak{g})$.

REMARK 5.7.4. Similarly to the case of $\mathfrak{sl}_2$ one can define the quantum group $U_v^{\mathrm{rat}}(\mathfrak{g})$ over $\mathbb{k}(q)$, the De Concini-Kac and Lusztig forms $U_v^{DK}(\mathfrak{g})$, $U_v^L(\mathfrak{g})$ over $\mathbb{k}[v, v^{-1}]$, the corresponding specialized versions $U_q^{DK}(\mathfrak{g})$, $U_q^L(\mathfrak{g})$ at $v = q$, and the small quantum group $u_q(\mathfrak{g})$ at an ℓ-th root of unity of dimension $\ell^{\dim(\mathfrak{g})}$. Moreover, these algebras reduce to the usual, divided power, and restricted enveloping algebras in positive characteristic, respectively. The details of these constructions are rather involved, and we refer the reader to the book [**Lus5**] for details.

5.8. Representations of quantum groups and quantum function algebras

Suppose that q is not a root of unity, and consider the representation theory of $U_q(\mathfrak{sl}_2)$. Given an integer λ, we can define the *Verma module* M_λ generated by the vector v_λ with defining relations $Ev_\lambda = 0$, $Kv_\lambda = q^\lambda v_\lambda$. This module has a basis $v_\lambda, Fv_\lambda, F^2v_\lambda, \ldots$. It is easy to show that M_λ is irreducible if and only if $\lambda \notin \mathbb{Z}_+$. If $\lambda \in \mathbb{Z}_+$ then M_λ has a submodule J_λ spanned by $F^j v_\lambda$ for $j > \lambda$, and the quotient $V_\lambda = M_\lambda/J_\lambda$ is irreducible, of dimension $\lambda+1$. The representations V_λ are deformations of finite dimensional representations of $\mathfrak{sl}_2$, and thus they satisfy the Clebsch-Gordan rule:

$$V_i \otimes V_j = \sum_{l=0}^{\min(i,j)} V_{i+j-2l}. \tag{5.16}$$

Thus, we have a semisimple tensor category with simple objects V_i ($V_0 = \mathbf{1}$), which categorifies the Clebsch-Gordan unital based ring (for any q which is not a root of unity).

In a similar way, if $\mathfrak{g}$ is a simple Lie algebra, and λ is an integral weight, then we can define the Verma module M_λ over $U_q(\mathfrak{g})$ generated by v_λ with relations $E_i v_\lambda = 0$, $K_i v_\lambda = q^{(\lambda,\alpha_i^\vee)} v_\lambda$, where $\alpha_i^\vee$ are the simple coroots. Then, if λ is dominant, then M_λ has a maximal proper submodule J_λ generated by the vectors $F_i^{(\lambda,\alpha_i^\vee)+1} v_\lambda$, and $M_\lambda/J_\lambda = V_\lambda$ is a finite dimensional irreducible representation. The representations V_λ have characters given by the classical Weyl character formula for $\mathfrak{g}$. They span a semisimple tensor category, whose fusion rules are the same as for the category $\mathsf{Rep}(\mathfrak{g})$.

REMARK 5.8.1. Unlike the case of the Lie algebra $\mathfrak{sl}_2$, for $U_q(\mathfrak{sl}_2)$ it is not quite true that the representations V_λ exhaust all the irreducible finite dimensional representations. However, this statement is almost true, in the following sense: any finite dimensional irreducible representation is a tensor product of V_λ with a 1-dimensional representation, and there are two 1-dimensional representations: the trivial one and χ given by $\chi(E) = \chi(F) = 0$, $\chi(K) = -1$.

A similar statement holds for any simple Lie algebra $\mathfrak{g}$. In this case, there are 2^r 1-dimensional representations, which map E_i and F_i to zero and K_i to signs ε_i.

Representations V_λ are called *representations of type I* (they correspond to the case when $\varepsilon_i = 1$ for all i).

EXERCISE 5.8.2. (i) Prove that for $U_q(\mathfrak{sl}_2)$, M_λ has a basis $F^i v_\lambda$, $i \geq 0$. Deduce from this the Poincare-Birkhoff-Witt theorem for $U_q(\mathfrak{sl}_2)$: the elements $F^i K^j E^r$, $i, r \geq 0$, $j \in \mathbb{Z}$, form a basis of $U_q(\mathfrak{sl}_2)$.

(ii) Find the center of $U_q(\mathfrak{sl}_2)$ when q is not a root of unity, and compute the action of the center on the representations V_λ.

Let us now describe the irreducible representations of the small quantum group $u_q(\mathfrak{sl}_2)$. It is easy to see that the representations V_λ for $\lambda = 0, 1, \ldots, \ell-1$ of $U_q(\mathfrak{sl}_2)$

are defined for q being an ℓ-th root of 1, factor through $u_q(\mathfrak{sl}_2)$ and are irreducible. Also, these representations are $\mathbb{Z}$-graded.

The fusion rules for V_λ can therefore be easily derived by looking at the $\mathbb{Z}$-grading. Namely, we have the usual Clebsch-Gordan rule (5.16), where if $m \geq \ell$ then $V_m := 2V_{m-\ell} + V_{2\ell-m-2}$. In particular, we have

$$V_1 V_i = V_{i+1} + V_{i-1}$$

for any $i \leq \ell - 2$, and

$$V_1 V_{\ell-1} = 2 + 2V_{\ell-2}.$$

For example, if $\ell = 3$, we obtain (using that $V_0 = 1$):

$$V_1^2 = 1 + V_2,\ V_1 V_2 = 2 + 2V_1,\ V_2^2 = 2 + 2V_1 + V_2.$$

Note that this is only the multiplication law in the Grothendieck ring; the underlying tensor category $\mathsf{Rep}\, u_q(\mathfrak{sl}_2)$ is not semisimple. Also, the Grothendieck ring of this category is a transitive unital $\mathbb{Z}_+$-ring with an involution, but not a based (or even a weak based) ring: the product $V_1 V_{\ell-1}$ involves two copies of 1, even though $V_{\ell-1} \neq V_1^*$ (note that in the tensor product $V_1 \otimes V_{\ell-1}$, the corresponding two copies of $\mathbf{1}$ are necessarily in the middle of the composition series).

If ℓ is a prime, then we can reduce modulo ℓ, and obtain the representation category of $U^{\mathrm{res}}(\mathfrak{sl}_2)$ in characteristic ℓ. This category has the same description of simple objects and the same fusion rules.

EXERCISE 5.8.3. Let ℓ be a positive integer.
(i) Show that $V_0, ..., V_{\ell-1}$ are the only irreducible representations of $u_q(\mathfrak{sl}_2)$.
(ii) Prove the tensor product rule for V_λ.
(iii) Do (i),(ii) for the Lie algebra $\mathfrak{sl}_2$ in characteristic ℓ and show that the answer is the same.

EXERCISE 5.8.4. Let q be a primitive root of unity of odd order $\ell \geq 3$.
(i) Show that E^ℓ, F^ℓ, K^ℓ are central elements of $U_q(\mathfrak{sl}_2)$.
(ii) Classify irreducible representations of $U_q(\mathfrak{sl}_2)$.
(iii) Describe the Grothendieck ring of $U_q(\mathfrak{sl}_2)$.
(iv) Do (i)-(iii) for $\mathfrak{sl}_2$ in characteristic ℓ.

Now let G be the simply connected algebraic group with Lie algebra $\mathfrak{g}$, and consider the semisimple tensor category $\mathcal{C}_q(G)$ spanned by the representations V_λ. We have a fiber functor $F : \mathcal{C}_q(G) \to \mathsf{Vec}$. Let $O_q(G) := \mathsf{Coend}(F)$. Then $O_q(G) = \oplus_\lambda V_\lambda \otimes V_\lambda^*$ as a vector space, and it is a Hopf algebra. More specifically, it is easy to see that $O_q(G) = U_q(\mathfrak{g})^*_{\mathcal{C}_q(G)}$, the algebra of linear functionals on $U_q(\mathfrak{g})$ that generate representations from $\mathcal{C}_q(G)$ under both left and right translations. Note that for $q = 1$ (i.e., when we work with the category $\mathcal{C}_1(G) := \mathsf{Rep}(G)$), the algebra $O_q(G)$ coincides with the usual Hopf algebra $O(G)$ of regular functions on G.

DEFINITION 5.8.5. The algebra $O_q(G)$ is called the *quantum function algebra* attached to G.

EXERCISE 5.8.6. Give an explicit description of the operations in the Hopf algebra $O_q(G) = \oplus_\lambda V_\lambda \otimes V_\lambda^*$ (choose bases in all V_λ, consider the corresponding basis in $O_q(G)$, and write the operations explicitly in this basis).

REMARK 5.8.7. Let us explain in more detail the motivation for the term "quantum Frobenius" for the map F_q from Section 5.6. The (finite) dual of F_q is a Hopf algebra map $F_q^* : O(SL_2) \to O_q(SL_2)$. Now assume that the order ℓ of q is a prime. Then, reducing modulo ℓ, we get a Hopf algebra map $F^* : O(SL_2)_\ell \to O(SL_2)_\ell$ in characteristic ℓ (since the reduction of q modulo ℓ is 1). One can show that F^* is the pullback for the usual Frobenius map $F : SL_2 \to SL_2$ in characteristic ℓ, which simply raises all the coordinates to the ℓ-th power.

5.9. Absence of primitive elements

We have seen that many interesting finite dimensional Hopf algebras contain nontrivial (g, h)-skew-primitive elements with $g \neq h$.

However, it follows from Theorem 4.4.1 that this is no longer the case for primitive elements.

COROLLARY 5.9.1. *A finite dimensional bialgebra H over a field of characteristic zero cannot contain nonzero primitive elements.*

PROOF. Apply Theorem 4.4.1 to the category $H-$comod and use Proposition 1.9.12. □

REMARK 5.9.2. Here is a "linear algebra" proof of this corollary. Let x be a nonzero primitive element of H. Then we have a family of grouplike elements $e^{stx} \in H((t))$, $s \in \Bbbk$, i.e., an infinite collection of characters of $H^*((t))$, which is impossible, as H is finite dimensional.

COROLLARY 5.9.3. *If H is a finite dimensional* commutative *Hopf algebra over an algebraically closed field $\Bbbk$ of characteristic 0, then $H = \mathsf{Fun}(G, k)$ for a unique finite group G.*

PROOF. Let $G = \mathrm{Spec}(H)$ (a finite group scheme), and $x \in T_1G = (\mathfrak{m}/\mathfrak{m}^2)^*$ where $\mathfrak{m}$ is the kernel of the counit. Then x is a linear function on $\mathfrak{m}$. Extend it to H by setting $x(1) = 0$. Then x s a derivation:

$$x(fg) = x(f)g(1) + f(1)x(g).$$

This implies that x is a primitive element in H^*. So by Corollary 5.9.1, $x = 0$. This implies that G is reduced at the point 1. By using translations, we see that G is reduced at all other points. So G is a finite group, and we are done. □

5.10. The Cartier-Gabriel-Kostant theorem

PROPOSITION 5.10.1. *If H is a Hopf algebra over a field of characteristic zero, then the natural map $\xi : U(\mathsf{Prim}(H)) \to H$ is injective.*

PROOF. By Proposition 1.13.8, it suffices to check the injectivity of ξ in degree 1 of the coradical filtration. Thus, it is enough to check that ξ is injective on primitive elements of $U(\mathsf{Prim}(H))$. But by Proposition 5.5.4, all of them lie in $\mathsf{Prim}(H)$, as desired. □

THEOREM 5.10.2. *Any cocommutative Hopf algebra H over an algebraically closed field of characteristic zero is of the form $\Bbbk G \ltimes U(\mathfrak{g})$, where $\mathfrak{g}$ is a Lie algebra, and G is a group acting on $\mathfrak{g}$.*

PROOF. Let G be the group of grouplike elements of H. Since H is cocommutative, it is pointed (see Definition 5.11.3), and $\mathsf{Ext}^1(g,h) = 0$ if $g, h \in G$, $g \neq h$. Hence the category $\mathcal{C} = H-$ comod splits into a direct sum of blocks $\mathcal{C} = \oplus_{g\in G}\mathcal{C}_g$, where $\mathcal{C}_g$ is the category of objects of $\mathcal{C}$ which have a filtration with successive quotients isomorphic to g. So $H = \oplus_{g\in G} H_g$, where $\mathcal{C}_g = H_g -$ comod, and $H_g = gH_1$. Moreover, $A = H_1$ is a Hopf algebra, and we have an action of G on A by Hopf algebra automorphisms.

Now let $\mathfrak{g} = \mathsf{Prim}(A) = \mathsf{Prim}(H)$. This is a Lie algebra, and the group G acts on it (by conjugation) by Lie algebra automorphisms. So we need just to show that the natural homomorphism $\psi : U(\mathfrak{g}) \to A$ is actually an isomorphism.

It is clear that any morphism of coalgebras preserves the coradical filtration, so we can pass to the associated graded morphism $\psi_0 : S\mathfrak{g} \to A_0$, where $A_0 = \mathrm{gr}(A)$. It is enough to check that ψ_0 is an isomorphism.

The morphism ψ_0 is an isomorphism in degrees 0 and 1, and by Proposition 5.10.1, it is injective. So we only need to show surjectivity.

We prove the surjectivity in each degree n by induction. To simplify notation, let us identify $S\mathfrak{g}$ with its image under ψ_0. Suppose that the surjectivity is known in all degrees below n. Let z be a homogeneous element in A_0 of degree n. Then it is easy to see from the counit axiom that

$$\Delta(z) - z \otimes 1 - 1 \otimes z = u \tag{5.17}$$

where $u \in S\mathfrak{g} \otimes S\mathfrak{g}$ is a symmetric element (as Δ is cocommutative).

Equation 5.17 implies that the element u satisfies the equation

$$(\Delta \otimes \mathsf{id})(u) + u \otimes 1 = (\mathsf{id} \otimes \Delta)(u) + 1 \otimes u. \tag{5.18}$$

LEMMA 5.10.3. *Let V be a vector space over a field $\Bbbk$ of characteristic zero. Let $u \in SV \otimes SV$ be a symmetric element satisfying equation (5.18). Then $u = \Delta(w) - w \otimes 1 - 1 \otimes w$ for some $w \in SV$.*

PROOF. Clearly, we may assume that V is finite dimensional. Regard u as a polynomial function on $V^* \times V^*$; our job is to show that

$$u(x, y) = w(x + y) - w(x) - w(y)$$

for some polynomial w.

If we regard u as a polynomial, equation (5.18) takes the form of the 2-cocycle condition

$$u(x + y, t) + u(x, y) = u(x, y + t) + u(y, t).$$

Thus u defines a group law on $U := V^* \oplus \Bbbk$, given by

$$(x, a) + (y, b) = (x + y, a + b + u(x, y)).$$

Clearly, we may assume that u is homogeneous, of some degree $d \neq 1$. Since u is symmetric, the group U is abelian. So in U we have

$$((x, 0) + (x, 0)) + ((y, 0) + (y, 0)) = ((x, 0) + (y, 0)) + ((x, 0) + (y, 0))$$

Computing the second component of both sides, we get

$$u(x, x) + u(y, y) + 2^d u(x, y) = 2u(x, y) + u(x + y, x + y).$$

So one can take $w(x) = (2^d - 2)^{-1} u(x, x)$, as desired. □

Now, applying Lemma 5.10.3, we get that there exists $w \in A_0$ such that $z - w$ is a primitive element, which implies that $z - w \in A_0$, so $z \in A_0$. □

REMARK 5.10.4. The Cartier-Gabriel-Kostant theorem implies that any cocommutative Hopf algebra over an algebraically closed field of characteristic zero in which the only grouplike element is 1 is of the form $U(\mathfrak{g})$, where $\mathfrak{g}$ is a Lie algebra (the Milnor-Moore theorem, [**MiM**]), in particular it is generated by primitive elements. The latter statement is false in positive characteristic. Namely, consider the commutative Hopf algebra $\mathbb{Q}[x, z]$ where x, z are primitive, and set $y = z + x^p/p$, where p is a prime. Then

$$\Delta(y) = y \otimes 1 + 1 \otimes y + \sum_{i=1}^{p-1} \frac{1}{p}\binom{p}{i} x^i \otimes x^{p-i}. \tag{5.19}$$

Since the numbers $\frac{1}{p}\binom{p}{i}$ are integers, this formula (together with $\Delta(x) = x\otimes 1+1\otimes x$, $S(x) = -x$, $S(y) = -y$) defines a Hopf algebra structure on $H = \Bbbk[x, y]$ for any field $\Bbbk$, in particular, one of characteristic p. But if $\Bbbk$ has characteristic p, then it is easy to see that H is not generated by primitive elements (namely, the element y is not in the subalgebra generated by them).

The Cartier-Gabriel-Kostant theorem implies that any affine pro-algebraic group scheme over a field of characteristic zero is in fact a pro-algebraic group. Namely, we have

COROLLARY 5.10.5. *Let H be a commutative Hopf algebra over a field $\Bbbk$ of characteristic zero. Then H has no nonzero nilpotent elements.*

PROOF. It is clear that H is a union of finitely generated Hopf subalgebras (generated by finite dimensional subcoalgebras of H), so we may assume that H is finitely generated. Let $\mathfrak{m}$ be the kernel of the counit of H, and $B = \cup_{n=1}^{\infty}(H/\mathfrak{m}^n)^*$ (i.e., B is the continuous dual of the formal completion of H near the ideal $\mathfrak{m}$). It is easy to see that B is a cocommutative Hopf algebra, and its only grouplike element is 1. So by the Cartier-Gabriel-Kostant theorem $B = U(\mathfrak{g})$, where $\mathfrak{g} = (\mathfrak{m}/\mathfrak{m}^2)^*$. This implies that $G = \mathrm{Spec}(H)$ is smooth at $1 \in G$, i.e., it is an algebraic group, as desired. □

REMARK 5.10.6. Note that Corollary 5.10.5 is a generalization of Corollary 5.9.1.

5.11. Pointed tensor categories and Hopf algebras

DEFINITION 5.11.1. A tensor category $\mathcal{C}$ is *pointed* if every simple object of $\mathcal{C}$ is invertible.

EXAMPLE 5.11.2. The category Vec_G^{ω} is a pointed tensor category. If G is a p-group and $\Bbbk$ has characteristic p, then $\mathrm{Rep}_{\Bbbk}(G)$ is pointed.

DEFINITION 5.11.3. A Hopf algebra H is *pointed* if it is pointed as a coalgebra.

Note that a Hopf algebra H is pointed if and only if the tensor category of finite dimensional H-comodules is pointed.

EXERCISE 5.11.4. Show that any cocommutative Hopf algebra, the Taft and Nichols Hopf algebras, as well as the quantum groups $U_q(\mathfrak{g})$ are pointed Hopf algebras.

REMARK 5.11.5. It follows from Section 4.12 that in a pointed Hopf algebra H the coradical filtration is a Hopf algebra filtration, i.e., $H_iH_j \subset H_{i+j}$ and $S(H_i) = H_i$, so $\mathrm{gr}(H)$ (the associated graded algebra of H under the coradical filtration) is a Hopf algebra.

In this situation, the Hopf algebra H is said to be a *lifting* of the coradically graded Hopf algebra $\mathrm{gr}(H)$.

EXAMPLE 5.11.6. The Taft algebra and the Nichols Hopf algebras are coradically graded. The associated graded Hopf algebra of $U_q(\mathfrak{g})$ is the Hopf algebra defined by the same relations as $U_q(\mathfrak{g})$ (see Section 5.7), except that the commutation relation between E_i and F_j is replaced with the condition that E_i and F_j commute (for all i, j). The same applies to the small quantum group $u_q(\mathfrak{sl}_2)$.

It is easy to see that any Hopf algebra generated by grouplike and skew-primitive elements is automatically pointed.

On the other hand, there exist pointed Hopf algebras which are not generated by grouplike and skew-primitive elements. Perhaps the simplest example of such a Hopf algebra is the commutative Hopf algebra H of regular functions on the Heisenberg group

$$\left\{ \begin{pmatrix} 1 & x & z \\ 0 & 1 & y \\ 0 & 0 & 1 \end{pmatrix} \mid x,\, y,\, z \in \mathbb{R} \right\}$$

with values in an algebraically closed field $\Bbbk$ of characteristic 0. It is easy to see that H is the polynomial algebra in generators x, y, z (entries of the matrix), so that x, y are primitive, and

$$\Delta(z) = z \otimes 1 + 1 \otimes z + x \otimes y.$$

Since the only grouplike element in H is 1, and the only skew-primitive elements are x, y, H is not generated by grouplike and skew-primitive elements.

However, one has the following conjecture, due to Andruskiewitsch and Schneider, see [**AndrS**].

CONJECTURE 5.11.7. Any finite dimensional pointed Hopf algebra over a field of characteristic zero is generated in degree 1 of its coradical filtration, i.e., by grouplike and skew-primitive elements.

It is easy to see that it is enough to prove this conjecture for coradically graded Hopf algebras. It holds in all known examples, and has been proved in the case when the group of grouplike elements is abelian ([**Ang2**]), and a number of other special cases.

The reason we discuss this conjecture here is that it is essentially a categorical statement. Let us make the following definition.

DEFINITION 5.11.8. We say that a tensor category $\mathcal{C}$ is *tensor-generated* by a collection of objects X_α if every object of $\mathcal{C}$ is a subquotient of a finite direct sum of tensor products of X_α.

PROPOSITION 5.11.9. *A pointed Hopf algebra H is generated by grouplike and skew-primitive elements if and only if the tensor category $H - \mathrm{comod}$ is tensor-generated by objects of length* 2.

PROOF. This follows from the fact that matrix elements of the tensor product of comodules V, W for H are products of matrix elements of V, W. □

Thus, one may generalize Conjecture 5.11.7 to the following conjecture about tensor categories.

CONJECTURE 5.11.10. Any finite pointed tensor category over a field of characteristic zero is tensor generated by objects of length 2.

As we have seen, this property fails for infinite categories, e.g., for the category of rational representations of the Heisenberg group. In fact, this is very easy to see categorically: the center of the Heisenberg group acts trivially on 2-dimensional representations, but it is not true for a general rational representation.

5.12. Quasi-bialgebras

Let us now discuss reconstruction theory for quasi-fiber functors. This leads to the notions of quasi-bialgebra and quasi-Hopf algebra.

DEFINITION 5.12.1. Let $\mathcal{C}$ be a tensor category over $\Bbbk$, and $(F, J) : \mathcal{C} \to \mathsf{Vec}$ be a quasi-fiber functor. (F, J) is said to be *normalized* if $J_{\mathbf{1}X} = J_{X\mathbf{1}} = \mathsf{id}_{F(X)}$ for all $X \in \mathcal{C}$.

DEFINITION 5.12.2. Two quasi-fiber functors (F, J_1) and (F, J_2) are said to be *twist equivalent* (by the twist $J_1^{-1}J_2$).

Since for a quasi-fiber functor (unlike a fiber functor), the isomorphism J is not required to satisfy any equations, it typically does not carry any valuable structural information, and thus it is more reasonable to classify quasi-fiber functors not up to isomorphism, but rather up to twist equivalence combined with isomorphism.

EXERCISE 5.12.3. Show that any quasi-fiber functor is equivalent to a normalized one.

Now let $\mathcal{C}$ be a finite tensor category over $\Bbbk$, and let (F, J) be a normalized quasi-fiber functor. Let $H = \mathsf{End}\, F$ be the corresponding finite dimensional algebra. Then H has a coproduct Δ and a counit ε defined exactly as in the case of a fiber functor, which are algebra homomorphisms. The only difference is that, in general, Δ is not coassociative, since J does not satisfy the monoidal structure axiom. Rather, there is an invertible element $\Phi \in H^{\otimes 3}$, defined by the commutative diagram

$$\begin{array}{ccc} (F(X)\otimes F(Y))\otimes F(Z) & \xrightarrow{\Phi_{F(X),F(Y),F(Z)}} & F(X)\otimes (F(Y)\otimes F(Z)) \\ {\scriptstyle J_{X,Y}\otimes \mathsf{id}_{F(Z)}}\downarrow & & \downarrow{\scriptstyle \mathsf{id}_{F(X)}\otimes J_{Y,Z}} \\ F(X\otimes Y)\otimes F(Z) & & F(X)\otimes F(Y\otimes Z) \\ {\scriptstyle J_{X\otimes Y,Z}}\downarrow & & \downarrow{\scriptstyle J_{X,Y\otimes Z}} \\ F((X\otimes Y)\otimes Z) & \xrightarrow{F(a_{X,Y,Z})} & F(X\otimes (Y\otimes Z)) \end{array} \tag{5.20}$$

for all $X, Y, Z \in \mathcal{C}$, and we have the following proposition.

PROPOSITION 5.12.4. *The following identities hold:*

$$(\mathrm{id}\otimes\Delta)(\Delta(h)) = \Phi(\Delta\otimes\mathrm{id})(\Delta(h))\Phi^{-1}, \quad h\in H, \tag{5.21}$$

$$(\mathrm{id}\otimes\mathrm{id}\otimes\Delta)(\Phi)(\Delta\otimes\mathrm{id}\otimes\mathrm{id})(\Phi) = (1\otimes\Phi)(\mathrm{id}\otimes\Delta\otimes\mathrm{id})(\Phi)(\Phi\otimes 1), \tag{5.22}$$

$$(\varepsilon\otimes\mathrm{id})(\Delta(h)) = h = (\mathrm{id}\otimes\varepsilon)(\Delta(h)), \tag{5.23}$$

$$(\mathrm{id}\otimes\varepsilon\otimes\mathrm{id})(\Phi) = 1\otimes 1. \tag{5.24}$$

PROOF. The first identity follows from the definition of Φ, the second one from the pentagon axiom for $\mathcal{C}$, the third one from the condition that (F, J) is normalized, and the fourth one from the triangle axiom and the condition that (F, J) is normalized. □

DEFINITION 5.12.5. An associative unital $\Bbbk$-algebra H equipped with unital algebra homomorphisms $\Delta : H \to H\otimes H$ (the comultiplication or coproduct) and $\varepsilon : H \to \Bbbk$ (the counit) and an invertible element $\Phi\in H^{\otimes 3}$ satisfying the identities of Proposition 5.12.4 is called a *quasi-bialgebra*. The element Φ is called the *associator* of H. A homomorphism of quasi-bialgebras is a unital algebra homomorphism that preserves the coproduct, counit, and associator.

Thus, the notion of a quasi-bialgebra is a generalization of the notion of a bialgebra; namely, a bialgebra is a quasi-bialgebra with $\Phi = 1$.[2]

For a quasi-bialgebra H, the tensor product of (left) H-modules V and W is an H-module via Δ, i.e., in the same way as for bialgebras. Also, it follows from (5.21) that for any H-modules U, V, W the mapping

$$a_{U,V,W} : (U\otimes V)\otimes W \cong U\otimes(V\otimes W) : u\otimes v\otimes w \mapsto \Phi(u\otimes v\otimes w) \tag{5.25}$$

is an H-module isomorphism. Axiom (5.23) implies that the natural maps $l_V = \mathrm{id} : \mathbf{1}\otimes V \xrightarrow{\sim} V$ and $r_V = \mathrm{id} : V\otimes\mathbf{1} \xrightarrow{\sim} V$ are also H-module isomorphisms. Finally, equations (5.22) and (5.24) say, respectively, that the pentagon axiom (2.2) and the triangle axiom (2.10) are satisfied for $\mathbf{Rep}(H)$. In other words, $\mathbf{Rep}(H)$ is a monoidal category.

DEFINITION 5.12.6. A *twist* for a quasi-bialgebra H is an invertible element $J\in H\otimes H$ such that $(\varepsilon\otimes\mathrm{id})(J) = (\mathrm{id}\otimes\varepsilon)(J) = 1$. Given a twist, we can define a new quasi-bialgebra H^J which is H as an algebra, with the same counit, the coproduct given by

$$\Delta^J(x) = J^{-1}\Delta(x)J,$$

and the associator given by

$$\Phi^J = (\mathrm{id}\otimes J)^{-1}(\mathrm{id}\otimes\Delta)(J)^{-1}\Phi(\Delta\otimes\mathrm{id})(J)(J\otimes\mathrm{id})$$

The algebra H^J is called *twist equivalent* to H, by the twist J.

It is easy to see that twist equivalent quasi-fiber functors produce twist-equivalent quasi-bialgebras, and vice versa.

Therefore, we have the following reconstruction theorem.

---

[2]However, note that Δ can be coassociative even if $\Phi\neq 1$.

THEOREM 5.12.7. *The assignments*

$$(5.26) \qquad (\mathcal{C}, F) \mapsto H = \mathsf{End}(F), \quad H \mapsto (\mathsf{Rep}(H), \text{Forget})$$

are mutually inverse bijections between (1) the set of monoidal equivalence classes of finite ring categories $\mathcal{C}$ over $\Bbbk$ with a quasi-fiber functor and (2) the set of equivalence classes of finite dimensional quasi-bialgebras H over $\Bbbk$ up to twist equivalence and isomorphism.

PROOF. Straightforward from the above. □

Also, we have the following proposition.

PROPOSITION 5.12.8. *If a finite ring category $\mathcal{C}$ with left duals*[3] *over $\Bbbk$ admits a quasi-fiber functor, then this functor is unique up to twisting (i.e., changing the quasi-tensor structure).*

PROOF. Let X_i, $i = 1, ..., n$ be the (representatives of isomorphism classes of) simple objects of $\mathcal{C}$. The functor F is exact, so it is determined up to an isomorphism by the numbers $d_i = \dim F(X_i)$. So we need to show that these numbers are uniquely determined by $\mathcal{C}$. This is a consequence of Propositions 4.5.4 and 3.3.6(3), since $d_i = \mathsf{FPdim}(X_i)$. □

EXERCISE 5.12.9. Suppose that in the situation of Exercise 5.2.7, the functor F is equipped with a normalized quasi-tensor structure J, i.e., an isomorphism

$$J : F(-) \otimes F(-) \to F(- \otimes -)$$

such that $J_{\mathbf{1}X} = J_{X\mathbf{1}} = \mathsf{id}_{F(X)}$. Show that this endows H with the structure of a quasi-bialgebra, such that (F, J) defines a monoidal equivalence $\mathcal{C} \to \mathbf{Rep}(H)$.

REMARK 5.12.10. Proposition 5.12.8 is false for infinite categories. For example, it is known that if $\mathcal{C} = \mathsf{Rep}(SL_2(\mathbb{C}))$, and $V \in \mathcal{C}$ is a 2-dimensional representation, then for any positive integer $n \geq 2$ there exists a fiber functor on $\mathcal{C}$ with $\dim F(V) = n$ (see [**Bi**]).

5.13. Quasi-bialgebras with an antipode and quasi-Hopf algebras

Now consider the situation of the previous section, and assume that a finite ring category $\mathcal{C}$ has left duals. In this case, by Proposition 4.2.9, the left dualization functor is exact; it is also faithful by Proposition 2.10.8. Therefore, the functor $F(V^*)^*$ is another quasi-fiber functor on $\mathcal{C}$. So by Proposition 5.12.8, this functor is isomorphic to F. Let us fix such an isomorphism $\xi = (\xi_V)$, $\xi_V : F(V) \to F(V^*)^*$. Then we have natural linear maps $\Bbbk \to F(V) \otimes F(V^*)$, $F(V^*) \otimes F(V) \to \Bbbk$ constructed as in Exercise 2.10.6, which can be regarded as linear maps $\hat{\alpha} : F(V) \to F(V^*)^*$ and $\hat{\beta} : F(V^*)^* \to F(V)$. Thus, the quasi-bialgebra $H = \mathsf{End}(F)$ has the following additional structures.

1. The elements $\alpha, \beta \in H$ such that for any $V \in \mathcal{C}$, $\alpha_V = \xi_V^{-1} \circ \hat{\alpha}_V$, $\beta_V = \hat{\beta}_V \circ \xi_V$.

2. The antipode $S : H \to H$, which is a unital algebra antihomomorphism such that if $\Delta(a) = \sum_i a_i^1 \otimes a_i^2$, $a \in H$, then

$$(5.27) \qquad \sum_i S(a_i^1)\alpha a_i^2 = \varepsilon(a)\alpha, \qquad \sum_i a_i^1 \beta S(a_i^2) = \varepsilon(a)\beta.$$

[3]By Proposition 4.2.10, such a category is always a tensor category (i.e., also admits right duals), but we don't use this here.

Namely, for $a \in H$ the element $S(a)$ acts on $F(V)$ by $\xi^{-1} \circ a^*_{F(V^*)} \circ \xi$.

Let us write the associator as $\Phi = \sum_i \Phi_i^1 \otimes \Phi_i^2 \otimes \Phi_i^3$ and its inverse as $\Phi^{-1} = \sum \bar{\Phi}_i^1 \otimes \bar{\Phi}_i^2 \otimes \bar{\Phi}_i^3$.

PROPOSITION 5.13.1. *One has*

$$\sum \Phi_i^1 \beta S(\Phi_i^2) \alpha \Phi_i^3 = 1, \qquad \sum S(\bar{\Phi}_i^1) \alpha \bar{\Phi}_i^2 \beta S(\bar{\Phi}_i^3) = 1. \tag{5.28}$$

PROOF. This follows directly from the duality axioms. □

DEFINITION 5.13.2. An *antipode* on a quasi-bialgebra H is a triple (S, α, β), where $S : H \to H$ is a unital antihomomorphism and $\alpha, \beta \in H$, satisfying identities (5.27) and (5.28).

A *quasi-Hopf algebra* is a quasi-bialgebra $(H, \Delta, \varepsilon, \Phi)$ for which there exists an antipode (S, α, β) such that S is bijective.

Thus, the notion of a quasi-Hopf algebra is a generalization of the notion of a Hopf algebra; namely, a Hopf algebra is a quasi-Hopf algebra with $\Phi = 1$, $\alpha = \beta = 1$.

We see that if in the above setting $\mathcal{C}$ is rigid (i.e., a tensor category), then $H = \mathsf{End}(F)$ is a finite dimensional quasi-Hopf algebra.

Conversely, if H is a quasi-bialgebra with an antipode (not necessarily finite dimensional), then the category $\mathcal{C} = \mathsf{Rep}(H)$ admits left duals. Indeed, the left dual module of an H-module V is defined as in the Hopf algebra case: it is the dual vector space V^* with the action of H given by

$$\langle h\phi, v \rangle = \langle \phi, S(h)v \rangle, \qquad v \in V, \phi \in V^*, h \in H.$$

Let $\sum v_i \otimes f_i$ be the image of id_V under the canonical isomorphism $\mathsf{End}(V) \xrightarrow{\sim} V \otimes V^*$. Then the evaluation and coevaluation maps are defined using the elements α and β:

$$\mathsf{ev}_V(f \otimes v) = f(\alpha v), \mathsf{coev}_V(1) = \sum \beta v_i \otimes f_i.$$

Axiom (5.27) is then equivalent to ev_V and coev_V being H-module maps. Equations (5.28) are equivalent, respectively, to axioms (2.43) and (2.44) of a left dual.

If S is invertible, then the right dualization functor is an equivalence of categories, so the representation category $\mathsf{Rep}(H)$ of a quasi-Hopf algebra H is rigid, i.e., is a tensor category.

REMARK 5.13.3. Proposition 4.2.10 implies that any finite dimensional quasi-bialgebra with an antipode is a quasi-Hopf algebra (i.e., the antipode is automatically invertible).

EXERCISE 5.13.4. Let $H := (H, \Delta, \varepsilon, \Phi, S, \alpha, \beta)$ be a quasi-bialgebra with an antipode, and $u \in H$ be an invertible element.

(i) Show that if one sets

$$\overline{S}(h) = uS(h)u^{-1}, \qquad \overline{\alpha} = u\alpha, \quad \text{and} \quad \overline{\beta} = \beta u^{-1} \tag{5.29}$$

then the triple $(\overline{S}, \overline{\alpha}, \overline{\beta})$ is an antipode.

(ii) Conversely, show that any $\overline{S}$, $\overline{\alpha}$, and $\overline{\beta}$ satisfying conditions (5.27) and (5.28) are given by formulas (5.29) for a uniquely defined u.

Hint: if H is finite dimensional, (ii) can be formally deduced from the uniqueness of the left dual in a tensor category up to a unique isomorphism. Use this approach to obtain the unique possible formula for u, and check that it does the job for any H.

REMARK 5.13.5. The non-uniqueness of S, α, and β observed in Exercise 5.13.4 reflects the freedom in choosing the isomorphism ξ.

EXAMPLE 5.13.6. (cf. Example 2.10.14) Let G be a finite group and let $\omega \in Z^3(G, \Bbbk^\times)$ be a normalized 3-cocycle, see (2.18). Consider the algebra $H = \mathsf{Fun}(G, \Bbbk)$ of $\Bbbk$-valued functions on G with the usual coproduct and counit. Set

$$\Phi = \sum \omega(f,g,h)p_f \otimes p_g \otimes p_h, \quad \alpha = \sum \omega(g, g^{-1}, g)p_g, \quad \beta = 1,$$

where p_g is the primitive idempotent of H corresponding to $g \in G$. It is straightforward to check that these data define a commutative quasi-Hopf algebra, which we denote $\mathsf{Fun}(G, \Bbbk)_\omega$. The tensor category $\mathsf{Rep}(\mathsf{Fun}(G, \Bbbk)_\omega)$ is obviously equivalent to Vec_G^ω.

It is easy to show that a twist of a quasi-bialgebra H with an antipode is again a quasi-bialgebra with an antipode (this reflects the fact that in the finite dimensional case, the existence of an antipode for H is the property of the category of finite dimensional representations of H). Indeed, if the twist J and its inverse have the form

$$J = \sum_i a_i \otimes b_i, \ J^{-1} = \sum_i a_i' \otimes b_i'$$

then H^J has an antipode (S^J, α^J, β^J) with $S^J = S$ and

$$\alpha^J = \sum_i S(a_i)\alpha b_i, \quad \beta^J = \sum_i a_i' \beta S(b_i').$$

Thus, we have the following reconstruction theorem.

THEOREM 5.13.7. *The assignments*

$$(\mathcal{C}, F) \mapsto H = \mathsf{End}(F), \quad H \mapsto (\mathsf{Rep}(H), \mathrm{Forget})$$

are mutually inverse bijections between tensor equivalence classes of finite tensor categories $\mathcal{C}$ admitting a quasi-fiber functor, and equivalence classes of finite dimensional quasi-Hopf algebras H over $\Bbbk$, up to twist equivalence and isomorphism.

EXERCISE 5.13.8. Generalize Example 4.5.5 to quasi-Hopf algebras and show that Frobenius-Perron dimensions in $\mathsf{Rep}(H)$, where H is a quasi-Hopf algebra, coincide with vector space dimensions.

REMARK 5.13.9. One can define the dual notions of a coquasi-bialgebra and coquasi-Hopf algebra, and prove the corresponding reconstruction theorems for tensor categories which are not necessarily finite. This is straightforward, but fairly tedious, and we will not do it here.

5.14. Twists for bialgebras and Hopf algebras

Let H be a bialgebra. We can regard it as a quasi-bialgebra with $\Phi = 1$. Let J be a twist for H.

DEFINITION 5.14.1. J is called a bialgebra twist if H^J is a bialgebra, i.e., $\Phi^J = 1$.

Thus, a bialgebra twist for H is an invertible element $J \in H \otimes H$ such that $(\varepsilon \otimes \mathsf{id})(J) = (\mathsf{id} \otimes \varepsilon)(J) = 1$, and J satisfies the *twist equation*

$$(\mathsf{id} \otimes \Delta)(J)(\mathsf{id} \otimes J) = (\Delta \otimes \mathsf{id})(J)(J \otimes \mathsf{id}). \tag{5.30}$$

EXERCISE 5.14.2. Show that if a bialgebra H has an antipode S, and J is a bialgebra twist for H, then the bialgebra H^J also has an antipode. Namely, let $J = \sum a_i \otimes b_i$, $J^{-1} = \sum a'_i \otimes b'_i$, and set $Q_J = \sum_i S(a_i)b_i$. Then Q_J is invertible with $Q_J^{-1} = \sum_i a'_i S(b'_i)$, and the antipode of H^J is defined by $S^J(x) = Q_J^{-1} S(x) Q_J$. In particular, if H is a Hopf algebra then so is H^J.

REMARK 5.14.3. Twisting does not change the category of H-modules as a monoidal category, and the existence of an antipode (for finite dimensional H) is a categorical property (existence of left duals). This yields the above formulas, and then one easily checks that they work for any H (not necessarily finite dimensional).

Any twist on a bialgebra H defines a fiber functor (Forget, J) on the category $\mathsf{Rep}(H)$. However, two different twists J_1, J_2 may define isomorphic fiber functors. It is easy to see that this happens if and only if there is an invertible element $v \in H$ such that

$$J_2 = \Delta(v) J_1 (v^{-1} \otimes v^{-1}).$$

In this case the twists J_1 and J_2 are called *gauge equivalent* by the gauge transformation v, and the bialgebras H^{J_1}, H^{J_2} are isomorphic (by conjugation by v). So, we have the following result.

PROPOSITION 5.14.4. *Let H be a finite dimensional Hopf algebra. Then $J \mapsto$ (Forget, J) is a bijection between:*

(1) *gauge equivalence classes of bialgebra twists for H, and*
(2) *fiber functors on $\mathcal{C} = \mathsf{Rep}(H)$, up to isomorphism.*

PROOF. By Proposition 5.12.8, any fiber functor on $\mathcal{C}$ is isomorphic to the forgetful functor Forget as an additive functor. So any fiber functor, up to an isomorphism, has the form (Forget, J), where J is a bialgebra twist. Now it remains to determine when (Forget, J_1) and (Forget, J_2) are isomorphic. Let v : (Forget, J_1) $\to$ (Forget, J_2) be an isomorphism. Then $v \in H$ is an invertible element, and it defines a gauge transformation mapping J_1 to J_2. $\square$

REMARK 5.14.5. By Proposition 2.6.1(ii), the isomorphism classes of fiber functors on Vec_G are in bijection with $H^2(G, \Bbbk^\times)$. This shows that there may exist non-isomorphic fiber functors on a given finite tensor category $\mathcal{C}$, defining isomorphic Hopf algebras. Indeed, all fiber functors on Vec_G yield the same Hopf algebra $\mathsf{Fun}(G, \Bbbk)$. (These fiber functors are, however, all equivalent to each other by composing them with monoidal autoequivalences of $\mathcal{C}$.)

Note that Vec_G^ω does not admit fiber functors for cohomologically nontrivial ω.

5.15. Bibliographical notes

5.1. Fiber functors were first discussed in Deligne-Milne [**DelM**], Section 3 (in the setting of symmetric categories). In the nonsymmetric situation they were considered by Drinfeld [**Dr3**], p.812, and Lyubashenko, [**Ly1**]. Quasi-fiber functors were essentially first considered by Drinfeld in [**Dr4**].

5.2, 5.3. Bialgebras and Hopf algebras were introduced by H. Hopf in topology. For a review discussing the theory of Hopf algebras and their applications see [**Car2**]. They were studied independently in early 1960s under the name of *ring groups* by Kac in [**KacG1**]. An early classical source on Hopf algebras is the textbook by Sweedler [**Sw**]. For a much more detailed treatment of Hopf algebras,

CHAPTER 6

Finite tensor categories

In this chapter we will undertake a systematic study of finite multitensor and tensor categories which are not necessarily semisimple. There are several interrelated motivations to consider nonsemisimple finite tensor categories:

(1) Representations of finite groups in positive characteristic.
(2) Finite dimensional Hopf algebras, in particular quantum groups $u_q(\mathfrak{g})$ at roots of unity, see Section 5.7.
(3) Logarithmic conformal field theories; they lead to nonsemisimple finite tensor categories, similarly to how rational conformal field theories lead to semisimple ones (see [**Gabe**]).
(4) Fusion categories of zero categorical dimension, see Section 7.21. Their dual tensor categories with respect to module categories (see Section 7.12) may be non-semisimple.

6.1. Properties of projective objects

In this section we will study general properties of finite multitensor and tensor categories.

Let $\mathcal{C}$ be a multitensor category. By Definition 1.8.6, every simple object X has a projective cover $P(X)$. The object $P(X)$ is unique up to a non-unique isomorphism.

In Section 1.8 we introduced the group $K_0(\mathcal{C})$ freely generated by isomorphism classes of indecomposable projective objects of $\mathcal{C}$ and the Cartan matrix C, see Definition 1.8.14. Recall that the entries of C are $[P(X) : Y]$, where X, Y are simple objects of $\mathcal{C}$.

Let $\{X_i\}_{i\in I}$ be the set of isomorphism classes of simple objects of $\mathcal{C}$, and let i^* and ${}^*i$ be such that $X_i^* \cong X_{i^*}$ and ${}^*X_i \cong X_{{}^*i}$. Let $N_{jk}^i = [X_j \otimes X_k : X_i]$ denote the multiplicity of X_i in the Jordan-Hölder series of $X_j \otimes X_k$. Let P_i denote the projective cover of X_i.

Let $\mathsf{Gr}(\mathcal{C})$ be the Grothendieck ring of $\mathcal{C}$, see Definition 4.5.2.

PROPOSITION 6.1.1. *Let $\mathcal{C}$ be a finite multitensor category. Then $K_0(\mathcal{C})$ is a $\mathsf{Gr}(\mathcal{C})$-bimodule.*

PROOF. This follows from the fact that the tensor product of a projective object with any object is projective, Proposition 4.2.12. □

Let us describe this bimodule explicitly.

PROPOSITION 6.1.2. *For any object Z of $\mathcal{C}$,*

$$P_i \otimes Z \cong \bigoplus_{j,k} N_{kj^*}^i [Z : X_j] P_k, \quad Z \otimes P_i \cong \bigoplus_{j,k} N_{{}^*jk}^i [Z : X_j] P_k.$$

PROOF. $\mathsf{Hom}_{\mathcal{C}}(P_i \otimes Z, X_k) = \mathsf{Hom}_{\mathcal{C}}(P_i, X_k \otimes Z^*)$, and the first formula follows from Proposition 4.2.12 and formula (1.7). The second formula is analogous. □

PROPOSITION 6.1.3. *Let P be a projective object in a multitensor category $\mathcal{C}$. Then P^* is also projective. Hence, any projective object in a multitensor category is also injective.*

PROOF. We need to show that the functor $\mathsf{Hom}_{\mathcal{C}}(P^*, -)$ is exact. This functor is isomorphic to $\mathsf{Hom}_{\mathcal{C}}(\mathbf{1}, P \otimes -)$. The functor $P \otimes -$ is exact, and by Proposition 4.2.12, any exact sequence splits after tensoring with P, as an exact sequence consisting of projective objects. □

REMARK 6.1.4. A finite abelian category $\mathcal{A}$ is called a *quasi-Frobenius* category if any projective object of $\mathcal{A}$ is injective, and vice versa. Proposition 6.1.3 says that any multitensor category is a quasi-Frobenius category. It is well known that any object of a quasi-Frobenius category admitting a finite projective resolution is projective (indeed, the last nonzero arrow of this resolution is an embedding of projective (= injective) modules and therefore is an inclusion of a direct summand. Hence the resolution can be replaced by a shorter one, and we can use induction in the length of the resolution). Thus, any quasi-Frobenius category is either semisimple or of infinite homological dimension.

REMARK 6.1.5. Proposition 6.1.3 implies that an indecomposable projective object P has a unique simple subobject, i.e., that the socle of P is simple.

Recall the notion of Frobenius-Perron dimension $\mathsf{FPdim}(X)$ of an object X of a finite tensor category, introduced in Section 4.5.

DEFINITION 6.1.6. For any finite tensor category $\mathcal{C}$ the virtual projective object

$$R_{\mathcal{C}} = \sum_{i \in I} \mathsf{FPdim}(X_i) P_i \in K_0(\mathcal{C}) \otimes_{\mathbb{Z}} \mathbb{C}. \tag{6.1}$$

will be called the *regular object* of $\mathcal{C}$.

DEFINITION 6.1.7. Let $\mathcal{C}$ be a finite tensor category. Then the *Frobenius-Perron* dimension of $\mathcal{C}$ is defined by

$$\mathsf{FPdim}(\mathcal{C}) := \mathsf{FPdim}(R_{\mathcal{C}}) = \sum_{i \in I} \mathsf{FPdim}(X_i) \, \mathsf{FPdim}(P_i). \tag{6.2}$$

REMARK 6.1.8. If $\mathcal{C}$ is semisimple (i.e., a fusion category) then $\mathsf{Gr}(\mathcal{C})$ is a fusion ring, and $\mathsf{FPdim}(\mathcal{C}) = \mathsf{FPdim}(\mathsf{Gr}(\mathcal{C}))$.

EXAMPLE 6.1.9. Let H be a finite dimensional quasi-Hopf algebra. Then $\mathsf{FPdim}(\mathsf{Rep}(H)) = \dim_{\Bbbk}(H)$.

EXERCISE 6.1.10. Let G be a group acting on a finite tensor category $\mathcal{C}$.

(i) Show that $\mathsf{FPdim}(\mathcal{C} \rtimes G) = |G| \, \mathsf{FPdim}(\mathcal{C})$, where $\mathcal{C} \rtimes G$ is the crossed product category introduced in Definition 4.15.5.
(ii) Prove that $\mathsf{FPdim}(\mathcal{C}^G) = |G| \, \mathsf{FPdim}(\mathcal{C})$.

PROPOSITION 6.1.11. *$Z \otimes R_{\mathcal{C}} = R_{\mathcal{C}} \otimes Z = \mathsf{FPdim}(Z) R_{\mathcal{C}}$ for all $Z \in \mathsf{Gr}(\mathcal{C})$, and hence the image of $R_{\mathcal{C}}$ in $\mathsf{Gr}(\mathcal{C}) \otimes_{\mathbb{Z}} \mathbb{C}$ is a regular element.*

PROOF. We have $\sum_i \mathsf{FPdim}(X_i) \dim_k \mathsf{Hom}_{\mathcal{C}}(P_i, Z) = \mathsf{FPdim}(Z)$ for any object Z of $\mathcal{C}$. Hence,

$$\begin{aligned}\sum_i \mathsf{FPdim}(X_i) \dim_{\Bbbk} \mathsf{Hom}_{\mathcal{C}}(P_i \otimes Z, Y)& \\ &= \sum_i \mathsf{FPdim}(X_i) \dim_{\Bbbk} \mathsf{Hom}_{\mathcal{C}}(P_i, Y \otimes Z^*) \\ &= \mathsf{FPdim}(Y \otimes Z^*) \\ &= \mathsf{FPdim}(Y)\,\mathsf{FPdim}(Z^*) \\ &= \mathsf{FPdim}(Y)\,\mathsf{FPdim}(Z) \\ &= \mathsf{FPdim}(Z) \sum_i \mathsf{FPdim}(X_i) \dim_{\Bbbk} \mathsf{Hom}_{\mathcal{C}}(P_i, Y).\end{aligned}$$

Now, $P_i \otimes Z$ are projective objects by Proposition 4.2.12. Hence, the formal sums $\sum_i \mathsf{FPdim}(X_i) P_i \otimes Z = R_{\mathcal{C}} \otimes Z$ and $\mathsf{FPdim}(Z) \sum_i \mathsf{FPdim}(X_i) P_i = \mathsf{FPdim}(Z) R_{\mathcal{C}}$ are linear combinations of P_j, $j \in I$ with the same coefficients. □

REMARK 6.1.12. We note the following inequality:

$$\mathsf{FPdim}(\mathcal{C}) \geq N\,\mathsf{FPdim}(P), \tag{6.3}$$

where N is the number of simple objects in $\mathcal{C}$, and P is the projective cover of the unit object $\mathbf{1}$. Indeed, for any simple object V the projective object $P(V) \otimes {}^*V$ has a nontrivial homomorphism to $\mathbf{1}$, and hence contains P. So

$$\mathsf{FPdim}(P(V))\,\mathsf{FPdim}(V) \geq \mathsf{FPdim}(P).$$

Adding these inequalities over all simple V, we get (6.3).

DEFINITION 6.1.13. A finite tensor category $\mathcal{C}$ is called *integral* if $\mathsf{Gr}(\mathcal{C})$ is integral in the sense of Definition 3.5.5.

PROPOSITION 6.1.14. *A finite tensor category $\mathcal{C}$ is integral if and only if $\mathcal{C}$ is equivalent to the representation category of a finite dimensional quasi-Hopf algebra.*

PROOF. The "if" part is clear from Proposition 4.5.7, see also Exercise 5.13.8. To prove the "only if" part, it is enough, in view of Theorem 5.13.7, to construct a quasi-fiber functor on $\mathcal{C}$. Define $P = \oplus_i \mathsf{FPdim}(X_i) P_i$, where X_i are the simple objects of $\mathcal{C}$, and P_i are their projective covers. Define $F = \mathsf{Hom}_{\mathcal{C}}(P, -)$. Obviously, F is exact and faithful, $F(\mathbf{1}) \cong \mathbf{1}$, and

$$\dim_{\Bbbk} F(X) = \mathsf{FPdim}(X) \tag{6.4}$$

for all $X \in \mathcal{C}$. Using Proposition 1.11.2, we continue the functors

$$F(- \otimes -) \quad \text{and} \quad F(-) \otimes F(-)$$

to the functors $\mathcal{C} \boxtimes \mathcal{C} \to \mathsf{Vec}$. Both of these functors are exact and take the same values on the simple objects of $\mathcal{C} \boxtimes \mathcal{C}$. Thus these functors are isomorphic and we are done. □

COROLLARY 6.1.15. *The assignment $H \mapsto \mathsf{Rep}(H)$ defines a bijection between finite dimensional quasi-Hopf algebras H over $\Bbbk$ (up to twist equivalence and isomorphism) and integral finite tensor categories over $\Bbbk$ (up to a tensor equivalence).*

THEOREM 6.1.16. *Let $\mathcal{C}$, $\mathcal{D}$ be finite tensor categories, and let $F : \mathcal{C} \to \mathcal{D}$ be a surjective quasi-tensor functor. Then F maps projective objects to projective ones.*

PROOF. The notion of a surjective functor was introduced in Definition 1.8.3. Let X be an object of $\mathcal{C}$. Let us write X as a direct sum of indecomposable objects (recall that by the Krull-Schmidt theorem such a representation is unique). Define the *projectivity defect* $p(X)$ of X to be the sum of Frobenius-Perron dimensions of all the non-projective summands in this sum. It is clear that $p(X \oplus Y) = p(X) + p(Y)$. Also, it follows from Proposition 4.2.12 that $p(X \otimes Y) \leq p(X)p(Y)$.

Let P_i be the indecomposable projective objects in $\mathcal{C}$. Let $P_i \otimes P_j \cong \oplus_k B_{ij}^k P_k$, and let B_i be the matrix with entries B_{ij}^k. Also, let $B = \sum B_i$. Obviously, B has strictly positive entries, and the Frobenius-Perron eigenvalue of B is $\sum_i \mathsf{FPdim}(P_i)$.

Let $F : \mathcal{C} \to \mathcal{D}$ be a surjective quasi-tensor functor between finite tensor categories. Let $p_j = p(F(P_j))$, and $\mathbf{p}$ be the vector with entries p_j. Then we get $p_i p_j \geq \sum_k B_{ij}^k p_k$, so $(\sum_i p_i)\mathbf{p} \geq B\mathbf{p}$. So, either p_i are all zero, or they are all positive, and the norm of B with respect to the norm $|x| = \sum p_i |x_i|$ is at most $\sum p_i$. Since $p_i \leq \mathsf{FPdim}(P_i)$, this implies $p_i = \mathsf{FPdim}(P_i)$ for all i (as the largest eigenvalue of B is $\sum_i \mathsf{FPdim}(P_i)$).

Assume the second option is the case. Then $F(P_i)$ does not contain nonzero projective objects as direct summands, and hence for any projective $P \in \mathcal{C}$, $F(P)$ cannot contain a nonzero projective object as a direct summand. However, let Q be a projective object of $\mathcal{D}$. Then, since F is surjective, there exists an object $X \in \mathcal{C}$ such that Q is a subquotient of $F(X)$. Since any X is a quotient of a projective object, and F is exact, we may assume that $X = P$ is projective. So Q occurs as a subquotient in $F(P)$. As Q is both projective and injective, it is actually a direct summand in $F(P)$. Contradiction.

Thus, $p_i = 0$ and $F(P_i)$ are projective. □

EXERCISE 6.1.17. A *quasi-Hopf subalgebra* of a quasi-Hopf algebra H is a unital subalgebra $K \subset H$ which is closed under the coproduct, counit, and antipode, and such that the associator and its inverse belong to $K^{\otimes 3}$. Show that a quasi-Hopf subalgebra of a semisimple quasi-Hopf algebra is semisimple (over a field of any characteristic).

Hint: Apply Theorem 6.1.16 to the restriction functor and the trivial representation.

6.2. Categorical freeness

Let $\mathcal{C}$, $\mathcal{D}$ be finite tensor categories and let $F : \mathcal{C} \to \mathcal{D}$ be a surjective quasi-tensor functor. Let $R_\mathcal{C}$, $R_\mathcal{D}$ denote the regular objects of $\mathcal{C}$ and $\mathcal{D}$, see Definition 6.1.6.

THEOREM 6.2.1. *We have*

$$F(R_\mathcal{C}) = \frac{\mathsf{FPdim}(\mathcal{C})}{\mathsf{FPdim}(\mathcal{D})} R_\mathcal{D}. \tag{6.5}$$

PROOF. By Theorem 6.1.16, $F(R_\mathcal{C})$ is a virtually projective object. Thus, $F(R_\mathcal{C})$ must be proportional to $R_\mathcal{D}$, since both (when written in the basis P_i) are eigenvectors of a matrix with strictly positive entries with its Frobenius-Perron eigenvalue, see Theorem 3.2.1(ii). (For this matrix we may take the matrix of multiplication by $F(X)$, where X is such that $F(X)$ contains as composition factors all simple objects of $\mathcal{D}$; such a matrix exists by the surjectivity of F). The coefficient is obtained by computing the Frobenius-Perron dimensions of both sides. □

COROLLARY 6.2.2. *In the above situation, one has* $\mathsf{FPdim}(\mathcal{C}) \geq \mathsf{FPdim}(\mathcal{D})$, *and* $\mathsf{FPdim}(\mathcal{D})$ *divides* $\mathsf{FPdim}(\mathcal{C})$ *in the ring of algebraic integers. In fact,*

$$\frac{\mathsf{FPdim}(\mathcal{C})}{\mathsf{FPdim}(\mathcal{D})} = \sum_i \mathsf{FPdim}(X_i) \dim_{\Bbbk} \mathsf{Hom}_{\mathcal{D}}(F(P_i), \mathbf{1}_{\mathcal{D}}), \tag{6.6}$$

where X_i *runs over simple objects of* $\mathcal{C}$.

PROOF. The statement is obtained by comparing the dimensions of $\mathsf{Hom}_{\mathcal{D}}(X, \mathbf{1}_{\mathcal{D}})$ where X is taken to be the left (respectively, right) side of (6.5). □

REMARK 6.2.3. Note that in the case when $\mathcal{C}$ is semisimple (i.e., a fusion category), and thus $\mathrm{Gr}(\mathcal{C})$ is a fusion ring, Corollary 6.2.2 reduces to Proposition 3.3.13(3).

LEMMA 6.2.4. *Let* $I : \mathcal{D} \to \mathcal{C}$ *be the right adjoint to* F. *For any object* $X \in \mathcal{D}$ *one has*

$$\mathsf{FPdim}(I(X)) = \frac{\mathsf{FPdim}(\mathcal{C})}{\mathsf{FPdim}(\mathcal{D})} \mathsf{FPdim}(X).$$

PROOF. We compute:

$$\begin{aligned}
\mathsf{FPdim}(I(X)) &= \sum_{V \in \mathcal{O}(\mathcal{C})} \mathsf{FPdim}(V)[I(X) : V] \\
&= \sum_{V \in \mathcal{O}(\mathcal{C})} \mathsf{FPdim}(V) \dim_{\Bbbk} \mathsf{Hom}_{\mathcal{C}}(P(V), I(X)) \\
&= \dim_{\Bbbk} \mathsf{Hom}_{\mathcal{C}}(R_{\mathcal{C}}, I(X)) \\
&= \frac{\mathsf{FPdim}(\mathcal{C})}{\mathsf{FPdim}(\mathcal{D})} \dim_{\Bbbk} \mathsf{Hom}_{\mathcal{D}}(R_{\mathcal{D}}, X) \\
&= \frac{\mathsf{FPdim}(\mathcal{C})}{\mathsf{FPdim}(\mathcal{D})} \mathsf{FPdim}(X),
\end{aligned}$$

where we used (1.7) and Theorem 6.2.1. Note that $\dim_{\Bbbk} \mathsf{Hom}_{\mathcal{D}}(R_{\mathcal{D}}, X)$ is well defined since $R_{\mathcal{D}}$ is a virtual projective object. □

Suppose now that $\mathcal{C}$ is integral, i.e., by Proposition 6.1.14, it is the representation category of a quasi-Hopf algebra H. In this case, $R_{\mathcal{C}}$ is an honest (not only virtual) projective object of $\mathcal{C}$, namely the free rank 1 module over H. Therefore, multiples of $R_{\mathcal{C}}$ are free H-modules of finite rank, and vice versa.

Then Theorem 6.1.16 and the fact that $F(R_{\mathcal{C}})$ is proportional to $R_{\mathcal{D}}$ implies the following categorical freeness result.

COROLLARY 6.2.5. *If* $\mathcal{C}$ *is integral, and* $F : \mathcal{C} \to \mathcal{D}$ *is a surjective quasi-tensor functor then* $\mathcal{D}$ *is also integral, and the object* $F(R_{\mathcal{C}})$ *is free of rank* $\frac{\mathsf{FPdim}(\mathcal{C})}{\mathsf{FPdim}(\mathcal{D})}$ *(which is an integer).*

PROOF. The Frobenius-Perron dimensions of simple objects of $\mathcal{D}$ are coordinates of the unique eigenvector of the positive integer matrix of multiplication by $F(R_{\mathcal{C}})$ with integer eigenvalue $\mathsf{FPdim}(\mathcal{C})$, normalized so that the component of $\mathbf{1}$ is 1. Thus, all coordinates of this vector are rational numbers, hence integers (because they are algebraic integers). This implies that the category $\mathcal{D}$ is integral. The second statement is clear from the above. □

COROLLARY 6.2.6. *A finite dimensional quasi-Hopf algebra H is a free module over its quasi-Hopf subalgebra B. In particular* $\dim H$ *is divisible by* $\dim B$.

For Hopf algebras, this is the famous theorem of Nichols and Zoeller, [**NicZ**].

COROLLARY 6.2.7. *Let $G := \mathbf{G}(H)$ be the group of grouplike elements of a finite dimensional Hopf algebra H. Then G is linearly independent in H, i.e., $\Bbbk G$ is a Hopf subalgebra of H. Moreover, the order of G divides the dimension of H. In particular, the order of any grouplike element of H divides the dimension of H.*

PROOF. Consider the natural Hopf algebra homomorphism $\phi : \Bbbk G \to H$. Let B be the image of ϕ. Then we have a Hopf algebra inclusion $\phi^* : B^* \to \mathsf{Fun}(G, \Bbbk)$. Thus, B^* is commutative and semisimple, so $B^* = \mathsf{Fun}(K, \Bbbk)$ for some finite group K, and we have a surjective group homomorphism $G \to K$. This homomorphism is in fact the restriction of ϕ, so it must be an isomorphism (as $\phi|_G$ is injective by definition). Thus ϕ is injective, and $B \cong \Bbbk G$. So by Corollary 6.2.6, the order of G divides $\dim H$. The last statement is just Lagrange's theorem for finite groups. □

6.3. Injective and surjective tensor functors

Let $F : \mathcal{C} \to \mathcal{D}$ be a tensor functor from a finite tensor category $\mathcal{C}$ to a tensor category $\mathcal{D}$. By Proposition 1.8.19, the image $\mathsf{Im}F$ is a finite abelian category. Also, $\mathsf{Im}F$ is clearly a tensor category. Thus, $\mathsf{Im}F$ is a finite tensor category which is a full tensor subcategory of $\mathcal{D}$.

The notions of injective and surjective additive functors were introduced in Definition 1.8.3.

PROPOSITION 6.3.1. *Let $F : \mathcal{C} \to \mathcal{D}$ be an injective tensor functor from a finite to a locally finite tensor category. Then F sends simple objects to simple ones.*

PROOF. Let Q be an indecomposable projective object in $\mathsf{Im}F$. Then there exists $X \in \mathcal{C}$ such that $Q = F(X)$. Indeed, we can find X' such that $F(X')$ contains Q as a subquotient. Since Q is both projective and injective by Proposition 6.1.3, Q is a direct summand of $F(X')$. Since F is fully faithful, there exists a direct summand X of X' such that $F(X) = Q$.

Now let $X \in \mathcal{C}$ be such that $F(X) = Q$ is indecomposable projective. Then X is indecomposable projective. Indeed, X is a quotient of some projective object P', so Q is a quotient of $F(P')$. Since Q is projective, it is a direct summand of $F(P')$. Since F is fully faithful, there is an indecomposable summand P'' of P' such that $F(P'') = Q$; clearly, P'' is isomorphic to X.

Thus the indecomposable projectives of $\mathsf{Im}F$ are precisely the images of indecomposable projectives in $\mathcal{C}$ under F (by Theorem 6.1.16 applied to $F : \mathcal{C} \to \mathsf{Im}F$). Let $X_i \in \mathcal{C}$ be the simple objects, and P_i the projective covers of X_i. Then $\dim \mathsf{Hom}(P_j, X_i) = \delta_{ij}$. Hence $\dim \mathsf{Hom}(F(P_j), F(X_i)) = \delta_{ij}$. Thus, the set of the composition factors of $F(X_i)$ contains exactly one simple object. In other words, $F(X_i)$ is simple. □

REMARK 6.3.2. Note that Proposition 6.3.1 is a statement solely about tensor categories, and it fails for non-tensor exact injective functors between finite abelian categories. For example, let $\mathcal{C}$ be the category of finite dimensional modules over the algebra of 2 by 2 matrices, and $\mathcal{D}$ be the category of finite dimensional modules over the subalgebra of upper triangular 2 by 2 matrices. Then the restriction functor

$F : \mathcal{C} \to \mathcal{D}$ is exact and injective, but it maps the simple 2-dimensional module to a non-simple module.

PROPOSITION 6.3.3. *Let $\mathcal{C}$ and $\mathcal{D}$ be finite tensor categories and let $F : \mathcal{C} \to \mathcal{D}$ be an injective tensor functor. Then* $\mathsf{FPdim}(\mathcal{C}) \leq \mathsf{FPdim}(\mathcal{D})$. *The equality is achieved if and only if F is an equivalence.*

PROOF. We may assume that $\mathcal{C}$ is a tensor subcategory of $\mathcal{D}$. Let X be a simple object of $\mathcal{C}$. Then by Proposition 6.3.1, X is also a simple object of $\mathcal{D}$. Let $P_{\mathcal{C}}(X)$, $P_{\mathcal{D}}(X)$ be the projective covers of X in $\mathcal{C}$, $\mathcal{D}$. Then we have projections $a_{\mathcal{C}} : P_{\mathcal{C}}(X) \to X$, $a_{\mathcal{D}} : P_{\mathcal{D}}(X) \to X$ (in $\mathcal{C}$, $\mathcal{D}$, respectively), and there exists a morphism $b : P_{\mathcal{D}}(X) \to P_{\mathcal{C}}(X)$ such that $a_{\mathcal{C}} \circ b = a_{\mathcal{D}}$. We claim that b is onto. Indeed, assume the contrary. Let L be a simple quotient of the cokernel of b, and f a projection $P_{\mathcal{C}}(X) \to L$. It is clear that $L = X$ and f is proportional to $a_{\mathcal{C}}$. But this is a contradiction, since $f \circ b = 0$.

Thus, $\mathsf{FPdim}(P_{\mathcal{D}}(X)) \geq \mathsf{FPdim}(P_{\mathcal{C}}(X))$. This implies the first statement of the proposition.

Let us now prove the second statement. The equality $\mathsf{FPdim}(\mathcal{C}) = \mathsf{FPdim}(\mathcal{D})$ implies that 1) all simple objects of $\mathcal{D}$ are also contained in $\mathcal{C}$, and 2) b is an isomorphism, i.e., $P_{\mathcal{C}}(X) = P_{\mathcal{D}}(X)$ (i.e., $\mathcal{C}$ is a Serre subcategory of $\mathcal{D}$, see Definition 4.14.1). This implies that $\mathcal{C} = \mathcal{D}$. □

PROPOSITION 6.3.4. *Let $\mathcal{C}$ and $\mathcal{D}$ be finite tensor categories and let $F : \mathcal{C} \to \mathcal{D}$ be a surjective tensor functor. Then* $\mathsf{FPdim}(\mathcal{C}) \geq \mathsf{FPdim}(\mathcal{D})$. *The equality is achieved if and only if F is an equivalence.*

PROOF. The first statement has already been proved, so let us prove the second one. Let X_i, P_i, $j = 1, \dots, n$ be the simple and projective objects in $\mathcal{C}$, and X'_j, P'_j, $j = 1, \dots, m$ the simple and projective objects in $\mathcal{D}$. Let d_i, d'_j be the Frobenius-Perron dimensions of X_i, X'_j. We have $F(X_i) = \sum_j a_{ij} X'_j$ (in the Grothendieck group $\mathsf{Gr}(\mathcal{C})$), and $F(P_i) = \oplus_j b_{ji} P'_j$ (in $K_0(\mathcal{C})$). From the first equation we get $d_i = \sum a_{ij} d'_j$, and since $F(R_{\mathcal{C}}) = R_{\mathcal{D}}$, from the second equation we get $\sum b_{ji} d_i = d'_j$, So if $\mathbf{d}$, $\mathbf{d}'$ are the vectors with entries d_i, d'_j, and A, B matrices with entries a_{ij}, b_{ji}, respectively, then $A\mathbf{d}' = \mathbf{d}$, $B\mathbf{d} = \mathbf{d}'$. Thus, $AB\mathbf{d} = \mathbf{d}$.

Now, the functor F is faithful. Hence, $\mathsf{Hom}_{\mathcal{D}}(F(P_i), F(X_i)) \neq 0$. But the dimension of this space is $\sum_j a_{ij} b_{ji} = (AB)_{ii}$. Hence the diagonal entries of AB are ≥ 1. Since $AB\mathbf{d} = \mathbf{d}$, the entries of $\mathbf{d}$ are positive, and the entries of AB are non-negative, we conclude that $AB = 1$.

We will now show that $n \geq m$. This will imply that $BA = 1$. Since $AB = 1$, for any i there exists a unique j such that $a_{ij} b_{ji} \neq 0$; call it $j(i)$. It suffices to show that for any j there exists i such that $j = j(i)$. Assume the contrary, i.e., some $j \neq j(i)$ for any i. Then $a_{ij} b_{jk} = 0$ for all i, k. Choose i so that X'_j is contained as a constituent in $F(X_i)$ (it must exist as F is surjective). Then $a_{ij} \neq 0$, so $b_{jk} = 0$ for all k. This means that P'_j is not a direct summand of $F(P_k)$ for any k, i.e., is not a subquotient of $F(Q)$ for any projective object Q. This is a contradiction with the surjectivity of F.

Thus $AB = 1$, $BA = 1$. This means that A is a permutation matrix, and $B = A^{-1}$. This easily implies that F is an equivalence. □

Now let $F : \mathcal{C} \to \mathcal{D}$ be any tensor functor between finite abelian categories. The functor F is naturally written as a composition of two tensor functors: $F = F_i \circ F_s$, where $F_s : \mathcal{C} \to \mathsf{Im}F$ is surjective, and $F_i : \mathsf{Im}F \to \mathcal{D}$ is injective. Clearly, F is surjective if and only if F_i is an equivalence, and F is injective if and only if F_s is an equivalence.

COROLLARY 6.3.5. *Let $F : \mathcal{C} \to \mathcal{D}$ be a tensor functor between finite tensor categories.*

(i) $\mathsf{FPdim}(\mathsf{Im}F) = \mathsf{FPdim}(\mathcal{C})$ *if and only if F is injective.*
(ii) $\mathsf{FPdim}(\mathsf{Im}F) = \mathsf{FPdim}(\mathcal{D})$ *if and only if F is surjective.*

PROOF. (i) follows from Proposition 6.3.3 and (ii) follows from Proposition 6.3.4. □

COROLLARY 6.3.6. *Suppose that a tensor functor $F : \mathcal{C} \to \mathcal{D}$ between finite tensor categories factors through a finite tensor category $\mathcal{E}$, such that $\mathsf{FPdim}(\mathcal{E}) < \min(\mathsf{FPdim}(\mathcal{C}), \mathsf{FPdim}(\mathcal{D}))$. Then F is neither surjective nor injective.*

PROOF. We have $F = F_1 \circ F_2$, $F_2 : \mathcal{C} \to \mathcal{E}$, $F_1 : \mathcal{E} \to \mathcal{D}$. Clearly, $\mathsf{Im}F$ is a tensor subcategory in $\mathsf{Im}F_1$, so $\mathsf{FPdim}(\mathsf{Im}F) \le \mathsf{FPdim}(\mathsf{Im}F_1) \le \mathsf{FPdim}(\mathcal{E})$. Thus, by Proposition 6.3.5, F is neither surjective nor injective. □

6.4. The distinguished invertible object

Let $\mathcal{C}$ be a finite tensor category with simple objects X_i, $i \in I$. Let P_i denote the projective cover of X_i. Since duals to projective objects are projective, we can define a map $D : I \to I$ such that $P_i^* = P_{D(i)}$. It is clear that $D^2(i) = i^{**}$.

Let 0 be the label for the unit object. Let $\rho = D(0)$. (In other words, ${}^*X_\rho$ is the socle of $P_0 = P(\mathbf{1})$). We have

$$\mathsf{Hom}_{\mathcal{C}}(P_i^*, X_j) = \mathsf{Hom}_{\mathcal{C}}(\mathbf{1}, P_i \otimes X_j) = \mathsf{Hom}_{\mathcal{C}}(\mathbf{1}, \oplus_k N_{kj^*}^i P_k).$$

This space has dimension $N_{\rho j^*}^i$. Thus we get

$$N_{\rho j^*}^i = \delta_{D(i),j}.$$

Let now X_ρ be the corresponding simple object. By Proposition 6.1.2, we have

$$X_\rho^* \otimes P_m \cong \oplus_k N_{\rho k}^m P_k \cong P_{D(m)^*}.$$

LEMMA 6.4.1. *X_ρ is an invertible object.*

PROOF. The last equation implies that the matrix of action of X_{ρ^*} on projectives is a permutation matrix. Hence, the Frobenius-Perron dimension of X_{ρ^*} is 1, hence, X_ρ is invertible by Corollary 3.3.10. □

LEMMA 6.4.2. *One has: $P_{D(i)} = P_{{}^*i} \otimes X_\rho$; $X_{D(i)} = X_{{}^*i} \otimes X_\rho$.*

PROOF. It suffices to prove the first statement. Therefore, our job is to show that $\dim_{\Bbbk} \mathsf{Hom}_{\mathcal{C}}(P_i^*, X_j) = \dim_{\Bbbk} \mathsf{Hom}_{\mathcal{C}}(P_{{}^*i}, X_j \otimes X_{\rho^*})$. The left hand side was computed before, it is $N_{\rho j^*}^i$. On the other hand, the right hand side is $N_{j,\rho^*}^{{}^*i}$ (we use that $\rho^* = {}^*\rho$ for an invertible object ρ). These numbers are equal by the properties of duality, so we are done. □

COROLLARY 6.4.3. *$P_{i^{**}} = X_\rho^* \otimes P_{{}^{**}i} \otimes X_\rho$ and $X_{i^{**}} = X_\rho^* \otimes X_{{}^{**}i} \otimes X_\rho$.*

PROOF. Again, it suffices to prove the first statement. We have

$$P_{i^{**}} = P_i^{**} = (P_{{}^*i} \otimes X_\rho)^* = X_\rho^* \otimes P_{{}^*i}^* = X_\rho^* \otimes P_{{}^{**}i} \otimes X_\rho$$

by Lemma 6.4.2. □

DEFINITION 6.4.4. X_ρ is called the *distinguished invertible object* of $\mathcal{C}$.

We see that for any i, the socle of P_i is $\hat{X}_i := X_\rho^* \otimes^{**} X_i = X_i^{**} \otimes X_\rho^*$. This implies the following result.

COROLLARY 6.4.5. *Any finite dimensional quasi-Hopf algebra H is a Frobenius algebra, i.e., H is isomorphic to H^* as a left H-module.*

PROOF. It is easy to see that a Frobenius algebra is a quasi-Frobenius algebra (i.e., a finite dimensional algebra for which projective and injective modules coincide) in which the socle of every indecomposable projective module has the same dimension as its cosocle (i.e., the simple quotient). As follows from the above, these conditions are satisfied for finite dimensional quasi-Hopf algebras (namely, the first condition follows from Proposition 6.1.3, and the second condition follows from the fact that X_ρ is 1-dimensional). □

REMARK 6.4.6. An alternative characterization of the distinguished object X_ρ will be given in Theorem 7.18.7.

6.5. Integrals in quasi-Hopf algebras and unimodular categories

DEFINITION 6.5.1. A *left integral* in an algebra H with a counit (i.e., a character) $\varepsilon : H \to \Bbbk$ is an element $I \in H$ such that $xI = \varepsilon(x)I$ for all $x \in H$. Similarly, a *right integral* in H is an element $I \in H$ such that $Ix = \varepsilon(x)I$ for all $x \in H$.

Note that any antiautomorphism of H preserving the counit (e.g., an antipode of a quasi-Hopf algebra) maps left integrals to right integrals, and vice versa.

REMARK 6.5.2. Let H be the convolution algebra of compactly supported distributions on a locally compact topological group (e.g., a Lie group) G. This algebra has a counit ε defined by $\varepsilon(\xi) = \xi(1)$. Let dg be a right-invariant Haar measure on G. Then the distribution $I(f) = \int_G f(g)dg$ (which is compactly supported, i.e., lies in H, if and only if G is compact) is a right integral in the sense that $I * \psi = \varepsilon(\psi)I$ for any $\psi \in H$ (it is also a left integral if and only if G is unimodular, which is always true if G is compact). This motivates the terminology.

Note that this example makes sense for a finite group G over any field $\Bbbk$. In this case, $H = \Bbbk G$, and $I = \sum_{g\in G} g$ is both a left and a right integral.

PROPOSITION 6.5.3. *Any finite dimensional quasi-Hopf algebra admits a unique nonzero left integral up to scaling and a unique nonzero right integral up to scaling.*

PROOF. It suffices to prove the statement for left integrals (for right integrals the statement is obtained by applying the antipode). A left integral is the same thing as a homomorphism of left H-modules $\Bbbk \to H$. Since H is Frobenius, this is the same as a homomorphism $\Bbbk \to H^*$, i.e., a homomorphism $H \to \Bbbk$. But such homomorphisms are just multiples of the counit. □

EXAMPLE 6.5.4. Let H be the Taft Hopf algebra of dimension n^2 (Example 5.5.6). Then the element $I_\ell := (1 + g + ... + g^{n-1})x^{n-1}$ is a left integral, and $I_r := x^{n-1}(1 + g + ... + g^{n-1})$ is a right integral (both are nonzero and not proportional to each other).

Note that the space of left integrals of an algebra H with a counit is a right H-module (indeed, if I is a left integral, then so is Iy for all $y \in H$). Thus, for finite dimensional quasi-Hopf algebras, we obtain a character $\alpha : H \to \Bbbk$, such that $Ix = \alpha(x)I$ for all $x \in H$. This character is called *the distinguished character* of H.

PROPOSITION 6.5.5. *Let H be a finite dimensional quasi-Hopf algebra, and $\mathcal{C} = \mathsf{Rep}(H)$. Then X_ρ coincides with the distinguished character α.*

PROOF. Let I be a nonzero left integral in H. We have $xI = \varepsilon(x)I$ and $Ix = \alpha(x)I$. This means that for any $V \in \mathcal{C}$, I defines a morphism from $V \otimes \alpha^{-1}$ to V given by $v \mapsto Iv$.

The element I belongs to the submodule P_i of H, whose socle is the trivial H-module. Thus, $P_i^* = P(\mathbf{1})$, and hence by Lemma 6.4.2, $i = \rho$. Thus, I defines a nonzero (but rank 1) morphism $P_\rho \otimes \alpha^{-1} \to P_\rho$. The image of this morphism, because of rank 1, must be $X_0 = \mathbf{1}$, so $\mathbf{1}$ is a quotient of $P_\rho \otimes \alpha^{-1}$, and hence α is a quotient of P_ρ. Thus, $\alpha = X_\rho$, and we are done. □

PROPOSITION 6.5.6. *The following conditions on a finite dimensional quasi-Hopf algebra H are equivalent:*

(i) *H is semisimple,*
(ii) *$\varepsilon(I) \neq 0$ (where I is a left integral in H),*
(iii) *$I^2 \neq 0$,*
(iv) *I can be normalized to be an idempotent.*

PROOF. (ii) implies (i): If $\varepsilon(I) \neq 0$ then $\Bbbk = \mathbf{1}$ is a direct summand in H as a left H-module. This implies that $\mathbf{1}$ is projective, hence $\mathsf{Rep}(H)$ is semisimple (Corollary 4.2.13).

(i) implies (iv): If H is semisimple, the integral is a multiple of the projector to the trivial representation, so the statement is obvious.

(iv) implies (iii): obvious.

(iii) implies (ii): clear, since $I^2 = \varepsilon(I)I$. □

DEFINITION 6.5.7. A finite tensor category $\mathcal{C}$ is *unimodular* if $X_\rho = \mathbf{1}$. A finite dimensional quasi-Hopf algebra H is unimodular if $\mathsf{Rep}(H)$ is a unimodular category, i.e., if left and right integrals in H coincide.

REMARK 6.5.8. This terminology is motivated by the notion of a unimodular Lie group, which is a Lie group on which a left invariant Haar measure is also right invariant, and vice versa.

REMARK 6.5.9. Every semisimple finite tensor category is automatically unimodular. In particular, if H is a semisimple quasi-Hopf algebra, then $\alpha = \varepsilon$, and any left integral in H is a right integral and vice versa.

EXERCISE 6.5.10.
(i) Let H be the Nichols Hopf algebra of dimension 2^{n+1} (Example 5.5.8). Find the projective covers of simple objects, the distinguished invertible object, and show that H is not unimodular. In particular, Sweedler's finite dimensional Hopf algebra is not unimodular.
(ii) Do the same if H is the Taft Hopf algebra (Example 5.5.6).
(iii) Let $H = u_{\mathsf{q}}(\mathfrak{sl}_2)$ be the small quantum group at a root of unity q of odd order (see Section 5.6). Show that H is unimodular, but H^* is not. Find the distinguished character of H^* (i.e., the distinguished grouplike element of H). What happens for the corresponding graded Hopf algebra $\mathrm{gr}(H)$?

6.6. Degeneracy of the Cartan matrix

Let $\mathcal{C}$ be a finite tensor category. Let X_i, P_i, $i \in I$ be the simple objects in $\mathcal{C}$ and their projective covers. Let $C_{ij} = [P_i : X_j]$ be the entries of the Cartan matrix of $\mathcal{C}$, see Definition 1.8.14.

THEOREM 6.6.1. *Suppose that $\mathcal{C}$ is not semisimple, and admits an isomorphism of additive functors $u : \mathrm{id}_{\mathcal{C}} \to **$. Then the Cartan matrix C is degenerate over the ground field $\Bbbk$.*

PROOF. Let $\dim(V) = \mathsf{Tr}^L_V(u)$ be the dimension function defined by the (left) categorical trace of u. This function is additive on exact sequences, so it is a linear functional on $\mathsf{Gr}(\mathcal{C})$.

On the other hand, the dimension of every projective object P with respect to this function is zero. Indeed, the dimension of P is the composition of maps

$$\mathbf{1} \xrightarrow{\mathsf{ev}_P} P \otimes P^* \xrightarrow{u \otimes \mathsf{id}_{P^*}} P^{**} \otimes P^* \xrightarrow{\mathsf{coev}_{P^*}} \mathbf{1}.$$

If this map is nonzero then $\mathbf{1}$ is a direct summand in $P \otimes P^*$, which is projective. Thus $\mathbf{1}$ is projective, So $\mathcal{C}$ is semisimple by Corollary 4.2.13. Contradiction.

Since the dimension of the unit object $\mathbf{1}$ is not zero, $\mathbf{1}$ is not a linear combination of projective objects in the Grothendieck group tensored with $\Bbbk$. This means that the linear endomorphism of $\mathsf{Gr}(\mathcal{C})$ given by C is not surjective. □

6.7. Bibliographical notes

6.1. For the results in this section, see [**EtO1**].

6.2. The results in this section were proved in [**EtO1**]. Corollary 6.2.5 for the representation category of a finite dimensional Hopf algebra is well known and much used; it is due to Nichols and Zoeller [**NicZ**]. Corollary 6.2.6 is due to Schauenburg [**Schau3**].

6.3. The results in this section are taken from [**EtO1**].

6.4. The distinguished character for finite tensor categories was first introduced and studied in [**EtO1**], and was studied further in [**ENO1**]. For quasi-Hopf algebras, it was studied in by Haussner and Nill [**HaN**]. For Hopf algebras, the distinguished character of H is the same as the distinguished grouplike element of H^*. It was introduced by Radford in [**Ra2**] and is discussed in detail in Hopf algebra textbooks, such as [**Ra5**].

6.5. For the theory of integrals in Hopf algebras see the original paper by Larson and Sweedler [**LaS**], and the books of Montgomery [**Mon**] and Radford [**Ra5**]. The theory of integrals for quasi-Hopf algebras was developed in [**HaN**].

6.6. Theorem 6.6.1 in the Hopf algebra case was proved by Lorenz [**Lor**]. Our proof in the categorical setting is analogous to his and is taken from [**EtO1**].

CHAPTER 7

Module categories

We have seen that the notion of a tensor category categorifies the notion of a ring. In a similar way, the notion of a module category categorifies the notion of a module over a ring. In this section we will develop a systematic theory of module categories over tensor categories. This theory is interesting by itself, but is also crucial for understanding the structure of tensor categories, similarly to how the study of modules is important in understanding the structure of rings.

We will begin with a discussion of module categories over general monoidal categories, and then pass to the $\Bbbk$-linear case.

7.1. The definition of a module category

Let $\mathcal{C} = (\mathcal{C}, \otimes, \mathbf{1}, a, l, r)$ be a monoidal category.

DEFINITION 7.1.1. A *left module category* over $\mathcal{C}$ is a category $\mathcal{M}$ equipped with an *action (or module product) bifunctor* $\otimes : \mathcal{C} \times \mathcal{M} \to \mathcal{M}$ and a natural isomorphism

$$m_{X,Y,M} : (X \otimes Y) \otimes M \xrightarrow{\sim} X \otimes (Y \otimes M), \qquad X, Y \in \mathcal{C},\ M \in \mathcal{M}, \tag{7.1}$$

called *module associativity constraint* such that the functor $M \mapsto \mathbf{1} \otimes M : \mathcal{M} \to \mathcal{M}$ is an autoequivalence, and the *pentagon diagram*:

$$\begin{array}{ccccc} & & ((X \otimes Y) \otimes Z) \otimes M & & \\ & \swarrow^{a_{X,Y,Z} \otimes \mathrm{id}_M} & & \searrow^{m_{X \otimes Y, Z, M}} & \\ (X \otimes (Y \otimes Z)) \otimes M & & & & (X \otimes Y) \otimes (Z \otimes M) \\ \downarrow{\scriptstyle m_{X, Y \otimes Z, M}} & & & & \downarrow{\scriptstyle m_{X,Y,Z \otimes M}} \\ X \otimes ((Y \otimes Z) \otimes M) & & \xrightarrow{\mathrm{id}_X \otimes m_{Y,Z,M}} & & X \otimes (Y \otimes (Z \otimes M)) \end{array} \tag{7.2}$$

is commutative for all objects X, Y, Z in $\mathcal{C}$ and M in $\mathcal{M}$.

Clearly, this definition categorifies the notion of a module over a monoid.

In a similar way one defines a *right* $\mathcal{C}$-module category. Namely, a right $\mathcal{C}$-module category is the same thing as a left $\mathcal{C}^{\mathrm{op}}$-module category. By a *module category* we will always mean a left module category unless otherwise specified.

Similarly to the case of monoidal categories, for any $\mathcal{C}$-module category $\mathcal{M}$, one has a canonical functorial *unit isomorphism of a module category*

$$l_M : \mathbf{1} \otimes M \xrightarrow{\sim} M, \tag{7.3}$$

called the *unit constraint*, and one can give the following equivalent definition of a module category, making this isomorphism a part of the data.

DEFINITION 7.1.2. A *left module category* over $\mathcal{C}$ is a category $\mathcal{M}$ equipped with a bifunctor $\otimes : \mathcal{C} \times \mathcal{M} \to \mathcal{M}$ and natural isomorphisms (7.1) and (7.3) such that the pentagon diagram (7.2) and the *triangle diagram*:

$$(7.4)\qquad \begin{array}{ccc} (X \otimes \mathbf{1}) \otimes M & \xrightarrow{m_{X,\mathbf{1},M}} & X \otimes (\mathbf{1} \otimes M) \\ & {\scriptstyle r_X \otimes \mathrm{id}_M} \searrow \quad \swarrow {\scriptstyle \mathrm{id}_X \otimes l_M} & \\ & X \otimes M & \end{array}$$

commute for all $X \in \mathcal{C}$, $M \in \mathcal{M}$.

We leave it to the reader to establish the equivalence of the two definitions; this is entirely parallel to the case of monoidal categories, cf. Section 2.1.

The following proposition gives an alternative definition of a module category. Let $\mathcal{M}$ be a category. Consider the category $\mathsf{End}(\mathcal{M})$ of endofunctors of $\mathcal{M}$. As we have seen in Example 2.3.12, $\mathsf{End}(\mathcal{M})$ is a monoidal category.

PROPOSITION 7.1.3. *There is a bijective correspondence between structures of a $\mathcal{C}$-module category on $\mathcal{M}$ and monoidal functors $F : \mathcal{C} \to \mathsf{End}(\mathcal{M})$.*

PROOF. Let $F : \mathcal{C} \to \mathsf{End}(\mathcal{M})$ be a monoidal functor with the monoidal structure $J_{X,Y} : F(X) \circ F(Y) \xrightarrow{\sim} F(X \otimes Y)$, see Definition 2.4.1. Set

$$X \otimes M := F(X)(M), \quad X \in \mathcal{C},\ M \in \mathcal{M},$$

and define the associativity constraint m of $\mathcal{M}$ using the monoidal structure of F:

$$m_{X,Y,M} : (X \otimes Y) \otimes M = F(X \otimes Y)(M) \xrightarrow{J_{X,Y}^{-1}} F(X)(F(Y)(M)) = X \otimes (Y \otimes M),$$

for all $X, Y \in \mathcal{C}$, $M \in \mathcal{M}$. Conversely, let $\mathcal{M}$ be a module category over $\mathcal{C}$. Then for any $X \in \mathcal{C}$ we have the functor $M \mapsto X \otimes M$ of left tensor multiplication by X. Thus we have a functor $F : \mathcal{C} \to \mathsf{End}(\mathcal{M})$. Using the associativity isomorphism m of $\mathcal{M}$, one defines a monoidal structure on F:

$$(J_{X,Y})_M : F(X)(F(Y)(M)) = X \otimes (Y \otimes M) \xrightarrow{m_{X,Y,M}^{-1}} (X \otimes Y) \otimes M = F(X \otimes Y)(M).$$

Note that under the above correspondence the hexagon diagram (2.23) for the monoidal structure on F corresponds to the pentagon diagram (7.2). This is because one of the sides of the hexagon (2.23) disappears due to the fact that the category $\mathsf{End}(\mathcal{M})$ is strict, so its associativity constraint is the identity. □

Clearly, Proposition 7.1.3 categorifies the fact in elementary algebra that a module over a ring is the same thing as a representation.

DEFINITION 7.1.4. A *module subcategory* $\mathcal{N}$ of a $\mathcal{C}$-module category $\mathcal{M}$ is a full subcategory $\mathcal{N} \subset \mathcal{M}$ which is closed under the action of $\mathcal{C}$.

REMARK 7.1.5. Let $\mathcal{C}$ be a rigid monoidal category, and let $\mathcal{M}$ be a right $\mathcal{C}$-module category. Let $\mathcal{M}^\vee$ be the category dual to $\mathcal{M}$. Then $\mathcal{M}^\vee$ is a left $\mathcal{C}$-module category with the $\mathcal{C}$-action $\odot : \mathcal{C} \times \mathcal{M}^\vee \to \mathcal{M}^\vee$ given by $X \odot M := M \otimes {}^*X$ and the associativity constraint given by

$$(X \otimes Y) \odot M = M \otimes {}^*(X \otimes Y) \cong M \otimes ({}^*Y \otimes {}^*X) \xrightarrow{\sim} (M \otimes {}^*Y) \otimes {}^*X = X \odot (Y \odot M),$$

for all $X, Y \in \mathcal{C}$, $M \in \mathcal{M}^\vee$, where the middle arrow is the associativity constraint of $\mathcal{M}$. Similarly, if $\mathcal{N}$ is a left $\mathcal{C}$-module category, then $\mathcal{N}^\vee$ is a right $\mathcal{C}$-module category, with the $\mathcal{C}$-action $\odot$ given by $N \odot X := X^* \otimes N$.

Analogously, for a right $\mathcal{C}$-module category $\mathcal{M}$ we can define the left module category ${}^{\vee}\mathcal{M}$, which is dual to $\mathcal{M}$ as a category, with the left action of $\mathcal{C}$ defined by $X \odot M := M \otimes X^*$, and likewise for left module categories. Then we have ${}^{\vee}(\mathcal{M}^{\vee}) \cong ({}^{\vee}\mathcal{M})^{\vee} \cong \mathcal{M}$, but $(\mathcal{M}^{\vee})^{\vee}$ (which is the twist of $\mathcal{M}$ by the tensor autoequivalence $X \mapsto X^{**}$ of $\mathcal{C}$) is, in general, not equivalent to $\mathcal{M}$ as a $\mathcal{C}$-module category.

PROPOSITION 7.1.6. *Let $\mathcal{C}$ be a rigid monoidal category and let $\mathcal{M}$ be a $\mathcal{C}$-module category. There is a canonical isomorphism*

$$\mathsf{Hom}_{\mathcal{M}}(X^* \otimes M, N) \xrightarrow{\sim} \mathsf{Hom}_{\mathcal{M}}(M, X \otimes N) \tag{7.5}$$

natural in $X \in \mathcal{C}$ and $M, N \in \mathcal{M}$.

PROOF. This is completely parallel to Proposition 2.10.8. □

Thus, if $\mathcal{C}$ is rigid, the endofunctor $M \mapsto X^* \otimes M : \mathcal{M} \to \mathcal{M}$ is a left adjoint to $M \mapsto X \otimes M$ and $M \mapsto {}^*X \otimes M$ is a right adjoint to $M \mapsto X \otimes M$.

Let us also define a notion of a bimodule category over a pair of monoidal categories.

DEFINITION 7.1.7. Let $\mathcal{C}, \mathcal{D}$ be monoidal categories. A $(\mathcal{C}, \mathcal{D})$-*bimodule category* is a category $\mathcal{M}$ that has left $\mathcal{C}$-module and right $\mathcal{D}$-module category structures with modules associativity constraints $m_{X,Y,M} : (X \otimes Y) \otimes M \xrightarrow{\sim} X \otimes (Y \otimes M)$ and $n_{M,W,Z} : M \otimes (W \otimes Z) \xrightarrow{\sim} (M \otimes W) \otimes Z$ respectively, compatible by a collection of natural isomorphisms $b_{X,M,Z} : (X \otimes M) \otimes Z \xrightarrow{\sim} X \otimes (M \otimes Z)$ called *middle associativity constraints* such that the diagrams

$$\begin{array}{ccccc}
 & & ((X \otimes Y) \otimes M) \otimes Z & & \\
 & \swarrow^{m_{X,Y,M} \otimes \mathsf{id}_Z} & & \searrow^{b_{X \otimes Y, M, Z}} & \\
(X \otimes (Y \otimes M)) \otimes Z & & & & (X \otimes Y) \otimes (M \otimes Z) \\
\downarrow{\scriptstyle b_{X, Y \otimes M, Z}} & & & & \downarrow{\scriptstyle m_{X,Y,M \otimes Z}} \\
X \otimes ((Y \otimes M) \otimes Z) & & \xrightarrow[\mathsf{id}_X \otimes b_{Y,M,Z}]{} & & X \otimes (Y \otimes (M \otimes Z))
\end{array}$$

and

$$\begin{array}{ccccc}
 & & X \otimes (M \otimes (W \otimes Z)) & & \\
 & \swarrow^{\mathsf{id}_X \otimes n_{M,W,Z}} & & \nwarrow^{b_{X,M,W \otimes Z}} & \\
X \otimes ((M \otimes W) \otimes Z) & & & & (X \otimes M) \otimes (W \otimes Z) \\
\uparrow{\scriptstyle b_{X, M \otimes W, Z}} & & & & \downarrow{\scriptstyle n_{X \otimes M, W, Z}} \\
(X \otimes (M \otimes W)) \otimes Z & & \xleftarrow[b_{X,M,W} \otimes \mathsf{id}_Z]{} & & ((X \otimes M) \otimes W) \otimes Z
\end{array}$$

commute for all $X, Y \in \mathcal{C}$, $Z, W \in \mathcal{D}$, and $M \in \mathcal{M}$.

7.2. Module functors

DEFINITION 7.2.1. Let $\mathcal{M}$ and $\mathcal{N}$ be two module categories over $\mathcal{C}$ with associativity constraints m and n, respectively. A *$\mathcal{C}$-module functor* from $\mathcal{M}$ to $\mathcal{N}$ consists of a functor $F : \mathcal{M} \to \mathcal{N}$ and a natural isomorphism

$$s_{X,M} : F(X \otimes M) \to X \otimes F(M), \quad X \in \mathcal{C},\ M \in \mathcal{M},$$

such that the following diagrams

(7.6)

$$\begin{array}{ccccc} & & F((X\otimes Y)\otimes M) & & \\ & \swarrow^{F(m_{X,Y,M})} & & \searrow^{s_{X\otimes Y,M}} & \\ F(X\otimes(Y\otimes M)) & & & & (X\otimes Y)\otimes F(M) \\ \downarrow{\scriptstyle s_{X,Y\otimes M}} & & & & \downarrow{\scriptstyle n_{X,Y,F(M)}} \\ X\otimes F(Y\otimes M) & & \xrightarrow{\mathrm{id}_X\otimes s_{Y,M}} & & X\otimes(Y\otimes F(M)) \end{array}$$

and

(7.7)

$$\begin{array}{ccc} F(\mathbf{1}\otimes M) & \xrightarrow{s_{\mathbf{1},M}} & \mathbf{1}\otimes F(M) \\ & {\scriptstyle F(l_M)}\searrow \quad \swarrow{\scriptstyle l_{F(M)}} & \\ & F(M) & \end{array}$$

commute for all $X, Y \in \mathcal{C}$ and $M \in \mathcal{M}$.

A $\mathcal{C}$-module *equivalence* $F : \mathcal{M} \to \mathcal{N}$ of $\mathcal{C}$-module categories is a module functor (F, s) from $\mathcal{M}$ to $\mathcal{N}$ such that F is an equivalence of categories.

Clearly, the notion of a $\mathcal{C}$-module functor categorifies that of a homomorphism of modules over a ring.

Let $\mathcal{M}_1$, $\mathcal{M}_2$ be two module categories over a monoidal category $\mathcal{C}$, and let $(F,\ s)$, $(G,\ t)$ be two module functors $\mathcal{M}_1 \to \mathcal{M}_2$.

DEFINITION 7.2.2. A morphism of $\mathcal{C}$-module functors from $(F,\ s)$ to $(G,\ t)$ is a natural transformation ν between F and G such that the following diagram commutes for any $X \in \mathcal{C}$ and $M \in \mathcal{M}$:

(7.8)

$$\begin{array}{ccc} F(X\otimes M) & \xrightarrow{s_{X,M}} & X\otimes F(M) \\ \downarrow{\scriptstyle \nu_{X\otimes M}} & & \downarrow{\scriptstyle \nu_X\otimes \mathrm{id}_M} \\ X\otimes G(M) & \xrightarrow{t_{X,M}} & G(X)\otimes M. \end{array}$$

It is easy to see that $\mathcal{C}$-module functors between $\mathcal{M}_1$ and $\mathcal{M}_2$ with module functor morphisms introduced above form a category.

EXERCISE 7.2.3. Let $\mathcal{M}$, $\mathcal{N}$, $\mathcal{L}$ be $\mathcal{C}$-module categories and let $F_1 : \mathcal{M} \to \mathcal{N}$ and $F_2 : \mathcal{N} \to \mathcal{L}$ be $\mathcal{C}$-module functors. Show that the composition $F_2 \circ F_1$ has a canonical structure of a $\mathcal{C}$-module functor.

REMARK 7.2.4. Similarly to the case of monoidal categories, one can prove an analog of Mac Lane's strictness theorem for module categories stating that positions of brackets are, essentially, immaterial, see Section 2.8. We leave it to the reader to state and prove this theorem. Hence, one can assume without loss of generality that $\mathbf{1} \otimes M = M$ and $l_M = \mathrm{id}_M$ for all $M \in \mathcal{M}$. We will often do so from now on.

REMARK 7.2.5. Note that in the strict case diagram (7.7) reduces to the condition that $s_{\mathbf{1},M} = \mathrm{id}_{F(M)}$.

7.3. Module categories over multitensor categories

We will be interested in module categories over multitensor categories (defined over a field $\Bbbk$), see Definition 4.1.1. In this case, we would like to consider module categories with an additional structure of an abelian category.

Let $\mathcal{C}$ be a multitensor category over $\Bbbk$.

DEFINITION 7.3.1. A *module category* over $\mathcal{C}$ (or $\mathcal{C}$-module category) is a locally finite abelian category $\mathcal{M}$ over $\Bbbk$ which is equipped with a structure of a $\mathcal{C}$-module category, such that the module product bifunctor $\otimes : \mathcal{C} \times \mathcal{M} \to \mathcal{M}$ is bilinear on morphisms and exact in the first variable.

EXERCISE 7.3.2. Show that the module product $\otimes : \mathcal{C} \times \mathcal{M} \to \mathcal{M}$ is always exact in the second variable.

Let $\mathrm{End}_l(\mathcal{M})$ be the category of left exact functors from $\mathcal{M}$ to $\mathcal{M}$. This is an abelian category. (Namely, if $\mathcal{M}$ is the category of finite dimensional comodules over a coalgebra C then $\mathrm{End}_l(\mathcal{M})$ is equivalent to a full subcategory of the category of C-bicomodules, via $F \mapsto F(C)$; note that $F(C)$ is well defined, since F, being left exact, commutes with direct limits, and thus extends to the ind-completion of $\mathcal{M}$).

PROPOSITION 7.3.3. *There is a bijection between structures of a $\mathcal{C}$-module category on $\mathcal{M}$ and tensor functors $F : \mathcal{C} \to \mathrm{End}_l(\mathcal{M})$.*[1]

PROOF. The proof is the same as that of Proposition 7.1.3. □

We will also need to consider module functors between module categories over multitensor categories. Unless otherwise specified, we will consider only *left exact* module functors, referring to them just as "module functors".

There is an obvious construction of the direct sum of module categories.

PROPOSITION 7.3.4. *Let $\mathcal{M}_1, \mathcal{M}_2$ be two module categories over $\mathcal{C}$. Then the category $\mathcal{M} = \mathcal{M}_1 \oplus \mathcal{M}_2$ with module product, associativity constraints, and the unit constraints being sums of those of $\mathcal{M}_1$ and $\mathcal{M}_2$ is a module category over $\mathcal{C}$.*

PROOF. Obvious. □

DEFINITION 7.3.5. The module category $\mathcal{M}$ is called the *direct sum* of the module categories $\mathcal{M}_1$ and $\mathcal{M}_2$.

DEFINITION 7.3.6. We will say that a module category $\mathcal{M}$ over $\mathcal{C}$ is *indecomposable* if it is not equivalent to a nontrivial direct sum of module categories (that is, with $\mathcal{M}_1, \mathcal{M}_2$ nonzero).

---

[1]The category $\mathrm{End}_{\mathcal{C}}(\mathcal{M})$ is not, in general, a multiring category, so we use the term "tensor functor" in a broader sense (meaning an exact monoidal functor). Also, note that for any $X \in \mathcal{C}$, $F(X)$ is dualizable, i.e., has left and right duals of all orders, so we also have that $F : \mathcal{C} \to \mathrm{End}_r(\mathcal{M})$, where $\mathrm{End}_r(\mathcal{M})$ is the category of right exact end functors of $\mathcal{M}$. In other words, $F(X)$ is exact for any $X \in \mathcal{C}$.

7.4. Examples of module categories

The following are some basic examples of module categories.

EXAMPLE 7.4.1. Any (multi)tensor category $\mathcal{C}$ is a module category over itself: the module product is the tensor product of $\mathcal{C}$ and $m = a$ is the associativity constraint of $\mathcal{C}$. This module category can be considered as a categorification of the regular representation of an algebra.

More generally, if $\mathcal{D} \subset \mathcal{C}$ is a (multi)tensor subcategory of $\mathcal{C}$ then $\mathcal{C}$ is a $\mathcal{D}$-module category.

EXAMPLE 7.4.2. Let $\mathcal{C}$ be a multitensor category. Then one considers $\mathcal{M} = \mathcal{C}$ as a module category over $\mathcal{C} \boxtimes \mathcal{C}^{\mathrm{op}}$ via $(X \boxtimes Y) \otimes Z := X \otimes Z \otimes Y$. (This can be extended to the entire category $\mathcal{C} \boxtimes \mathcal{C}^{\mathrm{op}}$ by resolving objects of this category by injective $\boxtimes$-decomposable objects). The associativity and unit constraints for this category are defined using associativity and unit constraints in $\mathcal{C}$. This module category corresponds to the algebra considered as a bimodule over itself.

EXERCISE 7.4.3. Let $\mathcal{C}, \mathcal{D}$ be multitensor categories. A $(\mathcal{C}, \mathcal{D})$-*bimodule category* is a module category over Deligne's tensor product $\mathcal{C} \boxtimes \mathcal{D}^{\mathrm{op}}$.

EXERCISE 7.4.4. Let $\mathcal{M}$ be a $(\mathcal{C}, \mathcal{D})$-bimodule category. Show that the dual category $\mathcal{M}^\vee$ is a $(\mathcal{D}, \mathcal{C})$-bimodule category. (The left and right module category structures on $\mathcal{M}^\vee$ were defined in Remark 7.1.5).

EXAMPLE 7.4.5. Let $\mathcal{C}$ be a multitensor category and let $\mathcal{C} = \bigoplus_{i,j} \mathcal{C}_{ij}$ be its decomposition into components (see Remark 4.3.4). Then obviously $\mathcal{C}_{ij}$ is a $(\mathcal{C}_{ii}, \mathcal{C}_{jj})$-bimodule category.

EXAMPLE 7.4.6. Let us study when the simplest category $\mathcal{M} = \mathsf{Vec}$ is a module category over a tensor category $\mathcal{C}$. Obviously $\mathsf{End}_{\mathcal{C}}(\mathcal{M}) = \mathsf{Vec}$ as a tensor category. Hence by Proposition 7.1.3 the structures of the module category over $\mathcal{C}$ on $\mathcal{M}$ are in a natural bijection with tensor functors $F : \mathcal{C} \to \mathsf{Vec}$, that is, with fiber functors. Thus the theory of module categories can be considered as an extension of the theory of fiber functors.

EXAMPLE 7.4.7. Let $F : \mathcal{C} \to \mathcal{D}$ be a tensor functor. Then $\mathcal{M} = \mathcal{D}$ has a structure of a module category over $\mathcal{C}$ with $X \otimes Y := F(X) \otimes Y$.

EXERCISE 7.4.8. Define the associativity and unit constraints in Example 7.4.7 using the tensor structure of the functor F and verify the axioms.

EXAMPLE 7.4.9. Let G be a finite group and let $L \subset G$ be a subgroup. Since the restriction functor $\mathrm{Res} : \mathsf{Rep}(G) \to \mathsf{Rep}(L)$ is a tensor functor, we conclude from Example 7.4.7 that $\mathsf{Rep}(L)$ is a module category over $\mathcal{C} = \mathsf{Rep}(G)$.

More generally, let $\psi \in Z^2(L, \Bbbk^\times)$ be a 2-cocycle on L. By definition, a *projective representation* of L on a vector space V with the *Schur multiplier* ψ is a map $\rho : L \to GL(V)$ such that $\rho(g) \circ \rho(h) = \psi(g, h)\rho(gh)$ for all $g, h \in L$. Let $\mathsf{Rep}_\psi(L)$ denote the abelian category of projective representations of L with the Schur multiplier ψ. These representations are the same as representations of the *twisted group algebra* $\Bbbk L_\psi$ with multiplication

$$g \cdot h = \psi(g, h)\, gh, \qquad g, h \in L.$$

The usual tensor product and usual associativity and unit constraints endow $\mathsf{Rep}_\psi(L)$ with the structure of a $\mathsf{Rep}(G)$-module category. We will see in Corollary 7.12.20 that all semisimple indecomposable $\mathsf{Rep}(G)$-module categories are of this form.

EXAMPLE 7.4.10. Let $\mathcal{C} = \mathsf{Vec}_G$, where G is a group. In this case, a module category $\mathcal{M}$ over $\mathcal{C}$ is an abelian category $\mathcal{M}$ with a collection of autoequivalences

$$F_g : M \mapsto \delta_g \otimes M : \mathcal{M} \to \mathcal{M},$$

together with a collection of tensor functor isomorphisms

$$\eta_{g,h} : F_g \circ F_h \to F_{gh}, \quad g, h \in G,$$

satisfying the 2-cocycle relation: $\eta_{gh,k} \circ \eta_{gh} = \eta_{g,hk} \circ \eta_{hk}$ as natural isomorphisms $F_g \circ F_h \circ F_k \xrightarrow{\sim} F_{ghk}$ for all $g, h, k \in G$.

Recall from Definition 2.7.1 that such data is called an *action* of G on $\mathcal{M}$. So, module categories over Vec_G is the same thing as abelian categories with an action of G.

Let us describe indecomposable semisimple Vec_G-module categories explicitly. In any such category $\mathcal{M}$ the set of simple objects is a transitive G-set $X = G/L$, where a subgroup $L \subset G$ is determined up to a conjugacy. Let us view the group of functions $\mathsf{Fun}(G/L, \Bbbk^\times)$ as the coinduced module $\mathrm{Coind}_L^G \Bbbk^\times$. The Vec_G-module associativity constraint on $\mathcal{M}$ defines a function

$$\Psi : G \times G \to \mathrm{Coind}_L^G \Bbbk^\times,$$

by $\Psi(x, y, b) = m_{x,y,y^{-1}x^{-1}b}$. It is easy to see that the pentagon axiom (7.2) implies that $\Psi \in Z^2(G, \mathrm{Coind}_L^G \Bbbk^\times)$. Clearly, the equivalence class of Ψ depends only on the cohomology class of M in $H^2(G, \mathrm{Coind}_L^G \Bbbk^\times)$. By Shapiro's Lemma the restriction map

$$Z^2(G, \mathrm{Coind}_L^G \Bbbk^\times) \to Z^2(L, \Bbbk^\times) : \Psi \mapsto \psi$$

induces an isomorphism $H^2(G, \mathrm{Coind}_L^G \Bbbk^\times) \xrightarrow{\sim} H^2(L, \Bbbk^\times)$.

Thus, an indecomposable Vec_G-module category is determined by a pair (L, ψ), where $L \subset G$ is a subgroup and $\psi \in H^2(L, \Bbbk^\times)$. Let $\mathcal{M}(L, \psi)$ denote the corresponding category.

EXERCISE 7.4.11. Show that Vec_G-module categories $\mathcal{M}(L, \psi)$ and $\mathcal{M}(L', \psi')$ are equivalent if and only if there is $g \in G$ such that $L' = gLg^{-1}$ and ψ' is cohomologous to ψ^g in $H^2(L', \Bbbk^\times)$, where $\psi^g(x, y) := \psi(gxg^{-1}, gyg^{-1})$ for all $x, y \in L$. Here we abuse notation and identify ψ and ψ' with cocycles representing them.

REMARK 7.4.12. Note that indecomposable $\mathsf{Rep}(G)$-module categories in Example 7.4.9 and indecomposable Vec_G-module categories in Example 7.4.10 are parametrized by the same set of data. We will see in Section 7.16 below that this is not merely a coincidence.

EXAMPLE 7.4.13. Here is an example which we consider as somewhat pathological with respect to finiteness properties: let $\mathcal{C} = \mathsf{Vec}$ and let $\mathcal{M} = \mathbf{Vec}$ be the category of all (possibly infinite dimensional) vector spaces. Then the usual tensor product and the usual associativity and unit constraints determine the structure of a module category over $\mathcal{C}$ on $\mathcal{M}$.

7.5. Exact module categories over finite tensor categories

Consider the simplest tensor category $\mathcal{C} = \mathsf{Vec}$. Let $\mathcal{M}$ be any locally finite abelian category over $\Bbbk$. Then $\mathcal{M}$ has a unique (up to equivalence) structure of a module category over $\mathcal{C}$. Thus in this case the understanding of all locally finite module categories over $\mathcal{C}$ is equivalent to the understanding of all $\Bbbk$-linear abelian categories. This seems to be too complicated even if we restrict ourselves only to finite categories. Thus in this section we introduce a much smaller class of module categories which is quite manageable (for example, this class admits an explicit classification for many interesting tensor categories $\mathcal{C}$) but on the other hand contains many interesting examples. Here is the main definition:

Definition 7.5.1. Let $\mathcal{C}$ be a multitensor category with enough projective objects. A locally finite module category $\mathcal{M}$ over $\mathcal{C}$ is called *exact* if for any projective object $P \in \mathcal{C}$ and any object $M \in \mathcal{M}$ the object $P \otimes M$ is projective in $\mathcal{M}$.

Exercise 7.5.2. Let $\mathcal{M}$ be an arbitrary module category over $\mathcal{C}$. Show that for any object $X \in \mathcal{C}$ and any projective object $Q \in \mathcal{M}$ the object $X \otimes Q$ is projective in $\mathcal{M}$.

It is immediate from the definition that any semisimple module category is exact (since any object in a semisimple category is projective).

Remark 7.5.3. We will see that the notion of an exact module category may be regarded as the categorical analog of the notion of a projective module in ring theory.

Example 7.5.4. Notice that in the category $\mathcal{C} = \mathsf{Vec}$ the object $\mathbf{1}$ is projective. Therefore for an exact module category $\mathcal{M}$ over $\mathcal{C}$ any object $M = \mathbf{1} \otimes M$ is projective. Hence an abelian category $\mathcal{M}$ considered as a module category over $\mathcal{C}$ is exact if and only if it is semisimple. Thus the exact module categories over Vec are classified by the cardinality of the set of the isomorphism classes of simple objects. More generally, the same argument shows that if $\mathcal{C}$ is semisimple (and hence $\mathbf{1}$ is projective) then any exact module category over $\mathcal{C}$ is semisimple. But the classification of exact module categories over non-semisimple categories $\mathcal{C}$ can be quite nontrivial.

Example 7.5.5. Since in a multitensor category, the tensor product of a projective object with any object is projective, any finite multitensor category $\mathcal{C}$ considered as a module category over itself (see Example 7.4.1) is exact. Also the category $\mathcal{C}$ considered as a module category over $\mathcal{C} \boxtimes \mathcal{C}^{\mathrm{op}}$ (see Example 7.4.2) is exact.

Example 7.5.6. Let $\mathcal{C}$ and $\mathcal{D}$ be finite multitensor categories and let $F : \mathcal{C} \to \mathcal{D}$ be a *surjective* tensor functor. Then the category $\mathcal{D}$ considered as a module category over $\mathcal{C}$ (see Example 7.4.7) is exact by Theorem 6.1.16.

Exercise 7.5.7. Show that the assumption that F is surjective is essential for Example 7.5.6.

7.6. First properties of exact module categories

Let $\mathcal{C}$ be a finite multitensor category.

LEMMA 7.6.1. *Let $\mathcal{M}$ be an exact $\mathcal{C}$-module category. Then the category $\mathcal{M}$ has enough projective objects.*

PROOF. Let P_0 denote the projective cover of the unit object in $\mathcal{C}$. Then the natural map $P_0 \otimes X \to \mathbf{1} \otimes X \simeq X$ is surjective for any $X \in \mathcal{M}$ since $\otimes$ is exact. Also $P_0 \otimes X$ is projective by the definition of an exact module category. □

COROLLARY 7.6.2. *Assume that an exact module category $\mathcal{M}$ over $\mathcal{C}$ has finitely many isomorphism classes of simple objects. Then $\mathcal{M}$ is finite.*

LEMMA 7.6.3. *Let $\mathcal{M}$ be an exact module category over $\mathcal{C}$. Let $P \in \mathcal{C}$ be projective and $X \in \mathcal{M}$. Then $P \otimes X$ is injective.*

PROOF. The functor $\mathsf{Hom}_{\mathcal{M}}(-, P \otimes X)$ is isomorphic to the functor $\mathsf{Hom}_{\mathcal{M}}(P^* \otimes -, X)$. The object P^* is projective by Proposition 6.1.3. Thus for any exact sequence

$$0 \to Y_1 \to Y_2 \to Y_3 \to 0$$

the sequence

$$0 \to P^* \otimes Y_1 \to P^* \otimes Y_2 \to P^* \otimes Y_3 \to 0$$

splits, and hence the functor $\mathsf{Hom}(P^* \otimes -, X)$ is exact. The lemma is proved. □

COROLLARY 7.6.4. *In an exact $\mathcal{C}$-module category any projective object is injective, and vice versa.*

PROOF. Any projective object X of $\mathcal{M}$ is a direct summand of the object of the form $P_0 \otimes X$ and thus is injective. □

REMARK 7.6.5. Corollary 7.6.4 says that any exact module category over a finite multitensor category (in particular, any finite multitensor category itself) is a quasi-Frobenius category, cf. Remark 6.1.4.

Let $\mathcal{M}$ be an exact $\mathcal{C}$-module category. Let $\mathcal{O}(\mathcal{M})$ denote the set of (isomorphism classes of) simple objects in $\mathcal{M}$. Let us introduce the following relation on $\mathcal{O}(\mathcal{M})$: two objects $X, Y \in \mathcal{O}(\mathcal{M})$ are related if Y appears as a subquotient of $L \otimes X$ for some $L \in \mathcal{C}$.

LEMMA 7.6.6. *The relation above is reflexive, symmetric and transitive.*

PROOF. Since $\mathbf{1} \otimes X = X$, we have the reflexivity. Let $X, Y, Z \in \mathcal{O}(\mathcal{M})$ and $L_1, L_2 \in \mathcal{C}$. If Y is a subquotient of $L_1 \otimes X$ and Z is a subquotient of $L_2 \otimes Y$ then Z is a subquotient of $(L_2 \otimes L_1) \otimes X$ (since $\otimes$ is exact), so we get the transitivity. Now assume that Y is a subquotient of $L \otimes X$. Then the projective cover $P(Y)$ of Y is a direct summand of $P_0 \otimes L \otimes X$; hence there exists $S \in \mathcal{C}$ such that $\mathsf{Hom}(S \otimes X, Y) \neq 0$ (for example $S = P_0 \otimes L$). Thus $\mathsf{Hom}_{\mathcal{M}}(X, S^* \otimes Y) = \mathsf{Hom}_{\mathcal{M}}(S \otimes X, Y) \neq 0$ and hence X is a subobject of $S^* \otimes Y$. Consequently, our relation is symmetric. □

Thus our relation is an equivalence relation. Hence $\mathcal{O}(\mathcal{M})$ is partitioned into equivalence classes,

$$\mathcal{O}(\mathcal{M}) = \bigsqcup_{i \in I} \mathcal{O}(\mathcal{M})_i.$$

For an equivalence class $i \in I$ let $\mathcal{M}_i$ denote the full subcategory of $\mathcal{M}$ consisting of objects whose simple subquotients lie in $\mathcal{O}(\mathcal{M})_i$. Clearly, $\mathcal{M}_i$ is a module subcategory of $\mathcal{M}$.

We have the following complete reducibility property for exact module categories.

PROPOSITION 7.6.7. *The module categories $\mathcal{M}_i$ are exact. The category $\mathcal{M}$ is the direct sum of its module subcategories $\mathcal{M}_i$, i.e.,*

$$\mathcal{M} = \bigoplus_{i \in I} \mathcal{M}_i. \tag{7.9}$$

PROOF. For any $X \in \mathcal{O}(\mathcal{M})_i$ its projective cover is a direct summand of $P_0 \otimes X$ and hence lies in the category $\mathcal{M}_i$. Hence the category $\mathcal{M}$ is the direct sum of its subcategories $\mathcal{M}_i$, and $\mathcal{M}_i$ are exact. $\square$

DEFINITION 7.6.8. We will call the indecomposable categories $\mathcal{M}_i$, $i \in I$, in (7.9) the *components* of $\mathcal{M}$.

A crucial property of exact module categories is the following

PROPOSITION 7.6.9. *Let $\mathcal{M}_1$ and $\mathcal{M}_2$ be two module categories over $\mathcal{C}$. Assume that $\mathcal{M}_1$ is exact. Then any additive module functor $F : \mathcal{M}_1 \to \mathcal{M}_2$ is exact.*

PROOF. Let $0 \to X \to Y \to Z \to 0$ be an exact sequence in $\mathcal{M}_1$. Assume that the sequence $0 \to F(X) \to F(Y) \to F(Z) \to 0$ is not exact. Then the sequence $0 \to P \otimes F(X) \to P \otimes F(Y) \to P \otimes F(Z) \to 0$ is also non-exact for any nonzero object $P \in \mathcal{C}$ since the functor $P \otimes -$ is exact and $P \otimes X = 0$ implies $X = 0$. In particular, we can take P to be projective. But then the sequence

$$0 \to P \otimes X \to P \otimes Y \to P \otimes Z \to 0$$

is exact and split and hence the sequence

$$0 \to F(P \otimes X) \to F(P \otimes Y) \to F(P \otimes Z) \to 0$$

is exact and we get a contradiction. $\square$

REMARK 7.6.10. We will see in Proposition 7.9.7 that Proposition 7.6.9 in fact characterizes exact module categories.

7.7. Module categories and $\mathbb{Z}_+$-modules

Recall that for any multitensor category $\mathcal{C}$ its Grothendieck ring $\mathsf{Gr}(\mathcal{C})$ is naturally a $\mathbb{Z}_+$-ring. The notion of a $\mathbb{Z}_+$-module over a $\mathbb{Z}_+$-ring was introduced in Section 3.4.

Let $\mathcal{M}$ be a module category over $\mathcal{C}$. Recall from Definition 1.5.8 that the Grothendieck group $\mathsf{Gr}(\mathcal{M})$ is a free abelian group with the basis given by the isomorphism classes of simple objects. It is easy to see that $\mathsf{Gr}(\mathcal{M})$ is a $\mathbb{Z}_+$-module over $\mathsf{Gr}(\mathcal{C})$. Obviously, the direct sum of module categories corresponds to the direct sum of $\mathbb{Z}_+$-modules.

EXAMPLE 7.7.1. There exists an indecomposable module category $\mathcal{M}$ over $\mathcal{C}$ such that $\mathsf{Gr}(\mathcal{M})$ is not indecomposable over $\mathsf{Gr}(\mathcal{C})$. E.g., take $\mathcal{C} = \mathsf{Vec}$ and $\mathcal{M}$ the category of modules over the algebra of upper-triangular 2×2 matrices.

Note, however, that, as follows immediately from Proposition 7.6.7, for an indecomposable exact module category $\mathcal{M}$ the $\mathbb{Z}_+$-module $\mathsf{Gr}(\mathcal{M})$ is indecomposable over $\mathsf{Gr}(\mathcal{C})$. In fact, even more is true.

PROPOSITION 7.7.2. *Let $\mathcal{M}$ be an indecomposable exact module category over $\mathcal{C}$. Then $\mathsf{Gr}(\mathcal{M})$ is an irreducible $\mathbb{Z}_+$-module over $\mathsf{Gr}(\mathcal{C})$.*

PROOF. By the way of contradiction, let A be a $\mathbb{Z}_+$-submodule of $\mathsf{Gr}(\mathcal{M})$. Let $A' \subset \mathsf{Gr}(\mathcal{M})$ be the subgroup spanned by simple objects of $\mathcal{M}$ not contained in A. We would like to show that A' is a $\mathbb{Z}_+$-submodule of $\mathsf{Gr}(\mathcal{M})$. Then $\mathsf{Gr}(\mathcal{M}) = A \oplus A'$, a contradiction. Let $N' \in A'$ be a simple object. It suffices to show that for each projective object Q of $\mathcal{C}$ the object $Q \otimes N'$ is in A'. Since $\mathcal{M}$ is exact, for every simple $N \in A$ its projective cover $P(N)$ also belongs to A. We have

$$[Q \otimes N' : N] = \dim_{\Bbbk} \mathsf{Hom}_{\mathcal{M}}(P(N), Q \otimes N') = \dim_{\Bbbk} \mathsf{Hom}_{\mathcal{M}}(Q^* \otimes P(N), N') = 0,$$

where we used Equations (1.7) and (7.5). Thus, $Q \otimes N' \in A'$. □

By Proposition 3.4.6, for a given finite multitensor category $\mathcal{C}$ there are only finitely many $\mathbb{Z}_+$-modules over $\mathsf{Gr}(\mathcal{C})$ which are of the form $\mathsf{Gr}(\mathcal{M})$ where $\mathcal{M}$ is an indecomposable exact module category over $\mathcal{C}$.

7.8. Algebras in multitensor categories

Let $\mathcal{C}$ be a multitensor category.

DEFINITION 7.8.1. An *algebra* in $\mathcal{C}$ is a triple (A, m, u), where A is an object of $\mathcal{C}$, and $m : A \otimes A \to A$ and $u : \mathbf{1} \to A$ are morphisms (called multiplication and unit, respectively) such that the following diagrams

(7.10)
$$\begin{array}{ccc} (A \otimes A) \otimes A & \xrightarrow{a_{A,A,A}} & A \otimes (A \otimes A) \\ \downarrow{\scriptstyle m \otimes \mathsf{id}_A} & & \downarrow{\scriptstyle \mathsf{id}_A \otimes m} \\ A \otimes A & & A \otimes A \\ & \searrow^{m} \quad \swarrow^{m} & \\ & A, & \end{array}$$

and

(7.11)
$$\begin{array}{ccc} \mathbf{1} \otimes A & \xrightarrow{l_A} & A \\ \downarrow{\scriptstyle u \otimes \mathsf{id}_A} & & \downarrow{\scriptstyle \mathsf{id}_A} \\ A \otimes A & \xrightarrow{m} & A, \end{array} \qquad \begin{array}{ccc} A \otimes \mathbf{1} & \xrightarrow{r_A} & A \\ \downarrow{\scriptstyle \mathsf{id}_A \otimes u} & & \downarrow{\scriptstyle \mathsf{id}_A} \\ A \otimes A & \xrightarrow{m} & A \end{array}$$

commute. Here, as usual, a, l, r denote the associativity and unit constraints of $\mathcal{C}$.

Axiom (7.10) represents associativity of multiplication m and (7.11) represents the properties of unit with respect to the multiplication. Of course, in the case when $\mathcal{C} = \mathbf{Vec}$, we get the definition of an associative algebra with unit, and in the case $\mathcal{C} = \mathsf{Vec}$ we get the definition of a finite dimensional associative algebra with unit.

REMARK 7.8.2. If $\mathcal{C}$ is not closed under direct limits (e.g., $\mathcal{C}$ is a multitensor category), one can generalize the above definition, allowing A to be an ind-object (i.e., "infinite dimensional"). However, we will mostly deal with algebras honestly in $\mathcal{C}$ (i.e., "finite dimensional"), and will make this assumption unless otherwise specified.

EXAMPLE 7.8.3. The following are simple examples of algebras in multitensor categories.

(1) In any multitensor category $\mathcal{C}$ the unit object $\mathbf{1}$ is an algebra.
(2) The algebra of functions $\mathsf{Fun}(G)$ on a finite group G (with values in the ground field $\Bbbk$) is an algebra in $\mathsf{Rep}(G)$ (here G acts on $\mathsf{Fun}(G)$ by right translations).
(3) An algebra in Vec_G is the same thing as a G-graded algebra. In particular, if L is a subgroup of G then the group algebra $\Bbbk L$ is an algebra in Vec_G.
(4) More generally, let ω be a 3-cocycle on G with values in $\Bbbk^\times$, and ψ be a 2-cochain of H such that $\omega = d_2\psi$. Then one can define the *twisted group algebra* $(kH)_\psi$ in Vec_G^ω, cf. Example 7.4.9, where

$$(kH)_\psi := \bigoplus_{h\in H} \delta_h$$

as an object of Vec_G^ω, and the multiplication $\delta_h \otimes \delta_{h'} \to \delta_{hh'}$ is the operation of multiplication by $\psi(h, h')$. If $\omega = 1$ (i.e., ψ is a 2-cocycle), the twisted group algebra is associative in the usual sense, and is a familiar object from group theory. However, if ω is nontrivial, this algebra is not associative in the usual sense, but is only associative in the tensor category Vec_G^ω.
(5) Let H be a bialgebra. An algebra in the category of H-comodules, or an H-*comodule algebra* A is an algebra which is also an H-comodule such that the multiplication $A \otimes A \to A$ and unit $\Bbbk \to A$ are H-comodule homomorphisms (equivalently, the comodule structure map $A \to H \otimes A$ is an algebra homomorphism). A special class of H-comodule algebras consists of *coideal subalgebras* of H, i.e., subalgebras $K \subset H$ such that $\Delta(K) \subset H \otimes K$, where Δ is the comultiplication of H.
(6) One defines an algebra in $\mathbf{Rep}(H)$, or an H-*module algebra* A as an algebra which is also an H-module such that the multiplication and unit of A are H-module homomorphisms. That is, the structure map

$$H \otimes A \to A : h \otimes a \mapsto h \cdot a$$

satisfies $m(\Delta(h) \cdot (a \otimes b)) = h \cdot (ab)$, where m is the multiplication map in A, and $h \cdot 1 = \varepsilon(h)1$ for all $h \in H$, $a, b \in A$.

EXAMPLE 7.8.4. Let $\mathcal{C}$ be a multitensor category and $X \in \mathcal{C}$. Then the object $A = X \otimes X^*$ has a natural structure of an algebra with unit $u = \mathsf{coev}_X$ and multiplication $m = \mathsf{id}_X \otimes \mathsf{ev}_X \otimes \mathsf{id}_{X^*}$. In particular for $X = \mathbf{1}$ we get a (trivial) structure of an algebra on $A = \mathbf{1}$.

Now we define modules over algebras:

DEFINITION 7.8.5. A *right module* over an algebra (A, m, u) (or, simply, *a right A-module*) in $\mathcal{C}$ is a pair (M, p), where M is an object in $\mathcal{C}$ and $p : M \otimes A \to M$ is a morphism such that the following diagrams

$$\begin{array}{ccc} (M \otimes A) \otimes A & \xrightarrow{a_{M,A,A}} & M \otimes (A \otimes A) \\ \downarrow{\scriptstyle p \otimes \mathsf{id}_A} & & \downarrow{\scriptstyle \mathsf{id}_M \otimes m} \\ M \otimes A & & M \otimes A \\ & {\scriptstyle p} \searrow \quad \swarrow {\scriptstyle p} & \\ & M & \end{array} \tag{7.12}$$

and

$$\begin{array}{ccc} M \otimes \mathbf{1} & \xrightarrow{r_M} & M \\ \downarrow{\scriptstyle \mathsf{id}_M \otimes u} & & \downarrow{\scriptstyle \mathsf{id}_M} \\ M \otimes A & \xrightarrow{p} & M \end{array} \tag{7.13}$$

commute.

We leave it to the reader to define subalgebras, ideals, homomorphisms of algebras and modules, etc. in the categorical setting.

REMARK 7.8.6. (i) One can similarly define the notion of a left A-module. If (M, p) is a right A-module then $({}^*M, q)$ is a left A-module with the structure morphism $q : A \otimes {}^*M \to M$ given by the image of $p : M \otimes A \to M$ under the isomorphism

$$\mathsf{Hom}_{\mathcal{C}}(M \otimes A,\, M) \cong \mathsf{Hom}_{\mathcal{C}}({}^*M,\, {}^*A \otimes {}^*M) \cong \mathsf{Hom}_{\mathcal{C}}(A \otimes {}^*M,\, {}^*M),$$

where the last isomorphism is (2.52). Similarly, if M is a left A-module then M^* is a right A-module.

(ii) Given A-modules M_1, M_2 in $\mathcal{C}$, module homomorphisms between them form a subspace of the vector space $\mathsf{Hom}_{\mathcal{C}}(M_1, M_2)$. We will denote this subspace by $\mathsf{Hom}_A(M_1, M_2)$. It is easy to see that a composition of homomorphisms is a homomorphism. Thus, right A-modules in $\mathcal{C}$ form a category $\mathsf{Mod}_{\mathcal{C}}(A)$.

EXERCISE 7.8.7. Check that $\mathsf{Mod}_{\mathcal{C}}(A)$ is an abelian category.

The following observations relate the categories $\mathsf{Mod}_{\mathcal{C}}(A)$ and module categories.

EXERCISE 7.8.8. Show that for any A-module (M, p) and any $X \in \mathcal{C}$ the object $X \otimes M$ also has a structure of an A-module given by the composition

$$(X \otimes M) \otimes A \xrightarrow{a_{X,M,A}} X \otimes (M \otimes A) \xrightarrow{\mathsf{id}_X \otimes p} X \otimes M.$$

Thus, we have a functor

$$\mathcal{C} \times \mathsf{Mod}_{\mathcal{C}}(A) \to \mathsf{Mod}_{\mathcal{C}}(A). \tag{7.14}$$

EXERCISE 7.8.9. (This exercise defines associativity and unit constraints for the category $\mathsf{Mod}_{\mathcal{C}}(A)$). Show that for any A-module (M, p) and objects $X, Y \in \mathcal{C}$ the associativity morphism $a_{X,Y,M} : (X \otimes Y) \otimes M \to X \otimes (Y \otimes M)$ is an isomorphism of A-modules. Similarly, the unit morphism $\mathbf{1} \otimes M \to M$ is an isomorphism of A-modules.

PROPOSITION 7.8.10. *The category* $\mathsf{Mod}_{\mathcal{C}}(A)$ *together with module product (7.14) and associativity and unit constraints is a left* $\mathcal{C}$*-module category.*

PROOF. The proof amounts to verification of axioms (7.2) and (7.4). This is left to the reader as an exercise. □

EXAMPLE 7.8.11. Let $A = X \otimes X^*$ be the algebra from Example 7.8.4. Then for each $Y \in \mathcal{C}$ the object $M_Y := Y \otimes X^*$ is an A-module via

$$\mathsf{id}_Y \otimes \mathsf{ev}_X \otimes \mathsf{id}_{X^*} : Y \otimes X^* \otimes X \otimes X^* \to Y \otimes X^*.$$

Furthermore the functor $Y \mapsto M_Y$ gives rise to a $\mathcal{C}$-module functor between the regular $\mathcal{C}$-module category $\mathcal{C}$ and $\mathsf{Mod}_{\mathcal{C}}(A)$. We will see below in Example 7.10.2 that this functor is an equivalence.

Note that for any algebra $A \in \mathcal{C}$ we have a $\mathcal{C}$-module functor assigning to $X \in \mathcal{C}$ the "free" A-module, namely, $X \mapsto X \otimes A : \mathcal{C} \to \mathsf{Mod}_{\mathcal{C}}(A)$. We also have the forgetful $\mathcal{C}$-module functor Forg : $\mathsf{Mod}_{\mathcal{C}}(A) \to \mathcal{C}$. The next Proposition shows that these functors are adjoints of each other.

LEMMA 7.8.12. *For any* $X \in \mathcal{C}$ *we have a natural isomorphism*

$$\mathsf{Hom}_A(X \otimes A, M) = \mathsf{Hom}_{\mathcal{C}}(X, \mathit{Forg}(M)). \tag{7.15}$$

PROOF. Define a linear map $\phi : \mathsf{Hom}_A(X \otimes A, M) \to \mathsf{Hom}_{\mathcal{C}}(X, \mathrm{Forg}(M))$ by defining $\phi(f)$ for $f \in \mathsf{Hom}_A(X \otimes A, M)$ as

$$\phi(f) : X \cong X \otimes \mathbf{1} \xrightarrow{\mathsf{id}_X \otimes u} X \otimes A \xrightarrow{f} M.$$

Define a linear map $\psi : \mathsf{Hom}_{\mathcal{C}}(X, \mathrm{Forg}(M)) \to \mathsf{Hom}_A(X \otimes A, M)$ by defining $\psi(g)$ for $g \in \mathsf{Hom}_{\mathcal{C}}(X, \mathrm{Forg}(M))$

$$\psi(g) : X \otimes A \xrightarrow{g \otimes \mathsf{id}_A} M \otimes A \xrightarrow{p} M,$$

where m and u are the multiplication and unit of A and $p : M \otimes A \to M$ is the module structure on M. It is easy to check that ϕ and ψ are inverses of each other. □

REMARK 7.8.13. In what follows we will usually abuse notation and denote $\mathrm{Forg}(M)$ simply by M.

EXERCISE 7.8.14. Show that for any $M \in \mathsf{Mod}_{\mathcal{C}}(A)$ there exists $X \in \mathcal{C}$ and a surjection $X \otimes A \to M$ (e.g., $X = M$ regarded as an object of $\mathcal{C}$).

EXERCISE 7.8.15. Assume that the category $\mathcal{C}$ has enough projective objects. Then the category $\mathsf{Mod}_{\mathcal{C}}(A)$ has enough projective objects.

EXERCISE 7.8.16. Assume that the category $\mathcal{C}$ is finite. Then the category $\mathsf{Mod}_{\mathcal{C}}(A)$ is finite.

Thus we get a general construction of module categories from algebras in the category $\mathcal{C}$. Not any module category over $\mathcal{C}$ is of the form $\mathsf{Mod}_{\mathcal{C}}(A)$: for $\mathcal{C} = \mathsf{Vec}$ the module category of all (possibly infinite dimensional) vector spaces (see Example 7.4.13) is not of this form. But note that for $\mathcal{C} = \mathsf{Vec}$ any finite module category is of the form $\mathsf{Mod}_{\mathcal{C}}(A)$ (just because every finite abelian category over $\Bbbk$ is equivalent to $\mathsf{Mod}(A)$ for some finite dimensional $\Bbbk$-algebra A). We will show later that all finite module categories over a finite $\mathcal{C}$ are of the form $\mathsf{Mod}_{\mathcal{C}}(A)$ for a suitable A. But of course different algebras A can give rise to the same module categories.

DEFINITION 7.8.17. We say that two algebras A and B in $\mathcal{C}$ are *Morita equivalent* if the module categories $\mathsf{Mod}_{\mathcal{C}}(A)$ and $\mathsf{Mod}_{\mathcal{C}}(B)$ are equivalent $\mathcal{C}$-module categories.

Note that in the case $\mathcal{C} = \mathsf{Vec}$ this definition specializes to the usual notion of Morita equivalence of finite dimensional algebras.

EXAMPLE 7.8.18. We will see later in Example 7.10.2 that all the algebras from Example 7.8.4 are Morita equivalent (cf. Example 7.8.11); moreover any algebra which is Morita equivalent to $A = \mathbf{1}$ is of the form $X \otimes X^*$ for a suitable $X \in \mathcal{C}$.

Not every module category of the form $\mathsf{Mod}_{\mathcal{C}}(A)$ is exact.

EXERCISE 7.8.19. Give an example of module category of the form $\mathsf{Mod}_{\mathcal{C}}(A)$ which is not exact.

DEFINITION 7.8.20. An algebra A in a multitensor category $\mathcal{C}$ is called *exact* if the module category $\mathsf{Mod}_{\mathcal{C}}(A)$ is exact.

It is obvious from the definition that the exactness is invariant under Morita equivalence.

We will need the notion of a tensor product over an algebra $A \in \mathcal{C}$.

DEFINITION 7.8.21. Let A be an algebra in $\mathcal{C}$ and let (M, p) be a right A-module, and (N, q) be a left A-module. A *tensor product* of M and N over A is the object $M \otimes_A N \in \mathcal{C}$ defined as the co-equalizer of the diagram

$$M \otimes A \otimes N \underset{\mathsf{id}_M \otimes q}{\overset{p \otimes \mathsf{id}_N}{\rightrightarrows}} M \otimes N \longrightarrow M \otimes_A N, \tag{7.16}$$

i.e., the cokernel of the morphism $p \otimes \mathsf{id}_N - \mathsf{id}_M \otimes q$.

EXERCISE 7.8.22. Prove that for a right A-module M and a left A-module N one has $M \otimes_A A \cong M$ and $A \otimes_A N \cong N$.

EXERCISE 7.8.23. Show that the functor $\otimes_A$ is right exact in each variable (that is, for fixed M, N, the functors $M \otimes_A -$ and $- \otimes_A N$ are right exact).

LEMMA 7.8.24. *Let A be an algebra in $\mathcal{C}$, and let M, N be left A-modules. There is a natural isomorphism*

$$\mathsf{Hom}_{\mathcal{C}}(M \otimes_A {}^*N, X) \cong \mathsf{Hom}_A(M, X \otimes N), \tag{7.17}$$

where ${}^*N$ is a left A-module by Remark 7.8.6(i).

PROOF. The required isomorphism is defined using the composition

$$\mathsf{Hom}_{\mathcal{C}}(M \otimes_A {}^*N, X) \to \mathsf{Hom}_{\mathcal{C}}(M \otimes {}^*N, X) \cong \mathsf{Hom}_{\mathcal{C}}(M, X \otimes N),$$

where the first arrow comes from (7.16). We leave it to the reader to show that the image of this composition is in $\mathsf{Hom}_A(M, X \otimes N) \subset \mathsf{Hom}_{\mathcal{C}}(M, X \otimes N)$ and that the resulting map $\mathsf{Hom}_{\mathcal{C}}(M \otimes_A {}^*N, X) \to \mathsf{Hom}_A(M, X \otimes N)$ is an isomorphism. $\square$

DEFINITION 7.8.25. Let A, B be two algebras in $\mathcal{C}$. An (A, B)-*bimodule* in $\mathcal{C}$ is a triple (M, p, q) where $M \in \mathcal{C}$ and $p : A \otimes M \to M$, $q : M \otimes B \to M$ such that

(1) The pair (M, p) is a left A-module in $\mathcal{C}$.
(2) The pair (M, q) is a right B-module in $\mathcal{C}$.

(3) The following diagram commutes:

$$
\begin{array}{ccccc}
(A \otimes M) \otimes B & & \xrightarrow{a_{A,M,B}} & & A \otimes (M \otimes B) \\
\downarrow{\scriptstyle p \otimes \mathrm{id}_B} & & & & \downarrow{\scriptstyle \mathrm{id}_A \otimes q} \\
M \otimes B & & & & A \otimes M \\
& \searrow^{q} & & \swarrow^{p} & \\
& & M. & &
\end{array}
\tag{7.18}
$$

A homomorphism of (A, B)-bimodules is a morphism in $\mathcal{C}$ which is a homomorphism of both left A-modules and right B-modules. Given (A, B)-bimodules M, N, let $\mathsf{Hom}_{A-B}(M, N)$ denote the space of bimodule homomorphisms from M to N. It is clear that (A, B)-bimodules in $\mathcal{C}$ and their homomorphisms form a category which we will denote $\mathsf{Bimod}_{\mathcal{C}}(A, B)$.

REMARK 7.8.26. Note that in the categorical setting, we cannot, in general, define (A, B)-bimodules as modules over $A \otimes B^{\mathrm{op}}$, since neither the opposite algebra nor the tensor product of algebras is defined in a general multitensor category.

We will say "A-bimodule" instead of "(A, A)-bimodule" and write $\mathsf{Bimod}_{\mathcal{C}}(A)$ instead of $\mathsf{Bimod}_{\mathcal{C}}(A, A)$.

EXERCISE 7.8.27. Let A, B, and C be algebras in $\mathcal{C}$. Let M be an (A, B)-bimodule and N be a (B, C)-bimodule. Show that $M \otimes_A N$ has a canonical structure of an (A, C)-bimodule.

EXERCISE 7.8.28. Let A, B, C, and D be algebras in $\mathcal{C}$. Let M be an (A, B)-bimodule, N be a (B, C)-bimodule, and P be a (C, D)-bimodule. Construct a natural (A, D)-bimodule associativity isomorphism

$$
a_{M,N,P} : (M \otimes_B N) \otimes_C P \xrightarrow{\sim} M \otimes_B (N \otimes_C P)
\tag{7.19}
$$

and bimodule unit isomorphisms $l_M : A \otimes_A M \xrightarrow{\sim} M$ and $r_M : M \otimes_B B \xrightarrow{\sim} M$. Prove the pentagon and triangle relations for these isomorphisms.

DEFINITION 7.8.29. An algebra A in a tensor category $\mathcal{C}$ is called *separable* if the multiplication morphism $m : A \otimes A \to A$ splits as a morphism of A-bimodules.

For example, an absolutely semisimple algebra over a field (i.e., one remaining semisimple after any field extension) is separable.

PROPOSITION 7.8.30. *Let A be a separable algebra in a fusion category $\mathcal{C}$. Then the category $\mathsf{Mod}_{\mathcal{C}}(A)$ is semisimple.*

PROOF. Note that A considered as a bimodule over itself is a direct summand of the A-bimodule $A \otimes A$. Thus any right A-module $M = M \otimes_A A$ is a direct summand of $M \otimes_A A \otimes A = M \otimes A$. It follows from Lemma 7.8.12 that the functor $\mathsf{Hom}_A(M \otimes A, -)$ is exact, i.e., the object $M \otimes A \in \mathsf{Mod}_{\mathcal{C}}(A)$ is projective. Thus any $M \in \mathsf{Mod}_{\mathcal{C}}(A)$ is projective, i.e., $\mathsf{Mod}_{\mathcal{C}}(A)$ is semisimple. $\square$

REMARK 7.8.31. Proposition 7.8.30 implies that a separable algebra in a fusion category is exact.

EXERCISE 7.8.32. Let A be an H-module algebra, i.e., an algebra in the category $\mathbf{Rep}(H)$, where H is a bialgebra. Define the *smash product* $A\#H$ to be the tensor product $A \otimes H$ with multiplication rule

$$(a \otimes h)(a' \otimes h') := a(h_1 \cdot a') \otimes h_2 h', \quad a, a' \in A, h, h' \in H,$$

where $\Delta(h) = h_1 \otimes h_2$ (Sweedler's notation).[2]

(i) Show that if $H = \Bbbk G$ for a group G then $A\#H$ is the usual semidirect product $A \rtimes G$.

(ii) Show that the category of A-modules in $\mathbf{Rep}(H)$ is equivalent to the category of $A\#H$-modules.

7.9. Internal Homs in module categories

An important technical tool in the study of module categories is the notion of internal Hom.

Let $\mathcal{C}$ be a finite multitensor category (the finiteness condition is not strictly necessary in this Section but simplifies the exposition). Let $\mathcal{M}$ be a $\mathcal{C}$-module category and fix objects $M_1, M_2 \in \mathcal{M}$. Consider the functor

$$X \mapsto \mathsf{Hom}_{\mathcal{M}}(X \otimes M_1, M_2) : \mathcal{C} \to \mathsf{Vec}. \tag{7.20}$$

This functor is left exact and thus is representable, i.e., there exists an object $\underline{\mathsf{Hom}}(M_1, M_2) \in \mathcal{C}$ and a natural isomorphism

$$\mathsf{Hom}_{\mathcal{M}}(X \otimes M_1, M_2) \cong \mathsf{Hom}_{\mathcal{C}}(X, \underline{\mathsf{Hom}}(M_1, M_2)). \tag{7.21}$$

REMARK 7.9.1. If we do not assume that the category $\mathcal{C}$ is finite, the functor above is still representable, but by an ind-object of $\mathcal{C}$. Working with ind-objects, one can extend the theory below to this more general case. We leave this for an interested reader.

DEFINITION 7.9.2. The object $\underline{\mathsf{Hom}}(M_1, M_2)$ representing the functor (7.20) is called the *internal Hom* from M_1 to M_2.

Note that by the Yoneda Lemma

$$(M_1, M_2) \mapsto \underline{\mathsf{Hom}}(M_1, M_2) : \mathcal{M}^{\mathrm{op}} \times \mathcal{M} \to \mathcal{C} \tag{7.22}$$

is a bifunctor.

EXERCISE 7.9.3. Show that bifunctor (7.22) is left exact in both variables.

LEMMA 7.9.4. *There are canonical natural isomorphims*

$$\mathsf{Hom}_{\mathcal{M}}(X \otimes M_1, M_2) \cong \mathsf{Hom}_{\mathcal{C}}(X, \underline{\mathsf{Hom}}(M_1, M_2)), \tag{7.23}$$

$$\mathsf{Hom}_{\mathcal{M}}(M_1, X \otimes M_2) \cong \mathsf{Hom}_{\mathcal{C}}(\mathbf{1}, X \otimes \underline{\mathsf{Hom}}(M_1, M_2)), \tag{7.24}$$

$$\underline{\mathsf{Hom}}(M_1, X \otimes M_2) \cong X \otimes \underline{\mathsf{Hom}}(M_1, M_2), \tag{7.25}$$

$$\underline{\mathsf{Hom}}(X \otimes M_1, M_2) \cong \underline{\mathsf{Hom}}(M_1, M_2) \otimes X^*. \tag{7.26}$$

[2] "Sweedler's notation" means that for the sake of brevity we abuse notation and write $\Delta(x) = x_1 \otimes x_2$ instead of the more appropriate $\Delta(x) = \sum_i x_{1i} \otimes x_{2i}$, implying summation on the right hand side, and similarly for triple coproduct, etc. This is a common convention in the theory of Hopf algebras, which does not cause confusion after some practice (one needs to remember that this does not mean that $\Delta(x)$ is a pure tensor). This convention is somewhat similar to Einstein's convention in tensor calculus – implied summation over repeated indices.

PROOF. Ismorphism (7.23) is just the definition of $\underline{\mathsf{Hom}}(M_1, M_2)$. Isomorphism (7.24) is the composition

$$\begin{aligned}\mathsf{Hom}_{\mathcal{M}}(M_1, X\otimes M_2) &\cong \mathsf{Hom}_{\mathcal{M}}(X^*\otimes M_1, M_2)\\ &\cong \mathsf{Hom}_{\mathcal{C}}(X^*, \underline{\mathsf{Hom}}(M_1, M_2))\\ &\cong \mathsf{Hom}_{\mathcal{C}}(\mathbf{1}, X\otimes \underline{\mathsf{Hom}}(M_1, M_2)),\end{aligned}$$

where we used adjunction (7.5). We get isomorphism (7.25) and (7.26) by the Yoneda Lemma from the following calculations:

$$\begin{aligned}\mathsf{Hom}_{\mathcal{C}}(Y, \underline{\mathsf{Hom}}(M_1, X\otimes M_2)) &\cong \mathsf{Hom}_{\mathcal{M}}(Y\otimes M_1, X\otimes M_2)\\ &\cong \mathsf{Hom}_{\mathcal{M}}(X^*\otimes (Y\otimes M_1), M_2)\\ &\cong \mathsf{Hom}_{\mathcal{M}}((X^*\otimes Y)\otimes M_1, M_2)\\ &\cong \mathsf{Hom}_{\mathcal{C}}(X^*\otimes Y, \underline{\mathsf{Hom}}(M_1, M_2))\\ &\cong \mathsf{Hom}_{\mathcal{C}}(Y, X\otimes \underline{\mathsf{Hom}}(M_1, M_2))\\ \mathsf{Hom}_{\mathcal{C}}(Y, \underline{\mathsf{Hom}}(X\otimes M_1, M_2)), &\cong \mathsf{Hom}_{\mathcal{M}}(Y\otimes (X\otimes M_1), M_2)\\ &\cong \mathsf{Hom}_{\mathcal{M}}((Y\otimes X)\otimes M_1, M_2)\\ &\cong \mathsf{Hom}_{\mathcal{C}}(Y\otimes X, \underline{\mathsf{Hom}}(M_1, M_2))\\ &\cong \mathsf{Hom}_{\mathcal{C}}(Y, \underline{\mathsf{Hom}}(M_1, M_2)\otimes X^*),\end{aligned}$$

where we used the definition of internal Hom and adjunctions (2.49), (2.50), and (7.5). □

COROLLARY 7.9.5. *Let M be a fixed object of a $\mathcal{C}$-module category $\mathcal{M}$. The assignment*

$$N \mapsto \underline{\mathsf{Hom}}(M, N) : \mathcal{M} \to \mathcal{C} \tag{7.27}$$

is a $\mathcal{C}$-module functor.

PROOF. The module functor structure for (7.27) is given by natural isomorphism (7.25) of Lemma 7.9.4. □

COROLLARY 7.9.6. *Assume that $\mathcal{M}$ is an exact module category. Then bifunctor (7.22) is exact in each variable.*

PROOF. This follows from Corollary 7.9.5 and Proposition 7.6.9. □

The mere definition of the internal Hom allows us to prove the following converse to Proposition 7.6.9.

PROPOSITION 7.9.7. (1) *Suppose that for a $\mathcal{C}$-module category $\mathcal{M}$ the bifunctor $\underline{\mathsf{Hom}}$ is exact in the second variable, i.e., for any object $N \in \mathcal{M}$ the functor $M \mapsto \underline{\mathsf{Hom}}(N, M) : \mathcal{M} \to \mathcal{C}$ is exact. Then $\mathcal{M}$ is exact.*
(2) *Let $\mathcal{M}_1, \mathcal{M}_2$ be nonzero $\mathcal{C}$-module categories. Assume that any module functor from $\mathcal{M}_1$ to $\mathcal{M}_2$ is exact. Then the module category $\mathcal{M}_1$ is exact.*

PROOF. (1) Let $P \in \mathcal{C}$ be any projective object. Then for any $N \in \mathcal{M}$ one has $\mathsf{Hom}_{\mathcal{M}}(P\otimes N, -) = \mathsf{Hom}_{\mathcal{C}}(P, \underline{\mathsf{Hom}}(N, -))$, and thus the functor $\mathsf{Hom}_{\mathcal{M}}(P\otimes N, -)$ is exact. By the definition of an exact module category, we are done.

(2) We claim that under our assumptions any module functor $F \in \mathsf{Fun}_{\mathcal{C}}(\mathcal{M}_1, \mathcal{C})$ is exact. Indeed, let $0 \neq M \in \mathcal{M}_2$. The functor $F(-)\otimes M \in \mathsf{Fun}_{\mathcal{C}}(\mathcal{M}_1, \mathcal{M}_2)$ is exact. Since $-\otimes M$ is exact, and $X\otimes M = 0$ implies $X = 0$, we see that F is exact.

In particular, we see that for any object $N \in \mathcal{M}_1$, the functor $\underline{\mathsf{Hom}}(N, -) : \mathcal{M}_1 \to \mathcal{C}$ is exact, since it is a module functor. Now (2) follows from (1). $\square$

EXAMPLE 7.9.8. It is instructive to calculate $\underline{\mathsf{Hom}}$ for the category $\mathsf{Mod}_{\mathcal{C}}(A)$. Let $M, N \in \mathsf{Mod}_{\mathcal{C}}(A)$. Then M^* has a natural structure of a left A^{**}-module and $\underline{\mathsf{Hom}}(M, N) = N \otimes^A M^*$ where

$$N \otimes^A M^* := (M \otimes_A {}^*N)^*$$

(recall that by Remark 7.8.6(i) ${}^*N$ is a left A-module). Thus $N \otimes^A M^*$ is naturally a subobject of $N \otimes M^*$ while $N \otimes_A {}^*M$ is a quotient of $N \otimes {}^*M$. We leave to the reader to state and prove the associativity properties of $\otimes^A$. One deduces from this description of $\underline{\mathsf{Hom}}$ that exactness of A is equivalent to biexactness of the product $\otimes^A$ (and to biexactness of $\otimes_A$).

For two objects M_1, M_2 of a module category $\mathcal{M}$ we have the canonical "evaluation" morphism

$$\mathsf{ev}_{M_1,M_2} : \underline{\mathsf{Hom}}(M_1, M_2) \otimes M_1 \to M_2 \tag{7.28}$$

obtained as the image of $\mathsf{id}_{\underline{\mathsf{Hom}}(M_1,M_2)}$ under natural isomorphism (7.21):

$$\mathsf{Hom}_{\mathcal{C}}(\underline{\mathsf{Hom}}(M_1, M_2), \underline{\mathsf{Hom}}(M_1, M_2)) \cong \mathsf{Hom}_{\mathcal{M}}(\underline{\mathsf{Hom}}(M_1, M_2) \otimes M_1, M_2).$$

Let M_1, M_2, M_3 be objects of $\mathcal{M}$. Consider the following canonical composition:

$$\begin{aligned}(\underline{\mathsf{Hom}}(M_2, M_3) \otimes \underline{\mathsf{Hom}}(M_1, M_2)) \otimes M_1 &\xrightarrow{\sim} \\ \underline{\mathsf{Hom}}(M_2, M_3) \otimes (\underline{\mathsf{Hom}}(M_1, M_2) \otimes M_1) &\xrightarrow{\mathsf{id}_{\underline{\mathsf{Hom}}(M_2,\, M_3)} \otimes\, \mathsf{ev}_{M_1,\, M_2}} \\ \underline{\mathsf{Hom}}(M_2, M_3) \otimes M_2 &\xrightarrow{\mathsf{ev}_{M_2,\, M_3}} M_3.\end{aligned}$$

By (7.21) this composition produces the following multiplication morphism:

$$\underline{\mathsf{Hom}}(M_2, M_3) \otimes \underline{\mathsf{Hom}}(M_1, M_2) \to \underline{\mathsf{Hom}}(M_1, M_3). \tag{7.29}$$

We also have for every $M \in \mathcal{M}$ a canonical unit morphism:

$$u_M : \mathbf{1} \to \underline{\mathsf{Hom}}(M, M) \tag{7.30}$$

obtained from (7.21) with $X = \mathbf{1}$ as the image of id_M.

EXERCISE 7.9.9. Check that multiplication (7.29) is associative and compatible with unit (7.30) and with natural isomorphisms of Lemma 7.9.4.

In particular, for every $M \in \mathcal{M}$ the object $\underline{\mathsf{Hom}}(M, M)$ has a canonical structure of an algebra in $\mathcal{C}$ and for every $N \in \mathcal{M}$ the object $\underline{\mathsf{Hom}}(M, N)$ is a right $\underline{\mathsf{Hom}}(M, M)$-module. Furthermore, the assignment

$$F : N \mapsto \underline{\mathsf{Hom}}(M, N) : \mathcal{M} \to \mathsf{Mod}_{\mathcal{C}}(\underline{\mathsf{Hom}}(M, M)) \tag{7.31}$$

is a $\mathcal{C}$-module functor (recall that $\mathsf{Mod}_{\mathcal{C}}(\underline{\mathsf{Hom}}(M, M))$ is a left $\mathcal{C}$-module category by Proposition 7.8.10). The $\mathcal{C}$-module functor structure of functor (7.31) is provided by isomorphism (7.25). We will see in Theorem 7.10.1 that (7.31) is an equivalence.

EXAMPLE 7.9.10. Let $\mathcal{C}$, $\mathcal{D}$ be multitensor categories, let $F : \mathcal{C} \to \mathcal{D}$ be a tensor functor, and let $I : \mathcal{D} \to \mathcal{C}$ be the right adjoint of F. We can view $\mathcal{D}$ as a $\mathcal{C}$-module category. Then $I(Y) = \underline{\mathsf{Hom}}(\mathbf{1}, Y)$, where $\mathbf{1}$ denotes the unit object of $\mathcal{D}$. Indeed, this is a direct consequence of the adjunction isomorphism

$$\mathsf{Hom}_{\mathcal{D}}(F(X) \otimes \mathbf{1}, Y) \cong \mathsf{Hom}_{\mathcal{C}}(X, I(Y)).$$

In particular, the object $I(\mathbf{1}) = \underline{\mathsf{Hom}}(\mathbf{1},\mathbf{1}) \in \mathcal{C}$ has a canonical structure of an algebra in $\mathcal{C}$ and $Y \mapsto I(Y) : \mathcal{D} \to \mathsf{Mod}_{\mathcal{C}}(I(\mathbf{1}))$ is a $\mathcal{C}$-module functor.

EXAMPLE 7.9.11. As a special case of Example 7.9.10, consider the situation when $\mathcal{C} = \mathsf{Rep}(H)$, the representation category of a finite dimensional Hopf algebra H, $\mathcal{D} = \mathsf{Vec}$, and $F : \mathsf{Rep}(H) \to \mathsf{Vec}$ is the fiber functor. Then $\underline{\mathsf{Hom}}(\Bbbk, \Bbbk) = H^*$, the Hopf algebra dual to H, viewed as an algebra in $\mathsf{Rep}(H)$, where H acts on H^* by right translations.

Let $\mathcal{C}$ be a multitensor category viewed as a $(\mathcal{C} \boxtimes \mathcal{C}^{\mathrm{op}})$-module category.

DEFINITION 7.9.12. We will call the algebra $A := \underline{\mathsf{Hom}}(\mathbf{1},\, \mathbf{1})$ in $\mathcal{C} \boxtimes \mathcal{C}^{\mathrm{op}}$ the *canonical algebra* of $\mathcal{C}$.

REMARK 7.9.13. If $\mathcal{C}$ is the representation category of a finite dimensional Hopf algebra H then $A = H^*$ viewed as an algebra in the category of H-bimodules, where H acts on H^* by left and right translations.

EXAMPLE 7.9.14. Suppose that $\mathcal{C}$ is a multifusion category and let $\mathcal{O}(\mathcal{C})$ denote the set of simple objects in $\mathcal{C}$. We claim that

$$A = \bigoplus_{X \in \mathcal{O}(\mathcal{C})} X \boxtimes {}^*X \tag{7.32}$$

as an object of $\mathcal{C} \boxtimes \mathcal{C}^{\mathrm{op}}$. Indeed, (7.32) is a consequence of the natural isomorphism

$$\begin{aligned} \mathsf{Hom}_{\mathcal{C}}(Y \otimes Z,\, \mathbf{1}) &\cong \mathsf{Hom}_{\mathcal{C}}(Y,\, Z^*) \\ &\cong \bigoplus_{X \in \mathcal{O}(\mathcal{C})} \mathsf{Hom}_{\mathcal{C}}(Y,\, X) \otimes \mathsf{Hom}_{\mathcal{C}}(X,\, Z^*) \\ &\cong \mathsf{Hom}_{\mathcal{C}\boxtimes\mathcal{C}^{\mathrm{op}}}\left(Y \boxtimes Z,\, \bigoplus_{X \in \mathcal{O}(\mathcal{C})} X \boxtimes {}^*X \right). \end{aligned}$$

Using (7.29) one explicitly describes the multiplication in the algebra A as follows. We have

$$A \otimes A = \bigoplus_{X,Y \in \mathcal{O}(\mathcal{C})} (X \otimes Y) \boxtimes {}^*(X \otimes Y).$$

For any $Z \in \mathcal{O}(\mathcal{C})$ the vector spaces $\mathsf{Hom}_{\mathcal{C}}(X \otimes Y,\, Z)$ and $\mathsf{Hom}_{\mathcal{C}}({}^*(X \otimes Y),\, {}^*Z) = \mathsf{Hom}(Z,\, X \otimes Y)$ are canonically dual to each other via the pairing

$$\mathsf{Hom}_{\mathcal{C}}(X \otimes Y,\, Z) \otimes \mathsf{Hom}_{\mathcal{C}}(Z,\, X \otimes Y) \to \mathsf{Hom}_{\mathcal{C}}(Z,\, Z) \cong \Bbbk,$$

hence there is a canonical morphism $(X \otimes Y) \boxtimes {}^*(X \otimes Y) \to Z \boxtimes {}^*Z$. Then the multiplication $A \otimes A \to A$ is the direct sum (over $Z \in \mathcal{O}(\mathcal{C})$) of all such morphisms.

7.10. Characterization of module categories in terms of algebras

Let $\mathcal{C}$ be a finite multitensor category.

THEOREM 7.10.1. *Let $\mathcal{M}$ be a $\mathcal{C}$-module category, and let $M \in \mathcal{M}$ be an object satisfying the following two conditions:*

(i) *The functor $\underline{\mathsf{Hom}}(M,\, -)$ is right exact (note that it is automatically left exact).*
(ii) *For any $N \in \mathcal{M}$ there exists $X \in \mathcal{C}$ and a surjection $X \otimes M \to N$.*

Let $A = \underline{\mathsf{Hom}}(M, M)$. Then the functor $F : \mathcal{M} \to \mathsf{Mod}_{\mathcal{C}}(A)$ defined in (7.31) is an equivalence of $\mathcal{C}$-module categories.

PROOF. We will proceed in steps by proving the following claims.

(1) *The map $F : \mathsf{Hom}_{\mathcal{M}}(N_1, N_2) \to \mathsf{Hom}_A(F(N_1), F(N_2))$ is an isomorphism for any $N_2 \in \mathcal{M}$ and N_1 of the form $X \otimes M$ with $X \in \mathcal{C}$, $M \in \mathcal{M}$.*

Indeed, by (7.25) we have $F(N_1) = \underline{\mathsf{Hom}}(M, X \otimes M) = X \otimes A$ and the claim follows from the following calculation:

$$\begin{aligned}\mathsf{Hom}_A(F(N_1), F(N_2)) &= \mathsf{Hom}_A(X \otimes A, F(N_2)) \cong \mathsf{Hom}_{\mathcal{C}}(X, F(N_2)) \\ &= \mathsf{Hom}_{\mathcal{C}}(X, \underline{\mathsf{Hom}}(M, N_2)) \cong \mathsf{Hom}_{\mathcal{M}}(X \otimes M, N_2) \\ &= \mathsf{Hom}_{\mathcal{M}}(N_1, N_2),\end{aligned}$$

where we used Lemma 7.8.12 and the definition (7.21) of internal Hom.

(2) *The map $F : \mathsf{Hom}_{\mathcal{M}}(N_1, N_2) \to \mathsf{Hom}_A(F(N_1), F(N_2))$ is an isomorphism for all objects $N_1, N_2 \in \mathcal{M}$.*

By condition (ii), there exist objects $X, Y \in \mathcal{C}$ and an exact sequence

$$Y \otimes M \to X \otimes M \to N_1 \to 0.$$

Since F is exact, the sequence

$$F(Y \otimes M) \to F(X \otimes M) \to F(N_1) \to 0$$

is exact. Since for each $N \in \mathcal{M}$ the functor $\mathsf{Hom}_{\mathcal{M}}(-, N)$ is left exact, the rows in the commutative diagram

$$\begin{array}{ccccccc} 0 \longrightarrow & \mathsf{Hom}_{\mathcal{M}}(N_1, N_2) & \longrightarrow & \mathsf{Hom}_{\mathcal{M}}(X \otimes M, N_2) & \longrightarrow & \mathsf{Hom}_{\mathcal{M}}(Y \otimes M, N_2) \\ & \downarrow{\scriptstyle F} & & \downarrow{\scriptstyle F} & & \downarrow{\scriptstyle F} \\ 0 \longrightarrow & \mathsf{Hom}_{\mathcal{M}}(F(N_1), F(N_2)) & \longrightarrow & \mathsf{Hom}_{\mathcal{M}}(F(X \otimes M), F(N_2)) & \longrightarrow & \mathsf{Hom}_{\mathcal{M}}(F(Y \otimes M), F(N_2)) \end{array}$$

are exact. Since by step (1) the second and third vertical arrows are isomorphisms, so is the first one.

(3) *The functor F is surjective on isomorphism classes of objects of $\mathsf{Mod}_{\mathcal{C}}(A)$.*

We know (see Exercise 7.8.14) that for any object $L \in \mathsf{Mod}_{\mathcal{C}}(A)$ there exists an exact sequence

$$Y \otimes A \xrightarrow{\tilde{f}} X \otimes A \to L \to 0$$

for some $X, Y \in \mathcal{C}$. Let $f \in \mathsf{Hom}(Y \otimes M, X \otimes M)$ be the image of $\tilde{f}$ under the isomorphism

$$\mathsf{Hom}_A(Y \otimes A, X \otimes A) \cong \mathsf{Hom}_A(F(Y \otimes M), F(X \otimes M)) \cong \mathsf{Hom}_{\mathcal{M}}(Y \otimes M, X \otimes M),$$

and let $N \in \mathcal{M}$ be the cokernel of f. Then $F(N) \cong L$ by the exactness of F.

Thus, F is an equivalence of categories. □

EXAMPLE 7.10.2. Let X be an object in $\mathcal{C}$ and let $A = X \otimes X^*$ be the algebra from Example 7.8.4. It follows from Theorem 7.10.1 that the assignment

$$Y \mapsto Y \otimes X^* : \mathcal{C} \to \mathsf{Mod}_{\mathcal{C}}(A)$$

is an equivalence of $\mathcal{C}$-module categories (here $\mathcal{C}$ is viewed as the regular $\mathcal{C}$-module category). Similarly, the assignment $Y \mapsto X \otimes Y$ is an equivalence between $\mathcal{C}$ and the category of left A-modules in $\mathcal{C}$.

We have two situations where condition (i) of Theorem 7.10.1 is satisfied:

(1) $\mathcal{M}$ is an arbitrary indecomposable $\mathcal{C}$-module category and $M \in \mathcal{M}$ is projective.
(2) $\mathcal{M}$ is an indecomposable exact $\mathcal{C}$-module category and $M \in \mathcal{M}$ is arbitrary.

EXERCISE 7.10.3. Check that in both of these cases $\underline{\mathsf{Hom}}(M, -)$ is exact.
Hint: in the first case first prove that $\underline{\mathsf{Hom}}(M, N)$ is a projective object of $\mathcal{C}$ for any $N \in \mathcal{M}$.

EXERCISE 7.10.4. Show that condition (2) above is equivalent to the fact that $[M]$ generates $\mathsf{Gr}(\mathcal{M})$ as a $\mathbb{Z}_+$-module over $\mathsf{Gr}(\mathcal{C})$.

COROLLARY 7.10.5. (i) *Let $\mathcal{M}$ be a finite module category over $\mathcal{C}$. Then there exists an algebra $A \in \mathcal{C}$ and a module equivalence $\mathcal{M} \cong \mathsf{Mod}_{\mathcal{C}}(A)$.*
(ii) *Let $\mathcal{M}$ be an exact module category over $\mathcal{C}$ and let $M \in \mathcal{M}$ be an object such that $[M]$ generates $\mathsf{Gr}(\mathcal{M})$ as a $\mathbb{Z}_+$-module over $\mathsf{Gr}(\mathcal{C})$. Then there is a module equivalence $\mathcal{M} \cong \mathsf{Mod}_{\mathcal{C}}(A)$ where $A = \underline{\mathsf{Hom}}(M, M)$.*

EXERCISE 7.10.6. Show that any indecomposable exact algebra A in a finite tensor category $\mathcal{C}$ can be written as $\underline{\mathsf{Hom}}(M, M)$, where $\mathcal{M}$ is an exact indecomposable $\mathcal{C}$-module category and $M \in \mathcal{M}$ a nonzero object. Moreover, show that the pair $(\mathcal{M}, M)$ is unique up to equivalence.
Hint: Take $\mathcal{M} = \mathsf{Mod}_{\mathcal{C}}(A)$ and $M = A$.

This implies the following corollary.

COROLLARY 7.10.7. *Let H be a finite dimensional Hopf algebra. Let $\mathcal{N}$ be the category of left H^*-modules in $\mathsf{Rep}(H)$, where H acts on H^* by right translations (Example 7.9.11). Then:*

(i) The functor $G : \mathsf{Vec} \to \mathcal{N}$ given by $G(V) = V \otimes H^*$ is an equivalence of categories.

(ii) The algebra $H^* \# H$ (see Exercise 7.8.32) is naturally isomorphic to the matrix algebra $\mathsf{End}_{\Bbbk}(H^*)$.

PROOF. Part (i) follows from Corollary 7.10.5(ii) with $\mathcal{M} = \mathsf{Vec}$, and part (ii) follows from part (i). □

DEFINITION 7.10.8. A left H^*-module in the category of left H-modules is called a *Hopf module* for H. [3]

Corollary 7.10.7(i) is called the *Fundamental Theorem for Hopf modules*; it states that any Hopf module for H is a multiple of H^* (see also Exercise 7.10.10(iii)).

DEFINITION 7.10.9. The algebra $H^* \# H$ is called *the Heisenberg double* of H.

EXERCISE 7.10.10. Let H be a finite dimensional Hopf algebra over an algebraically closed field $\Bbbk$ of any characteristic. Consider the linear map $\eta : H^* \otimes H \to H$ given by $\eta(f \otimes h) = f(S^{-1}(h_1))h_2$, where $\Delta(h) = h_1 \otimes h_2$.

(i) Show that η is an action of H^* on the vector space H.

---

[3] In **[Mon]**, Definition 1.9.1, a Hopf module is defined as left H^*-module (or, equivalently, a right H-comodule) in the category of *right* H-modules. But this is equivalent to our setting by using the antipode.

(ii) Show that η is H-invariant (where H^* is regarded as a left H-module via right translations, and H via left translations). Deduce that η equips H with the structure of a Hopf module for H.

(iii) Show that H is isomorphic to H^* as a Hopf module for H (use the Fundamental theorem). Moreover, show that any Hopf module isomorphism $\phi : H \to H^*$ maps $1 \in H$ to a nonzero left integral $\lambda \in H^*$. Thus, any Hopf module for H is a multiple of H, which is a more usual formulation of the Fundamental Theorem for Hopf modules.

(iv) Show that $\phi(h)(x) = \lambda(xh)$, $h, x \in H$. Deduce that the pairing $(x, h) := \lambda(xh)$ on H is non-degenerate for any nonzero left integral λ of H^*.

(v) Deduce that if I is a nonzero left or right integral of H, then $\lambda(I) \neq 0$.

(vi) Show that if I is a left integral of H then for each $x \in H$ one has

$$\lambda(S^{-1}(I))x = \lambda(S^{-1}(I_1)x)I_2,$$

and

$$\lambda(I)x = \lambda(xI_2)S^{-1}(I_1),$$

where $\Delta(I) = I_1 \otimes I_2$. Deduce that $\lambda(S^{-1}(I)) = \lambda(I)$ (set $x = I$ in the first identity). [4] Deduce that

$$xI_2 \otimes S^{-1}(I_1) = I_2 \otimes S^{-1}(I_1)x$$

and

$$xI_1 \otimes S(I_2) = I_1 \otimes S(I_2)x, \ x \in H$$

(multiply these identities on the left by $a \in H$ and apply λ to the first component).

(vii) Let $A : H \to H$ be a linear operator. Show that in the notation of (vi), if $\lambda(I) = 1$, then $\mathsf{Tr}(A) = \lambda(A(S^{-1}(I_1))I_2)$. (Choose a basis $\{x_k\}$ of H and write $\mathsf{Tr}(A)$ as $\sum_k x_k^*(Ax_k)$.)

(viii) Prove that if I is a left integral in H and $\lambda(I) = 1$ then $\mathsf{Tr}(S^2) = \varepsilon(I)\lambda(1)$. (Use (vii)). Prove the same statement if I is a right integral in H (replace I with $S(I)$ and use (vi)).

(ix) Deduce the Larson-Radford theorem: H is semisimple and cosemisimple if and only if $\mathsf{Tr}(S^2)$ is nonzero.

(x) Show that η commutes with the left action of H^{cop} defined by $h \circ x = xS(h)$ for $x \in H$, and $(h \circ f)(x) = f(S^{-1}(h)x)$ for $f \in H^*$ (where S is the antipode of H).

REMARK 7.10.11. The Fundamental Theorem for Hopf modules (see Exercise 7.10.10(iii)) also holds in the infinite dimensional situation, when H is replaced by any Hopf algebra K, and the category $\mathbf{Rep}(H)$ is replaced by the category of K-modules. Specifically, a Hopf module for K is a left K-module and a right K-comodule M such that $\tau(xm) = \Delta(x)\tau(m)$, where $\tau : M \to M \otimes K$ is the coaction map (i.e., τ is a module homomorphism). Namely, this theorem states that any Hopf module for K is a multiple of K, with the left multiplication action of K, and $\tau = \Delta$.

EXERCISE 7.10.12. Prove the Fundamental Theorem for Hopf modules in the general case, by generalizing the finite dimensional proof.

---

[4]Note that in general $\lambda(S(I)) \neq \lambda(I)$, as $S^2(I) \neq I$. Indeed, for the Taft algebra (Example 5.5.6), $S^2(I) = q^{-1}I$.

7.11. Categories of module functors

Let $\mathcal{M}_1$, $\mathcal{M}_2$ be two module categories over a multitensor category $\mathcal{C}$.

The category of $\mathcal{C}$-module functors between $\mathcal{M}_1$ and $\mathcal{M}_2$ was introduced in Section 7.2. This category is very difficult to work with (consider the case $\mathcal{C} = \mathsf{Vec!}$) and we are going to consider its subcategory which is more manageable. Let $\mathsf{Fun}_{\mathcal{C}}(\mathcal{M}_1, \mathcal{M}_2)$ denote the full subcategory of the category of module functors consisting of *right exact* module functors (which are not necessarily left exact). First of all, this category can be described in down-to-earth terms:

PROPOSITION 7.11.1. *Assume that* $\mathcal{M}_1 \simeq \mathsf{Mod}_{\mathcal{C}}(A)$ *and* $\mathcal{M}_2 \simeq \mathsf{Mod}_{\mathcal{C}}(B)$ *for some algebras* $A, B \in \mathcal{C}$. *The category* $\mathsf{Fun}_{\mathcal{C}}(\mathcal{M}_1, \mathcal{M}_2)$ *is equivalent to the category of* $A-B$*-bimodules via the functor*

$$M \mapsto (- \otimes_A M) : \mathsf{Bimod}_{\mathcal{C}}(A,\, B) \to \mathsf{Fun}_{\mathcal{C}}(\mathcal{M}_1, \mathcal{M}_2). \tag{7.33}$$

In particular, $\mathsf{Fun}_{\mathcal{C}}(\mathcal{M}_1, \mathcal{M}_2)$ *is an abelian category.*

PROOF. The proof repeats the standard proof from ring theory in the categorical setting. □

In a similar way one can show that the category of left exact module functors is abelian (using Hom_A instead of $\otimes_A$).

We would like now to construct new tensor categories in the following way. Given a $\mathcal{C}$-module category $\mathcal{M}$, the category $\mathsf{Fun}_{\mathcal{C}}(\mathcal{M}, \mathcal{M})$ is a monoidal category with the tensor product being composition of functors.

Note that in general the category $\mathsf{Fun}_{\mathcal{C}}(\mathcal{M}, \mathcal{M})$ is not rigid (consider the case $\mathcal{C} = \mathsf{Vec!}$). Thus to get a good theory (and examples of new tensor categories), we restrict ourselves to the case of *exact* module categories, see Definition 7.5.1. In this case we can say much more about the categories $\mathsf{Fun}_{\mathcal{C}}(\mathcal{M}, \mathcal{M})$ than in general.

Let $\mathcal{M}_1$ and $\mathcal{M}_2$ be two exact module categories over $\mathcal{C}$. Note that the category $\mathsf{Fun}_{\mathcal{C}}(\mathcal{M}_1, \mathcal{M}_2)$ coincides with the category of the additive module functors from $\mathcal{M}_1$ to $\mathcal{M}_2$ by Proposition 7.6.9.

EXERCISE 7.11.2. Any object of $\mathsf{Fun}_{\mathcal{C}}(\mathcal{M}_1, \mathcal{M}_2)$ is of finite length.

LEMMA 7.11.3. *Let* $\mathcal{M}_1, \mathcal{M}_2, \mathcal{M}_3$ *be exact module categories over* $\mathcal{C}$. *The bifunctor of composition* $\mathsf{Fun}_{\mathcal{C}}(\mathcal{M}_2, \mathcal{M}_3) \times \mathsf{Fun}_{\mathcal{C}}(\mathcal{M}_1, \mathcal{M}_2) \to \mathsf{Fun}_{\mathcal{C}}(\mathcal{M}_1, \mathcal{M}_3)$ *is biexact.*

PROOF. This directly follows Proposition 7.6.9. □

Another immediate consequence of Proposition 7.6.9 is the following:

LEMMA 7.11.4. *Let* $\mathcal{M}_1, \mathcal{M}_2$ *be exact module categories over* $\mathcal{C}$. *Any functor* $F \in \mathsf{Fun}_{\mathcal{C}}(\mathcal{M}_1, \mathcal{M}_2)$ *has left and right adjoints.*

COROLLARY 7.11.5. *Let* $\mathcal{M}_1, \mathcal{M}_2$ *be exact module categories over* $\mathcal{C}$. *Any functor* $F \in \mathsf{Fun}_{\mathcal{C}}(\mathcal{M}_1, \mathcal{M}_2)$ *maps projective objects to projective objects.*

In view of Example 7.5.6 this Corollary is a generalization of Theorem 6.1.16 (but this does not give a new proof of Theorem 6.1.16).

PROPOSITION 7.11.6. *Let* $\mathcal{C}$ *be a finite multitensor category. Then category* $\mathsf{Fun}_{\mathcal{C}}(\mathcal{M}_1, \mathcal{M}_2)$ *is finite.*

PROOF. We will use Theorem 7.10.1. There are $\mathcal{C}$-module equivalences $\mathcal{M}_1 \cong \mathsf{Mod}_{\mathcal{C}}(A_1)$ and $\mathcal{M}_2 \cong \mathsf{Mod}_{\mathcal{C}}(A_2)$ for some algebras $A_1, A_2 \in \mathcal{C}$. By Proposition 7.11.1 the category $\mathsf{Fun}_{\mathcal{C}}(\mathcal{M}_1, \mathcal{M}_2)$ is equivalent to the category of (A_1, A_2)-bimodules. But this category clearly has enough projective objects: for any projective $P \in \mathcal{C}$ the bimodule $A_1 \otimes P \otimes A_2$ is projective. □

7.12. Dual tensor categories and categorical Morita equivalence

Let $\mathcal{C}$ be a multitensor category, let $\mathcal{M}, \mathcal{N}$ be exact $\mathcal{C}$-module categories, and let $F : \mathcal{M} \to \mathcal{N}$ be a $\mathcal{C}$-module functor. The right adjoint $G : \mathcal{N} \to \mathcal{M}$ of F has a natural structure of a $\mathcal{C}$-module functor. Its module structure is defined by the natural isomorphism

$$\mathsf{Hom}_{\mathcal{M}}(M, G(X \otimes N)) \cong \mathsf{Hom}_{\mathcal{N}}(F(M), X \otimes N) \cong \mathsf{Hom}_{\mathcal{N}}({}^*X \otimes F(M), N) \xrightarrow{\sim} \mathsf{Hom}_{\mathcal{N}}(F({}^*X \otimes M), N) \cong \mathsf{Hom}_{\mathcal{M}}(M, X \otimes G(N)),$$

where the first and the last isomorphisms come from Proposition 7.1.6 combined with the adjunction between F and G and the middle arrow comes from the $\mathcal{C}$-module structure of F. Similarly, the left adjoint of F has a structure of a $\mathcal{C}$-module functor.

EXERCISE 7.12.1. Show that that the category $\mathsf{Fun}_{\mathcal{C}}(\mathcal{M}, \mathcal{M})$ is a rigid monoidal category (namely, the left and right duals are left and right adjoint functors and the evaluation and coevaluation morphisms are the counit $FG \to \mathsf{id}_{\mathcal{M}}$ and the unit $\mathsf{id}_{\mathcal{M}} \to GF$ of the adjunction, respectively).

DEFINITION 7.12.2. We denote category $\mathsf{Fun}_{\mathcal{C}}(\mathcal{M}, \mathcal{M})$ as $\mathcal{C}^*_{\mathcal{M}}$ and call it the *dual tensor category* to $\mathcal{C}$ *with respect to* $\mathcal{M}$.

By Proposition 7.11.6, $\mathcal{C}^*_{\mathcal{M}}$ is a finite multitensor category. Moreover, it is a tensor category if and only if $\mathcal{M}$ is indecomposable.

EXAMPLE 7.12.3. Let $\mathcal{C}$ be a multitensor category viewed as the regular module category over itself, see Example 7.4.1. It is clear that every $\mathcal{C}$-module endofunctor of $\mathcal{C}$ is of the form $(- \otimes X)$ for some $X \in \mathcal{C}$ (the module structure on this functor is given by the associativity constraint of $\mathcal{C}$) and that $X \mapsto (- \otimes X) : \mathcal{C}^{\mathrm{op}} \to \mathcal{C}^*_{\mathcal{C}}$ is a tensor equivalence.

REMARK 7.12.4. The notion of the dual category is a categorical version of notion of the endomorphism ring of a module, i.e., the centralizer algebra. We will see that it gives rise to many new examples of tensor categories.

REMARK 7.12.5. Let $A \in \mathcal{C}$ be an algebra such that $\mathcal{M} = \mathsf{Mod}_{\mathcal{C}}(A)$. By Proposition 7.11.1 the category $\mathcal{C}^*_{\mathcal{M}}$ is identified with the category $\mathsf{Bimod}_{\mathcal{C}}(A)^{\mathrm{op}}$ of A-bimodules with opposite tensor product (because A-bimodules act naturally on $\mathsf{Mod}_{\mathcal{C}}(A)$ from the right).

Recall from (7.9) that an exact $\mathcal{C}$-module category $\mathcal{M}$ is decomposed into the sum of its components:

$$\mathcal{M} = \bigoplus_{i \in I} \mathcal{M}_i.$$

LEMMA 7.12.6. *The unit object* $\mathbf{1} \in \mathcal{C}^*_{\mathcal{M}}$ *is a direct sum of the projectors to the subcategories* $\mathcal{M}_i$. *Each such projector is a simple object.*

Proof. The first statement is clear. For the second statement it is enough to consider the case when $\mathcal{M}$ is indecomposable. Let F be a nonzero module subfunctor of the identity functor. Then $F(X) \neq 0$ for any $X \neq 0$. Hence $F(X) = X$ for any simple $X \in \mathcal{M}$ and thus $F(X) = X$ for any $X \in \mathcal{M}$ since F is exact. $\square$

Thus, the category $\mathcal{C}^*_{\mathcal{M}}$ is a finite multitensor category. In particular, if $\mathcal{M}$ is indecomposable then $\mathcal{C}^*_{\mathcal{M}}$ is a finite tensor category.

Note that by definition $\mathcal{M}$ is a left module category over $\mathcal{C}^*_{\mathcal{M}}$.

Lemma 7.12.7. *The module category $\mathcal{M}$ over $\mathcal{C}^*_{\mathcal{M}}$ is exact.*

Proof. Note that $\mathcal{C}^*_{\mathcal{M}} \cong \mathsf{Bimod}_{\mathcal{C}}(A)^{\mathrm{op}}$ by Remark 7.12.5. Any projective object in the category of A-bimodules is a direct summand of the object of the form $A \otimes P \otimes A$ for some projective $P \in \mathcal{C}$. Now for any $M \in \mathsf{Mod}_{\mathcal{C}}(A)$ one has that $M \otimes_A A \otimes P \otimes A = (M \otimes P) \otimes A$ is projective by exactness of the category $\mathsf{Mod}_{\mathcal{C}}(A)$. $\square$

Example 7.12.8. It is instructive to consider the internal Hom for the category $\mathsf{Mod}_{\mathcal{C}}(A)$ considered as a right module category over $\mathcal{C}^{*\mathrm{op}}_{\mathcal{M}} = \mathsf{Bimod}_{\mathcal{C}}(A)$ (cf. Example 7.9.8). We leave to the reader to check that $\underline{\mathsf{Hom}}_{\mathcal{C}^{*\mathrm{op}}_{\mathcal{M}}}(M, N) = {}^*M \otimes N$ (by Remark 7.8.6(i) ${}^*M$ is a left A-module and so ${}^*M \otimes N$ has an obvious structure of an A-bimodule). In particular, $B := \underline{\mathsf{Hom}}_{\mathcal{C}^{*\mathrm{op}}_{\mathcal{M}}}(A, A) = {}^*A \otimes A$ is an algebra in the category of A-bimodules. Thus B is an algebra in the category $\mathcal{C}$ and it follows from (7.29) that its multiplication comes from the evaluation morphism $\mathsf{ev}_{^*A} : A \otimes {}^*A \to \mathbf{1}$, cf. Example 7.8.4. Moreover, there is an embedding of algebras

$$A \cong \mathbf{1} \otimes A \xrightarrow{\mathsf{coev}_{^*A} \otimes \mathsf{id}_A} {}^*A \otimes A \otimes A \xrightarrow{\mathsf{id}_{^*A} \otimes m} {}^*A \otimes A = B, \tag{7.34}$$

where $m : A \otimes A \to A$ is the multiplication of A, and the A-bimodule structure of B comes from the left and right multiplication by A.

By Lemma 7.12.7 for any exact $\mathcal{C}$-module category $\mathcal{M}$ the category $(\mathcal{C}^*_{\mathcal{M}})^*_{\mathcal{M}}$ is well defined. There is an obvious tensor functor $\mathsf{can} : \mathcal{C} \to (\mathcal{C}^*_{\mathcal{M}})^*_{\mathcal{M}}$.

Definition 7.12.9. A module category $\mathcal{M}$ over $\mathcal{C}$ is *faithful* if each $\mathbf{1}_i \in \mathcal{C}$ acts by a nonzero functor in $\mathcal{M}$.

It is clear that $\mathcal{M}$ is faithful if and only if any nonzero object of $\mathcal{C}$ acts by a nonzero functor in $\mathcal{M}$. Note also that if $\mathcal{C}$ is an indecomposable multitensor category then any nonzero exact module category $\mathcal{M}$ over $\mathcal{C}$ is faithful.

Remark 7.12.10. In what follows we will always assume that module categories are faithful. Many statements for faithful module categories have obvious extensions to the non-faithful case.

Theorem 7.12.11. *Suppose that $\mathcal{M}$ is faithful. Then the functor*

$$\mathsf{can} : \mathcal{C} \to (\mathcal{C}^*_{\mathcal{M}})^*_{\mathcal{M}} \tag{7.35}$$

is a tensor equivalence of categories.

Proof. By Theorem 7.10.1 there is an algebra A in $\mathcal{C}$ such that $\mathcal{M} \cong \mathsf{Mod}_{\mathcal{C}}(A)$. The category $\mathcal{C}^*_{\mathcal{M}}$ is identified with the category $\mathsf{Bimod}_{\mathcal{C}}(A)^{\mathrm{op}}$. The category $(\mathcal{C}^*_{\mathcal{M}})^*_{\mathcal{M}}$ is identified with the category of B-bimodules in the category of A-bimodules (here B is the same as in Example 7.12.8 and is considered as an algebra in the category of A-modules). But this latter category is tautologically identified with the category of B-bimodules (here we view B as an algebra in $\mathcal{C}$) since for any B-module one reconstructs the A-module structure via the algebra embedding (7.34). It follows from Example 7.10.2 that any B-bimodule is of the form ${}^*A \otimes X \otimes A$ and it is easy to see that $\mathsf{can}(X) = {}^*A \otimes X \otimes A$. This is clearly an equivalence, using the faithfulness of $\mathcal{M}$. □

REMARK 7.12.12. Theorem 7.12.11 categorifies the classical "double centralizer theorem" for projective modules, which says that if A is a finite dimensional algebra and P is a faithful projective A-module then the centralizer of $\mathsf{End}_A(P)$ in $\mathsf{End}_{\Bbbk}(P)$ is A.

COROLLARY 7.12.13. *Assume that $\mathcal{C}$ is a finite tensor (not only multitensor) category. Then an exact module category $\mathcal{M}$ over $\mathcal{C}$ is indecomposable over $\mathcal{C}^*_{\mathcal{M}}$.*

PROOF. This is an immediate consequence of Theorem 7.12.11 and Lemma 7.12.6. □

Let $\mathcal{C}$ be a multitensor category, and $\mathcal{M}$ be a fixed module category over $\mathcal{C}$. For any other module category $\mathcal{M}_1$ over $\mathcal{C}$ the category $\mathsf{Fun}_{\mathcal{C}}(\mathcal{M}_1, \mathcal{M})$ has an obvious structure of a module category over $\mathcal{C}^*_{\mathcal{M}} = \mathsf{Fun}_{\mathcal{C}}(\mathcal{M}, \mathcal{M})$.

PROPOSITION 7.12.14. *The module category $\mathsf{Fun}_{\mathcal{C}}(\mathcal{M}_1, \mathcal{M})$ over $\mathcal{C}^*_{\mathcal{M}}$ is exact.*

PROOF. Assume that $\mathcal{M} = \mathsf{Mod}_{\mathcal{C}}(A)$ and $\mathcal{M}_1 = \mathsf{Mod}_{\mathcal{C}}(A_1)$ for some algebras A and A_1 in $\mathcal{C}$. Identify $\mathcal{C}^*_{\mathcal{M}}$ with the category of A-bimodules and $\mathsf{Fun}_{\mathcal{C}}(\mathcal{M}_1, \mathcal{M})$ with the category of (A_1, A)-bimodules. Any projective object of $\mathsf{Bimod}_{\mathcal{C}}(A)$ is a direct summand of an object of the form $A \otimes P \otimes A$ for some projective $P \in \mathcal{C}$. Let M be an (A_1, A)-bimodule, then $M \otimes_A (A \otimes P \otimes A) \cong M \otimes P \otimes A$. By Lemma 7.8.12

$$\mathsf{Hom}_{A_1-A}(M \otimes P \otimes A, -) \cong \mathsf{Hom}_{A_1}(M \otimes P, -),$$

so it is enough to check that $M \otimes P$ is a projective left A_1-module. This is equivalent to $(M \otimes P)^*$ being injective (since $N \mapsto N^*$ is an equivalence between the category of left A-modules and the category of right A-modules). But $(M \otimes P)^* = P^* \otimes M^*$, and the result follows from projectivity of P^* and Lemma 7.6.3. □

REMARK 7.12.15. Let $\mathcal{C}$ be a multitensor category. The collection of all exact $\mathcal{C}$-module categories forms a 2-category $\mathsf{Mod}(\mathcal{C})$ (the 1-morphisms in $\mathsf{Mod}(\mathcal{C})$ are $\mathcal{C}$-module functors, and 2-morphisms are natural transformations of $\mathcal{C}$-module functors, see Definition 7.2.2).

THEOREM 7.12.16. *Let $\mathcal{M}$ be a faithful exact module category over $\mathcal{C}$. The 2-functor*

$$\mathcal{N} \mapsto \mathsf{Fun}_{\mathcal{C}}(\mathcal{M}, \mathcal{N}) : \mathsf{Mod}(\mathcal{C}) \to \mathsf{Mod}((\mathcal{C}^*_{\mathcal{M}})^{\mathrm{op}}) \tag{7.36}$$

is a 2-equivalence.

PROOF. This is similar to the proof of Theorem 7.12.11 and is left to the reader. □

DEFINITION 7.12.17. Let $\mathcal{C}$, $\mathcal{D}$ be tensor categories. We will say that $\mathcal{C}$ and $\mathcal{D}$ are *categorically Morita equivalent* if there is an exact $\mathcal{C}$-module category $\mathcal{M}$ and a tensor equivalence $\mathcal{D}^{\mathrm{op}} \cong \mathcal{C}^*_{\mathcal{M}}$.

Note that by Remark 7.12.5 the category $\mathcal{D}$ is categorically Morita equivalent to $\mathcal{C}$ if and only if there exists an exact algebra $A \in \mathcal{C}$ such that $\mathcal{D}$ is tensor equivalent to $\mathsf{Bimod}_{\mathcal{C}}(A)$.

PROPOSITION 7.12.18. *Categorical Morita equivalence is an equivalence relation.*

PROOF. The reflexivity of categorical Morita equivalence follows from Example 7.12.3. Its symmetry is proved in Theorem 7.12.11.

So it remains to prove that categorical Morita equivalence is transitive. Let $A \in \mathcal{C}$ be an exact algebra. Note that an algebra $B \in \mathsf{Bimod}_{\mathcal{C}}(A)$ is the same as an algebra $B \in \mathcal{C}$ endowed with a homomorphism $A \to B$. It is a tautology that a B-bimodule in the category $\mathsf{Bimod}_{\mathcal{C}}(A)$ is the same as a B-bimodule in the category $\mathcal{C}$, and this correspondence is compatible with the tensor product over B. Thus, in order to prove the transitivity, we just need to show that an exact algebra $B \in \mathsf{Bimod}_{\mathcal{C}}(A)$ is exact when considered as an algebra in $\mathcal{C}$.

Let $P \in \mathcal{C}$ be a projective object and let $M \in \mathsf{Mod}_{\mathcal{C}}(B)$. We need to show that $P \otimes M \in \mathsf{Mod}_{\mathcal{C}}(B)$ is projective. Note that $A \otimes P \otimes A \in \mathsf{Bimod}_{\mathcal{C}}(A)$ is projective and $A \otimes M$ is a B-module in the category $\mathsf{Bimod}_{\mathcal{C}}(A)$. Since the algebra B considered as an algebra in $\mathsf{Bimod}_{\mathcal{C}}(A)$ is exact, we see that $A \otimes P \otimes A \otimes_A A \otimes M = A \otimes P \otimes A \otimes M$ is projective in the category of B-modules in $\mathsf{Bimod}_{\mathcal{C}}(A)$, which is the same as the category $\mathsf{Bimod}_{\mathcal{C}}(A, B)$. We claim that $A \otimes P \otimes A \otimes M$ is projective when considered as an object of $\mathsf{Mod}_{\mathcal{C}}(B)$. Indeed,

$$\begin{aligned}&\mathsf{Hom}_B(A \otimes P \otimes A \otimes M, -)\\ &\qquad = \mathsf{Hom}_B(P \otimes A \otimes M, {}^*A \otimes -) = \mathsf{Hom}_{A-B}(A \otimes P \otimes A \otimes M, {}^*A \otimes -),\end{aligned}$$

so the functor $\mathsf{Hom}_B(A \otimes P \otimes A \otimes M, -)$ is isomorphic to a composition of exact functors. Finally, P is a subobject of $A \otimes P \otimes A$, hence P is a direct summand of $A \otimes P \otimes A$ by Proposition 6.1.3. Hence, the B-module $P \otimes M$ is a direct summand of the B-module $A \otimes P \otimes A \otimes M$, so it is projective. □

Let us consider examples of categorical Morita equivalence.

EXAMPLE 7.12.19. Let G be a finite group and let $\mathcal{C} = \mathsf{Vec}_G$ be the category of G-graded vector spaces. The category Vec is an exact Vec_G-module category via the forgetful tensor functor $\mathsf{Vec}_G \to \mathsf{Vec}$. Let us determine the dual category $(\mathsf{Vec}_G)^*_{\mathsf{Vec}}$. By definition, a Vec_G-module endofunctor F of Vec consists of a vector space $V := F(\Bbbk)$ and a collection of isomorphisms

$$s_g \in \mathsf{Hom}_{\mathsf{Vec}}(F(\delta_g \otimes \Bbbk), \delta_g \otimes F(\Bbbk)) = \mathsf{End}_{\Bbbk}(V),\ g \in G.$$

It follows from axiom (7.6) in Definition 7.2.1 of module functor that the map $g \mapsto s_g : G \to GL(V)$ is a representation of G on V. Conversely, any such representation determines a Vec_G-module endofunctor of Vec. It is easy to check that homomorphisms of representations are precisely morphisms between the corresponding module functors. Thus, $(\mathsf{Vec}_G)^*_{\mathsf{Vec}} \cong \mathsf{Rep}(G)$, i.e., the categories Vec_G and $\mathsf{Rep}(G)$ are categorically Morita equivalent, cf. Remark 4.15.7.

COROLLARY 7.12.20. *Let G be a finite group. Any indecomposable exact $\mathsf{Rep}(G)$-module category is equivalent to $\mathsf{Rep}_\psi(L)$ for some subgroup $L \subset G$ and a 2-cocycle $\psi \in Z^2(L, \Bbbk^\times)$.*

PROOF. This follows from Theorem 7.12.16, Example 7.12.19, and Example 7.4.10. □

Corollary 7.12.20 implies the classification of twists J (see Definition 5.14.1) for the group algebra $\Bbbk G$ of a finite group G, obtained by Movshev, [**Mov**]. Namely, let J be a twist for $\Bbbk G$. Then J defines a Hopf algebra $H = \Bbbk G^J$ and hence a fiber functor $F_J : \mathsf{Rep}(G) \to \mathsf{Vec}$, which is the forgetful functor with the tensor structure defined by J. Such a functor endows Vec with a structure of a module category over $\mathsf{Rep}(G)$. By Corollary 7.12.20, this module category is of the form $\mathsf{Rep}_\psi(L)$ for some (L, ψ). Moreover, since $\mathsf{Rep}_\psi(L) = \mathsf{Vec}$, the twisted group algebra $\Bbbk L_\psi$ must be simple.

DEFINITION 7.12.21. A 2-cocycle ψ on a finite group L is said to be *non-degenerate* if the twisted group algebra $\Bbbk L_\psi$ is simple. A group L which admits a nondegenerate 2-cocycle is called a group of *central type*.

EXERCISE 7.12.22. (i) Let $L = A \oplus A^\vee$, where A is a finite abelian group, and $\psi((a_1, f_1), (a_2, f_2)) = f_1(a_2)$, where $a_i \in A$, $f_i \in A^\vee$. Show that ψ is a nondegerate 2-cocycle, so L is of central type.

(ii) More generally, let a finite group K act on an abelian group A, let $L = K \ltimes A^\vee$, and let $\pi : K \to A$ be a bijective 1-cocycle (this reduces to (i) in the case $K = A$, $\pi = \mathrm{id}$). Show that L is a group of central type, with $\psi((k_1, f_1), (k_2, f_2)) = f_1(\pi(k_2))$.

Hint: Construct an irreducible projective action of L on $\mathsf{Fun}(A, \Bbbk)$ by $(k\phi)(a) = \phi(k^{-1}a + \pi(k))$, $k \in K$, and $(f\phi)(a) = f(a)\phi(a)$, $f \in A^\vee$.

(iii) Show that if ψ is a nondegenerate 2-cocycle on L then $\Bbbk L_\psi$ is isomorphic to the regular representation of L (where L acts on $\Bbbk L_\psi$ by conjugation).

Hint: Let V be the irreducible $\Bbbk L_\psi$-module, and $\hat{L}$ be a central extension of L by a finite cyclic group acting linearly in V, so that $\Bbbk L_\psi = V \otimes V^*$ as a $\hat{L}$-module. Show that $g \in \hat{L}$ has zero trace in V unless g projects to $1 \in L$. Deduce the same statement about the trace of g in $V \otimes V^*$, and conclude the required statement.

(iv) Prove the converse to (iii).

Hint: Use that $\Bbbk L_\psi$ is semisimple (why?) and the center of $\Bbbk L_\psi$ carries a trivial action of L.

(v) Show that if ψ is a 2-cocycle on L then the function $\chi_x(y) := \psi(x, y)/\psi(y, x)$ is a character of the centralizer Z_x of x in L, and ψ is nondegenerate if and only if χ_x is nontrivial for every $x \neq 1$.

Hint (for the second statement): Compute the trace of the conjugation operator by x on $\Bbbk L_\psi$, using the basis of group elements. Then use (iii), (iv).

(vi) Let $\mathfrak{l}$ be a nilpotent finite dimensional Lie algebra over a prime field $\mathbb{F}_p$, $p > 2$, whose nilpotency index (i.e., length of the lower central series) is $< p$. In this case, we can define the group $L = \exp(\mathfrak{l})$, which is the same as $\mathfrak{l}$ as a set, with multiplication defined using the Campbell-Hausdorff series

$$x * y = CH(x, y) = x + y + \frac{1}{2}[x, y] + \ldots$$

(this is well defined since the length is $< p$). Now assume that $\mathfrak{l}$ is a nilpotent Lie algebra of length $< p - 1$ which is quasi-Frobenius, i.e., carries a non-degenerate 2-cocycle $\omega \in \wedge^2\mathfrak{l}^*$ (i.e., $\omega([x, y], z) + \omega([y, z], x) + \omega([z, x], y) = 0$). Show that the group $L = \exp(\mathfrak{l})$ is of central type, and compute its nondegenerate 2-cocycle.

Hint: Use that 2-cocycles correspond to central extensions, for both groups and Lie algebras.

(vii) Give examples of nonabelian nilpotent quasi-Frobenius Lie algebras.

It is clear that the order of a group of central type is a square. A much deeper result is the following theorem of Howlett and Isaacs (conjectured by Iwahori and Matsumoto in 1964):

THEOREM 7.12.23. [**HoI**] *Any group of central type is solvable.*

The proof is based on the method of Liebler and Yellen, which uses the classification of finite simple groups. [5]

Thus, we see that the subgroup L obtained from a twist J must be a group of central type (in particular, its order is a square and it is solvable), and J gives rise to a non-degenerate 2-cocycle ψ on L. Conversely, given a central type subgroup $L \subset G$ and a non-degenerate 2-cocycle ψ on L, the algebra $\Bbbk L_\psi$ is simple, so $\mathsf{Rep}_\psi(L)$ is a module category over $\mathsf{Rep}(G)$ equivalent to Vec, which gives rise to a fiber functor on $\mathsf{Rep}(G)$ and thus a twist J. It is easy to see that this correspondence is a bijection between equivalence classes. Thus, we have the following classification of twists in $\Bbbk G$.

COROLLARY 7.12.24. *The above assignments $J \mapsto (L, \psi)$, $(L, \psi) \mapsto J$ are mutually inverse bijections between the gauge equivalence classes of twists J for $\Bbbk G$ and conjugacy classes of pairs (L, ψ), where L is a subgroup of G and ψ is a non-degenerate 2-cocycle on L.*

For more details, see [**Mov**], [**EtG3**].

EXAMPLE 7.12.25. This is a generalization of Example 7.12.19. Let G be a finite group with an action $g \mapsto T_g$ on a finite tensor category $\mathcal{C}$. Recall the equivariantization category $\mathcal{C}^G$ introduced in Section 2.7 and the crossed product category $\mathcal{C} \rtimes G$ introduced in Definition 4.15.5.

Note that $\mathcal{C}$ has a natural structure of a $\mathcal{C} \rtimes G$-module category:

$$(X \boxtimes g) \otimes V = T_g(X \otimes V),$$

for all objects X, V in $\mathcal{C}$ and $g \in G$.

We claim that the dual category $(\mathcal{C} \rtimes G)^*_{\mathcal{C}}$ is equivalent to $(\mathcal{C}^G)^{\mathrm{op}}$, i.e., that $\mathcal{C} \rtimes G$ and $\mathcal{C}^G$ are categorically Morita equivalent. (Note that for $\mathcal{C} = \mathsf{Vec}$ this reduces to Example 7.12.19).

Indeed, let $F \in (\mathcal{C} \rtimes G)^*_{\mathcal{C}}$ be a $(\mathcal{C} \rtimes G)$-module endofunctor of $\mathcal{C}$. In particular, F is a $\mathcal{C}$-module functor, hence $F(V) = V \otimes X$ for some X in $\mathcal{C}$. It is straightforward to check (and the reader is invited to do so) that a $(\mathcal{C} \rtimes G)$-module functor structure on the latter functor is the same thing as a G-equivariant structure on X.

EXAMPLE 7.12.26. Here is another generalization of Example 7.12.19. Let H be a finite dimensional Hopf algebra. The category Vec is a $\mathsf{Rep}(H)$-module category via the fiber functor $F : \mathsf{Rep}(H) \to \mathsf{Vec}$, see Section 5.1. Let us determine the corresponding dual category. Any $\Bbbk$-linear additive endofunctor E of Vec is completely determined by a vector space V such that $E(\Bbbk) = V$. The $\mathsf{Rep}(H)$-module functor structure on E gives rise to a natural isomorphism

$$F(X) \otimes_\Bbbk V \xrightarrow{\sim} F(X) \otimes_\Bbbk V, \qquad X \in \mathcal{C},$$

[5] Our terminology is slightly different from that of Howlett and Isaacs. Namely, they use the term "central type" for the central extension of L by a cyclic group obtained from the cocycle ψ.

and, hence to a homomorphism $\pi_V : V \to \mathsf{End}(F) \otimes_{\Bbbk} V$. As we know from the reconstruction theory (see Sections 5.2 and 5.3) there is a canonical Hopf algebra structure on $\mathsf{End}(F)$. Furthermore, the module functor axioms (7.6) and (7.7) are equivalent to π_V being an $\mathsf{End}(F)$-comodule structure on V. By Theorem 5.2.3 there is a Hopf algebra isomorphism $\mathsf{End}(F) \cong H$. So V is an H-comodule, and hence, an H^*-module. It follows that $\mathsf{Rep}(H)^*_{\mathsf{Vec}} \cong \mathsf{Rep}(H^*)$. Thus, the notion of categorical duality extends that of duality for Hopf algebras (namely, $\mathsf{Rep}(H)$ is categorically Morita equivalent to $\mathsf{Rep}(H^{*\mathrm{cop}})$, where "cop" denotes the opposite coproduct).

EXERCISE 7.12.27. Fill the details in Example 7.12.26. In particular, prove that π_V is indeed a $\mathsf{End}(F)$-comodule structure on V and that the corresponding functor $\mathsf{Rep}(H)^*_{\mathsf{Vec}} \to \mathsf{Rep}(H^*)$ is tensor.

Let $\mathcal{M}$ be an exact module category over $\mathcal{C}$. For $X, Y \in \mathcal{M}$ we have two notions of internal Hom: one taking values in $\mathcal{C}$ and another taking values in $\mathcal{C}^*_{\mathcal{M}}$, denoted by $\underline{\mathsf{Hom}}_{\mathcal{C}}$ and $\underline{\mathsf{Hom}}_{\mathcal{C}^*_{\mathcal{M}}}$, respectively. The following consequence of calculations in Examples 7.9.8 and 7.12.8 is very useful.

PROPOSITION 7.12.28. *Let $X, Y, Z \in \mathcal{M}$. There is a canonical isomorphism*

$$\underline{\mathsf{Hom}}_{\mathcal{C}}(X, Y) \otimes Z \cong {}^*\underline{\mathsf{Hom}}_{\mathcal{C}^*_{\mathcal{M}}}(Z, X) \otimes Y. \tag{7.37}$$

PROOF. By Theorem 7.12.11 it suffices to find a canonical isomorphism

$$^*\underline{\mathsf{Hom}}_{\mathcal{C}}(Z, X) \otimes Y \cong \underline{\mathsf{Hom}}_{\mathcal{C}^*_{\mathcal{M}}}(X, Y) \otimes Z, \quad X,\, Y,\, Z \in \mathcal{M}. \tag{7.38}$$

This isomorphism is constructed as follows. Choose an algebra A such that $\mathcal{M} = \mathsf{Mod}_{\mathcal{C}}(A)$. By Example 7.9.8 the left hand side of (7.38) is $^*(X \otimes^A Z^*) \otimes Y \cong {}^*((Z \otimes_A {}^*X)^*) \otimes Y \cong (Z \otimes_A {}^*X) \otimes Y$. On the other hand, by Example 7.12.8 the right hand side of (7.38) is $Z \otimes_A ({}^*X \otimes Y)$. Thus, the associativity constraint provides a canonical isomorphism (7.38). Note that the inverse to this isomorphism is the image of id_Y under the following morphism

$$\begin{aligned}
\mathsf{Hom}_{\mathcal{M}}(Y, Y) &\to \mathsf{Hom}_{\mathcal{M}}(\underline{\mathsf{Hom}}_{\mathcal{C}^*_{\mathcal{M}}}(X, Y) \otimes X,\, Y) \to \\
&\mathsf{Hom}_{\mathcal{M}}(\underline{\mathsf{Hom}}_{\mathcal{C}^*_{\mathcal{M}}}(X, Y) \otimes \underline{\mathsf{Hom}}_{\mathcal{C}}(Z, X) \otimes Z, Y) \cong \\
&\mathsf{Hom}_{\mathcal{M}}(\underline{\mathsf{Hom}}_{\mathcal{C}}(Z, X) \otimes \underline{\mathsf{Hom}}_{\mathcal{C}^*_{\mathcal{M}}}(X, Y) \otimes Z, Y) \cong \\
&\qquad \mathsf{Hom}_{\mathcal{M}}(\underline{\mathsf{Hom}}_{\mathcal{C}^*_{\mathcal{M}}}(X, Y) \otimes Z, {}^*\underline{\mathsf{Hom}}_{\mathcal{C}}(Z, X) \otimes Y),
\end{aligned}$$

where the arrows are given by the evaluation morphism (7.28). Thus, isomorphism (7.37) does not depend on the choice of the algebra A. □

REMARK 7.12.29. The isomorphism (7.37) is $\mathcal{C}$-equivariant in the following sense. Given $W \in \mathcal{C}$, the diagram

$$\begin{array}{ccc}
\underline{\mathsf{Hom}}_{\mathcal{C}}(X, W \otimes Y) \otimes Z & \xrightarrow{\sim} & {}^*\underline{\mathsf{Hom}}_{\mathcal{C}^*_{\mathcal{M}}}(Z, X) \otimes (W \otimes Y) \\
\downarrow \wr & & \downarrow \wr \\
W \otimes \underline{\mathsf{Hom}}_{\mathcal{C}}(X, Y) \otimes Z & \xrightarrow{\sim} & W \otimes {}^*\underline{\mathsf{Hom}}_{\mathcal{C}^*_{\mathcal{M}}}(Z, X) \otimes Y,
\end{array} \tag{7.39}$$

where the horizontal arrows are given by isomorphism (7.37), the left vertical arrow is given by isomorphism (7.25), and the right vertical arrow comes from the

$\mathcal{C}^*_{\mathcal{M}}$-module functor structure on W (since $\mathcal{C}$ is identified with $(\mathcal{C}^*_{\mathcal{M}})^*_{\mathcal{M}}$ by Theorem 7.12.11).

7.13. The center construction

The notion of the center of a monoidal category categorifies that of the center of a ring.

Let $\mathcal{C}$ be a monoidal category with the associativity constraint

$$a_{X,Y,Z} : (X \otimes Y) \otimes Z \xrightarrow{\sim} X \otimes (Y \otimes Z), \quad X, Y, Z \in \mathcal{C}.$$

DEFINITION 7.13.1. The *center* of $\mathcal{C}$ is the category $\mathcal{Z}(\mathcal{C})$ defined as follows. The objects of $\mathcal{Z}(\mathcal{C})$ are pairs (Z, γ), where $Z \in \mathcal{C}$ and

$$\gamma_X : X \otimes Z \xrightarrow{\sim} Z \otimes X, \quad X \in \mathcal{C} \tag{7.40}$$

is a natural isomorphism such that the following diagram

(7.41)

$$\begin{array}{ccccc} & X\otimes(Z\otimes Y) & \xrightarrow{\alpha^{-1}_{X,Z,Y}} & (X\otimes Z)\otimes Y & \\ {\scriptstyle \mathrm{id}_X\otimes\gamma_Y}\nearrow & & & & \searrow{\scriptstyle \gamma_X\otimes\mathrm{id}_Y} \\ X\otimes(Y\otimes Z) & & & & (Z\otimes X)\otimes Y, \\ {\scriptstyle\alpha^{-1}_{X,Y,Z}}\searrow & & & & \nearrow{\scriptstyle \alpha^{-1}_{Z,X,Y,}} \\ & (X\otimes Y)\otimes Z & \xrightarrow[\gamma_{X\otimes Y}]{} & Z\otimes(X\otimes Y) & \end{array}$$

is commutative for all $X, Y \in \mathcal{C}$.

A morphism from (Z, γ) to (Z', γ') is a morphism $f \in \mathsf{Hom}_{\mathcal{C}}(Z, Z')$ such that for each $X \in \mathcal{C}$ we have $(f \otimes \mathsf{id}_X) \circ \gamma_X = \gamma'_X \circ (\mathsf{id}_X \otimes f)$.

It turns out that $\mathcal{Z}(\mathcal{C})$ is a monoidal category with tensor product defined as follows. If (Z, γ) and (Z', γ') are objects in $\mathcal{Z}(\mathcal{C})$ then

$$(Z, \gamma) \otimes (Z', \gamma') := (Z \otimes Z', \tilde{\gamma}),$$

where $\tilde{\gamma}_X : X \otimes (Z \otimes Z') \xrightarrow{\sim} (Z \otimes Z') \otimes X$, $X \in \mathcal{C}$, is defined by the following commutative diagram:

$$\begin{array}{ccccc} X\otimes(Z\otimes Z') & \xrightarrow{\alpha^{-1}_{X,Z,Z'}} & (X\otimes Z)\otimes Z' & \xrightarrow{\gamma_X\otimes\mathrm{id}_{Z'}} & (Z\otimes X)\otimes Z' \\ \downarrow{\scriptstyle\tilde{\gamma}_X} & & & & \downarrow{\scriptstyle\alpha_{Z,X,Z'}} \\ (Z\otimes Z')\otimes X & \xleftarrow{\alpha^{-1}_{Z,Z',X}} & Z\otimes(Z'\otimes X) & \xleftarrow{\mathrm{id}_Z\otimes\gamma'_X} & Z\otimes(X\otimes Z'). \end{array} \tag{7.42}$$

The unit object of $\mathcal{Z}(\mathcal{C})$ is $(\mathbf{1}, r^{-1}l)$, where r and l are the unit constraints.

EXERCISE 7.13.2. Show that the associativity constraint of $\mathcal{C}$ is a morphism in $\mathcal{Z}(\mathcal{C})$ and so the latter is indeed a monoidal category.

If $Z \in \mathcal{C}$ has a left dual Z^* then (Z, γ) has a left dual $(Z^*, \overline{\gamma})$, where $\overline{\gamma}_X := (\gamma^{-1}_{{}^*X})^*$. Similarly for right duals. Thus, if $\mathcal{C}$ is rigid then so is $\mathcal{Z}(\mathcal{C})$.

REMARK 7.13.3. Isomorphisms (7.40) equip $\mathcal{Z}(\mathcal{C})$ with an additional structure called *braiding*. Tensor categories with such structures are called *braided*. We study braided categories in Chapter 8.

REMARK 7.13.4. Note that there is an obvious forgetful monoidal functor

$$F : \mathcal{Z}(\mathcal{C}) \to \mathcal{C} : (Z, \gamma) \mapsto Z. \tag{7.43}$$

EXERCISE 7.13.5. Show that $(Z, \gamma) \mapsto (Z, \gamma^{-1}) : \mathcal{Z}(\mathcal{C}) \to \mathcal{Z}(\mathcal{C}^{\mathrm{op}})$ is a tensor equivalence.

EXERCISE 7.13.6. Show that if $\mathcal{C}$ is a pivotal category, then so is its center $\mathcal{Z}(\mathcal{C})$.

Hint: Check that if $Z \in \mathcal{Z}(\mathcal{C})$ and a is a pivotal structure on $\mathcal{C}$, then $\gamma_{Z^{**}} \circ (a_Z \otimes \mathsf{id}) = (\mathsf{id} \otimes a_Z) \circ \gamma_Z$. Use functoriality of $\gamma_{Z,X}$ in X, the fact that $\gamma_{Z,Y}^{**} = \gamma_{Z^{**},Y^{**}}$ (the definition of $\gamma_{Z^{**}}$), tensoriality of a_X and functoriality of a_X with respect to $X \in \mathcal{C}$.

EXERCISE 7.13.7. Show that the center of an indecomposable multitensor category is a tensor category (i.e., show that it is abelian, and $\mathsf{End}(\mathbf{1}) = \Bbbk$).

Let $\mathcal{C}$ be a finite multitensor category. Then $\mathcal{C}$ is an exact module category over $\mathcal{C} \boxtimes \mathcal{C}^{\mathrm{op}}$, i.e., a $\mathcal{C}$-bimodule category.

PROPOSITION 7.13.8. *There is a canonical equivalence* $(\mathcal{C} \boxtimes \mathcal{C}^{\mathrm{op}})^*_{\mathcal{C}} \cong \mathcal{Z}(\mathcal{C})$. *In particular, the center of a finite multitensor category is finite.*

PROOF. Let $F : \mathcal{C} \to \mathcal{C}$ be a $\mathcal{C}$-bimodule endofunctor. Since F is, in particular, a left $\mathcal{C}$-module functor, we have $F = - \otimes Z$ for some $Z \in \mathcal{C}$. Since F is also a right $\mathcal{C}$-module functor, we must have a natural isomorphism

$$(X \otimes Y) \otimes Z = F(X \otimes Y) \xrightarrow{\sim} F(X) \otimes Y = (X \otimes Z) \otimes Y, \quad X, Y \in \mathcal{C}.$$

Taking $X = \mathbf{1}$ we obtain a natural isomorphism

$$\gamma : (- \otimes Z) \xrightarrow{\sim} (Z \otimes -). \tag{7.44}$$

The compatibility conditions (7.41) for the components $\gamma_X : X \otimes Z \xrightarrow{\sim} Z \otimes X$ come from the axiom (7.6) of a module functor. Finally, one checks that composition of $\mathcal{C}$-bimodule endofunctors of $\mathcal{C}$ corresponds to the tensor product of objects of the center and that this correspondence is a tensor functor. □

EXERCISE 7.13.9. Verify the last statement in the proof of Proposition 7.13.8.

REMARK 7.13.10. Proposition 7.13.8 shows that the above notion of the center of $\mathcal{C}$ categorifies the notion of the center of a ring. Indeed, the center of a ring R is isomorphic to the ring of R-bimodule endomorphisms of R.

COROLLARY 7.13.11. *The forgetful functor* (7.43) *is surjective (in the sense of Definition* 1.8.3*).*

PROOF. This is an immediate consequence of Corollary 7.12.13. □

7.14. The quantum double construction for Hopf algebras

In this section we will explain what the center construction gives, in more explicit terms, for the category of representations of a finite dimensional Hopf algebra. It turns out that in this case the center construction reduces to Drinfeld's quantum double construction.

Namely, let H be a finite dimensional Hopf algebra with coproduct Δ, counit ε and antipode S, and let $\mathcal{C}$ be the category $\mathsf{Rep}(H)$ of finite dimensional representations of H. Then by the reconstruction theory for Hopf algebras, we have a natural forgetful functor $F : \mathcal{C} \to \mathsf{Vec}$, such that $H = \mathsf{End}\, F$. Now, $\mathcal{Z}(\mathcal{C})$ is a finite tensor category with a natural tensor functor $\mathcal{Z}(\mathcal{C}) \to \mathcal{C}$. Composing this functor with F, we get a tensor functor $\widetilde{F} : \mathcal{Z}(\mathcal{C}) \to \mathsf{Vec}$. Thus we have a finite dimensional Hopf algebra $D := \mathsf{End}\, \widetilde{F}$.

DEFINITION 7.14.1. The Hopf algebra D is called the *quantum double* of H and is denoted by $D(H)$.

Let us describe the structure of $D(H)$ more explicitly. For this purpose, let $Z \in \mathcal{Z}(\mathcal{C})$. Then Z is a representation of H with some additional structure. Let us spell this structure out. The additional structure is a map $\gamma_X : X \otimes Z \to Z \otimes X$ for each $X \in \mathcal{C}$ such that diagram (7.41) commutes. We have $\gamma_X = \sigma \circ R_{XZ}$, where $R_{XZ} : X \otimes Z \to X \otimes Z$ is a linear isomorphism, and σ is the permutation of components. The map R_{XZ} is functorial in X, so it suffices to know it for $X = H$ being the free rank 1 H-module. Now, R_{HZ} commutes with right multiplications by elements of H in the first component, so we can view R_{HZ} as an element of $H \otimes \mathsf{End}_{\Bbbk}(Z)$.

EXERCISE 7.14.2. Show that the commutative diagram for γ is equivalent to the equation $R^{13}_{HZ} R^{23}_{HZ} = (\Delta \otimes \mathsf{id})(R_{HZ})$.

Define a linear map $\rho : H^* \otimes Z \to Z$ by the formula $\rho(f \otimes z) = (f \otimes \mathsf{id})(R_{HZ})z$. Exercise 7.14.2 implies that ρ is an action of H^* on Z. Thus, by definition, R_{HZ} is the action on $H \otimes Z$ of the canonical element of $H \otimes H^*$, $R := \sum_i h_i \otimes h_i^*$, where h_i is a basis of H and h_i^* is the dual basis of H^* (here h_i^* acts via ρ). Note that $(\Delta \otimes \mathsf{id})(R) = R^{13} R^{23}$.

PROPOSITION 7.14.3. *Let $H^{*\mathrm{cop}}$ be the Hopf algebra H^* with the opposite coproduct (so that the antipode of $H^{*\mathrm{cop}}$ is inverse to the antipode S of H^*). Then R is invertible in $H \otimes H^{*\mathrm{cop}}$ with*

$$R^{-1} = (S \otimes \mathsf{id})(R) = (\mathsf{id} \otimes S^{-1})(R). \tag{7.45}$$

EXERCISE 7.14.4. Prove Proposition 7.14.3.

Thus, we have an action of the free product [6] $H * H^{*\mathrm{cop}}$ on Z. However, there are certain commutation relations between H and $H^{*\mathrm{cop}}$.

To write these relations down, note that γ_X is an H-morphism, which implies that $R_{HZ}\Delta(a) = \Delta^{op}(a) R_{HZ}$ for $a \in H$. Thus, $R^{12}(\Delta \otimes \mathsf{id})(R) = (\Delta^{op} \otimes \mathsf{id})(R) R^{12}$. Thus, we have the *quantum Yang-Baxter equation* in $H \otimes Z \otimes H^{*\mathrm{cop}}$:

$$R^{12} R^{13} R^{23} = R^{23} R^{13} R^{12}.$$

Using (7.45), this can be rewritten as

$$R^{12} (\mathsf{id} \otimes S^{-1})(R)^{23} = (\mathsf{id} \otimes S^{-1})(R)^{13} (\mathsf{id} \otimes S^{-1})(R)^{23} R^{12} R^{13}$$

In components, this looks like

$$\sum_{i,j} h_i \otimes h_i^* h_j \otimes S^{-1}(h_j^*) = \sum_{p,q,r,s} h_p h_r h_s \otimes h_q h_r^* \otimes S^{-1}(h_p^*) S^{-1}(h_q^*) h_s^*.$$

[6] Recall that for two unital algebras A_1, A_2, the *free product* $A_1 * A_2$ is the unital algebra such that $\mathsf{Hom}(A_1 * A_2, B) = \mathsf{Hom}(A_1, B) \times \mathsf{Hom}(A_2, B)$ for any unital algebra B. If $A_i = TV_i/(R_i)$, $i = 1, 2$, where TV_i is the tensor algebra of a space V_i, and $R_i \subset TV_i$ is the space of relations, then $A_1 * A_2 = T(V_1 \oplus V_2)/(R_1 \oplus R_2)$. It is easy to see that if A_1, A_2 are bialgebras or Hopf algebras then so is $A_1 * A_2$, with the coproduct, counit and antipode induced by those on the factors.

Taking the inner product with h_i^* in the first component and with $S^{-1}(h_j)$ in the third component, and using that S preserves the inner product between H and $H^{*\mathrm{cop}}$, we obtain

$$h_i^* h_j = \sum_{p,q,r,s} (h_i^*, h_p h_r h_s)(S^{-1}(h_p^*)S^{-1}(h_q^*)h_s^*, S^{-1}(h_j))h_q h_r^*$$

Using the coproduct rule for the R-matrix, this can be written as

$$h_i^* h_j = \sum_q (S(h_{i1}^*)h_q^* h_{i3}^*, h_j)h_q h_{i2}^*.$$

where we use Sweedler's notation $(\Delta \otimes \mathsf{id})(\Delta(h_i^*)) = h_{i1}^* \otimes h_{i2}^* \otimes h_{i3}^*$ (i.e., a summation is implied).

Finally, we can rewrite this as

$$fh = (f_1, S^{-1}(h_1))(f_3, h_3)h_2 f_2, \tag{7.46}$$

where $h \in H$, $f \in H^{*\mathrm{cop}}$, and $(\Delta \otimes \mathsf{id})(\Delta(h)) = h_1 \otimes h_2 \otimes h_3$ (again Sweedler's notation).

DEFINITION 7.14.5. The quotient of the free product $H * H^{*\mathrm{cop}}$ by the relation (7.46) is called the quantum double (or *Drinfeld double*) of H and denoted by $D(H)$.

Thus, we see that each object of $\mathcal{Z}(\mathcal{C})$ is naturally a module over $D(H)$. Conversely, it is clear that any $D(H)$-module Z is naturally an object of $\mathcal{Z}(\mathcal{C})$. Thus, we obtain

PROPOSITION 7.14.6. *The category $\mathcal{Z}(\mathcal{C})$ is naturally equivalent to the category* $\mathsf{Rep}(D(H))$ *of finite dimensional $D(H)$-modules.*

EXERCISE 7.14.7. Show that the commutation relation (7.46) of $D(H)$ can be rewritten as

$$(f_2, S^{-1}(h_2))f_1 h_1 = (f_1, S^{-1}(h_1))h_2 f_2,$$

and as

$$hf = (f_1, h_1)(f_3, S^{-1}(h_3))f_2 h_2.$$

The properties of $D(H)$ are summarized in the following proposition.

PROPOSITION 7.14.8. *(i) The multiplication map $H \otimes H^{*\mathrm{cop}} \to D(H)$ is an isomorphism of vector spaces; in other words, the algebra $D(H)$ is $H \otimes H^{*\mathrm{cop}}$ with $H, H^{*\mathrm{cop}}$ being subalgebras, and the multiplication defined by*

$$h_p h_i^* \cdot h_j h_q^* = (h_{i1}^*, S^{-1}(h_{j1}))(h_{i3}^*, h_{j3})h_p h_{j2} h_{i2}^* h_q^*. \tag{7.47}$$

In particular, $\dim D(H) = (\dim H)^2$.

(ii) Relation (7.46) defines a Hopf ideal in $H * H^{*\mathrm{cop}}$, and thus $D(H)$ is a Hopf algebra with the coproduct and antipode induced by those of H and $H^{*\mathrm{cop}}$.

(iii) The equivalence of Proposition 7.14.6 *is naturally an equivalence of tensor categories. This equivalence identifies $D(H)$ with the Hopf algebra* $D := \mathsf{End}\,\widetilde{F}$.

EXERCISE 7.14.9. Prove Proposition 7.14.8.

Hint: Show that the product defined by (7.47) is associative and compatible with the coproduct.

REMARK 7.14.10. It is essential that we use the opposite coproduct in H^* in (ii); with the usual coproduct, this construction does not work.

EXERCISE 7.14.11. Let ℓ be an odd integer ≥ 2, and H be the Taft algebra of dimension ℓ^2, see Example 5.5.6. Show that the double $D(H)$ is isomorphic to the tensor product $u_q(\mathfrak{sl}_2) \otimes \Bbbk\mathbb{Z}/\ell\mathbb{Z}$ as a Hopf algebra.

REMARK 7.14.12. A similar result holds for the small quantum group $u_q(\mathfrak{g})$ of any simple Lie algebra $\mathfrak{g}$. Namely, let $u_q(\mathfrak{b})$ denote the subalgebra of $u_q(\mathfrak{g})$ generated by the elements K_i and E_i for all i. Then under appropriate conditions on the root of unity q (namely, if the order of q is coprime to the determinant of the Cartan matrix of $\mathfrak{g}$), one has $D(u_q(\mathfrak{b})) = u_q(\mathfrak{g}) \otimes \Bbbk T$, where T is the group generated by the K_i.

EXERCISE 7.14.13. Find an explicit presentation by generators and relations for the double $D(H)$ for the Nichols Hopf algebra H, see Example 5.5.8.

7.15. Yetter-Drinfeld modules

Now let $\mathcal{C} = K-\mathsf{comod}$, where K is a Hopf algebra not assumed to be finite dimensional. Let us generalize the description of the center $\mathcal{Z}(\mathcal{C})$ from the previous section to this case (the case of the previous section is $K = H^*$, $\dim H < \infty$). To do so, we just need to rewrite the compatibility condition between the actions of H and H^* on an object $Z \in \mathcal{Z}(\mathcal{C})$ in terms of a coaction $\tau : Z \to K \otimes Z$ and an action $\eta : K \otimes Z \to Z$. If we do this, the compatibility relation takes the form:

$$\tau(kz) = k_1 z_1 S(k_3) \otimes k_2 z_2, \tag{7.48}$$

where we use Sweedler's notation $(\Delta \otimes 1)(\Delta(k)) = k_1 \otimes k_2 \otimes k_3$ and $\tau(z) = z_1 \otimes z_2$ (in both cases, implicit summation is assumed).

EXERCISE 7.15.1. Verify relation (7.48).
Hint: use Exercise 7.14.7.

DEFINITION 7.15.2. A Yetter-Drinfeld module over K is a left K-module Z equipped with a left K-comodule structure $\tau : Z \to K \otimes Z$ such that relation (7.48) is satisfied.

The tensor product of Yetter-Drinfeld modules is defined as the tensor product of K^{cop}-modules and of K-comodules. This endows the category of Yetter-Drinfeld modules with a monoidal structure. Moreover, it is easy to see that the category $YD(K)$ of finite dimensional Yetter-Drinfeld modules over K is a tensor category.
Thus, we obtain

PROPOSITION 7.15.3. *Let K be a Hopf algebra. The center of the category $K-\mathsf{comod}$ is naturally equivalent to the category $YD(K)$ as a tensor category.*

EXERCISE 7.15.4. Fill in the details in the proof of Proposition 7.15.3.

7.16. Invariants of categorical Morita equivalence

Let $\mathcal{C}$ be a finite multitensor category and let $\mathcal{M}$ be an exact faithful $\mathcal{C}$-module category. Consider $\mathcal{M}$ as a $(\mathcal{C}\boxtimes\mathcal{C}^*_{\mathcal{M}})$-module category. Clearly, this module category is exact. We have the following generalization of Proposition 7.13.8.

THEOREM 7.16.1. *The category $(\mathcal{C} \boxtimes \mathcal{C}^*_{\mathcal{M}})^*_{\mathcal{M}}$ is canonically equivalent to $\mathcal{Z}(\mathcal{C})$.*

PROOF. An object of $(\mathcal{C} \boxtimes \mathcal{C}_{\mathcal{M}}^{*})_{\mathcal{M}}^{*}$ is, in particular, a $\mathcal{C}_{\mathcal{M}}^{*}$-module functor, and so by Theorem 7.12.11 it is isomorphic to the functor $Z \otimes -$ of multiplication by some object $Z \in \mathcal{C}$. The $(\mathcal{C} \boxtimes \mathcal{C}_{\mathcal{M}}^{*})$-module structure on this functor yields a natural isomorphism of the following endofunctors of $\mathcal{M}$:

$$X \otimes F(Z \otimes -) \cong Z \otimes (X \otimes F(-)), \quad X \in \mathcal{C},\ F \in \mathcal{C}_{\mathcal{M}}^{*}. \tag{7.49}$$

This, in turn, gives rise to an isomorphism between $\mathcal{C}_{\mathcal{M}}^{*}$-module functors $(Z \otimes X) \otimes -$ and $(X \otimes Z) \otimes -$. By Theorem 7.12.11 $X \mapsto (X \otimes -) : \mathcal{C} \mapsto (\mathcal{C}_{\mathcal{M}}^{*})_{\mathcal{M}}^{*}$ is an equivalence, therefore we get a natural isomorphism $\gamma_X : X \otimes Z \xrightarrow{\sim} Z \otimes X$. The compatibility condition of isomorphism (7.49) coming from axioms of a module functor translates to diagram (7.41) from Definition 7.13.1 of the center of $\mathcal{C}$. Thus, the functorial isomorphism $\gamma = \{\gamma_X\}$ makes Z an object of the center of $\mathcal{C}$.

Conversely, given a central object (Z, γ) there is a natural $(\mathcal{C} \boxtimes \mathcal{C}_{\mathcal{M}}^{*})$-module functor structure on $Z \otimes -$. It is clear that the above assignments are tensor functors quasi-inverse to each other. □

COROLLARY 7.16.2. *There is a canonical tensor equivalence*

$$\mathcal{Z}(\mathcal{C}) \cong \mathcal{Z}(\mathcal{C}_{\mathcal{M}}^{*}). \tag{7.50}$$

PROOF. In view of Theorem 7.12.11 the equivalence in Theorem 7.16.1 is symmetric in $\mathcal{C}$ and $\mathcal{C}_{\mathcal{M}}^{*}$. Thus both $\mathcal{Z}(\mathcal{C})$ and $\mathcal{Z}(\mathcal{C}_{\mathcal{M}}^{*})$ are canonically equivalent to $(\mathcal{C} \boxtimes \mathcal{C}_{\mathcal{M}}^{*})_{\mathcal{M}}^{*}$. □

REMARK 7.16.3. Let A be an algebra in $\mathcal{C}$ such that $\mathcal{M} \cong \mathsf{Mod}_{\mathcal{C}}(A)$, see Theorem 7.10.1. Then tensor equivalence (7.50) can be described as follows. For any $(Z, \gamma) \in \mathcal{Z}(\mathcal{C})$ the free right A-module $Z \otimes A$ has a structure of a left A-module via

$$A \otimes (Z \otimes A) \xrightarrow{a_{A,Z,A}^{-1}} (A \otimes Z) \otimes A \xrightarrow{\gamma_A \otimes \mathrm{id}_A} (Z \otimes A) \otimes A \xrightarrow{a_{Z,A,A}} Z \otimes (A \otimes A) \xrightarrow{\mathrm{id}_Z \otimes m} Z \otimes A,$$

where a denotes the associativity constraint of $\mathcal{C}$ and $m : A \otimes A \to A$ is the multiplication of A. It is easy to check that (Z, γ) is an A-bimodule. Furthermore, it has a structure of a central A-bimodule via

$$X \otimes_A (Z \otimes A) \cong X \otimes Z \xrightarrow{\gamma_X} Z \otimes X \cong (Z \otimes A) \otimes_A X, \quad X \in \mathsf{Bimod}_{\mathcal{C}}(A). \tag{7.51}$$

Equivalence (7.50) is identified with

$$Z \mapsto Z \otimes A : \mathcal{Z}(\mathcal{C}) \to \mathcal{Z}(\mathsf{Bimod}_{\mathcal{C}}(A)). \tag{7.52}$$

Note that $\mathsf{Bimod}_{\mathcal{C}}(A) \cong (\mathcal{C}_{\mathcal{M}}^{*})^{\mathrm{op}}$ by Remark 7.12.5 and $\mathcal{Z}((\mathcal{C}_{\mathcal{M}}^{*})^{\mathrm{op}}) \cong \mathcal{Z}(\mathcal{C}_{\mathcal{M}}^{*})$ as tensor categories by Exercise 7.13.5.

REMARK 7.16.4. We will see in Proposition 8.5.3 that (7.52) is in fact an equivalence of *braided* tensor categories.

As we have seen in Proposition 7.13.8, the dual category of $(\mathcal{C} \boxtimes \mathcal{C}^{\mathrm{op}})$ with respect to $\mathcal{C}$ is $\mathcal{Z}(\mathcal{C})$, the center of $\mathcal{C}$. Let $I : \mathcal{C} \to \mathcal{Z}(\mathcal{C})$ be the right adjoint of the forgetful functor $F : \mathcal{Z}(\mathcal{C}) \to \mathcal{C}$ (7.43). We have canonical algebras $I(\mathbf{1}) \in \mathcal{Z}(\mathcal{C})$ and $A := \underline{\mathsf{Hom}}_{\mathcal{C} \boxtimes \mathcal{C}^{\mathrm{op}}}(\mathbf{1}, \mathbf{1}) \in \mathcal{C} \boxtimes \mathcal{C}^{\mathrm{op}}$, see Example 7.9.10 and Definition 7.9.12.

PROPOSITION 7.16.5. *Let $\mathcal{C}$, F, and I be as above.*

(i) *$FI(\mathbf{1})$ is isomorphic to the image of ${}^*A$ under the tensor product functor $\otimes : \mathcal{C} \boxtimes \mathcal{C}^{\mathrm{op}} \to \mathcal{C}$.*
(ii) *If $\mathcal{C}$ is a tensor category then* $\mathsf{FPdim}(I(\mathbf{1})) = \mathsf{FPdim}(\mathcal{C})$.

PROOF. Taking $X = Y = Z = \mathbf{1} \in \mathcal{C}$ in Proposition 7.12.28, we have

$$FI(\mathbf{1}) = \underline{\mathsf{Hom}}_{\mathcal{Z}(\mathcal{C})}(\mathbf{1}, \mathbf{1}) \otimes \mathbf{1} = {}^*A \otimes \mathbf{1}, \tag{7.53}$$

which implies (i).

To prove (ii), observe that (7.53) implies $\mathsf{FPdim}(I(\mathbf{1})) = \mathsf{FPdim}(A)$. Let $\mathcal{O}(\mathcal{C})$ denote the set of simple objects of $\mathcal{C}$. The simple objects of $\mathcal{C} \boxtimes \mathcal{C}^{\mathrm{op}}$ are of the form $X \boxtimes Y$ where $X, Y \in \mathcal{O}(\mathcal{C})$ and their projective covers are $P(X) \boxtimes P(Y)$. Hence

$$\begin{aligned}
\mathsf{FPdim}(A) &= \sum_{X,Y \in \mathcal{O}(\mathcal{C})} \mathsf{FPdim}(X)\,\mathsf{FPdim}(Y)[A : X \boxtimes Y] \\
&= \sum_{X,Y \in \mathcal{O}(\mathcal{C})} \mathsf{FPdim}(X)\,\mathsf{FPdim}(Y) \dim_{\Bbbk} \mathsf{Hom}_{\mathcal{C}\boxtimes\mathcal{C}^{\mathrm{op}}}(P(X) \boxtimes P(Y),\, A) \\
&= \sum_{X,Y \in \mathcal{O}(\mathcal{C})} \mathsf{FPdim}(X)\,\mathsf{FPdim}(Y) \dim_{\Bbbk} \mathsf{Hom}_{\mathcal{C}}(P(X) \otimes P(Y),\, \mathbf{1}) \\
&= \sum_{X,Y \in \mathcal{O}(\mathcal{C})} \mathsf{FPdim}(X)\,\mathsf{FPdim}(Y) \dim_{\Bbbk} \mathsf{Hom}_{\mathcal{C}}(P(X), P(Y)^*) \\
&= \sum_{X,Y \in \mathcal{O}(\mathcal{C})} \mathsf{FPdim}(X)\,\mathsf{FPdim}(Y)\,[P(Y)^* : X] \\
&= \sum_{Y \in \mathcal{O}(\mathcal{C})} \mathsf{FPdim}(Y)\,\mathsf{FPdim}(P(Y)^*) = \mathsf{FPdim}(\mathcal{C}),
\end{aligned}$$

as required. $\square$

THEOREM 7.16.6. *For any finite tensor category* $\mathsf{FPdim}(\mathcal{Z}(\mathcal{C})) = \mathsf{FPdim}(\mathcal{C})^2$.

PROOF. The forgetful functor $F : \mathcal{Z}(\mathcal{C}) \to \mathcal{C}$ is surjective by Lemma 7.13.11. Hence,

$$\mathsf{FPdim}(I(\mathbf{1})) = \frac{\mathsf{FPdim}(\mathcal{Z}(\mathcal{C}))}{\mathsf{FPdim}(\mathcal{C})}$$

by Lemma 6.2.4. On the other hand, we have $\mathsf{FPdim}(I(\mathbf{1})) = \mathsf{FPdim}(\mathcal{C})$ by Proposition 7.16.5(i). $\square$

COROLLARY 7.16.7. *For any exact indecomposable* $\mathcal{C}$*-module category* $\mathcal{M}$ *we have* $\mathsf{FPdim}(\mathcal{C}) = \mathsf{FPdim}(\mathcal{C}^*_{\mathcal{M}})$.

PROOF. By Theorems 7.16.1 and 7.16.6 we have

$$\mathsf{FPdim}(\mathcal{C})^2 = \mathsf{FPdim}(\mathcal{Z}(\mathcal{C})) = \mathsf{FPdim}(\mathcal{Z}(\mathcal{C}^*_{\mathcal{M}})) = \mathsf{FPdim}(\mathcal{C}^*_{\mathcal{M}})^2$$

and numbers $\mathsf{FPdim}(\mathcal{C})$ and $\mathsf{FPdim}(\mathcal{C}^*_{\mathcal{M}})$ are positive. $\square$

EXERCISE 7.16.8. Let $\mathcal{C}$ be a fusion category, and $\mathcal{M}$ an indecomposable semisimple $\mathcal{C}$-module category with the simple objects M_i. Recall that the Frobenius-Perron dimensions $\mathsf{FPdim}(M)$ for $M \in \mathcal{M}$ are defined up to scaling by the condition that $\mathsf{FPdim}(X \otimes M) = \mathsf{FPdim}(X)\,\mathsf{FPdim}(M)$ for any $X \in \mathcal{C}$. Define a (positive) normalization of these dimensions, $\mathsf{FPdim}_c(M)$, which we will call the *canonical normalization*, by the condition

$$\sum_i \mathsf{FPdim}_c(M_i)^2 = \mathsf{FPdim}(\mathcal{C}),$$

Show that for the canonical normalization,

$$\mathsf{FPdim}(\underline{\mathsf{Hom}}(M, M')) = \mathsf{FPdim}_c(M)\,\mathsf{FPdim}_c(M')$$

for any $M, M' \in \mathcal{M}$ (and the canonical normalization is determined by this property).

Hint: consider the matrix of the action of the regular virtual object $R_{\mathcal{C}}$ in the basis M_i. Show that the entries of this matrix are $\mathsf{FPdim}_c(M_i)\,\mathsf{FPdim}_c(M_j)$, and use this to deduce the required statement.

EXERCISE 7.16.9. Let $\mathcal{C}$ be a fusion category, and A be an indecomposable semisimple algebra in $\mathcal{C}$ (i.e., an algebra whose category $\mathcal{M}$ of left modules in $\mathcal{C}$ is indecomposable as a right $\mathcal{C}$-module category and semisimple).

(i) Consider the dual category $\mathcal{C}^*_{\mathcal{M}}$, i.e., the category of A-bimodules in $\mathcal{C}$. Let X_j be the simple objects of $\mathcal{C}^*_{\mathcal{M}}$. For $X \in \mathcal{C}^*_{\mathcal{M}}$, let $d(X)$ be the dimension of X as an object of $\mathcal{C}$. Show that $d(X) = \mathsf{FPdim}(A)\,\mathsf{FPdim}(X)$, and thus

$$\sum_j d(X_j)^2 = \mathsf{FPdim}(A)^2\,\mathsf{FPdim}(\mathcal{C}).$$

Hint: use that A is the unit object in $\mathcal{C}^*_{\mathcal{M}}$, and Corollary 7.16.7.

(ii) Let M_i be the simple objects of $\mathcal{M}$. Show that

$$\sum_i \mathsf{FPdim}(M_i)^2 = \mathsf{FPdim}(A)\,\mathsf{FPdim}(\mathcal{C}).$$

Hint: Show that $\underline{\mathsf{Hom}}(A, M) = M$, where on the right hand side M is regarded as an object of $\mathcal{C}$. Deduce that $\mathsf{FPdim}_c(A)\,\mathsf{FPdim}_c(M) = \mathsf{FPdim}(M)$, where $\mathsf{FPdim}_c(M)$ is defined in Exercise 7.16.8. Applying this to $M = A$, deduce that $\mathsf{FPdim}_c(A) = \mathsf{FPdim}(A)^{1/2}$ and hence $\mathsf{FPdim}(M) = \mathsf{FPdim}(A)^{1/2}\,\mathsf{FPdim}_c(M)$.

(iii) Let V be an object of $\mathcal{C}$. Show that

$$\dim_{\Bbbk} \mathsf{Hom}_{\mathcal{C}}(V \otimes V^*, A) \geq \frac{\mathsf{FPdim}(A)\,\mathsf{FPdim}(V)^2}{\mathsf{FPdim}(\mathcal{C})}.$$

In particular, taking $V = \mathbf{1}$, we get

$$\dim_{\Bbbk} \mathsf{Hom}_{\mathcal{C}}(\mathbf{1}, A) \geq \frac{\mathsf{FPdim}(A)}{\mathsf{FPdim}(\mathcal{C})}.$$

Hint: Let $M_i \otimes V = \oplus_j a_{ij} M_j$, $a_{ij} \in \mathbb{Z}_{\geq 0}$. Let $A = \oplus_i k_i M_i$ as a left A-module. Show that $\mathsf{End}_A(A \otimes V) = \mathsf{Hom}_{\mathcal{C}}(V \otimes V^*, A)$, and $\dim \mathsf{End}_A(A \otimes V) = \sum_j (\sum_i k_i a_{ij})^2$. On the other hand, denoting $\mathsf{FPdim}(M_i)$ by m_i, show that

$$\sum_{i,j} k_i a_{ij} m_j = \mathsf{FPdim}(A)\,\mathsf{FPdim}(V) \text{ and } \sum_j m_j^2 = \mathsf{FPdim}(A)\,\mathsf{FPdim}(\mathcal{C})$$

(use (ii)). Then use the Cauchy-Schwarz inequality.

(iv) Let H be a semisimple (quasi-) Hopf algebra, and A be a finite dimensional H-module algebra. Show that if A is semisimple as an algebra in the category of H-modules (i.e., $A\#H$ is a semisimple algebra), then for any finite dimension H-module V, one has

$$\dim \mathsf{Hom}_H(V \otimes V^*, A) \geq \frac{\dim(A)\dim(V)^2}{\dim(H)}.$$

In particular, $\dim(A^H) \geq \frac{\dim(A)}{\dim(H)}$, where A^H is the algebra of H-invariants in A.

(v) Let H be a semisimple Hopf algebra, and A be a finite dimensional H-module algebra. Show that if A is semisimple in the usual sense, then it is also semisimple in the category of H-modules.

Hint: Let $\mathcal{M}$ be the category of A-modules in $\mathcal{C}$, and $\mathcal{M}'$ be the category of ordinary A-modules (in Vec). Then $\mathcal{M}'$ is a module category over the tensor category $\mathcal{C}'$ of H^*-modules, which is Morita equivalent to $\mathcal{C}$. The category $\mathcal{M}'$ is semisimple, hence exact. Show that $\mathcal{M}'$ corresponds to $\mathcal{M}$ under the Morita equivalence between $\mathcal{C}'$ and $\mathcal{C}$, and deduce that $\mathcal{M}$ is exact, hence semisimple.

(vi) Deduce that (iv) holds for any H-module algebra A which is semisimple in the usual sense.

(vii) Does the converse to (v) hold over a field of any characteristic?

Hint: Consider $A = H^*$ with action of H by left translations, where H is the function algebra on a finite group in characteristic p.

(viii) Show that the converse to (v) holds if H is cosemisimple.

(ix) Show that (vi) fails if H is not semisimple; i.e., there are actions of finite dimensional nonsemisimple H on a matrix algebra A such that $\dim(A^H)$ is much less than $\dim(A)/\dim(H)$. Namely, let H be the Nichols Hopf algebra of dimension 16 over $\mathbb{C}$ generated by a grouplike element g such that $g^2 = 1$ and skew-primitive elements x_i, $i = 0, 1, 2$, such that $gx_i = -x_i g$, $x_i x_j = -x_j x_i$, $x_i^2 = 0$, and $\Delta(x_i) = 1 \otimes x_i + x_i \otimes g$. Let B be a matrix algebra. Show that there is a right B-linear left action of H on $B \oplus B$ with right B-basis e_1, e_2 by

$$g(e_1) = e_1,\ g(e_2) = -e_2,\ x_i(e_1) = 0,\ x_0(e_2) = e_1, x_1(e_2) = xe_1,\ x_2(e_2) = ye_1,$$

where $x, y \in B$ are any elements. Now consider the corresponding adjoint action of H on $A := \mathrm{End}(B^2)_B = \mathrm{Mat}_2(B)$, via

$$a \circ M = a_1 M S(a_2)$$

(using Sweedler's notation). Show that A^H is the set of matrices $b \cdot \mathrm{Id}$, where $b \in Z_{x,y}$, and $Z_{x,y}$ is the centralizer of x, y in B.

Now take $x, y \in B$ to be an irreducible (i.e., generating) pair of elements (clearly, generic x, y satisfy this condition). Show that $Z_{x,y} = \mathbb{C}$, so the invariants A^H in A are trivial, and hence $\dim A^H = 1 < \dim(A)/\dim(H) = \dim(B)/4$ if $\dim(B) > 4$.

7.17. Duality for tensor functors and Lagrange's Theorem

Let $\mathcal{C}, \mathcal{D}$ be finite multitensor categories, let $\mathcal{M}$ be an exact faithful $\mathcal{D}$-module category, and $F : \mathcal{C} \to \mathcal{D}$ be a tensor functor (i.e., we require that $F(\mathbf{1}) = \mathbf{1}$). Then $\mathcal{M}$ is a module category over $\mathcal{C}$ which is, obviously, not always exact (e.g., consider the case when $\mathcal{C}$ is trivial and $\mathcal{M} = \mathcal{D}$).

DEFINITION 7.17.1. The pair $(F, \mathcal{M})$ is called an *exact pair* if $\mathcal{M}$ is an exact $\mathcal{C}$-module category.

Suppose $(F, \mathcal{M})$ is an exact pair. There is an obvious tensor functor

$$F^* : \mathcal{D}^*_{\mathcal{M}} \to \mathcal{C}^*_{\mathcal{M}}$$

(note that F endows a $\mathcal{D}$-module endofunctor of $\mathcal{M}$ with the structure of a $\mathcal{C}$-module endofunctor) and $(F^*, \mathcal{M})$ is an exact pair.

DEFINITION 7.17.2. We will call the above functor $F^* : \mathcal{D}^*_{\mathcal{M}} \to \mathcal{C}^*_{\mathcal{M}}$ the *dual* functor to F. We will also call the exact pair $(F^*, \mathcal{M})$ *dual* to the pair $(F, \mathcal{M})$, and write $(F^*, \mathcal{M}) = (F, \mathcal{M})^*$.

Clearly, for any exact pair T, one has $T^{**} = T$.

For simplicity, below we will consider only exact pairs in which $\mathcal{C}$ and $\mathcal{D}$ are tensor (i.e., not just multitensor) categories, and $\mathcal{M}$ is indecomposable over $\mathcal{C}$ (the class of such pairs is obviously stable under dualization). We note, however, that the results below can be extended to the general case.

DEFINITION 7.17.3. An exact pair $(F, \mathcal{M})$ is *surjective* if F is surjective, and *injective* if F is injective.

THEOREM 7.17.4. *The dualization map takes surjective exact pairs into injective ones, and vice versa.*

PROOF. Let $T = (F : \mathcal{C} \to \mathcal{D}, \mathcal{M})$ be an exact pair. Denote $\mathsf{FPdim}(\mathcal{C}) = c$ and $\mathsf{FPdim}(\mathcal{D}) = d$. Then $\mathsf{FPdim}(\mathcal{C}^*_{\mathcal{M}}) = c$ and $\mathsf{FPdim}(\mathcal{D}^*_{\mathcal{M}}) = d$ by Corollary 7.16.7.

Assume first that F is injective, but F^* is not surjective. Then by Proposition 6.3.5, $\mathsf{FPdim}(\mathsf{Im}F^*) < c$. Since F is injective, we also have $\mathsf{FPdim}(\mathsf{Im}F^*) < d$ (as $c \le d$). The functor F factors through $\mathcal{E} = (\mathrm{Im}F^*)^*_{\mathcal{M}}$ (it is not difficult to show that $\mathcal{M}$ is exact and indecomposable over $\mathsf{Im}F^*$, so $\mathcal{E}$ is a finite tensor category). Since $\mathsf{FPdim}(\mathcal{E}) = \mathsf{FPdim}(\mathsf{Im}F^*) < \min(c, d)$, by Proposition 6.3.6, F is not injective. Contradiction.

Assume now that F is surjective, but F^* is not injective. Then by Proposition 6.3.5, $\mathsf{FPdim}(\mathsf{Im}F^*) < d$. Since F is surjective, we also have $\mathsf{FPdim}(\mathsf{Im}F^*) < c$. The functor F factors through $\mathcal{E} = (\mathrm{Im}F^*)^*_{\mathcal{M}}$. Since $\mathsf{FPdim}(\mathcal{E}) = \mathsf{FPdim}(\mathsf{Im}F^*) < \min(c, d)$, by Proposition 6.3.6, F is not surjective. Contradiction. □

Let $\mathcal{C}$ be a multitensor category and let

$$\mathcal{C} = \bigoplus_{ik} \mathcal{C}_{ik}$$

be its decomposition into a direct sum of component subcategories, see (4.1). Note that each $\mathcal{C}_{ii}$ is a tensor category and $\mathcal{C}_{ik}$ is a $\mathcal{C}_{ii} - \mathcal{C}_{kk}$ bimodule category.

PROPOSITION 7.17.5. *We have $(\mathcal{C}_{ii})^*_{\mathcal{C}_{ik}} \cong \mathcal{C}^{\mathrm{op}}_{kk}$, i.e., tensor categories $\mathcal{C}_{ii}$ and $\mathcal{C}_{kk}$ are categorically Morita equivalent for all i, k.*

PROOF. For each k let $\mathcal{M}_k := \oplus_j \mathcal{C}_{jk}$. The regular $\mathcal{C}$-module category decomposes as $\mathcal{C} = \oplus_k \mathcal{M}_k$. Then $\mathcal{C}^*_{\mathcal{C}} = \oplus_{kl} \mathsf{Fun}_{\mathcal{C}}(\mathcal{M}_k, \mathcal{M}_l)$. It follows from Example 7.12.3 that $\mathcal{C}^*_{\mathcal{M}_k} \cong \mathcal{C}^{\mathrm{op}}_{kk}$. Let $\tilde{\mathcal{C}} := \oplus_i \mathcal{C}_{ii}$ be the "diagonal" subcategory of $\mathcal{C}$. It is clear that $\tilde{\mathcal{C}}^*_{\mathcal{M}_k} = \oplus_i (\mathcal{C}_{ii})^*_{\mathcal{C}_{ik}}$.

By Theorem 7.17.4 the dual functor $\mathcal{C}^*_{\mathcal{M}_k} \to \tilde{\mathcal{C}}^*_{\mathcal{M}_k}$ is surjective. This means that for all i, k the obvious tensor functor $\mathcal{C}^{op}_{kk} \to (\mathcal{C}_{ii})^*_{\mathcal{C}_{ik}}$ is surjective. In view of Proposition 6.3.4 this proves the result. □

Next, we prove Lagrange's theorem for subcategories of finite tensor categories.

THEOREM 7.17.6. *Let $\mathcal{D}$ be a finite tensor category, and $\mathcal{C} \subset \mathcal{D}$ be a tensor subcategory. Then the ratio $\mathsf{FPdim}(\mathcal{D})/\mathsf{FPdim}(\mathcal{C})$ is an algebraic integer.*

PROOF. Consider the natural embedding $F : \mathcal{C} \boxtimes \mathcal{D}^{\mathrm{op}} \to \mathcal{D} \boxtimes \mathcal{D}^{\mathrm{op}}$. Consider $\mathcal{M} = \mathcal{D}$ as a module category over $\mathcal{D} \boxtimes \mathcal{D}^{\mathrm{op}}$. It is easy to check that the pair $(F, \mathcal{M})$ is exact, and $\mathcal{M}$ is indecomposable over $\mathcal{C} \boxtimes \mathcal{D}^{\mathrm{op}}$. Thus, Theorem 7.17.4 applies, and the functor $F^* : (\mathcal{D} \otimes \mathcal{D}^{\mathrm{op}})^*_{\mathcal{M}} = \mathcal{Z}(\mathcal{D}) \to (\mathcal{C} \otimes \mathcal{D}^{\mathrm{op}})^*_{\mathcal{M}}$ is surjective. The Frobenius-Perron dimension of the first category is $\mathsf{FPdim}(\mathcal{D})^2$ and the Frobenius-Perron dimension of the second one is $\mathsf{FPdim}(\mathcal{C})\,\mathsf{FPdim}(\mathcal{D})$. By Corollary 6.2.2 the ratio of these dimensions is an algebraic integer. $\square$

REMARK 7.17.7. In the case when $\mathcal{D} = \mathsf{Vec}_G$, this theorem reduces to classical Lagrange's theorem, saying that the order of any subgroup of G divides the order of G.

7.18. Hopf bimodules and the Fundamental Theorem

Let $\mathcal{C}$, $\mathcal{D}$ be finite tensor categories over a field $\Bbbk$. Recall from Proposition 4.6.1 that the category $\mathcal{C} \boxtimes \mathcal{D}$ has a natural structure of a tensor category.

Let $\mathcal{C}^{\mathrm{op}}$ denote the opposite tensor category, see Definition 2.1.5. The category $\mathcal{C}$ has a natural structure of an exact $\mathcal{C}$-bimodule category. We will denote the corresponding action by

$$(X, V) \mapsto X \circ V, \quad X \in \mathcal{C} \boxtimes \mathcal{C}^{\mathrm{op}},\, V \in \mathcal{C}.$$

In Definition 7.9.12 we introduced the canonical algebra $A := \underline{\mathsf{Hom}}(\mathbf{1}, \mathbf{1}) \in \mathcal{C} \boxtimes \mathcal{C}^{\mathrm{op}}$.

DEFINITION 7.18.1. The category of right A-modules in $\mathcal{C} \boxtimes \mathcal{C}^{\mathrm{op}}$ will be called the category of *Hopf bimodules* in $\mathcal{C}$.

Clearly, this definition is a generalization of the definition of a Hopf module.

Let $\mathcal{H}$ denote the category of Hopf bimodules in $\mathcal{C}$.

Observe that $(X, Y) \mapsto \mathsf{Hom}_{\mathcal{C}}(X, Y^*)^*$ is an additive bifunctor from $\mathcal{C}^{\mathrm{op}} \times \mathcal{C}$ to Vec. Hence it defines an additive functor $H_{\mathcal{C}} : \mathcal{C}^{\mathrm{op}} \boxtimes \mathcal{C} \to \mathsf{Vec}$. Therefore, one can define another tensor product $\odot$ on $\mathcal{C} \boxtimes \mathcal{C}^{\mathrm{op}}$ by

$$(\mathsf{id}_{\mathcal{C}} \boxtimes H_{\mathcal{C}}) \boxtimes \mathsf{id}_{\mathcal{C}^{\mathrm{op}}} : (\mathcal{C} \boxtimes \mathcal{C}^{\mathrm{op}}) \boxtimes (\mathcal{C} \boxtimes \mathcal{C}^{\mathrm{op}}) \cong (\mathcal{C} \boxtimes (\mathcal{C}^{\mathrm{op}} \boxtimes \mathcal{C})) \boxtimes \mathcal{C}^{\mathrm{op}} \to \mathcal{C} \boxtimes \mathcal{C}^{\mathrm{op}}, \tag{7.54}$$

where we implicitly used a natural action of Vec on $\mathcal{C}$.

REMARK 7.18.2. Another way to define the tensor product $\odot$ is the following: one identifies $\mathcal{C}^{\mathrm{op}}$ with the dual category $\mathcal{C}^{\vee}$ via the functor $X \mapsto X^*$. Then the category $\mathcal{C} \boxtimes \mathcal{C}^{\mathrm{op}}$ identifies with the category of right exact functors from $\mathcal{C}$ to itself, and $\odot$ corresponds to the composition of functors. Under this identification, the object $A \in \mathcal{C} \boxtimes \mathcal{C}^{\mathrm{op}}$ representing the functor

$$X \boxtimes Y \mapsto \mathsf{Hom}(X \otimes Y, \mathbf{1}) = \mathsf{Hom}(X, Y^*) : \mathcal{C} \boxtimes \mathcal{C}^{\mathrm{op}} \to \mathsf{Vec}$$

corresponds to the identity functor (this is why we use $X \mapsto X^*$ and not $X \mapsto {}^*X$ to identify $\mathcal{C}^{\mathrm{op}}$ and $\mathcal{C}^{\vee}$). In particular, A is the unit object for $\odot$.

EXAMPLE 7.18.3. Let H be a finite dimensional Hopf algebra and let $\mathcal{C} = \mathsf{Rep}(H)$. We have $\mathcal{C} \boxtimes \mathcal{C}^{\mathrm{op}} = \mathsf{Rep}(H \otimes H^{\mathrm{cop}})$.

Then $A = H^*$ is an $H \otimes H^{\mathrm{cop}}$-module algebra via the action $((x \otimes y)\phi)(a) = \phi(S^{-1}(y)ax)$, where S^{-1} is the inverse antipode of H, i.e., the antipode of H^{cop} (cf. Example 7.9.11).

Furthermore, $\mathcal{H}$ is the category of H^*-modules in $\mathsf{Rep}(H \otimes H^{\mathrm{cop}})$ (equivalently, it is the category of H-comodules in the category of H-bimodules) and $\odot$ is dual to the usual bimodule tensor product (i.e., $M \odot N = (N^* \otimes_H M^*)^*$ where the star denotes the dual vector space).

EXERCISE 7.18.4. Prove statements in Example 7.18.3.

PROPOSITION 7.18.5. *Let $\mathcal{C}$ be a finite tensor category.*

(a) *$\mathcal{H}$ is a tensor category and functors $V \mapsto (V \boxtimes \mathbf{1}) \otimes A$ and $V \mapsto (\mathbf{1} \boxtimes V) \otimes A$ from $\mathcal{C}$ to $\mathcal{H}$ are equivalences of tensor categories.*

(b) *There is a natural isomorphism of tensor functors*

$$\rho_V : (V \boxtimes \mathbf{1}) \otimes A \cong (\mathbf{1} \boxtimes V) \otimes A. \tag{7.55}$$

PROOF. (a) For all objects V in $\mathcal{C}$ we have

$$\begin{aligned} \underline{\mathsf{Hom}}(\mathbf{1}, V) &= \underline{\mathsf{Hom}}(\mathbf{1}, (V \boxtimes \mathbf{1}) \circ \mathbf{1}) \simeq (V \boxtimes \mathbf{1}) \otimes \underline{\mathsf{Hom}}(\mathbf{1}, \mathbf{1}) = (V \boxtimes \mathbf{1}) \otimes A, \\ \underline{\mathsf{Hom}}(\mathbf{1}, V) &= \underline{\mathsf{Hom}}(\mathbf{1}, (\mathbf{1} \boxtimes V) \circ \mathbf{1}) \cong (\mathbf{1} \boxtimes V) \circ \underline{\mathsf{Hom}}(\mathbf{1}, \mathbf{1}) = (\mathbf{1} \boxtimes V) \otimes A. \end{aligned}$$

It follows from Theorem 7.10.1 that $(- \boxtimes \mathbf{1}) \otimes A$ is an equivalence between $\mathcal{C}$ and $\mathcal{H}$. To see that it is tensor, observe that under the identification in Remark 7.18.2 (i) the functor $(- \boxtimes \mathbf{1}) \otimes A$ (and, similarly, $(\mathbf{1} \boxtimes -) \otimes A$) sends $V \in \mathcal{C}$ to the functor $V \otimes -$ from $\mathcal{C}$ to itself. Thus, the associativity constraint in the category $\mathcal{C}$ gives rise to a tensor structure on these functors.

(b) The tensoriality of the natural isomorphism (7.55) is obvious from the description in (a). It is also equivalent to commutativity of the following diagram

$$\begin{array}{ccc} ((V \boxtimes \mathbf{1}) \otimes A) \odot ((W \boxtimes \mathbf{1}) \otimes A) & \longrightarrow & ((V \otimes W) \boxtimes \mathbf{1}) \otimes A \\ \big\downarrow {\scriptstyle \rho_V \odot \rho_W} & & \big\downarrow {\scriptstyle \rho_{V \otimes W}} \\ ((\mathbf{1} \boxtimes V) \otimes A) \odot ((\mathbf{1} \boxtimes W) \otimes A) & \longrightarrow & (\mathbf{1} \boxtimes (V \otimes W)) \otimes A. \end{array} \tag{7.56}$$

□

REMARK 7.18.6. Another way to state Proposition 7.18.5 (b) is to say that the following diagram commutes:

$$\begin{array}{ccc} (V \boxtimes \mathbf{1}) \otimes (W \boxtimes \mathbf{1}) \otimes A & \longrightarrow & ((V \otimes W) \boxtimes \mathbf{1}) \otimes A \\ \big\downarrow {\scriptstyle \mathsf{id} \otimes \rho_W} & & \big\downarrow {\scriptstyle \rho_{V \otimes W}} \\ (V \boxtimes \mathbf{1}) \otimes (\mathbf{1} \boxtimes W) \otimes A & & (\mathbf{1} \boxtimes (V \otimes W)) \otimes A \\ \big\downarrow & & \big\uparrow \\ (\mathbf{1} \boxtimes W) \otimes (V \boxtimes \mathbf{1}) \otimes A & \xrightarrow{\mathsf{id} \otimes \rho_V} & (\mathbf{1} \boxtimes W) \otimes (\mathbf{1} \boxtimes V) \otimes A. \end{array} \tag{7.57}$$

Let $A = \underline{\mathsf{Hom}}(\mathbf{1}, \mathbf{1})$ and let M be a left A-module. By Remark 7.8.6(i) the object M^* has a natural structure of a right A-module with the action given by

$$M^* \otimes A \xrightarrow{p^* \otimes \mathsf{id}_A} M^* \otimes A^* \otimes A \xrightarrow{\mathsf{id}_{M^*} \otimes \mathsf{coev}_A} M^*, \tag{7.58}$$

where $p : A \otimes M \to M$ is the left action of A on M and coev_A is the coevaluation morphism of A. In particular, A^* has a canonical structure of a Hopf bimodule. Thus, according to Proposition 7.18.5(a), there exists a unique up to an isomorphism object $D \in \mathcal{C}$ such that

$$(D \boxtimes \mathbf{1}) \otimes A \cong A^* \tag{7.59}$$

as Hopf bimodules. Moreover, isomorphism (7.59) is unique up to scaling. It follows immediately from the definition that the Frobenius-Perron dimension (see Section 3.3) of D equals to 1 and thus D is invertible. Note that its isomorphism class in $\mathcal{C}$ is canonically defined.

Recall that in Section 6.4 we defined the distinguished invertible object X_ρ of $\mathcal{C}$ as the dual of the socle of the projective cover of $\mathbf{1}$, see Definition 6.4.4.

THEOREM 7.18.7. *The object $D \in \mathcal{C}$ is isomorphic to the dual of the distinguished invertible object: $D \cong X_\rho^*$.*

PROOF. Let I be a set indexing the isomorphism classes of simple objects in $\mathcal{C}$; for $\alpha \in I$ let $X_\alpha, P_\alpha, I_\alpha$ denote a simple object corresponding to α, its projective cover, and its injective hull. We will assume that $0 \in I$ and $X_0 = \mathbf{1}$. Let i run through I. We are going to compute $\dim_{\Bbbk} \mathsf{Hom}_{\mathcal{C}\boxtimes\mathcal{C}^{\mathrm{op}}}(P_0 \boxtimes X_i, A^*)$ in two ways.
First calculation:

$$\begin{aligned}\dim_{\Bbbk} \mathsf{Hom}_{\mathcal{C}\boxtimes\mathcal{C}^{\mathrm{op}}}(P_0 \boxtimes X_i, A^*) &= \dim_{\Bbbk} \mathsf{Hom}_{\mathcal{C}\boxtimes\mathcal{C}^{\mathrm{op}}}(P_0 \boxtimes X_i, (D \boxtimes \mathbf{1}) \otimes A) \\ &= \dim_{\Bbbk} \mathsf{Hom}_{\mathcal{C}}(P_0 \otimes X_i, D) = \begin{cases} 1 & \text{if } X_i = D, \\ 0 & \text{otherwise.} \end{cases}\end{aligned}$$

Second calculation:

$$\begin{aligned}\dim_{\Bbbk} \mathsf{Hom}_{\mathcal{C}\boxtimes\mathcal{C}^{\mathrm{op}}}(P_0 \boxtimes X_i, A^*) &= \dim_{\Bbbk} \mathsf{Hom}_{\mathcal{C}\boxtimes\mathcal{C}^{\mathrm{op}}}(A, {}^*(P_0 \boxtimes X_i)) \\ &= \dim_{\Bbbk} \mathsf{Hom}_{\mathcal{C}\boxtimes\mathcal{C}^{\mathrm{op}}}((P_0 \boxtimes \mathbf{1}) \otimes A, \mathbf{1} \boxtimes X_i^*).\end{aligned}$$

Let us look closely at the object $(P_0 \boxtimes \mathbf{1}) \otimes A$ in $\mathcal{C} \boxtimes \mathcal{C}^{\mathrm{op}}$.

LEMMA 7.18.8. *The object $(P_0 \boxtimes \mathbf{1}) \otimes A$ is injective.*

PROOF. Observe that the functor

$$\mathsf{Hom}(X \boxtimes Y, (P_0 \boxtimes \mathbf{1}) \otimes A) = \mathsf{Hom}((P_0^* \otimes X) \boxtimes Y, A) = \mathsf{Hom}(P_0^* \otimes X \otimes Y, \mathbf{1})$$

is exact in both variables X, Y since $P_0^* \otimes X \otimes Y$ is injective, see Proposition 4.2.12. Thus the functor $\mathsf{Hom}(-, (P_0 \boxtimes \mathbf{1}) \otimes A)$ is exact. The lemma is proved. □

We continue the proof of the Theorem. By Lemma 7.18.8,

$$(P_0 \boxtimes \mathbf{1}) \otimes A = \oplus_{\alpha,\beta \in I} M_{\alpha\beta} I_\alpha \boxtimes I_\beta$$

for some non-negative integer multiplicities $M_{\alpha\beta}$. We have

$$\begin{aligned}M_{\alpha\beta} &= \dim_{\Bbbk} \mathsf{Hom}_{\mathcal{C}\boxtimes\mathcal{C}^{\mathrm{op}}}(X_\alpha \boxtimes X_\beta, (P_0 \boxtimes \mathbf{1}) \otimes A) \\ &= \dim_{\Bbbk} \mathsf{Hom}_{\mathcal{C}}(P_0^* \otimes X_\alpha \otimes X_\beta, \mathbf{1}) \\ &= \dim_{\Bbbk} \mathsf{Hom}_{\mathcal{C}}(X_\alpha \otimes X_\beta, P_0) \\ &= [X_\alpha \otimes X_\beta : X_\rho^*],\end{aligned}$$

where $[X : X_i]$ denotes the multiplicity of a simple object X_i in the Jordan-Hölder series of X. To calculate $\dim_{\Bbbk} \mathsf{Hom}_{\mathcal{C}\boxtimes\mathcal{C}^{\mathrm{op}}}((P_0 \boxtimes \mathbf{1}) \otimes A, \mathbf{1} \boxtimes X_i^*)$, it is enough to consider the summands with $I_\alpha = P_0$. In this case $X_\alpha = X_\rho^*$ and $[X_\alpha \otimes X_\beta : X_\rho^*] = [X_\beta : \mathbf{1}]$. Thus

$$\dim_{\Bbbk} \mathsf{Hom}_{\mathcal{C}\boxtimes\mathcal{C}^{\mathrm{op}}}((P_0 \boxtimes \mathbf{1}) \otimes A, \mathbf{1} \boxtimes X_i^*) = \begin{cases} 1 & \text{if } I_0 \text{ covers } X_i^*, \\ 0 & \text{otherwise.} \end{cases} \tag{7.60}$$

Since $I_0 = P_0^*$ covers X_ρ, the Theorem follows. □

PROPOSITION 7.18.9. *Let $f : A \to A^{**}$ be a morphism in $\mathcal{C} \boxtimes \mathcal{C}^{\mathrm{op}}$. Assume that $\mathsf{Tr}^L(f) \neq 0$. Then the category $\mathcal{C}$ is semisimple.*

PROOF. By definition $\mathsf{Tr}^L(f)$ is the following morphism :

$$\mathsf{Tr}^L(f) : \mathbf{1} \boxtimes \mathbf{1} \xrightarrow{\mathsf{coev}_A} A \otimes A^* \xrightarrow{f \otimes \mathsf{id}_A} A^{**} \otimes A^* \xrightarrow{\mathsf{ev}_{A^*}} \mathbf{1} \boxtimes \mathbf{1}. \tag{7.61}$$

In particular, if $\mathsf{Tr}^L(f) \neq 0$ then $\mathbf{1}$ is a direct summand of $A \otimes A^*$. Hence $P_0 \boxtimes \mathbf{1}$ is a direct summand of $(P_0 \boxtimes \mathbf{1}) \otimes A \otimes A^*$. By Lemma 7.18.8 $(P_0 \boxtimes \mathbf{1}) \otimes A$ is projective and therefore $(P_0 \boxtimes \mathbf{1}) \otimes A \otimes A^*$ is projective. Thus $P_0 \boxtimes \mathbf{1}$ is projective and consequently $\mathbf{1}$ is projective. Hence $\mathcal{C}$ is semisimple. □

As a corollary we obtain a classical theorem of Larson and Radford in Hopf algebra theory.

COROLLARY 7.18.10. *Let H be a finite dimensional Hopf algebra with the antipode S. If $\mathsf{Tr}_H(S^2) \neq 0$ then H is semisimple and cosemisimple.*

The converse to this theorem also holds and is proved below in Proposition 8.20.17.

Note that both directions of this theorem can be proved in another way (Exercise 7.10.10).

The next example illustrates the above constructions when $\mathcal{C} = \mathsf{Rep}(H)$, the category of finite dimensional representations of a finite dimensional Hopf algebra H.

EXAMPLE 7.18.11. For $\mathcal{C} = \mathsf{Rep}(H)$ we have $\mathcal{C} \boxtimes \mathcal{C}^{\mathrm{op}} = \mathsf{Rep}(H \otimes H^{\mathrm{cop}})$. As it was explained in Example 7.18.3, the algebra A in this case is isomorphic to H^*, with the action of $H \otimes H^{\mathrm{cop}}$ defined by $((x \otimes y)f)(a) = f(S^{-1}(y)ax)$. Similarly to Exercise 7.10.10, H is a Hopf bimodule for H^*. So by Proposition 7.18.5 there is an isomorphism of Hopf bimodules $\phi : H \to (\mathbf{1} \boxtimes D) \otimes H^*$. Let $\phi(1) = \lambda$. Then $f \circ 1 = f(1)$, so $f\lambda = f(1)\lambda$, hence λ is a nonzero left integral for H^*. Moreover, the H-invariance of ϕ implies that $\phi(h)(z) = \lambda(zh)$ for any $h, z \in H$. Thus, the compatibility condition of ϕ with the action of H^{cop} reads:

$$\phi(hS(b))(z) = \alpha^{-1}(b_1)\lambda(S^{-1}(b_2)zh),$$

where $\alpha = D^*$ is the distinguished character of H.

Thus, replacing b with $S(x)$ and zh with y, we obtain the following proposition, well known in Hopf algebra theory.

PROPOSITION 7.18.12. *If λ is a left integral of a Hopf algebra H, and α is its distinguished character, then*

$$\lambda(yS^2(x)) = \alpha(x_2)\lambda(x_1 y), \quad x, y \in H,$$

and

$$\lambda(yS^2(x_1))\alpha^{-1}(x_2) = \lambda(xy), \quad x, y \in H.$$

In particular, if H is unimodular (i.e., $\alpha = 1$) then

$$\lambda(yS^2(x)) = \lambda(xy), \quad x, y \in H.$$

REMARK 7.18.13. Recall that if B is a *Frobenius algebra* with the trace functional λ (i.e., the bilinear form $(x, y) \mapsto \lambda(xy)$ is non-degenerate) then there exists a unique automorphism $\sigma : B \to B$, called the *Nakayama automorphism*, such that $\lambda(xy) = \lambda(y\sigma(x))$. Proposition 7.18.12 says that the Nakayama automorphism of the Frobenius algebra H is $a \mapsto S^2(a_1)\alpha^{-1}(a_2)$, and it is S^2 in the unimodular case.

PROPOSITION 7.18.14. *Let H be a finite dimensional Hopf algebra such that H^* is unimodular (e.g., H is cosemisimple), and let I be a left integral of H. Suppose $\Delta(I) = \sum_i a_i \otimes b_i$. Then $\Delta(I) = \sum_i S^2(b_i) \otimes a_i$.*

PROOF. This follows from Proposition 7.18.12 by taking the dual. □

PROPOSITION 7.18.15. *(see* [**Mon**], *Corollary 10.3.13) Let H be a semisimple and cosemisimple Hopf algebra (over a field of any characteristic). Then the double $D(H)$ is semisimple.*

PROOF. Consider the $D(H)$-module $M := D(H) \otimes_{H^*} \Bbbk$. This module is projective, since $\Bbbk$ is a projective H^*-module (as H^* is semisimple). It suffices to show that the trivial $D(H)$-module $\Bbbk$ is a direct summand in M. Then it would follow that $\Bbbk$ is a projective $D(H)$-module, which would imply that any $D(H)$-module X is projective (since $X = X \otimes \Bbbk$, see Proposition 4.2.12).

Note that since $D(H) = H \otimes H^*$ as a space, we have a natural identification $M = H$ as a left H-module. We have a natural $D(H)$-module map $M = H \to \Bbbk$, which sends 1 to 1. Clearly, this map is (up to normalization) simply the counit ε of H.

Let I be a left integral of H. We claim I is a $D(H)$-invariant vector in M. Clearly, it is H-invariant (by the definition of the integral), so we only need to show that it is H^*-invariant. Let $f \in H^*$. Then by formula (7.46), we have

$$fI = (f_1, S^{-1}(I_1))(f_3, I_3)I_2 f_2. \tag{7.62}$$

But by Proposition 7.18.14, $I_1 \otimes I_2 \otimes I_3 = S^2(I_3) \otimes I_1 \otimes I_2$. Also, f_2 may be replaced by $\varepsilon(f_2)$, since we are tensoring over H^* with $\Bbbk$. Thus, equation (7.62) can be rewritten as

$$fI = (f_1, S(I_3))(f_2, I_2)I_1 = (f, I_2 S(I_3))I_1 = (f, 1)I = \varepsilon_{H^*}(f)I. \tag{7.63}$$

This implies the claim.

Since $\varepsilon(I) \neq 0$ (as H is semisimple), we see that we have a sequence of $D(H)$-maps $\Bbbk \to M \to \Bbbk$ such that the composition is nonzero. Thus, $\Bbbk$ is a direct summand in M, and we are done. □

REMARK 7.18.16. Another proof of Theorem 7.18.15 is obtained from Exercise 7.10.10. Namely, in this exercise we sketched a proof of the Larson-Radford theorem that H is semisimple and cosemisimple if and only if $\mathsf{Tr}_H(S^2) \neq 0$. But $\mathsf{Tr}_{D(H)}(S^2) = \mathsf{Tr}_H(S^2)\,\mathsf{Tr}_{H^{*\mathrm{op}}}(S^2)$, so we see that if H is semisimple and cosemisimple then $D(H)$ is semisimple.

7.19. Radford's isomorphism for the fourth dual

The classical formula of Radford [**Ra2**] expresses the fourth power of the antipode of a finite dimensional Hopf algebra H in terms of distinguished grouplike elements of H and H^*. In particular, some power of the fourth duality functor $V \mapsto V^{****}$ on $\mathsf{Rep}(H)$ is isomorphic, as a tensor functor, to the identity functor of $\mathsf{Rep}(H)$. Below we obtain a categorical version of this result.

Let $\mathcal{C}$ be a finite tensor category, let $A := \underline{\mathsf{Hom}}(\mathbf{1}, \mathbf{1})$ in $\mathcal{C} \boxtimes \mathcal{C}^{\mathrm{op}}$ be the canonical algebra introduced in Definition 7.9.12, and let D be the dual distinguished invertible object of $\mathcal{C}$ introduced in (7.59).

THEOREM 7.19.1. *There is a natural isomorphism of tensor functors:*

$$\delta_V : V^{**} \xrightarrow{\sim} D \otimes {}^{**}V \otimes D^{-1}. \tag{7.64}$$

PROOF. Isomorphism (7.59) produces a canonical isomorphism of algebras

$$A^{**} = (D \boxtimes \mathbf{1}) \otimes A \otimes (D \boxtimes \mathbf{1})^*. \tag{7.65}$$

We will identify these algebras using this isomorphism.

Recall that we have a tensor isomorphism (7.55) $\rho_V : (V \boxtimes \mathbf{1}) \otimes A \cong (\mathbf{1} \boxtimes V) \otimes A$. Its double dual $\rho_V^{**} : (V^{**} \boxtimes \mathbf{1}) \otimes A^{**} \cong (\mathbf{1} \boxtimes {}^{**}V) \otimes A^{**}$ is also tensor (i.e., the diagram analogous to (7.57) commutes).

Thus we have a tensor isomorphism of right A-modules

$$\tilde{\rho}_V : (V^{**} \boxtimes \mathbf{1}) \otimes (D \boxtimes \mathbf{1}) \otimes A \cong (\mathbf{1} \boxtimes {}^{**}V) \otimes (D \boxtimes \mathbf{1}) \otimes A$$

defined by $\tilde{\rho}_V \otimes \mathsf{id}_{(D \boxtimes \mathbf{1})^*} = \rho_V^{**}$.

Now define $\tilde{\delta}_V$ as the following composition:

$$(V^{**} \boxtimes \mathbf{1}) \otimes A \cong ((V^{**} \otimes D^*) \boxtimes \mathbf{1}) \otimes (D \boxtimes \mathbf{1}) \otimes A \xrightarrow{\tilde{\rho}_{V \otimes D^*}}$$

$$(\mathbf{1} \boxtimes ({}^{**}V \otimes D^*)) \otimes (D \boxtimes \mathbf{1}) \otimes A \cong (D \boxtimes \mathbf{1}) \otimes (\mathbf{1} \boxtimes ({}^{**}V \otimes D^*)) \otimes A \xrightarrow{\mathsf{id} \otimes \rho_{{}^{**}V \otimes D^*}}$$

$$(D \boxtimes \mathbf{1}) \otimes (({}^{**}V \otimes D^*) \boxtimes \mathbf{1}) \otimes A \cong ((D \otimes {}^{**}V \otimes D^*) \boxtimes \mathbf{1}) \otimes A.$$

Obviously, the isomorphism $\tilde{\delta}_V$ is tensor (again, the diagram analogous to (7.57) commutes).

Finally, define the isomorphism $\delta_V : V^{**} \to D \otimes {}^{**}V \otimes D^*$ by the condition $\delta_V \otimes \mathsf{id}_A = \tilde{\delta}_V$. Since $\tilde{\delta}_V$ is a morphism of right A-modules, Proposition 7.18.5(a) implies that δ_V is well defined. Since $\tilde{\delta}_V$ is tensor isomorphism of tensor functors, δ_V is a tensor isomorphism as well. □

COROLLARY 7.19.2. *There is a positive integer N such that the Nth powers of the tensor functors $-^{**}$ and ${}^{**}-$ are naturally isomorphic.*

PROOF. Since $\mathcal{C}$ has finitely many non-isomorphic invertible objects, there exists N such that $D^{\otimes N} \cong \mathbf{1}$. □

COROLLARY 7.19.3. *Let $\mathcal{C}$ be a fusion category. There is a canonical isomorphism of tensor functors $V^{**} \xrightarrow{\sim} {}^{**}V$. Equivalently, there is a canonical isomorphism of tensor functors $\mathsf{id}_{\mathcal{C}} \xrightarrow{\sim} -^{****}$.*

PROOF. By Remark 6.5.9 a fusion category is unimodular. Therefore we have that $D \cong \mathbf{1}$ by Theorem 7.18.7 and the statement follows by Theorem 7.19.1, where D is the dual distinguished invertible object of $\mathcal{C}$ introduced in (7.59). □

REMARK 7.19.4. Let $\mathcal{C}$ be a pivotal fusion category with pivotal structure a. Clearly, $a^2 : \mathsf{id}_{\mathcal{C}} \xrightarrow{\sim} -^{****}$ is an isomorphism of tensor functors. Therefore, a^2 and the canonical isomorphism of tensor functors $\mathsf{id}_{\mathcal{C}} \xrightarrow{\sim} -^{****}$ differ by an element from the group $\mathsf{Aut}_{\otimes}(\mathsf{id}_{\mathcal{C}})$.

EXAMPLE 7.19.5. Let $\mathcal{C}$ be the category of finite dimensional comodules over $u_q(\mathfrak{sl}_2)$ (see Section 5.6). This category is an example of a finite tensor category in which there are objects V such that V^{**} is not isomorphic to V. Namely, in this category, the functor $V \mapsto V^{**}$ is defined by the squared antipode S^2, which is conjugation by K: $S^2(x) = KxK^{-1}$. Now, we have $\mathsf{Ext}^1(1, K) = Y = \langle E, FK \rangle$, a 2-dimensional space. The set of isomorphism classes of nontrivial extensions of 1 by K is therefore the projective line $\mathbb{P}Y$. The operator of conjugation by K acts on Y with eigenvalues q^2, q^{-2}, hence nontrivially on $\mathbb{P}Y$. Thus for a generic extension V, the object V^{**} is not isomorphic to V.

Note also that in the category $\mathcal{C}$, $V^{**} \cong V$ if V is simple. This clearly has to be the case in any tensor category where all simple objects are invertible. We showed in Proposition 4.8.1 that this is the case in any semisimple tensor category. An example of a finite tensor category in which V^{**} is not always isomorphic to V even for simple V is the category of representations of the Hopf algebra A_q defined in Section 3.1 of [**Ra1**].[7]

7.20. The canonical Frobenius algebra of a unimodular category

Let $\mathcal{C}$ be a unimodular multitensor category (see Definition 6.5.7), and let $A := \underline{\mathsf{Hom}}(\mathbf{1}, \mathbf{1})$ in $\mathcal{C} \boxtimes \mathcal{C}^{\mathrm{op}}$ be the canonical algebra introduced in Definition 7.9.12. It was explained in Section 7.19 that in this case there is an isomorphism of A-modules,

$$\phi : A^* \xrightarrow{\sim} A. \tag{7.66}$$

This ϕ is unique up to an automorphism of $\mathbf{1}$. That is, if $\phi_1, \phi_2 : A^* \cong A$ are two isomorphisms of A-modules then $\phi_2 = ((\alpha \boxtimes \mathbf{1}) \otimes A) \circ \phi_1$, where $\alpha \in \mathsf{Aut}_{\mathcal{C}}(\mathbf{1})$.

PROPOSITION 7.20.1. *Let $\mathcal{C}$ be a unimodular multitensor category, let A be its canonical algebra with multiplication $m : A \otimes A \to A$ and unit $e : \mathbf{1} \boxtimes \mathbf{1} \to A$, and let ϕ be an isomorphism defined in* (7.66). *The comultiplication defined by*

$$\Delta := (\phi \otimes \phi) \circ m^* \circ \phi^{-1} : A \to A \otimes A \tag{7.67}$$

is coassociative, i.e., $(\Delta \otimes \mathsf{id}_A) \circ \Delta = (\mathsf{id}_A \otimes \Delta) \circ \Delta$, and is an A-bimodule morphism, i.e., $\Delta \circ m = (m \otimes \mathsf{id}_A) \circ (\mathsf{id}_A \otimes \Delta) = (\mathsf{id}_A \otimes m) \circ (\Delta \otimes \mathsf{id}_A)$. Furthermore, the morphism

$$\varepsilon := e^* \circ \phi^{-1} : A \to \mathbf{1} \boxtimes \mathbf{1} \tag{7.68}$$

is a counit for Δ, i.e., $(\varepsilon \otimes \mathsf{id}_A) \circ \Delta = \mathsf{id}_A = (\mathsf{id}_A \otimes \varepsilon) \circ \Delta$.

PROOF. This is completely parallel to the classical case (when $\mathcal{C} = \mathsf{Vec}$). □

EXERCISE 7.20.2. Prove Proposition 7.20.1.

DEFINITION 7.20.3. An algebra A in a tensor category $\mathcal{D}$ is called *Frobenius* if it has a structure of a coassociative and counital coalgebra in $\mathcal{D}$ with a comultiplication Δ which is a homomorphism of A-bimodules.

COROLLARY 7.20.4. *Let $\mathcal{C}$ be a unimodular multitensor category. Then the canonical algebra A of $\mathcal{C}$ is a Frobenius algebra in $\mathcal{C} \boxtimes \mathcal{C}^{\mathrm{op}}$ with the comultiplication and counit given by formulas* (7.67) *and* (7.68).

REMARK 7.20.5. The isomorphism type of a Frobenius algebra $(A, m, e, \Delta, \varepsilon)$ is well defined. Namely, suppose another choice of an isomorphism (7.66) gives a coalgebra structure (Δ', ε') on A. Then there is an algebra automorphism $\tau : A \to A$ such that $\Delta' = (\tau^{-1} \otimes \tau^{-1})\Delta\tau$ and $\varepsilon' = \varepsilon\tau$. Note that by Proposition 7.18.5 each such automorphism is of the form $\tau = (\alpha \boxtimes \mathbf{1}) \otimes A$ for some $\alpha \in \mathsf{Aut}_{\mathcal{C}}(\mathbf{1})$.

---

[7] We are very grateful to J. Cuadra for this reference.

EXAMPLE 7.20.6. Keep the setting of Example 7.18.11, and assume in addition that H is unimodular, i.e., $\alpha = \varepsilon$. Then H^* has another comultiplication

$$\delta(\psi) = \psi\lambda_1 \otimes S(\lambda_2), \qquad \psi \in H^* \tag{7.69}$$

that turns it into a Frobenius algebra.

EXERCISE 7.20.7. Show that δ is $H \otimes H^{\mathrm{cop}}$-linear (i.e., is a morphism in $\mathsf{Rep}(H \otimes H^{\mathrm{cop}})$) and is an H^*-bimodule morphism. (Use Proposition 7.18.12 and Exercise 7.10.10(vi)). What is the counit for δ?

Thus, we see that H^* is the canonical Frobenius algebra in $\mathsf{Rep}(H \otimes H^{\mathrm{cop}})$.

7.21. Categorical dimension of a multifusion category

Let $\mathcal{C}$ be a multifusion category. Let $\mathcal{O}(\mathcal{C})$ denote the set of simple objects of $\mathcal{C}$. Choose a natural isomorphism $\psi_X : X \xrightarrow{\sim} X^{**}$ (which exists by Proposition 4.8.1).

Recall from Definition 4.7.1 that the left categorical trace of an isomorphism $a : X \xrightarrow{\sim} X^{**}$ is defined by

$$\mathsf{Tr}^L(a) : \mathbf{1} \xrightarrow{\mathsf{coev}_X} X \otimes X^* \xrightarrow{a \otimes \mathsf{id}_{X^*}} X^{**} \otimes X^* \xrightarrow{\mathsf{ev}_{X^*}} \mathbf{1}.$$

see (4.8). We identify $\mathsf{Tr}^L(a)$ with an element of $\mathsf{End}_{\mathcal{C}}(\mathbf{1})$ (this element is a scalar when $\mathcal{C}$ is a fusion category).

Let $\mathbf{1} = \oplus_i \mathbf{1}_i$ be the decomposition of the unit object of $\mathcal{C}$ and let $\mathcal{C} = \oplus_{ij} \mathcal{C}_{ij}$ be the decomposition of $\mathcal{C}$ into the sum of component subcategories, where $\mathcal{C}_{ij} = \mathbf{1}_i \otimes \mathcal{C} \otimes \mathbf{1}_j$, see (4.1).

For any simple X in $\mathcal{C}_{ij}$ the scalar

$$|X|^2 := \mathsf{Tr}^L(\psi_X)\, \mathsf{Tr}^L((\psi_X^{-1})^*) \in \mathsf{End}_{\mathcal{C}}(\mathbf{1}_i) \otimes_{\Bbbk} \mathsf{End}_{\mathcal{C}}(\mathbf{1}_j) \cong \Bbbk \tag{7.70}$$

does not depend on the choice of ψ_X.

REMARK 7.21.1. Note that $|X|^2 \neq 0$ by Proposition 4.8.4.

DEFINITION 7.21.2. For any simple object X in $\mathcal{C}$ the scalar $|X|^2$ from (7.70) is called the *squared norm* of X.

DEFINITION 7.21.3. *The categorical dimension of a multifusion category $\mathcal{C}$ is*

$$\dim(\mathcal{C}) := \sum_{X \in \mathcal{O}(\mathcal{C})} |X|^2. \tag{7.71}$$

We also denote $\dim(\mathcal{C}_{ij}) := \sum_{X \in \mathcal{O}(\mathcal{C}_{ij})} |X|^2$.

EXAMPLE 7.21.4. Let H be a semisimple Hopf algebra. Then the categorical dimension of $\mathcal{C} = \mathsf{Rep}(H)$ is

$$\dim(\mathsf{Rep}(H)) = \mathsf{Tr}_H(S^2), \tag{7.72}$$

where S is the antipode of H. Indeed, since $X \cong X^{**}$ for every simple H-module X, it follows that S^2 is an inner automorphism, i.e., $S^2(x) = gxg^{-1}$ for some $g \in H$. Then $\mathsf{Tr}_H(S^2) = \sum_X \mathsf{Tr}_X(g)\, \mathsf{Tr}_{X^*}((g^{-1})^*) = \sum_X |X|^2 = \dim(\mathsf{Rep}(H))$.

Recall the description of the canonical algebra of $\mathcal{C}$:

$$A = \underline{\mathsf{Hom}}_{\mathcal{C} \boxtimes \mathcal{C}^{\mathrm{op}}}(\mathbf{1}, \mathbf{1})$$

from Example 7.9.14.

Let us describe the canonical isomorphism $A \xrightarrow{\sim} A^{**}$. We have canonically

$$A^{**} \cong \bigoplus_{X \in \mathcal{O}(\mathcal{C})} X^{**} \boxtimes {}^{***}X \cong \bigoplus_{X \in \mathcal{O}(\mathcal{C})} X \boxtimes {}^{*****}X.$$

Thus to specify a morphism $A \to A^{**}$ in $\mathcal{C} \boxtimes \mathcal{C}^{\mathrm{op}}$ is the same as to specify a collection of morphisms $\psi_X : {}^*X \to {}^{*****}X$ in $\mathcal{C}$.

LEMMA 7.21.5. *The canonical isomorphism $A \xrightarrow{\sim} A^{**}$ corresponds to the collection of isomorphisms*

$$\psi_X : {}^*X \to {}^{*****}X \qquad X \in \mathcal{O}(\mathcal{C})$$

characterized by the following property: for any isomorphism $\phi_X : {}^*X \to {}^{***}X$ one has $\mathsf{Tr}^L(\phi_X^{-1}) \,\mathsf{Tr}^L(\phi_X \circ \psi_X^{-1}) = |X|^2$.

PROOF. The statement is immediate from definitions since the isomorphism $A \xrightarrow{\sim} A^{**}$ is the composition of an isomorphism $A \xrightarrow{\sim} A^*$ and of the inverse of the dual of this isomorphism. □

COROLLARY 7.21.6. *The trace of the canonical isomorphism $A \cong A^{**}$ is equal to $\oplus_{ij} \dim(\mathcal{C}_{ij}) \,\mathsf{id}_{\mathbf{1}_i \boxtimes \mathbf{1}_j}$. When $\mathcal{C}$ is a fusion category, this trace is equal to $\dim(\mathcal{C})$.*

Let us relate the canonical isomorphism $\delta_X : X^{**} \cong {}^{**}X$ from Theorem 7.19.1 with squared norms of simple objects of $\mathcal{C}$.

THEOREM 7.21.7. *Let $X \in \mathcal{O}(\mathcal{C})$ be a simple object of $\mathcal{C}$. The canonical isomorphism $\delta_X : X^{**} \cong {}^{**}X$ can be characterized in the following way: for any isomorphism $\phi_X : X^{**} \to X$ one has $\mathsf{Tr}^L(\phi_X^{-1}) \,\mathsf{Tr}^L(\phi_X \circ \delta_X^{-1}) = |X|^2$.*

PROOF. Recall that A represents the functor $X \boxtimes Y \mapsto \mathsf{Hom}_{\mathcal{C}}(X \otimes Y, \mathbf{1})$ and A^{**} represents the functor $X \boxtimes Y \mapsto \mathsf{Hom}_{\mathcal{C}}({}^{**}X \otimes Y^{**}, \mathbf{1}) = \mathsf{Hom}_{\mathcal{C}}(X \otimes Y^{****}, \mathbf{1})$. It follows immediately from definitions that the canonical isomorphism $A \to A^{**}$ corresponds to the natural transformation

$$X \boxtimes Y \xrightarrow{\mathsf{id} \boxtimes \delta_{Y^{**}}^{-1}} X \boxtimes Y^{****}.$$

Now the result is an immediate consequence of Lemma 7.21.5. □

COROLLARY 7.21.8. *Let $V \in \mathcal{C}$ be an object and let $\phi_V : V \to {}^{**}V$ be a morphism. Then $\mathsf{Tr}^L(\phi_V^*) = \mathsf{Tr}^L(\phi_V \circ \delta_V^{-1})$.*

PROOF. It is enough to prove the statement for $V = X$, $X \in \mathcal{O}(\mathcal{C})$ and any isomorphism $\phi_X : X \to {}^{**}X$. But this is an immediate consequence of Theorem 7.21.7. □

Let $\mathcal{C}$ be a multifusion category. It is not known, in general, if $\mathcal{C}$ admits a pivotal structure. The following *pivotalization* construction will be used in the proof of Theorem 7.21.12 below.

By Corollary 7.19.3, there exists a natural isomorphism

$$g_X : X \xrightarrow{\sim} X^{****}$$

between the identity and the fourth duality tensor autoequivalences of $\mathcal{C}$. Since the second duality functor $X \mapsto X^{**}$ is a tensor autoequivalence of $\mathcal{C}$, we have an action of $\mathbb{Z}/2\mathbb{Z}$ on $\mathcal{C}$.

Let $\tilde{\mathcal{C}} := \mathcal{C}^{\mathbb{Z}/2\mathbb{Z}}$ be the corresponding equivariantization. Explicitly, simple objects of $\tilde{\mathcal{C}}$ are pairs (X, f), where X is a simple object of $\mathcal{C}$, and $f : X \to X^{**}$ is an isomorphism, such that $f^{**}f = g$. It is easy to see that this fusion category has a canonical pivotal structure.

DEFINITION 7.21.9. The category $\tilde{\mathcal{C}}$ will be called the *pivotalization* of $\mathcal{C}$.

COROLLARY 7.21.10. *The category $\tilde{\mathcal{C}}$ is spherical, that is,* $\dim(X) = \dim(X^*)$ *for any $X \in \tilde{\mathcal{C}}$.*

PROOF. Follows from Corollary 7.21.8. □

REMARK 7.21.11. We have $\dim(\tilde{\mathcal{C}}) = 2\dim(\mathcal{C})$ and $\mathsf{FPdim}(\tilde{\mathcal{C}}) = 2\,\mathsf{FPdim}(\mathcal{C})$.

THEOREM 7.21.12. *Let $\mathbb{k}$ be an algebraically closed field of characteristic* 0. *For any indecomposable multifusion category $\mathcal{C}$ over $\mathbb{k}$ one has* $\dim(\mathcal{C}_{ij}) \neq 0$ *for all component subcategories $\mathcal{C}_{ij}$ (so* $\dim(\mathcal{C}) \neq 0$ *for any fusion category $\mathcal{C}$). If $\mathbb{k} = \mathbb{C}$ then $|X|^2 > 0$ for all $X \in \mathcal{O}(\mathcal{C})$; in particular* $\dim(\mathcal{C}) \geq 1$, *and is* > 1 *for any non-trivial $\mathcal{C}$.*

PROOF. It suffices to prove the statement in the case when $\mathbb{k} = \mathbb{C}$ since $\mathcal{C}$ is always defined over a finitely generated subfield $\mathbb{k}'$ of $\mathbb{k}$ which can be embedded into $\mathbb{C}$.

First, consider the special case when $\mathcal{C}$ is pivotal. Let $a_X : X \xrightarrow{\sim} X^{**}$ be the pivotal structure of $\mathcal{C}$.

For any $X \in \mathcal{O}(\mathcal{C})$ let $d_X := \dim_a(X)$. Then $|X|^2 = d_X d_{X^*}$ and

$$d_X d_Y = \sum_{Z \in \mathcal{O}(\mathcal{C})} N_{XY}^Z d_Z,$$

where the numbers N_{XY}^Z are multiplicities coming from the tensor product decomposition $X \otimes Y \cong \oplus_{Z \in \mathcal{O}(\mathcal{C})} N_{XY}^Z Z$.

Let N_X be the matrix such that $(N_X)_{YZ} = N_{XY}^Z$, and let $\mathbf{d}$ be the vector with components d_X, $X \in \mathcal{O}(\mathcal{C})$. This vector is clearly nonzero (e.g., $d_\mathbf{1} \neq 0$), and we have

$$N_X \mathbf{d} = d_X \mathbf{d}.$$

Thus, $\mathbf{d}$ is an eigenvector of N_X with the eigenvalue d_X. Hence, $\mathbf{d}$ is an eigenvector of the non-negative matrix $N_X N_{X^*} = N_X N_X^t$. When $\mathbb{k} = \mathbb{C}$, the corresponding eigenvalue $|X|^2$ is non-negative by Remark 7.21.1. Consequently, $\dim(\mathcal{C})$ is non-negative.

Now let us prove the statement in a general case. Let $\tilde{\mathcal{C}}$ be the pivotalization of $\mathcal{C}$, see Definition 7.21.9. For any simple object X in $\tilde{\mathcal{C}}$ we have $|X|^2 > 0$. The forgetful tensor functor

$$F : \tilde{\mathcal{C}} \to \mathcal{C}, \quad (X, f) \mapsto X,$$

obviously preserves squared norms. Hence, $|V|^2 > 0$ for simple objects V in $\mathcal{C}$ (since for any V there exists $f : V \to V^{**}$ such that $f^{**}f = g$), which completes the proof. □

DEFINITION 7.21.13. Let $\mathbb{k}_{alg} \subset \mathbb{k}$ denote the subfield of algebraic numbers in $\mathbb{k}$. We say that $\alpha \in \mathbb{k}$ is *totally positive* if for any embedding $i : \mathbb{k}_{alg} \hookrightarrow \mathbb{C}$ we have $i(\alpha) > 0$.

In particular, a real algebraic number is totally positive if all its Galois conjugates are positive.

PROPOSITION 7.21.14. *Each* $|X|^2$ *is totally positive. Also,* $\dim(\mathcal{C})-1$ *is totally non-negative.*

PROOF. This follows from the proof of Theorem 7.21.12. □

Recall that actions of groups on tensor categories and the corresponding equivariantizations were defined in Sections 2.7 and 4.15. In the next Proposition we compute the categorical dimension of the equivariantized category.

PROPOSITION 7.21.15. *Let* G *be a finite group acting on a multifusion category* $\mathcal{C}$. *We have*

$$\dim(\mathcal{C}^G) = |G| \dim(\mathcal{C}). \tag{7.73}$$

PROOF. Let X be a simple object of $\mathcal{C}$ and $G_X \subset G$ the stabilizer of its isomorphism class. Recall from Remark 4.15.8 that the set of isomorphism classes of simple objects of $\mathcal{C}^G$ whose image under the forgetful functor $\mathcal{C}^G \to \mathcal{C}$ contains X is in natural bijection with the set $\mathrm{Irr}^1(\tilde{G}_X)$ of irreducible finite dimensional representations of $\tilde{G}_X$ such that each $\lambda \in \Bbbk^\times \subset \tilde{G}_X$ acts as multiplication by λ. (Here $\tilde{G}_X$ is a central extension of G.)

For $V \in \mathrm{Irr}^1(\tilde{G}_X)$ let X_V denote the corresponding simple object of $\mathcal{C}^G$. We will show that

$$|X_V|^2 = [G : G_X]^2 \dim_{\Bbbk}(V)^2 \, |X|^2 \tag{7.74}$$

To prove (7.74), notice that there exists a pair (ψ, ψ^G) of natural (but not necessarily tensor) isomorphisms $\psi_X : X \to X^{**}$, $X \in \mathcal{C}$ and $\psi^G_Y : Y \to Y^{**}$, $Y \in \mathcal{C}^G$ such that $F(\psi^G) = \psi$ where $F : \mathcal{C}^G \to \mathcal{C}$ is the forgetful functor (such pairs are in bijection with G-invariant functions $\mathcal{O}(\mathcal{C}) \to \Bbbk^\times$). For such a pair we clearly have $\mathsf{Tr}^{L,R}(\psi^G_{X_V}) = \mathsf{Tr}^{L,R}(\psi_{F(X_V)}) = [G : G_X] \dim_{\Bbbk}(V) \, \mathsf{Tr}^{L,R}(\psi_X)$, so (7.74) follows from (7.70).

Now the formula

$$\sum_{V \in \mathrm{Irr}^1(\tilde{G}_X)} \dim_{\Bbbk}(V)^2 = |G_X|.$$

along with equation (7.74) imply the result. □

Let $\mathcal{C}$ be a multifusion category and let $A = \underline{\mathsf{Hom}}_{\mathcal{C}\boxtimes\mathcal{C}^{\mathrm{op}}}(\mathbf{1}, \mathbf{1})$ be the canonical algebra associated to $\mathcal{C}$, see Definition 7.9.12. Recall from Proposition 7.20.1 that there is a Frobenius algebra structure $(A, m, e, \Delta, \varepsilon)$ on A (the definitions of Δ and ε use an isomorphism $\phi : A^* \xrightarrow{\sim} A$ of A-modules). Using this structure, we can give an alternative definition of a categorical dimension of $\mathcal{C}$, cf. Definition 7.21.3.

PROPOSITION 7.21.16. *The endomorphism*

$$\mathsf{Dim} = \varepsilon \circ m \circ \Delta \circ e : \mathbf{1} \boxtimes \mathbf{1} \to \mathbf{1} \boxtimes \mathbf{1} \tag{7.75}$$

does not depend on the choice of an isomorphism $\phi : A^* \xrightarrow{\sim} A$.

PROOF. Let (Δ', ε') be the comultiplication and counit of A resulting from a different choice of $\phi : A \to A^*$. Then, according to Remark 7.20.5 there is an endomorphism τ of the algebra A such that

$$\varepsilon' m \Delta' e = \varepsilon \tau m(\tau^{-1} \otimes \tau^{-1}) \Delta \tau e = \varepsilon m \Delta e, \tag{7.76}$$

as required. □

DEFINITION 7.21.17. We will call $\mathsf{Dim} \in \mathsf{Aut}_{\mathcal{C}\boxtimes\mathcal{C}^{\mathrm{op}}}(\mathbf{1}\boxtimes\mathbf{1})$ the *dimension endomorphism* of $\mathcal{C}$.

PROPOSITION 7.21.18. *Let $\mathcal{C} = \oplus_{ij}\mathcal{C}_{ij}$ be a multifusion category. Then*

$$\mathsf{Dim} = \oplus_{ij} \dim(\mathcal{C}_{ij})\, \mathsf{id}_{\mathbf{1}_i\boxtimes\mathbf{1}_j}. \tag{7.77}$$

PROOF. Let $\phi : A^* \xrightarrow{\sim} A$ be isomorphism (7.66). We will identify Dim with the trace of the canonical isomorphism $\phi^*\phi^{-1} : A \xrightarrow{\sim} A^{**}$. We have

$$\begin{aligned} \mathsf{ev}_A &= \varepsilon \circ m \circ (\phi \otimes \mathsf{id}_A) : A^* \otimes A \to \mathbf{1}\boxtimes\mathbf{1} \\ \mathsf{coev}_A &= (\mathsf{id}_A \otimes \phi^{-1}) \circ \Delta \circ e : \mathbf{1}\boxtimes\mathbf{1} \to A \otimes A^*. \end{aligned}$$

Evaluation and coevaluation axioms (2.43) and (2.44) follow from Proposition 7.20.1. We thus have a commutative diagram

$$\begin{array}{ccccccc} \mathbf{1}\boxtimes\mathbf{1} & \xrightarrow{\mathsf{coev}_A} & A\otimes A^* & \xrightarrow{\phi^*\phi^{-1}\otimes\mathsf{id}_{A^*}} & A^{**}\otimes A^* & \xrightarrow{\mathsf{ev}_{A^*}} & \mathbf{1}\boxtimes\mathbf{1}. \\ & \searrow^{e} & & \searrow^{\mathsf{id}_A\otimes\phi} \quad \swarrow^{\phi(\phi^*)^{-1}\otimes\phi} & & \nearrow^{\varepsilon} & \\ & A & \xrightarrow[\Delta]{} & A\otimes A & \xrightarrow[m]{} & A & \end{array} \tag{7.78}$$

The composition in the top row is $\mathsf{Tr}^L(\phi^*\phi^{-1})$ and the composition through the bottom row is Dim, so the result follows from Corollary 7.21.6. $\square$

COROLLARY 7.21.19. *Let $\mathcal{C}$ be a multifusion category and let $m : A \otimes A \to A$ and $\Delta : A \to A \otimes A$ be the multiplication and comultiplication of its canonical Frobenius algebra A. Then $m \circ \Delta : A \to A$ is an automorphism.*

PROOF. One can assume without loss of generality that $\mathcal{C}$ is indecomposable. In view of the Frobenius property of A (see Proposition 7.20.1) we have

$$m \circ \Delta = m \circ (\mathsf{id}_A \otimes (m \circ \Delta \circ e)) : A \otimes (\mathbf{1}\boxtimes\mathbf{1}) \cong A \to A,$$

where $e : \mathbf{1}\boxtimes\mathbf{1} \to A$ is the unit of A, so it suffices to check that $m\circ\Delta\circ e : \mathbf{1}\boxtimes\mathbf{1} \to A$ is injective. This immediately follows from Proposition 7.21.18. $\square$

When $\mathcal{C}$ is a fusion category, the endomorphism (7.75) can be identified with a scalar, which is nonzero by Theorem 7.21.12.

7.22. Davydov-Yetter cohomology and deformations of tensor categories

Let $\mathcal{C}$, $\mathcal{C}'$ be multitensor categories over a field $\Bbbk$, and $F : \mathcal{C} \to \mathcal{C}'$ be a tensor functor.

For a non-negative integer n, let $\mathcal{C}^n$ denote the n-th Cartesian power of $\mathcal{C}$. In particular, $\mathcal{C}^0$ has one object $\emptyset$ and one morphism (identity). Define the functor $T_n : \mathcal{C}^n \to \mathcal{C}$ by $T_n(X_1, ..., X_n) := X_1\otimes...\otimes X_n$. In particular, $T_0 : \mathcal{C}^0 \to \mathcal{C}$ is defined by $T_0(\emptyset) = \mathbf{1}$, and $T_1 = \mathsf{id}$. Let $C^n_F(\mathcal{C}) = \mathrm{End}(T_n\circ F^{\otimes n})$ (so e.g., $C^0_F(\mathcal{C}) = \mathsf{End}(\mathbf{1}_{\mathcal{C}'})$). We define a differential $d : C^n_F(\mathcal{C}) \to C^{n+1}_F(\mathcal{C})$ by the formula

$$df = \mathsf{id}\otimes f_{2,...,n+1} - f_{12,...,n+1} + f_{1,23,...,n+1} - ... + (-1)^n f_{1,...,nn+1} + (-1)^{n+1} f_{1,...,n}\otimes\mathsf{id},$$

where, for instance, $f_{12,3,...,n+1}$ is the endomorphism of the product of n objects $F(X_1)\otimes F(X_2), F(X_3), ..., F(X_{n+1})$, and we use the identification $F(X_1)\otimes F(X_2) \cong F(X_1 \otimes X_2)$ defined by the tensor structure on F.

It is easy to show that $d^2 = 0$. Thus $(C^\bullet_F(\mathcal{C}), d)$ is a complex.

DEFINITION 7.22.1. We will call the cohomology groups $H_F^n(\mathcal{C}), n \geq 0$ of this complex the *Davydov-Yetter cohomology groups* of $\mathcal{C}$ *with respect to* F. In the important special case $\mathcal{C} = \mathcal{C}'$, $F = \mathsf{id}$, we will denote this cohomology simply by $H^i(\mathcal{C})$ and call it the *Davydov-Yetter cohomology* of $\mathcal{C}$.

The motivation for considering this cohomology theory is that it describes deformations of multitensor categories and tensor functors.

EXERCISE 7.22.2. (i) Show that if $\mathcal{C}$ is an indecomposable multitensor category then $H_F^0(\mathcal{C}) = \Bbbk$.

(ii) Show that $H_F^1(\mathcal{C})$ is the Lie algebra of derivations (i.e., infinitesimal automorphisms) of F as a tensor functor; in particular, if $\Bbbk$ is of characteristic zero, then $H_F^1(\mathcal{C}) = \mathrm{Lie}(\mathsf{Aut}_\otimes(F))$.

(iii) Show that $H_F^2(\mathcal{C})$ parametrizes additively trivial first order deformations of F as a tensor functor modulo equivalence, and $H_F^3(\mathcal{C})$ is the obstruction space for such deformations. Thus, if $H_F^3(\mathcal{C}) = 0$, then such deformations of F are unobstructed, and the universal (additively trivial) formal deformation of F is parametrized by $H_F^2(\mathcal{C})$.

(iv) Show that $H^3(\mathcal{C})$ parametrizes additively trivial first order deformations of $\mathcal{C}$ as a multitensor category modulo equivalence, and $H^4(\mathcal{C})$ is the obstruction space for such deformations. Thus, if $H^4(\mathcal{C}) = 0$, then such deformations of $\mathcal{C}$ are unobstructed, and the universal (additively trivial) formal deformation of $\mathcal{C}$ is parametrized by $H^3(\mathcal{C})$.

(v) Compute explicitly the complex $C^\bullet(\mathcal{C}_n)$, where $\mathcal{C}_n$ is the category of bimodules over the algebra $\Bbbk^n$ (i.e., the category of $n \times n$-matrices of vector spaces), and compute $H^i(\mathcal{C}_n)$.

PROPOSITION 7.22.3. *If* $\mathcal{C} = A-\mathsf{comod}$ *for a Hopf algebra* A*, and* F *is the forgetful functor, then* $C_F^i(\mathcal{C}) = ((A^{\otimes i})^*)_A$*, where the subscript* A *means the space of elements commuting with the diagonal coaction of* A*, and the differential is the usual Hochschild differential (for trivial coefficients) restricted to invariants. In particular, if* A *is cocommutative, then* $H_F^i(\mathcal{C})$ *is the usual Hochschild cohomology* $HH^i(A, \Bbbk)$ *of* A *with trivial coefficients.*

EXERCISE 7.22.4. Prove Proposition 7.22.3.

EXAMPLE 7.22.5. Let G be a finite group, let $\Bbbk$ be an algebraically closed field (of any characteristic), and $\mathcal{C} = \mathsf{Vec}_G$ be the category of G-graded finite dimensional vector spaces over $\Bbbk$. Then $H^i(\mathcal{C}) = HH^i(\Bbbk G, \Bbbk)$, the Hochschild cohomology of the group algebra $\Bbbk G$ with trivial coefficients. In other words, $H^i(\mathcal{C}) = H^i(G, \Bbbk)$, the group cohomology of G with coefficients in $\Bbbk$.

EXAMPLE 7.22.6. Let G be a reductive algebraic group over $\Bbbk$ (of characteristic zero) with Lie algebra $\mathfrak{g}$. Let $\mathsf{Rep}(G)$ denote the category of algebraic representations of G. Then $H^i(\mathsf{Rep}(G)) = (\Lambda^i \mathfrak{g})^G$ for all i. Indeed, $\mathsf{Rep}(G)$ is the category of finite dimensional $O(G)$-comodules (where $O(G)$ is the Hopf algebra of regular functions on G). Therefore,

$$C^n(\mathsf{Rep}(G)) = (O(G^n)^*)^G,$$

where G acts diagonally by conjugation. Since G is reductive, the cohomology of this complex is the G-invariants in the Hochschild cohomology of $O(G)$ with coefficients in the trivial representation (corresponding to $1 \in G$). This cohomology is well known to be $\Lambda\mathfrak{g}$, so the answer is $(\Lambda\mathfrak{g})^G$, as desired. Thus, according to Exercise 7.22.2, if G is a simple algebraic group, then there are no nontrivial derivations or tensor structure deformations of the identity functor, but there exists a unique (up to scaling) first order deformation of the associativity constraint, corresponding to a basis element in $(\Lambda^3\mathfrak{g})^G$. It is easy to guess that this deformation comes from an actual deformation, namely from the deformation of $O(G)$ to the quantum group $O_q(G)$ (with $q = \exp(\hbar^{1/2})$, where $\hbar$ is the deformation parameter).

Let $A = \underline{\mathsf{Hom}}_{\mathcal{C}\boxtimes\mathcal{C}^{\mathrm{op}}}(\mathbf{1},\mathbf{1})$ be the canonical algebra associated to $\mathcal{C}$, see Definition 7.9.12. Let us compare the Davydov-Yetter complex of a tensor category $\mathcal{C}$ with the Hochschild complex of A.

Consider the Hochschild complex of A with coefficients in A. Let $C^n(A) = \mathsf{Hom}_{\mathcal{C}\boxtimes\mathcal{C}^{\mathrm{op}}}(A^{\otimes n}, A)$, $n = 0, 1, \ldots$, where $A^0 = \mathbf{1}\boxtimes\mathbf{1}$. The differential in this case is given by

$$(7.79)\quad d_n(f) = \mathrm{id}_A \otimes f + \sum_{i=1}^{n} (-1)^i f \circ (A^{\otimes^{i-1}} \otimes m \otimes A^{\otimes^{n-i}}) + (-1)^{n+1} f \otimes \mathrm{id}_A,$$

for all $f \in C^n(A)$, where $m : A \otimes A \to A$ is the multiplication of A. Let $HH^n(A)$, $n = 0, 1, \ldots$ denote the n-th cohomology group of the Hochschild complex.

PROPOSITION 7.22.7. *The cochain complexes $C^\bullet(\mathcal{C})$ and $C^\bullet(A)$ are isomorphic.*

PROOF. We may assume that $\mathcal{C}$ is strict. For $n \geq 2$ let $\iota^n_{k\,l} : \mathcal{C}\boxtimes\mathcal{C} \to \mathcal{C}^{\boxtimes n}$ be the inclusion coming from $V \boxtimes W \mapsto \mathbf{1}\boxtimes\cdots\boxtimes V\boxtimes\cdots\boxtimes W\boxtimes\cdots\mathbf{1}$ with V in the kth factor and W in the lth factor, $V, W \in \mathcal{C}$. Let $\otimes^n$ denotes the extension of the n-tuple tensor product to $\mathcal{C}^{\boxtimes n}$. Define

$$(7.80)\qquad A(n) := (\otimes^n \boxtimes \mathrm{id}_{\mathcal{C}} \boxtimes \cdots \boxtimes \mathrm{id}_{\mathcal{C}})(\iota^{2n}_{1\,2n}(A) \otimes \cdots \otimes \iota^{2n}_{n\,n+1}(A)) \in \mathcal{C}^{\boxtimes n+1},$$

where $\otimes^n \boxtimes \mathrm{id}_{\mathcal{C}} \boxtimes \cdots \boxtimes \mathrm{id}_{\mathcal{C}} : \mathcal{C}^{\boxtimes 2n} \to \mathcal{C}^{\boxtimes n+1}$. Let $A(0) = \mathbf{1}$ and $A(1) = A$. Then $A(n)$, $n \geq 0$, is an object of $\mathcal{C}^{\boxtimes(n+1)}$ representing the contravariant functor $\mathsf{Hom}_{\mathcal{C}}(\otimes^{n+1}(-), \mathbf{1}) : \mathcal{C}^{\boxtimes(n+1)} \to \mathsf{Vec}$, i.e., there is a natural isomorphism

$$(7.81)\qquad \mathsf{Hom}_{\mathcal{C}}(\otimes^{n+1}(X), \mathbf{1}) \cong \mathsf{Hom}_{\mathcal{C}^{\boxtimes(n+1)}}(X, A(n)), \qquad X \in \mathcal{C}^{\boxtimes(n+1)}.$$

EXAMPLE 7.22.8. Let $\mathcal{C}$ be a fusion category. In this case, we have $A = \oplus_{V\in\mathcal{O}(\mathcal{C})} V \boxtimes V^*$. Hence, $A(n) = \oplus_{V_1,\ldots,V_n\in\mathcal{O}(\mathcal{C})}(V_1 \otimes \ldots \otimes V_n) \boxtimes V_n^* \boxtimes \ldots \boxtimes V_1^*$, and (7.81) is easily verified.

As a consequence of (7.81), we have isomorphisms:

$$(7.82)\qquad \mathsf{End}_{\mathcal{C}^{\boxtimes(n+1)}}(A(n)) \cong \mathsf{Hom}_{\mathcal{C}}(\otimes^{n+1}(A(n)), \mathbf{1}) \cong \mathsf{Hom}_{\mathcal{C}\boxtimes\mathcal{C}^{\mathrm{op}}}(A^{\otimes n}, A).$$

For every $\eta \in \mathsf{End}(\otimes^n)$ let $I(\eta) = (\eta \boxtimes \mathrm{id}^{\boxtimes n})_{A(n)} : A(n) \to A(n)$ (that is, we take action on $A(n)$ of the natural endomorphism $\eta\boxtimes\mathrm{id}^{\boxtimes n}$ of the functor $\otimes^n \boxtimes \mathrm{id}_{\mathcal{C}}^{\boxtimes n} : \mathcal{C}^{\boxtimes 2n} \to \mathcal{C}^{\boxtimes n+1}$).

We claim that the map $I : \mathsf{End}(\otimes^n) \to \mathsf{End}_{\mathcal{C}^{\boxtimes(n+1)}}(A(n))$ is an isomorphism. Namely, its inverse J is given as follows. For each V in $\mathcal{C}$ let $F_V = \mathsf{Hom}_{\mathcal{C}}(\mathbf{1}, -\otimes V) : \mathcal{C} \to \mathsf{Vec}$, and for any $f \in \mathsf{End}_{\mathcal{C}^{\boxtimes(n+1)}}(A(n))$ define a natural endomorphism $J(f)$ of $\otimes^n$ by

$$J(f)_{V_1\boxtimes\cdots\boxtimes V_n} = (\mathrm{id}_{\mathcal{C}} \boxtimes F_{V_1} \boxtimes \cdots \boxtimes F_{V_n})(f).$$

Note that

$$(\mathrm{id}_{\mathcal{C}} \boxtimes F_{V_1} \boxtimes \cdots \boxtimes F_{V_n})(A(n)) \cong V_1 \otimes \cdots \otimes V_n$$

via the isomorphism (7.82).

The fact that the above assignments commute with the differentials follows from observing that the multiplication $m : A \otimes A \to A$ corresponds to the identity endomorphism of $\otimes$. □

7.23. Weak Hopf algebras

As we have seen, Hopf algebras arise through reconstruction theory from tensor categories $\mathcal{C}$ with a fiber functor $F : \mathcal{C} \to \mathsf{Vec}$, or, equivalently, with an action of $\mathcal{C}$ on the category Vec. However, we know that a tensor category (even a fusion category) does not always admit such an action, and thus is not always the category of modules (or comodules) over a Hopf algebra. On the other hand, a fusion category $\mathcal{C}$ always admits a tensor functor $F : \mathcal{C} \to R-$ bimod, where R is a semisimple $\Bbbk$-algebra, or, equivalently, an action on R-mod (which is a direct sum of several copies of Vec). In other words, any fusion category admits a nontrivial semisimple module category $\mathcal{M}$ (for example, $\mathcal{M} = \mathcal{C}$). This gives rise to a question what algebraic structure generalizing Hopf algebras arises from a multitensor category $\mathcal{C}$ with such a functor F. This structure is called a *weak Hopf algebra.*

To obtain the definition of a weak Hopf algebra, assume for simplicity that $\mathcal{C}$ is finite, and consider the algebra $A = \mathsf{End}_\Bbbk(F)$, i.e., the endomorphism algebra of the composition of F with the forgetful functor from R-bimod to Vec. Then, similarly to the reconstruction theory for Hopf algebras, we can analyze the structures induced on A by the multitensor category structure on $\mathcal{C}$ and the structure of a tensor functor on F. This leads to the following definition.

DEFINITION 7.23.1 ([**BohNS**]). A *weak Hopf algebra* is a vector space A with the structures of an associative algebra $(A, m, 1)$ with a multiplication $m : A \otimes_\Bbbk A \to A$ and unit $1 \in A$ and a coassociative coalgebra (A, Δ, ϵ) with a comultiplication $\Delta : A \to A \otimes_\Bbbk A$ and counit $\epsilon : A \to \Bbbk$ such that:

(i) The comultiplication Δ is a (not necessarily unit-preserving) homomorphism of algebras:

$$\Delta(hg) = \Delta(h)\Delta(g), \qquad h, g \in A; \tag{7.83}$$

(ii) The unit and counit satisfy the following identities:

$$(\Delta \otimes \mathsf{id})\Delta(1) = (\Delta(1) \otimes 1)(1 \otimes \Delta(1)) = (1 \otimes \Delta(1))(\Delta(1) \otimes 1), \tag{7.84}$$

$$\epsilon(fgh) = \epsilon(fg_1)\,\epsilon(g_2h) = \epsilon(fg_2)\,\epsilon(g_1h), \tag{7.85}$$

for all $f, g, h \in A$.

(iii) There is a linear map $S : A \to A$, called an *antipode*, such that

$$m(\mathsf{id} \otimes S)\Delta(h) = (\epsilon \otimes \mathsf{id})(\Delta(1)(h \otimes 1)), \tag{7.86}$$

$$m(S \otimes \mathsf{id})\Delta(h) = (\mathsf{id} \otimes \epsilon)((1 \otimes h)\Delta(1)), \tag{7.87}$$

$$S(h) = S(h_1)h_2S(h_3), \tag{7.88}$$

for all $h \in A$.

Here we use Sweedler's notation for the comultiplication: $\Delta(c) = c_1 \otimes c_2$.

Axioms (7.84) and (7.85) above are analogous to the usual bialgebra axioms of Δ being a unit preserving map and ϵ being an algebra homomorphism. Axioms (7.86) and (7.87) generalize the properties of the antipode in a Hopf algebra with respect to the counit. Also, it is possible to show that given (7.83) - (7.87), axiom (7.88) is equivalent to S being both an anti-algebra and anti-coalgebra map.

EXERCISE 7.23.2. (i) Show that similarly to Hopf algebras, the antipode of a finite dimensional weak Hopf algebra is bijective (see [**BohNS**, 2.10]).

(ii) Show that a weak Hopf algebra is a Hopf algebra if and only if the comultiplication is unit-preserving and if and only if ϵ is a homomorphism of algebras.

(iii) Show that when $\dim_{\Bbbk} A < \infty$, there is a natural weak Hopf algebra structure on the dual vector space A^* given by

$$\phi\psi(h) = (\phi \otimes \psi)(\Delta(h)), \tag{7.89}$$

$$\Delta(\phi)(h \otimes g) = \phi(hg) \tag{7.90}$$

$$S(\phi)(h) = \phi(S(h)), \tag{7.91}$$

for all $\phi, \psi \in A^*$, $h, g \in A$. The unit of A^* is ϵ and the counit is $\phi \mapsto \phi(1)$.

For simplicity from this point on let us consider only finite dimensional weak Hopf algebras.

The linear maps defined in (7.86) and (7.87) are called the *target* and *source counital maps* and are denoted ε_t and ε_s respectively :

$$\varepsilon_t(h) = \epsilon(1_1 h)1_2, \qquad \varepsilon_s(h) = 1_1\epsilon(h1_2), \tag{7.92}$$

for all $h \in A$.

EXERCISE 7.23.3. Show that the images of the counital maps

$$A_t = \varepsilon_t(A), \qquad A_s = \varepsilon_s(A) \tag{7.93}$$

are separable subalgebras of A.

The subalgebras A_t and A_s are called the target and source *bases* or *counital subalgebras* of A.

EXERCISE 7.23.4. Show that the subalgebras A_t and A_s commute with each other; moreover

$$\begin{aligned} A_t &= \{(\phi \otimes \mathsf{id})\Delta(1) \mid \phi \in A^*\} = \{h \in A \mid \Delta(h) = \Delta(1)(h \otimes 1)\}, \\ A_s &= \{(\mathsf{id} \otimes \phi)\Delta(1) \mid \phi \in A^*\} = \{h \in A \mid \Delta(h) = (1 \otimes h)\Delta(1)\}, \end{aligned}$$

i.e., A_t (respectively, A_s) is generated by the right (respectively, left) tensor factors of $\Delta(1)$ in the shortest possible presentation of $\Delta(1)$ in $A \otimes_{\Bbbk} A$. Show that $S(A_t) = A_s$ and $S(A_s) = A_t$.

EXERCISE 7.23.5. Show that the category $\mathsf{Rep}(A)$ of finite dimensional left A-modules is a rigid monoidal category. Namely, the tensor product of two A-modules V and W is $V \otimes_{A_t} W$ with the A-module structure defined via Δ, and the unit object $\mathbf{1}$ of $\mathsf{Rep}(A)$ is the target counital algebra A_t with the action $h \cdot z = \varepsilon_t(hz)$ for all $h \in A$, $z \in A_t$.

EXERCISE 7.23.6. For any algebra B denote by $Z(B)$ the center of B. Show that the unit object of $\mathsf{Rep}(A)$ is simple if and only if $Z(A) \cap A_t = \Bbbk$.

If the condition of Exercise 7.23.6 is satisfied, we will say that A is *connected*. We will say that A is *coconnected* if A^* is connected, and that A is *biconnected* if it is both connected and coconnected.

If $p \neq 0$ is an idempotent in $A_t \cap A_s \cap Z(A)$, then A is the direct sum of weak Hopf algebras pA and $(1-p)A$. Consequently, we say that A is *indecomposable* if $A_t \cap A_s \cap Z(A) = \Bbbk 1$.

EXERCISE 7.23.7. Show that $\mathsf{Rep}(A)$ is indecomposable if and only if A is indecomposable, and $\mathsf{Rep}(A)$ is a tensor category if and only if A is connected.

Every weak Hopf algebra A contains a canonical *minimal* weak Hopf subalgebra $A_{\min}$ generated, as an algebra, by A_t and A_s [**Nik1**, Section 3]. In other words, $A_{\min}$ is the minimal weak Hopf subalgebra of A that contains 1. Obviously, A is an ordinary Hopf algebra if and only if $A_{\min} = \Bbbk 1$.

EXERCISE 7.23.8. (i) Classify minimal weak Hopf algebras over $\Bbbk$, i.e., those for which $A = A_{\min}$ (see [**Nik1**, Proposition 3.4]).

(ii) Show that the restriction of S^2 on $A_{\min}$ is always an inner automorphism of $A_{\min}$ (see [**Nik1**]).

We will say that a weak Hopf algebra A is *regular* if

$$S^2|_{A_{\min}} = \mathsf{id}\,. \tag{7.94}$$

This property has an easy categorical interpretation.

EXERCISE 7.23.9. Let $\mathbf{1} = A_t$ be the trivial A-module. Show that A is regular if and only if the canonical isomorphism $\mathbf{1} \to \mathbf{1}^{**}$ is the identity map.

REMARK 7.23.10. It was shown in [**NikV1**, 6.1] that every weak Hopf algebra can be obtained as a twisting of some regular weak Hopf algebra with the same algebra structure.

Now let us return to studying the algebra $A = \mathsf{End}_{\Bbbk}(F)$, where $F : \mathcal{C} \to R-\text{bimod}$ is a tensor functor from a finite multitensor category $\mathcal{C}$ to the category of R-bimodules. That is, the category R-mod is a faithful semisimple module category over $\mathcal{C}$.

PROPOSITION 7.23.11. *([**Sz**]) The algebra A has a structure of a weak Hopf algebra with base R, and $\mathcal{C}$ is equivalent, as a tensor category, to the category $\mathsf{Rep}(A)$ of finite dimensional representations of A.*

EXERCISE 7.23.12. Prove Proposition 7.23.11.

Hint: In order to lift the naturally existing map $A \to A \otimes_R A$ to a weak Hopf algebra coproduct $A \to A \otimes_{\Bbbk} A$, one needs to use a separability idempotent $\Delta(1) \in R \otimes R^{\text{op}}$, i.e., a splitting of the multiplication $R \otimes R \to R$ as an R-bimodule map. The structure of a weak Hopf algebra on A depends on the choice of this idempotent.

EXERCISE 7.23.13. Let $\mathcal{C} = \mathsf{Vec}$, and $F(V) = V \otimes R$. Describe the weak Hopf algebras $A = \mathsf{End}_{\Bbbk} F$ and A^* explicitly, for all choices of the separability idempotent.

EXERCISE 7.23.14. Show that if $\Delta(1)$ is symmetric then the weak Hopf algebra $A = \mathsf{End}_{\Bbbk} F$ is regular. Deduce that any multifusion category is the representation category of a regular weak Hopf algebra A. Moreover, one can find A such that the base A_t of A is commutative.

The language of weak Hopf algebras is convenient to visualize various categorical constructions in linear-algebraic terms. One of them is that of a dual category. Indeed, let $\mathcal{C}$ be the category of representations of a weak Hopf algebra A. Let R be the base of A_t of A. Then we have a natural fiber functor from $\mathcal{C}$ to the category of R-bimodules – the forgetful functor. This functor defines a natural structure of a module category over $\mathcal{C}$ on the category $\mathcal{M} = R - \text{bimod}$.

EXERCISE 7.23.15. Show that the dual category $\mathcal{C}^*_{\mathcal{M}}$ is simply the representation category of the dual weak Hopf algebra A^{*cop} with the opposite coproduct (see [**Os1**]). Interpret categorically the notions of a coconnected and a biconnected weak Hopf algebra.

Note that as we just showed, this example is general, in the sense that any faithful module category over a multifusion category can be obtained in this way.

Thus, we see that a finite multitensor category $\mathcal{C}$ is equivalent to the representation category of a weak Hopf algebra if and only if $\mathcal{C}$ admits a semisimple faithful module category.

EXERCISE 7.23.16. This exercise constructs an example of a tensor category without a faithful semisimple module category, in characteristic p. We do not know an example in characteristic zero (but expect it exists).

(i) Let $\mathrm{char}\Bbbk = p$, and $H = \Bbbk[x, y, z]/(x^p, y^p, z^p)$. Put a Hopf algebra structure on H by declaring x, y, z to be primitive elements. Show that the element

$$\Phi := \exp(x \otimes y \otimes z) := \sum_{i=0}^{p-1} \frac{x^i \otimes y^i \otimes z^i}{i!}$$

defines the structure of a quasi-Hopf algebra on H (with the same coproduct).

(ii) Show that the quasi-Hopf algebra (H, Φ) is not twist equivalent to a Hopf algebra. Deduce that the category $\mathsf{Rep}(H)$ does not admit a semisimple faithful module category (use that $\mathsf{Rep}(H)$ has a unique simple object, which is $\mathbf{1}$).

Thus, we see that a tensor functor $F : \mathcal{C} \to R-\mathrm{bimod}$ does not always exist. However, it turns out that the situation improves if we look for quasi-tensor (not necessarily tensor) functors.

EXERCISE 7.23.17. Show that for any finite tensor category there exists a quasi-tensor functor $F : \mathcal{C} \to R-\mathrm{bimod}$ for some semisimple R (which may be taken to be commutative).

Hint: One can take R commutative with $\dim_\Bbbk R$ equal to the number of simple objects in $\mathcal{C}$. Mimick the proof of Proposition 6.1.14.

This raises the question what algebraic structure is induced on $\mathsf{End}_\Bbbk F$ when F is only a quasi-tensor (not necessarily a tensor) functor. This structure is called a *weak quasi-Hopf algebra*, and in principle it allows one to speak about any finite tensor category in explicit linear-algebraic terms. However, this structure is so cumbersome that it seems better not to consider it, and instead to use the language of tensor categories, which is the point of view of this book.

7.24. Bibliographical notes

7.1. As far as we know, the definition of a module category first appeared in Bernstein's lectures [**Ber**] and in the work of Crane and Frenkel [**CraF**]. The theory of module categories was further developed in [**Os1**]. The notion of a module category is implicitly present in Boundary Conformal Field Theory, see [**BehPPZ, FuS, PZ**] and in the theory of weak Hopf algebras [**BohNS**]. Many examples of module categories (without using this name) were studied in Operator Algebras Theory, see e.g. [**BocEK2, Oc2**].

7.2. The definition of a module functor between $\mathcal{C}$-module categories is taken from [**Os1**].

7.3. Module categories over multitensor categories were first considered in [**ENO2**] and [**EtO1**]. The notion of indecomposable module category was introduced in [**Os1**].

7.4. Example 7.4.6 is from [**Os1**, Section2]. Indecomposable module categories over $\mathsf{Rep}(G)$ were classified in [**Os1**, Theorem 3.2]. Indecomposable module categories over Vec_G^ω as well as tensor functors between such categories were classified in [**Os2**].

7.5. The notion of an exact module category was introduced in [**EtO1**].

7.6. All statements of this Section are taken from [**EtO1**, Section 3].

7.7. Proposition 7.7.2 is proved in [**EtO1**, Corollary 3.9]. It suggests the following approach to the classification of exact module categories over a multitensor category $\mathcal{C}$. First classify irreducible $\mathbb{Z}_+$-modules over $\mathsf{Gr}(\mathcal{C})$ (this is a combinatorial part), and then try to find all possible categorifications of a given $\mathbb{Z}_+$-module (this is a categorical part). Both these problems are quite nontrivial and interesting. This approach was employed in [**EtKh, KirO**], where indecomposable module categories over the fusion category associated to $\widehat{\mathfrak{sl}}_2$ were classified.

7.8. The notion of an algebra in a multitensor category $\mathcal{C}$ is introduced in [**Os1**, Section 3.1]. The category $\mathsf{Mod}_{\mathcal{C}}(A)$ was studied in the same paper.

7.9. Our discussion of internal Homs follows [**Os1**, Section 3.2] and [**EtO1**, Section 3.2]. Construction similar to that of the canonical algebra $\underline{\mathsf{Hom}}(\mathbf{1}, \mathbf{1})$ from Example 7.9.12 appeared in the work of Lyubashenko [**Ly2**] and in [**BesKLT**]. The canonical algebra is also related to Popa's symmetric enveloping algebra [**Po1**] in subfactor theory. Our exposition follows [**ENO1**], in particular, the multiplication of A (Example 7.9.14) is described in [**ENO1**, Section 7].

7.10. Theorem 7.10.1 is the main result of [**Os1**]. The proof is taken from [**Os1**, Theorem 3.1]. This result is a special case of the Barr-Beck Theorem in category theory ([**Mac2**]). We leave it to the interested reader to deduce Theorem 7.10.1 from the Barr-Beck Theorem. The Fundamental Theorem for Hopf modules is a classical result in Hopf algebra theory, see, e.g., textbooks [**Mon, Sw**].

7.11. Discussion of categories of module functors follows [**EtO1**] and [**Os1**].

7.12. Our exposition follows [**EtO1**, Section 3.3]. Theorem 7.12.11 is the main result of [**Os1**]. In the works of Fuchs, Runkel, and Schweigert [**FuRS2**] and Müger [**Mu2**] the theory of Morita equivalence was developed in terms of Frobenius algebras in a tensor category (without using module categories). An alternative proof of Proposition 7.12.18 is given in [**Mu2**, Proposition 4.6]. It also follows from [**Mu2**] that from an exact $\mathcal{C}$-module category $\mathcal{M}$ one can form a 2-category (see Section 2.12) with two objects $\mathfrak{A}$, $\mathfrak{B}$ such that $\mathsf{End}(\mathfrak{A}) = \mathcal{C}$, $\mathsf{End}(\mathfrak{B}) = (\mathcal{C}_{\mathcal{M}}^*)^{\mathrm{op}}$, $\mathsf{Hom}(\mathfrak{A}, \mathfrak{B}) = \mathcal{M}$, and $\mathsf{Hom}(\mathfrak{B}, \mathfrak{A}) = \mathsf{Fun}_{\mathcal{C}}(\mathcal{M}, \mathcal{C}) = \mathcal{M}^\vee$. In this language Proposition 7.12.28 expresses the associativity of the composition of Hom's.

7.13. The center construction is due to Drinfeld (unpublished) and appears in the work of Majid [**Maj1**] and Joyal and Street [**JoyS2**]. The braiding on the center (which is, of course, is its most important feature) will be considered in Section 8.5. Proposition 7.13.8 was proved in [**Os2**, Proposition 2.5].

7.14. The notion of the quantum double of a Hopf algebra is due to Drinfeld [**Dr3**]. Interpretation of the center of $\mathsf{Rep}(H)$, where H is a finite-dimensional Hopf algebra, as the representation category of $D(H)$ can be found in Kassel's textbook [**Kas**, Section XII.5].

7.15. Yetter-Drinfeld modules appeared in the work of Yetter [**Ye1**] under the name of *crossed bimodules*. Their relation with representations of $D(H)$ is discussed in Montgomery's textbook [**Mon**, Section 10.6].

7.16. Most of the results of this Section are taken from [**EtO1**, Section 3.4]. E.g., Theorem 7.16.1 is [**EtO1**, Theorem 3.34] and Theorem 7.16.6 is [**EtO1**, Theorem 3.42].

7.17. We follow [**EtO1**, Sections 3.5 and 3.6]. The analog of Theorem 7.17.6 for $\mathcal{D} = \mathsf{Rep}(H)$, where H is a finite dimensional Hopf algebra, follows from the work of Nichols and Zoeller [**NicZ**]. For quasi-Hopf algebras this is due to Schauenburg [**Schau4**].

7.18. The notions of Hopf modules and bimodules over Hopf algebras and their generalizations (quasi-Hopf algebras, weak Hopf algebras) were considered in [**LaS, Schau2, HaN, BohNS**].

The Fundamental Theorem for Hopf bimodules was proved by Schauenburg in [**Schau2**] for Hopf algebras and by Hausser and Nill [**HaN**] for quasi-Hopf algebras. The categorical analog of Hopf bimodules was introduced in [**ENO1**]. Proposition 7.18.5 provides a categorical version of the Fundamental Theorem for Hopf bimodules and generalizes the results of [**Schau2, HaN**].

Corollary 7.18.10 generalizes a classical result in Hopf algebra theory by Larson and Radford [**LaR2**].

7.19. In the case of braided tensor categories results of this Section were obtained the paper [**BesKLT**] by Bespalov, Kerler, Lyubashenko, and Turaev. For the representation theory of the Yangian in Example 7.19.5 see the book by Chari and Pressley, [**ChP**, 12.1].

7.20. Frobenius algebras in tensor categories were considered by Müger [**Mu2**] and by Schweigert and Fuchs [**SchwF**]. The proof of Corollary 7.20.4 follows the one of Abrams for classical Frobenius algebras, see [**Ab**].

7.21. The squared norm of simple object of a multifusion category $\mathcal{C}$ and categorical dimension of $\mathcal{C}$ were introduced by Müger [**Mu2**]. The description of categorical dimension in terms of the canonical Frobenius algebra was given in [**ENO1**]. Formula (7.72) can be established using the Larson-Radford trace formula for Hopf algebras [**LaR2**]. Theorem 7.21.12 is proved in [**ENO2**].

7.22. The cohomology of a tensor category with respect to a tensor functor was defined by Davydov [**Da1**] and independently by Yetter in [**Ye3, Ye4**] (motivated by the previous work [**CraY**]). Proposition 7.22.3 was proved in [**Da1**]. As usual, low dimensional Davydov-Yetter cohomology groups have an independent meaning [**Da1, Ye3, Ye4**] (see Exercise 7.22.2). The group $H^1_F(\mathcal{C})$ classifies derivations of F as a tensor functor. The group $H^2_F(\mathcal{C})$ classifies first order deformations of the tensor structure on the functor F. The group $H^3(\mathcal{C})$ classifies first order deformations of the associativity constraint in $\mathcal{C}$, i.e., of the structure of $\mathcal{C}$ itself. As usual, obstructions to these deformations live in the cohomology groups one degree higher.

On the other hand, it is known since the work of Gerstenhaber [**Ger**] that the deformation theory of an associative algebra is described by its Hochschild cohomology. A similarity between these two theories was observed in [**Ye3**]. Our Proposition 7.22.7 makes this similarity formal by means of the canonical Frobenius algebra A.

7.23. Weak Hopf algebras were introduced by Böhm, Nill, and Szlachanyi in [**BohNS**]. See the survey [**NikV1**] for more about them.

7.25. Other results

7.25.1. Classification of module categories. It is interesting to classify indecomposable exact module categories over a given tensor category. We have seen above how this problem is solved for group-theoretical fusion categories, but it has also been solved in many other cases, including nonsemisimple ones. Let us describe some of these results in more detail.

In [**EtO1**], indecomposable exact module categories are classified for the category $\mathsf{Rep}_{\Bbbk}(G)$, where G is a finite group and $\Bbbk$ has positive characteristic, as well as for finite supergroups G, and also for the category $\mathsf{Rep}(H)$, where H is the Taft Hopf algebra. These results were extended to representation categories of some classes of pointed Hopf algebras, as well as basic Hopf and quasi-Hopf algebras by Galindo, Garcia Iglesias, and Mombelli, see [**Mom1, Mom2, GarM, GalM**], based on previous work of Andruskiewitsch and Mombelli [**AndrM**]. In [**Gel2**], indecomposable exact module categories are classified for the category $\mathsf{Rep}_{\Bbbk}(G)$, where G is a *finite group scheme* and $\Bbbk$ has positive characteristic.

There are also some results on classification of indecomposable exact module categories over infinite tensor categories. An interesting special case of this problem is the problem of classifying fiber functors on representation categories of groups. We have seen above how this problem is solved for $\mathcal{C} = \mathsf{Rep}(G)$, where G is a finite group: in this case fiber functors correspond to twists, which are classified in Corollary 7.12.24 and correspond to groups of central type. However, if G is an affine algebraic group of positive dimension, the situation is more complicated. First of all, in this case, there may exist fiber functors $F : \mathcal{C} \to \mathsf{Vec}$ such that $\dim F(V) \neq \dim V$. For instance, when $G = SL_2$, and more generally for the category $O_q(SL_2)$ – comod, such fiber functors are classified in [**Bi**] (see Remark 5.12.10), and are related to quantum groups of the nondegenerate bilinear form (which are the corresponding Hopf algebras $\mathsf{Coend}(F)$, see [**DuL**]).

A generalization of this result to semisimple module categories (i.e., classification of such module categories over $O_q(SL_2)$ – comod) can be found in [**EtO2**] in the generic case, and in [**KirO**] and [**Os5**] in the root of unity case (in particular, for Verlinde categories, see Example 8.18.5). These works describe the variety parametrizing the module categories which categorify a given module M_Γ over the Grothendieck ring $\mathrm{Gr}(\mathcal{C})$, defined by the condition that the matrix of multiplication by the 2-dimensional representation is the incidence matrix A_Γ of a graph Γ. In particular, these works show that if Γ is a tree, then the categorification exists if and only if $-q - q^{-1}$ is an eigenvalue of A_Γ with an eigenvector having no nonzero entries. This exhibits special properties of quantum SL_2 when q are special algebraic numbers (such that $-q - q^{-1}$ are eigenvalues of A_Γ). Also, this shows that the case of Dynkin diagrams of type ADE (i.e., the case when the eigenvalues of A_Γ are in $(-2, 2)$) corresponds to the case when q is a root of unity (in particular, Verlinde categories), which is the result of [**KirO**] and [**Os5**].

For other algebraic groups, much less is known, even in the case of fiber functors. Namely, let's say that a fiber functor F on $\mathsf{Rep}(G)$ is classical if $\dim F(V) = \dim V$ for any $V \in \mathsf{Rep}(G)$; this is automatic for finite G by the Frobenius-Perron theorem, but, as we have just seen, may not hold for algebraic groups. It is not hard to show that classical fiber functors correspond to twists, i.e. elements in $O(G \times G)^*$ satisfying the twist axioms. However, the classification of twists is known in very few cases. E.g., for $G = SL_3$, see [**Oh1, Oh2**], for unipotent groups see [**EtG9**] (which gives a classification of fiber functors in terms of Lie bialgebras), and for nilpotent groups see [**Gel3**].

7.25.2. Subfactors. A *subfactor* is an inclusion $N \subset M$ of factors, i.e., von Neumann algebras with trivial centers. An important numerical characteristic of a subfactor is its *index* [**Jon1**]. This notion is similar to that of an index of a subgroup in a group (indeed, the latter can be recovered from groups acting on factors). However, there is an important difference. A famous result of Jones [**Jon1**] states that the index takes values in the set $\{4cos^2\frac{\pi}{n} \mid n \geq 3\} \cup [4, +\infty)$ and, conversely, that every such value is an index of some subfactor. The study of subfactors led Jones to his celebrated discovery of link invariants [**Jon2**].

There are many invariants (algebraic, combinatorial, analytic) describing a subfactor: Ocneanu's paragroups [**Oc1, Oc2**], Popa's λ-lattices [**Po2**], Jones's planar algebras [**Jon3**]. Finite depth subfactors can be described using weak Hopf algebras [**NikV2, NikV3**], cf. Definition 7.23.1. The subfactor literature is quite extensive, one can find many references in the recent survey by Jones, Morrison, and Snyder [**JMS**].

The relation between subfactors and tensor categories is as follows. Given a finite index subfactor $N \subset M$ there is a semisimple tensor category $\mathcal{C}$ of N-bimodules generated by the N-bimodule M (so the simple objects of $\mathcal{C}$ are simple N-bimodules contained in $M^{\otimes n}$, $n \geq 1$). The subfactor is said to have *finite depth* if $\mathcal{C}$ is a fusion category. Let $\mathcal{M}$ be the $\mathcal{C}$-module category of N-M bimodules generated by $X := M$. Thus, a finite depth subfactor determines a triple $(\mathcal{C}, \mathcal{M}, X)$ consisting of a fusion category, a module category over it, and a simple object of a module category.

Conversely, starting with a triple $(\mathcal{C}, \mathcal{M}, X)$ as above one can construct either a λ-lattice [**X1**] or a planar algebra [**Gh**] and, hence, a subfactor of a hyperfinite II_1-factor. In particular, one can construct a subfactor from any simple object in a fusion category [**Wen2**]. Note that the index of the resulting subfactor is equal to $\mathsf{FPdim}(\underline{\mathsf{Hom}}(X, X))$.

Among new fusion categories and their module categories constructed directly from subfactors are "exotic" examples of Asaeda and Haagerup [**AsH**], Bigelow, Peters, Morrison, and Snyder [**BiMPS**], and Izumi [**I1**]. These categories do not appear from quantum groups and affine Lie algebras in any known way. Furthermore, the classification of subfactors of index ≤ 5 is now complete [**JMS**].

In [**Oc3**] Ocneanu (partially) classified, in our terminology, module categories over the fusion category of representations of quantum $SU(N)$ at roots of unity.

In [**MorS**] examples of fusion categories that are not defined over cyclotomic integers were constructed.

Overall, interaction between subfactor theory and the theory of tensor categories proved to be very beneficial for both areas.

CHAPTER 8

Braided categories

8.1. Definition of a braided category

The notion of a braided monoidal category is a categorification of that of a commutative monoid. Similarly to the pentagon axiom for the associativity constraint in Definition 2.1.1, the commutativity constraint is required to have certain coherence properties.

DEFINITION 8.1.1. A *braiding* (or a *commutativity constraint*) on a monoidal category $\mathcal{C}$ is a natural isomorphism $c_{X,Y} : X \otimes Y \xrightarrow{\sim} Y \otimes X$ such that the hexagonal diagrams

(8.1)

$$\begin{array}{ccccc} & X \otimes (Y \otimes Z) & \xrightarrow{c_{X,Y\otimes Z}} & (Y \otimes Z) \otimes X & \\ {\scriptstyle a_{X,Y,Z}} \nearrow & & & & \searrow {\scriptstyle a_{Y,Z,X}} \\ (X \otimes Y) \otimes Z & & & & Y \otimes (Z \otimes X) \\ {\scriptstyle c_{X,Y}\otimes \mathrm{id}_Z} \searrow & & & & \nearrow {\scriptstyle \mathrm{id}_Y \otimes c_{X,Z}} \\ & (Y \otimes X) \otimes Z & \xrightarrow[a_{Y,X,Z}]{} & Y \otimes (X \otimes Z) & \end{array}$$

and

(8.2)

$$\begin{array}{ccccc} & (X \otimes Y) \otimes Z & \xrightarrow{c_{X\otimes Y,Z}} & Z \otimes (X \otimes Y) & \\ {\scriptstyle a^{-1}_{X,Y,Z}} \nearrow & & & & \searrow {\scriptstyle a^{-1}_{Z,X,Y}} \\ X \otimes (Y \otimes Z) & & & & (Z \otimes X) \otimes Y \\ {\scriptstyle \mathrm{id}_X \otimes c_{Y,Z}} \searrow & & & & \nearrow {\scriptstyle c_{X,Z}\otimes \mathrm{id}_Y} \\ & X \otimes (Z \otimes Y) & \xrightarrow[a^{-1}_{X,Z,Y}]{} & (X \otimes Z) \otimes Y & \end{array}$$

commute for all objects X, Y, Z in $\mathcal{C}$.

DEFINITION 8.1.2. A *braided* monoidal category is a pair consisting of a monoidal category and a braiding.

REMARK 8.1.3. Note that the same monoidal category can have several different structures of a braided category, see e.g., Section 8.4.

DEFINITION 8.1.4. Let $\mathcal{C}$ be a monoidal category equipped with a braiding $c_{X,Y} : X \otimes Y \xrightarrow{\sim} Y \otimes X$. We define the *reverse* braiding on $\mathcal{C}$ by

$$c'_{X,Y} := c^{-1}_{Y,X}. \tag{8.3}$$

We will call the corresponding braided category the *reverse* category and denote it $\mathcal{C}^{\text{rev}}$.

EXERCISE 8.1.5. Check the braiding axioms for c'.

EXERCISE 8.1.6. Show that for any object X in a braided monoidal category $\mathcal{C}$ one has

$$l_X \circ c_{X,\mathbf{1}} = r_X, \quad r_X \circ c_{\mathbf{1},X} = l_X, \quad \text{and} \quad c_{\mathbf{1},X} = c_{X,\mathbf{1}}^{-1}. \tag{8.4}$$

DEFINITION 8.1.7. Let $\mathcal{C}^1$ and $\mathcal{C}^2$ be braided monoidal categories whose braidings are denoted c^1 and c^2, respectively. A monoidal functor (F, J) from $\mathcal{C}^1$ to $\mathcal{C}^2$ is called *braided* if the following diagram commutes:

$$\begin{array}{ccc} F(X)\otimes F(Y) & \xrightarrow{c^2_{F(X),F(Y)}} & F(Y)\otimes F(X) \\ \Big\downarrow{\scriptstyle J_{X,Y}} & & \Big\downarrow{\scriptstyle J_{Y,X}} \\ F(X\otimes Y) & \xrightarrow{F(c^1_{X,Y})} & F(Y\otimes X) \end{array} \tag{8.5}$$

for all objects X, Y in $\mathcal{C}^1$.

A *braided monoidal equivalence* of braided monoidal categories is a braided monoidal functor which is also an equivalence of categories.

REMARK 8.1.8. Note that a monoidal functor is a functor with an additional *structure*, while for a monoidal functor to be braided is a *property*.

EXERCISE 8.1.9. Let $\mathcal{C}$ be a braided monoidal category with braiding c. Let $\mathcal{C}^{\text{op}}$ be the opposite monoidal category of $\mathcal{C}$, see Definition 2.1.5. Show that c defines a braiding on $\mathcal{C}^{\text{op}}$ and that the identity functor has a natural structure of a braided monoidal equivalence $\mathcal{C} \xrightarrow{\sim} \mathcal{C}^{\text{op}}$.

Recall that by Theorem 2.8.5 any monoidal category is equivalent to a strict category.

PROPOSITION 8.1.10. *Let $\mathcal{C}$ be a strict monoidal category with braiding c. For all $X, Y, Z \in \mathcal{C}$ the braiding satisfies the following* Yang-Baxter equation*:*

(8.6)

$$(c_{Y,Z}\otimes \mathsf{id}_X)\circ(\mathsf{id}_Y \otimes c_{X,Z})\circ(c_{X,Y}\otimes \mathsf{id}_Z) = (\mathsf{id}_Z \otimes c_{X,Y})\circ(c_{X,Z}\otimes \mathsf{id}_Y)\circ(\mathsf{id}_X \otimes c_{Y,Z}).$$

PROOF. Since $\mathcal{C}$ is strict, the hexagonal diagrams (8.1) and (8.2) become triangles. Consider the diagram

(8.7)

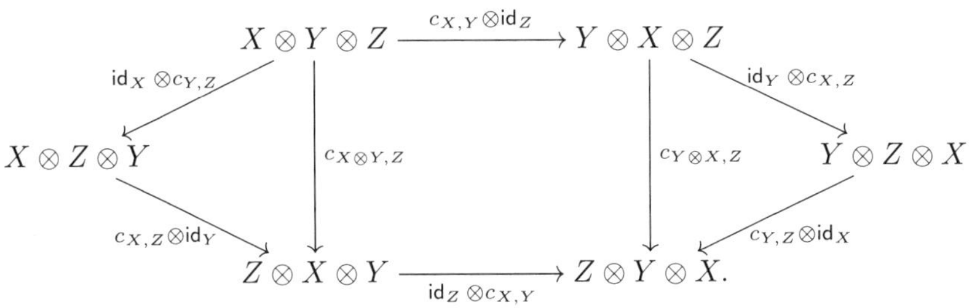

The two triangles are (8.1) and (8.2). The square in the middle commutes by naturality of c. Hence, the perimeter of (8.7) commutes. The two compositions along the perimeter are precisely the two sides of (8.6). $\square$

EXERCISE 8.1.11. State and prove an analog of Proposition 8.1.10 for non-strict categories.

DEFINITION 8.1.12. A braided monoidal category $\mathcal{C}$ is called *symmetric* if

$$c_{Y,X} \circ c_{X,Y} = \mathsf{id}_{X \otimes Y}$$

for all objects $X, Y \in \mathcal{C}$.

We will discuss symmetric tensor categories in more detail in Section 9.9.

8.2. First examples of braided categories and functors

Many examples of monoidal categories from Section 2.3 admit a natural braiding.

EXAMPLE 8.2.1. The categories **Set**, **Vec**, $\mathsf{Rep}(G)$ are braided with the braiding being the transposition of factors. For an abelian group G the category Vec_G is braided. Similarly, for a commutative ring R the category $\mathbf{Mod}_R$ is braided.

Note that all braidings in Example 8.2.1 are *symmetric*. Here is another important example of a symmetric category.

EXAMPLE 8.2.2. Let $\Bbbk$ be a field of characteristic $\neq 2$. Let $\mathcal{C}$ be the category $\mathsf{Vec}_{\mathbb{Z}/2\mathbb{Z}}$ of $\mathbb{Z}/2\mathbb{Z}$-graded vector spaces. Define a braiding on $\mathcal{C}$ by the formula $c_{X,Y}(x \otimes y) = (-1)^{\deg(x)\deg(y)} y \otimes x$ for homogeneous vectors x, y. It is easy to check that this endows $\mathcal{C}$ with the structure of a symmetric fusion category. This category is called *the category of super-vector spaces* and denoted sVec. This category is ubiquitous in many fields of mathematics, especially in homological algebra and its applications.

EXAMPLE 8.2.3. Let G be an abelian group, let $\mu : G \times G \to \Bbbk^\times$ be a 2-cocycle, and let $F_{\mathsf{id},\mu} : \mathsf{Vec}_G \to \mathsf{Vec}_G$ be the functor constructed in Example 2.6. The functor $F_{\mathsf{id},\mu}$ is braided if and only if $\mu(g, h) = \mu(h, g)$ for all $g, h \in G$ (note that in this case μ is cohomologically trivial). This example is generalized in Section 8.4 below.

The next example explains the name "braiding".

EXAMPLE 8.2.4. (**The category of braids.**) Recall that in Example 2.3.14 we introduced the notion of a tangle. *Braids* form a special class of tangles. Namely, a *braid* on n strands is a tangle obtained from n disjoint closed unit intervals (i.e., no circles allowed) such for each $t \in [0, 1]$ exactly one point of each interval belongs to $\mathbb{R}^2 \times \{t\}$. Clearly, braids form a monoidal subcategory $\mathcal{B}$ of the category $\mathcal{T}$ of tangles. There is a braiding of $\mathcal{T}$ given by (isotopy classes of) braids $c_{m,n} : m + n \to n + m$ that can be visualized as m strands "passing over" n strands.

Isotopy classes of braids on n strands form a group B_n, called the *braid group*, under the operation of composition. It is well-known (see e.g., [**KassT**]) that for $n \geq 2$ this group B_n is isomorphic to the abstract group with generators $\sigma_1, \dots, \sigma_{n-1}$ and relations

(i) $\sigma_i \sigma_j = \sigma_j \sigma_i \quad$ when $\quad |i - j| > 1$,
(ii) $\sigma_i \sigma_{i+1} \sigma_i = \sigma_{i+1} \sigma_i \sigma_{i+1}$.

Namely, one takes $\sigma_i = \mathsf{id}^{\otimes(i-1)} \otimes c_{1,1} \otimes \mathsf{id}^{\otimes(n-i-1)}$. For $n \geq 3$ this group is non-abelian.

REMARK 8.2.5. It follows from Proposition 8.1.10 that for any object V in a strict braided monoidal category $\mathcal{C}$ there is a group homomorphism

$$B_n \to \mathsf{Aut}_{\mathcal{C}}(V^{\otimes n}) : \sigma_i \mapsto \mathsf{id}_{V^{\otimes(i-1)}} \otimes c_{V,V} \otimes \mathsf{id}_{V^{\otimes(n-i-1)}}, \quad i = 1, \ldots, n-1.$$

If $\mathcal{C}$ is $\Bbbk$-linear then this is a representation of B_n over $\Bbbk$. Similarly, the pure braid group PB_n, which is the kernel of the natural homomorphism $B_n \to S_n$, acts naturally on $V_1 \otimes ... \otimes V_n$ for any objects $V_1, ..., V_n \in \mathcal{C}$.

REMARK 8.2.6. The braid group B_n can be interpreted as the fundamental group of the configuration space X_n of n unordered distinct points in $\mathbb{R}^2$. Here X_n is the quotient of the space $\{(z_1, z_2, \ldots, z_n) \mid z_i \neq z_j \text{ for } i \neq j\}$ by the natural action of S_n (with the quotient topology).

EXERCISE 8.2.7. Let $\mathcal{C}$ be a braided tensor category (not necessarily strict), and let $X_1, \ldots, X_n \in \mathcal{C}$. Let P_1, P_2 be any parenthesized products of $X_1, ..., X_n$ (in any orders) with arbitrary insertions of unit objects $\mathbf{1}$. Let $f = f_C : P_1 \to P_2$ be an isomorphism, obtained as a composition C of associativity, braiding, and unit isomorphisms and their inverses possibly tensored with identity morphisms. Explain how C defines a braid b_C. Show that if $b_C = b_{C'}$ in B_n then $f_C = f_{C'}$. This statement is called *Mac Lane's braided coherence theorem.*

8.3. Quasitriangular Hopf algebras

Let H be a Hopf algebra, and $\mathcal{C} = \mathbf{Rep}H$ be the monoidal category of H-modules. Assume that $\mathcal{C}$ is a braided category, with braiding $c = (c_{X,Y})$. Let $c^{\vee}_{H,H} = \sigma \circ c_{H,H} : H \otimes H \to H \otimes H$, where σ is the permutation of components. By functoriality of c, $c^{\vee}_{H,H}$ commutes with right multiplication by elements of $H \otimes H$, so it is the operator of left multiplication by a uniquely determined invertible element $R \in H \otimes H$. The axioms of a braided structure imply that

$$(8.8)\quad (\Delta \otimes \mathsf{id})(R) = R^{13}R^{23}, \ (\mathsf{id} \otimes \Delta)(R) = R^{13}R^{12}, \Delta^{\mathrm{op}}(h) = R\Delta(h)R^{-1}, \ h \in H,$$

where $\Delta^{\mathrm{op}} := \sigma \circ \Delta$ is the opposite coproduct.

This motivates the following definition.

DEFINITION 8.3.1. (*Drinfeld*) A *quasitriangular* Hopf algebra is a pair (H, R), where H is a Hopf algebra over $\Bbbk$, and $R \in H \otimes H$ (the *R-matrix*) is an invertible element satisfying relations (8.8). The element R is called the *universal R-matrix* of H.

REMARK 8.3.2. It immediately follows from the axioms that the R-matrix R of a quasitriangular Hopf algebra (H, R) satisfies the *quantum Yang-Baxter equation*

$$R^{12}R^{13}R^{23} = R^{23}R^{13}R^{12}.$$

Also, if (H, R) is a quasitriangular Hopf algebra then so is $(H, (R^{21})^{-1})$, and the braided category $\mathbf{Rep}(H, (R^{21})^{-1})$ is obtained from $\mathbf{Rep}(H, R)$ by inverting the braiding.

If the braiding c is symmetric, i.e., $c_{Y,X}c_{X,Y} = \mathsf{id}_{X,Y}$, then R satisfies the equation $R^{-1} = R^{21}$, called *the unitarity condition.* This motivates the following definition.

DEFINITION 8.3.3. If (H, R) is a quasitriangular Hopf algebra, and $R^{-1} = R^{21}$ then the R-matrix R is called *unitary* and (H, R) is called a *triangular* Hopf algebra.

EXAMPLE 8.3.4. If H is cocommutative then $R = 1 \otimes 1$ is a triangular structure on H. Thus, quasitriangular Hopf algebras generalize cocommutative Hopf algebras.

Conversely, it is easy to see that if (H, R) is a quasitriangular Hopf algebra then the category $\mathcal{C}$ of H-modules is a braided category, which is symmetric if (H, R) is triangular. Thus, we have a bijection between braidings on $\mathbf{Rep}(H)$ and quasitriangular structures on H, which restricts to a bijection between symmetric braidings and triangular structures.

Note that a given Hopf algebra may have no quasitriangular structures or can have several different quasitriangular structures.

EXAMPLE 8.3.5. Let $H := \mathsf{Fun}(G)$ be the (commutative) Hopf algebra of functions on a noncommutative finite group G with values in $\Bbbk$. Then H has no quasitriangular structures, since the category $\mathsf{Rep}(H)$ does not admit a braiding (as $X \otimes Y$ is in general not isomorphic to $Y \otimes X$ in this category).

EXAMPLE 8.3.6. Assume that $\Bbbk$ has characteristic $\neq 2$. Let g be the generator of $\mathbb{Z}/2\mathbb{Z}$. The group algebra $\Bbbk\mathbb{Z}/2\mathbb{Z}$ has two triangular R-matrices: $1 \otimes 1$ and $\frac{1}{2}(1 \otimes 1 + 1 \otimes g + g \otimes 1 - g \otimes g)$. The symmetric categories attached to these R-matrices are actually not equivalent: the first one is the category of representations of $\mathbb{Z}/2\mathbb{Z}$ (or, equivalently, the category $\mathsf{Vec}_{\mathbb{Z}/2\mathbb{Z}}$), and the second one is the category of super-vector spaces sVec.

EXAMPLE 8.3.7. The Sweedler Hopf algebra (see Example 5.5.6) has a family of triangular structures R_α parametrized by elements of $\Bbbk$:

$$R_\alpha = \frac{1}{2}(1 \otimes 1 + 1 \otimes g + g \otimes 1 - g \otimes g) + \frac{\alpha}{2}(x \otimes x + x \otimes gx - gx \otimes x + gx \otimes gx).$$

One of the main examples of a quasitriangular Hopf algebra is the Drinfeld double $D(H)$ of a finite dimensional Hopf algebra H, defined in Section 7.14. Namely, we have the following proposition, which follows from the material in Section 7.14.

PROPOSITION 8.3.8. *The quantum double $D(H) = H \otimes H^{*\mathrm{cop}}$ of a finite dimensional Hopf algebra H is a quasitriangular Hopf algebra, with the universal R-matrix $R = \sum h_i \otimes h_i^*$, where $\{h_i\}$ is a basis of H, and $\{h_i^*\}$ is the dual basis of $H^{*\mathrm{cop}}$. Moreover, the multiplication on $D(H)$ is the unique extension of the multiplication in H and $H^{*\mathrm{cop}}$ which makes R a quasitriangular structure (for the coproduct induced by the one from H and $H^{*\mathrm{cop}}$).*

It is easy to show that $D := D(H)$ is a *factorizable* quasitriangular Hopf algebra, i.e., the associated *Drinfeld-Reshetikhin map* ([**Dr5, Res**])

$$D^* \to D, \ f \mapsto (\mathsf{id} \otimes f)(R^{21}R)$$

is an isomorphism of vector spaces (see also Exercise 8.6.4 below). In particular, $(D(H), R)$ is not triangular when $\dim(H) > 1$.

EXAMPLE 8.3.9. Let G be a finite group. Then the underlying algebra of the Drinfeld double $D(G) := D(\Bbbk G)$ of $\Bbbk G$ is the semidirect product $\mathsf{Fun}(G, \Bbbk) \rtimes \Bbbk G$, where G acts on $\mathsf{Fun}(G, \Bbbk)$ by conjugation, and the universal R-matrix is $R = \sum_{g \in G} g \otimes \delta_g$, where δ_g is the delta-function at g.

EXAMPLE 8.3.10. From Example 7.14.11, we see that the small quantum group $u_q(\mathfrak{sl}_2)$ for a root of unity q of odd order ℓ is a quasitriangular Hopf algebra (as it is a tensor factor, hence a quotient of the double of the Taft algebra).

EXERCISE 8.3.11. Show that the R-matrix of $u_q(\mathfrak{sl}_2)$ coming from Example 7.14.11 is given by the following explicit formula:

$$R = R_0 \sum_{n=0}^{\ell-1} q^{n(n-1)/2} \frac{(q-q^{-1})^n E^n \otimes F^n}{[n]_q!},$$

with

$$R_0 = \sum_{i,j \in \mathbb{Z}/\ell\mathbb{Z}} q^{ij/2} \mathbf{1}_i \otimes \mathbf{1}_j,$$

where $\mathbf{1}_j$ are the idempotents of $\Bbbk[K]/(K^\ell - 1) = \Bbbk\mathbb{Z}/\ell\mathbb{Z}$ defined by $K^s \mathbf{1}_j = q^s \mathbf{1}_j$. (Here $q^{1/2}$ stands for the square root of q of odd order).

REMARK 8.3.12. Similarly, the quantum group $u_q(\mathfrak{g})$ is quasitriangular for any simple Lie algebra $\mathfrak{g}$.

EXERCISE 8.3.13. Assume that q is not a root of unity. Show that the category of finite dimensional type I representations of $U_q(\mathfrak{sl}_2)$ (see Section 5.8) is braided, with braiding defined by the R-matrix

$$R = R_0 \sum_{n=0}^{\infty} q^{n(n-1)/2} \frac{(q-q^{-1})^n E^n \otimes F^n}{[n]_q!},$$

with

$$R_0 = \sum_{i,j \in \mathbb{Z}} q^{ij/2} \mathbf{1}_i \otimes \mathbf{1}_j,$$

where $\mathbf{1}_j$ is the projection operator to weight j (and we fix a square root of q). Namely, if V, W are such representations, then this sum terminates when applied to $V \otimes W$, and gives rise to a well defined linear operator $V \otimes W \to V \otimes W$.

Note that this is not an honest quasitriangular structure on $U_q(\mathfrak{sl}_2)$, since the sum is infinite; in particular, it does not define a braiding on the category of all (not necessarily finite dimensional) representations of $U_q(\mathfrak{sl}_2)$.

Even though we have a bijection between braidings on the category of H-modules and quasitriangular structures on H, two non-isomorphic quasitriangular Hopf algebras may have equivalent braided categories of modules. Namely, this happens when the two Hopf algebras are related by a twist (see (5.30)).

PROPOSITION 8.3.14. *If (H, R) is a (quasi)triangular Hopf algebra and J is a twist for H, then $(H^J, R^J := (J^{21})^{-1} R J)$ is a quasi(triangular) Hopf algebra, and the categories of modules over (H, R) and (H^J, R^J) are naturally equivalent as a braided categories.*

EXERCISE 8.3.15. Prove Proposition 8.3.14.

EXAMPLE 8.3.16. If (H, R) is quasitriangular then R is a twist for $(H^{\mathrm{cop}}, R^{21})$ and $((H^{\mathrm{cop}})^R, (R^{21})^R) = (H, R)$.

EXAMPLE 8.3.17. Let G be a finite group and let J be a twist for $\Bbbk G$. Then $(\Bbbk G^J, (1 \otimes 1)^J = (J^{21})^{-1} J)$ is a triangular Hopf algebra.

REMARK 8.3.18. One can also define the notion of a (quasi)triangular quasi-Hopf algebra, which is the structure on the endomorphism algebra of a quasi-fiber functor on a braided monoidal category. This very important notion was defined and studied by Drinfeld in [**Dr4, Dr6**], and discussed in more detail in textbooks, such as [**Kas, EtS**]. We will not discuss this notion here.

Coquasitriangular Hopf algebras are duals to quasitriangular Hopf algebras and thus generalize commutative Hopf algebras.

Suppose that (A, R) is a finite dimensional quasitriangular Hopf algebra, and $H = A^*$. Then $R \in A \otimes A$ induces a bilinear form $H \otimes H \to \Bbbk$ (which we will also denote by R), and the properties of $R \in A \otimes A$ can be rewritten in terms of this form. This motivates the following definition.

DEFINITION 8.3.19. A *coquasitriangular* Hopf algebra is a pair (H, R), where H is a Hopf algebra over $\Bbbk$ and $R : H \otimes H \to \Bbbk$ (the *R-form*) is a convolution-invertible bilinear form on H satisfying the following axioms:

$$R(h, lg) = \sum R(h_1, g)R(h_2, l), \ \ R(gh, l) = \sum R(g, l_1)R(h, l_2)$$

and

$$\sum R(h_1, g_1)h_2g_2 = \sum g_1h_1R(h_2, g_2) \ \ (h, g, l \in H).$$

If $\sum R(h_1, g_1)R(g_2, h_2) = \varepsilon(g)\varepsilon(h)$ then (H, R) is called *cotriangular*.

EXAMPLE 8.3.20. Let (H, R) be a finite dimensional quasitriangular Hopf algebra. Viewing R as a linear map $R : H^* \otimes H^* \to \Bbbk$, it is straightforward to verify that (H^*, R) is coquasitriangular, and is cotriangular if (H, R) is triangular.

Note that if (H, R) is coquasitriangular, then the category $\mathcal{C}$ of H-comodules is a braided category, with braiding $c_{X,Y} = R^{13} \circ (\rho_X \otimes \rho_Y)$, where $\rho_X : X \to H \otimes X$, $\rho_Y : Y \to H \otimes Y$ are the coactions of H on X and Y. Conversely, it is easy to show that if $\mathcal{C}$ is braided, then H acquires a natural coquasitriangular structure. So, we have a bijection between braided structures on $\mathcal{C}$ and coquasitriangular structures on H up to twisting, which restricts to a bijection between symmetric structures on $\mathcal{C}$ and triangular structures on H up to twisting.

EXAMPLE 8.3.21. Any commutative Hopf algebra is cotriangular with the R-form $R = \varepsilon \otimes \varepsilon$.

EXAMPLE 8.3.22. (*Abelian groups*) Let A be an abelian group equipped with a bilinear form (i.e., a bicharacter) $R : A \times A \to \Bbbk^\times$. Let us extend R to a bilinear form on the group algebra $\Bbbk A$ by linearity. Then $(\Bbbk A, R)$ is a coquasitriangular Hopf algebra. If R is symmetric, $(\Bbbk A, R)$ is cotriangular.

Note that when A is a finite group, we may view R as an element of the algebra $\Bbbk A^\vee \otimes \Bbbk A^\vee$, where $A^\vee$ is the character group of A. We thus have that $(\Bbbk A^\vee, R)$ is a quasitriangular Hopf algebra.

EXERCISE 8.3.23. Let $G = SL_2$, and $O_q(G)$ be the corresponding quantum function algebra (See Section 5.8). Show that the R-matrix of 8.3.13 defines a coquasitriangular structure on $O_q(G)$, which gives rise to the braided structure on its category of comodules given in Exercise 8.3.13.

REMARK 8.3.24. One can define the notion of a *Hopf 2-cocycle* dual to the notion of a twist (see Definition 5.14.1) in an obvious way. If (H, R) is a co(quasi)triangular Hopf algebra and J is a Hopf 2-cocycle for H, then $(H^J, R^J := (J^{21})^{-1} * R * J)$ is a co(quasi)triangular Hopf algebra. We leave it to the reader to work out the details of this.

EXERCISE 8.3.25. Let $\mathcal{C}$ be a strict monoidal category. One says that $\mathcal{C}$ is a *coboundary* category if it is equipped with a functorial in $X, Y \in \mathcal{C}$ collection of morphisms $b_{X,Y} : X \otimes Y \to Y \otimes X$ such that

$$b_{X,Y} \circ b_{Y,X} = \mathsf{id}_{X \otimes Y},$$

and

$$(b_{Y,Z} \otimes \mathsf{id}_X) \circ b_{X,Y\otimes Z} = (\mathsf{id}_Z \otimes b_{X,Y}) \circ b_{X\otimes Y,Z}.$$

The collection $(b_{X,Y})$ is called the coboundary structure of $\mathcal{C}$.

(i) Extend this definition to not necessarily strict categories (so that it is invariant under equivalence of monoidal categories). In other words, write down the axioms of a coboundary category including the associativity morphisms.

(ii) Show that a coboundary structure on $\mathcal{C}$ is the same thing as a tensor structure c on the identity functor $\mathsf{id} : \mathcal{C} \to \mathcal{C}^{\mathrm{op}}$ such that $c^2 = \mathsf{id}$.

(iii) A Hopf algebra H is called *coboundary* if its category of representations is equipped with a coboundary structure. Show that H is coboundary if and only if it is equipped with a twist $R \in H \otimes H$ such that $R^{21} R = 1 \otimes 1$.

(iv) Show that any symmetric monoidal category is coboundary in a natural way (with $b_{X,Y} = c_{X,Y}$).

(v) (The Drinfeld construction). Let $\mathcal{C}$ be a braided category linear over $\Bbbk[[\hbar]]$ (with char$\Bbbk = 0$), and assume that the braiding satisfies $c_{Y,X} \circ c_{X,Y} = \mathsf{id}_{X\otimes Y} + O(\hbar)$ (such a category is called quasisymmetric). Show that the map

$$b_{X,Y} = c_{X,Y} \circ (c_{Y,X} \circ c_{X,Y})^{-1/2},$$

where by $(1+t)^{-1/2}$ we mean the Taylor series of this function at $t = 0$, is a coboundary structure on $\mathcal{C}$.

(vi) Show that the category of type I representations of $U_q(\mathfrak{sl}_2)$, where q is not a root of unity, admits a coboundary structure (imitate the construction of (v)).

REMARK 8.3.26. Similarly, for any simple Lie algebra $\mathfrak{g}$ the category of representations of $U_q(\mathfrak{g})$ of type I is a coboundary category. Moreover, when $q \to 0$, this category degenerates into the category of crystals, which is a semisimple monoidal category with the same Grothendieck ring. This category is not rigid or braided (these structures have a singularity at $q = 0$), but it remains coboundary in the limit. See [**HenK**] for more details.

EXERCISE 8.3.27. Define the cactus group Cact_n to be the group generated by elements S_{km}, $1 \le k < m \le n$, with defining relations

$$S_{km}^2 = 1, \; S_{km}S_{pq} = S_{pq}S_{km} \text{ if } [k,m] \cap [p,q] = \emptyset,$$

$$S_{km}S_{pq} = S_{k+m-q,k+m-p}S_{km} \text{ if } [k,m] \supset [p,q].$$

(i) Show that there exists a surjective homomorphism $\mathrm{Cact}_n \to S_n$ that sends S_{km} to permutations σ_{km} which reverse the interval $[k,m]$ and do not move other elements of $[1,n]$ (the kernel PCact_n of this homomorphism is called the pure cactus group). Describe the groups Cact_n and PCact_n for $n \le 3$. What about $n = 4$? (*Hint:* Show that PCact_4 is the fundamental group of the closed non-orientable surface of Euler characteristic -3).

(ii) Show that if X is any object of a coboundary category $\mathcal{C}$, then $X^{\otimes n}$ carries a natural action of Cact_n, sending S_{km} to the order reversal morphism acting on the k-th through m-th copies of X (there is a unique way to define such a morphism using the coboundary structure). For instance,

$$S_{k,k+1} \mapsto \mathsf{id}_X^{\otimes k-1} \otimes b_{X,X} \otimes \mathsf{id}_X^{\otimes n-k-1}.$$

Deduce that for any $X_1, ..., X_n \in \mathcal{C}$, the tensor product $X_1 \otimes ... \otimes X_n$ carries an action of the pure cactus group PCact_n.

(iii) Let $\widehat{PB}_n$ denote the prounipotent completion of the pure braid group PB_n over $\mathbb{Q}$. (see Example 5.4.4). Construct a homomorphism $\phi_n : \mathrm{PCact}_n \to \widehat{PB}_n$ such that for any quasisymmetric braided category $\mathcal{C}$ as in Exercise 8.3.25(v), the pure cactus group action on $X_1 \otimes ... \otimes X_n$ corresponding to the coboundary structure on $\mathcal{C}$ is the pullback of the pure braid group action corresponding to the braided structure on $\mathcal{C}$ via ϕ_n.

REMARK 8.3.28. The group PCact_n is the fundamental group of the real locus $\overline{M}_{0,n}(\mathbb{R})$ of the Deligne-Mumford compactification of the moduli space of genus zero curves with $n+1$ labeled marked points (points of this space are tree-like configurations of circles with marked points on them, which is the motivation for the term "Cactus group"). This space is a compact connected manifold, non-orientable for $n \geq 4$, which is known to be $K(\pi, 1)$ (a nontrivial geometric result). The group Cact_n is the orbifold fundamental group of $\overline{M}_{0,n}(\mathbb{R})/S_n$ (the moduli space of curves with n unlabeled points and one labelled point at infinity). See [**HenK**],[**EHKR**] and references therein for more details.

8.4. Pre-metric groups and pointed braided fusion categories

Let $\Bbbk$ be an algebraically closed field of characteristic 0. Let G be an abelian group. By a *quadratic form* on G (with values in $\Bbbk^\times$) we will mean a map $q : G \to \Bbbk^\times$ such that $q(g) = q(g^{-1})$ and the symmetric function

$$b(g, h) := \frac{q(gh)}{q(g)q(h)} \tag{8.9}$$

is a bicharacter, i.e., $b(g_1g_2, h) = b(g_1, h)b(g_2, h)$ for all $g, g_1, g_2, h \in G$. We will say that q is *non-degenerate* if the associated bicharacter b is non-degenerate.

The simplest way to construct a quadratic form is to start with a bicharacter $B : G \times G \to \Bbbk^\times$ and set

$$q(g) := B(g, g),\ g \in G. \tag{8.10}$$

When $|G|$ is odd, every quadratic form on G can be represented like this. However, this is not the case in general. Counterexample: $G = \mathbb{Z}/2\mathbb{Z}$ with $q(n) = i^{n^2}$, where $\Bbbk = \mathbb{C}$ and $i = \sqrt{-1}$.

DEFINITION 8.4.1. A *pre-metric group* is a pair (G, q) where G is a finite abelian group and $q : G \to \Bbbk^\times$ is a quadratic form. A *metric group* is a pre-metric group such that q is non-degenerate.

An *orthogonal homomorphism* between pre-metric groups (G, q) and (G', q') is a homomorphism $f : G \to G'$ such that $q' \circ f = q$. Pre-metric groups and orthogonal homomorphisms form a category.

The relation between braided fusion categories and pre-metric groups is as follows.

Let $\mathcal{C}$ be a pointed braided fusion category. Then isomorphism classes of simple objects of $\mathcal{C}$ form a finite abelian group G (i.e., $\mathcal{C} \cong \mathrm{Vec}_G^\omega$ as a tensor category). For $g \in G$ let $q(g) \in \Bbbk^\times$ denote the braiding $c_{X,X} \in \mathrm{Aut}_\mathcal{C}(X \otimes X) = \Bbbk^\times$, where X is a simple object of $\mathcal{C}$ whose isomorphism class equals g.

LEMMA 8.4.2. *The function $q : G \to \Bbbk^\times$ is a quadratic form.*

PROOF. By Theorem 2.8.5 we can assume that $\mathcal{C}$ is strict. In this case

$$q(gh) = q(g)q(h)b(g, h), \quad g, h \in G,$$

where $b(g, h) = c_{Y,X} \circ c_{X,Y} \in \mathrm{Aut}_\mathcal{C}(X \otimes Y) \in \Bbbk^\times$, where the isomorphism classes of X and Y are, respectively, g and h. It follows easily from the hexagon axiom (8.1) that $b : G \times G \to \Bbbk^\times$ is a bicharacter. □

EXERCISE 8.4.3. Let $\mathcal{C}$ be a skeletal pointed braided fusion category with the group of simple objects G. Then the associativity isomorphism in $\mathcal{C}$ is determined by a function $\omega : G \times G \times G \to \Bbbk^\times$, and the braiding is determined by a function $c : G \times G \to \Bbbk^\times$. Prove that the axioms of a braided fusion category are equivalent to the following equations:

(8.11)
$$\begin{aligned}
&\omega(g_1g_2, g_3, g_4)\omega(g_1, g_2, g_3g_4) = \omega(g_1, g_2, g_3)\omega(g_1, g_2g_3, g_4)\omega(g_2, g_3, g_4),\\
&\omega(g_2, g_3, g_1)c(g_1, g_2g_3)\omega(g_1, g_2, g_3) = c(g_1, g_3)\omega(g_2, g_1, g_3)c(g_1, g_2),\\
&\omega(g_3, g_1, g_2)^{-1}c(g_1g_2, g_3)\omega(g_1, g_2, g_3)^{-1} = c(g_1, g_3)\omega(g_1, g_3, g_2)^{-1}c(g_2, g_3),\\
&g_1, g_2, g_3, g_4 \in G.
\end{aligned}$$

Let $Z^3_{ab}(G, \Bbbk^\times)$ be the set of all pairs of functions (ω, c) satisfying (8.11). Observe that $Z^3_{ab}(G, \Bbbk^\times)$ is an abelian group with respect to the pointwise multiplication. The elements of this group are called *abelian cocycles* on the group G.

EXERCISE 8.4.4. Let $\omega \equiv 1$, that is $\omega(g_1, g_2, g_3) = 1$ for all $g_1, g_2, g_3 \in G$. Prove that the pair (ω, c) is an abelian cocycle if and only if c is a bicharacter on the group G.

EXERCISE 8.4.5. Let $\mathcal{C}$ be the skeletal pointed braided fusion category described by an abelian group G endowed with an abelian cocycle $(\omega, c) \in Z^3_{ab}(G, \Bbbk^\times)$. Show that $q(g) = c(g, g)$ for any $g \in G$. Use this to give an alternative proof of Lemma 8.4.2.

For a homomorphism $f : G_1 \to G_2$ and $(\omega, c) \in Z^3_{ab}(G_2, \Bbbk^\times)$ we define $f^*(\omega, c) = (\omega \circ f, c \circ f) \in Z^3_{ab}(G_1, \Bbbk^\times)$. It is clear that f^* is a homomorphism of abelian groups.

EXERCISE 8.4.6. Let $\mathcal{C}_1$ and $\mathcal{C}_2$ be the skeletal pointed braided fusion categories described by abelian groups G_1 and G_2 endowed with abelian cocycles $(\omega_1, c_1) \in Z^3_{ab}(G_1, \Bbbk^\times)$ and $(\omega_2, c_2) \in Z^3_{ab}(G_2, \Bbbk^\times)$. Prove that a braided tensor functor $F : \mathcal{C}_1 \to \mathcal{C}_2$ is determined by a homomorphism $f : G_1 \to G_2$ and a function $k : G_1 \times G_1 \to \Bbbk^\times$ satisfying

$$\begin{aligned}
&\omega(g_1, g_2, g_3) = k(g_2, g_3)k(g_1g_2, g_3)^{-1}k(g_1, g_2g_3)k(g_1, g_2)^{-1},\\
&c(g_1, g_2) = k(g_1, g_2)k(g_2, g_1)^{-1}.
\end{aligned} \tag{8.12}$$

where $(\omega, c) = (\omega_1, c_1)^{-1} f^*(\omega_2, c_2)$.

Let F_1 and F_2 be two such functors corresponding to pairs (f_1, k_1) and (f_2, k_2). Then F_1 and F_2 are not tensor isomorphic unless $f_1 = f_2$. If $f_1 = f_2$ then the tensor isomorphisms between F_1 and F_2 are in bijection with functions $\lambda : G \to \Bbbk^\times$ such that $k_2(g_1, g_2) = k_1(g_1, g_2)\lambda(g_1g_2)\lambda(g_1)^{-1}\lambda(g_2)^{-1}$, i.e., with solutions of the equation $d_2\lambda = k_2/k_1$.

For an abelian group G let $B^3_{ab}(G, \Bbbk^\times) \subset Z^3_{ab}(G, \Bbbk^\times)$ be the subgroup of *abelian coboundaries*, that is, of the abelian cocycles defined by (8.12) with $f = \mathsf{id}$ for all functions $k : G \times G \to \Bbbk^\times$.

DEFINITION 8.4.7. The group $H^3_{ab}(G, \Bbbk^\times) := Z^3_{ab}(G, \Bbbk^\times)/B^3_{ab}(G, \Bbbk^\times)$ is called the *abelian cohomology group* of G with coefficients in $\Bbbk^\times$.

The group $\mathsf{Aut}(G)$ acts on $H^3_{ab}(G, \Bbbk^\times)$ in an obvious way.

EXERCISE 8.4.8. Show that the braided equivalence classes of pointed braided fusion categories such that the group of isomorphism classes of simple objects is identified with G are in bijection with $H^3_{ab}(G, \Bbbk^\times)$. Thus the braided equivalence classes of pointed braided fusion categories such that the group of isomorphism classes of simple objects is isomorphic to G are in bijection with $H^3_{ab}(G, \Bbbk^\times)/\mathsf{Aut}(G)$.

Let $\mathrm{Quad}(G)$ be the group of quadratic forms with values in $\Bbbk^\times$ on a finite abelian group G. It is easy to check (and it follows from the discussion above) that the homomorphism $H^3_{ab}(G, \Bbbk^\times) \to \mathrm{Quad}(G)$, $(\omega, c) \mapsto q(g) = c(g, g)$ is well defined. The following result is due to Eilenberg and Mac Lane. For our proof we will need some results which will be proved later.

THEOREM 8.4.9. *The above homomorphism $H^3_{ab}(G, \Bbbk^\times) \to \mathrm{Quad}(G)$ is an isomorphism.*

PROOF. Let us show that this homomorphism is injective. Let $\mathcal{C}$ be a skeletal pointed braided fusion category corresponding to an abelian cocycle (ω, c) such that $q(g) = c(g, g) \equiv 1$. Then Corollary 9.9.24 states that there exists a braided tensor functor $\mathcal{C} \to \mathsf{Vec}$. Note that the category Vec is equivalent to the skeletal category with trivial group G_2 (see Exercise 8.4.6) and $(\omega_2, c_2) = (1, 1)$. Thus Exercise 8.4.6 implies that (ω, c) lies in $B^3_{ab}(G, \Bbbk^\times) \subset Z^3_{ab}(G, \Bbbk^\times)$. Hence our homomorphism is injective.

To prove surjectivity, we need to show that for any $q \in \mathrm{Quad}(G)$ there exists a pointed braided fusion category such that the corresponding form is q. In full generality this will be done in Example 8.23.10. For now we note that by Exercises 8.4.4 and 8.4.5 this result holds for quadratic forms q of the form $B(g, g)$ where $B : G \times G \to \Bbbk^\times$ is a bicharacter (possibly non-symmetric). By Exercise 8.4.10 this implies the surjectivity for groups G of odd order. □

EXERCISE 8.4.10. Prove that for an abelian group of odd order any quadratic form is of the form $B(g, g)$ for some bicharacter B.

Theorem 8.4.9 and Exercise 8.4.8 show that for any pre-metric group (G, q) there exists a unique up to a braided equivalence pointed braided fusion category $\mathcal{C}(G, q)$ such that the group of isomorphism classes of simple objects is G and the associated quadratic form is q.

EXERCISE 8.4.11. Let $\omega \in H^3(G, \Bbbk^\times)$ be the associator of the fusion category $\mathcal{C}(G, q)$.

(i) Show that ω is trivial if and only if there exists a bicharacter $B : G \times G \to \Bbbk^\times$ such that $q(x) = B(x, x)$ for all $x \in G$.

(ii) Show that ω is trivial if the category $\mathcal{C}(G, q)$ is symmetric (see Definition 8.1.12).

(iii) More generally, show that ω is trivial if and only if the quadratic form q has the following property: for any element $x \in G$ of order 2^n the order of $q(x) \in \Bbbk^\times$ does not exceed 2^n.

The categories $\mathcal{C}(G, q)$ exhaust all braided equivalence classes of pointed braided fusion categories. Next we describe all braided tensor functors between these categories. Consider the category whose objects are pointed braided fusion categories and morphisms are isomorphism classes of braided tensor functors (thus this category is a truncation of the 2-category where 2-morphisms are isomorphisms of tensor functors).

It is clear that a braided tensor functor between pointed braided fusion categories determines an orthogonal homomorphism between the corresponding pre-metric groups. We thus have a functor

$$F : (\text{pointed braided fusion categories}) \to (\text{pre-metric groups}). \tag{8.13}$$

THEOREM 8.4.12. *The above functor F is an equivalence.*

PROOF. We already know that the functor F is essentially surjective.

Let $f : (G_1, q_1) \to (G_2, q_2)$ be a morphism of pre-metric groups. Let us find abelian cocycles $(\omega_1, c_1) \in Z^3_{ab}(G_1, \Bbbk^\times)$ and $(\omega_2, c_2) \in Z^3_{ab}(G_2, \Bbbk^\times)$ such that $q_1(g) = c_1(g, g)$ and $q_2(g) = c_2(g, g)$. Then by the definition of an orthogonal homomorphism the abelian cocycle $(\omega, c) = (\omega_1, c_1)^{-1} f^*(\omega_2, c_2)$ satisfies $c(g, g) \equiv 1$. Thus by Theorem 8.4.9 we have $(\omega, c) \in B^3_{ab}(G_1, \Bbbk^\times)$. So there exists a function $k : G_1 \times G_1 \to \Bbbk^\times$ such that (8.12) holds. Hence by Exercise 8.4.6 there exists a braided tensor functor $\mathcal{C}(G_1, q_1) \to \mathcal{C}(G_2, q_2)$ lifting the morphism f. Therefore the functor F is full, that is, surjective on morphisms.

Finally, we need to show that the functor F is faithful, that is, injective on morphisms. Let k_1 and k_2 be two functions $G_1 \times G_1 \to \Bbbk^\times$ satisfying (8.12). Then the function $k(g_1, g_2) := k_1(g_1, g_2)^{-1} k_2(g_1, g_2)$ satisfies

$$\begin{aligned} &k(g_2, g_3) k(g_1 g_2, g_3)^{-1} k(g_1, g_2 g_3) k(g_1, g_2)^{-1} = 1, \\ &k(g_1, g_2) = k(g_2, g_1). \end{aligned} \tag{8.14}$$

In other words, the function k is a symmetric 2-cocycle on the group G_1 with values in $\Bbbk^\times$. It follows from Exercise 8.4.13 that there exists $\lambda : G_1 \to \Bbbk^\times$ such that $k(g_1, g_2) = \lambda(g_1 g_2) \lambda(g_1)^{-1} \lambda(g_2)^{-1}$. Thus

$$k_2(g_1, g_2) = k_1(g_1, g_2) \lambda(g_1 g_2) \lambda(g_1)^{-1} \lambda(g_2)^{-1}$$

and thus the functors corresponding to k_1 and k_2 are isomorphic. □

EXERCISE 8.4.13. Let G be a finite abelian group and let $k : G \times G \to \Bbbk^\times$ be a symmetric 2-cocycle, that is, a function satisfying (8.14). Show that k is trivial, that is, there exists a function $\lambda : G \to \Bbbk^\times$ such that $k(g_1, g_2) = \lambda(g_1 g_2) \lambda(g_1)^{-1} \lambda(g_2)^{-1}$. (*Hint:* consider the central extension of G by $\Bbbk^\times$ determined by k).

8.5. The center as a braided category

We introduced the center $\mathcal{Z}(\mathcal{C})$ of a monoidal category $\mathcal{C}$ in Section 7.13, see Definition 7.13.1. Namely, objects of $\mathcal{Z}(\mathcal{C})$ are pairs (Z, γ), where Z is an object of $\mathcal{C}$ and

$$\gamma_X : X \otimes Z \xrightarrow{\sim} Z \otimes X, \quad X \in \mathcal{C}$$

is a natural isomorphism satisfying compatibility conditions (7.41).

PROPOSITION 8.5.1. *$\mathcal{Z}(\mathcal{C})$ is a braided monoidal category with the associativity constraint given by that of $\mathcal{C}$ and braiding given by*

$$c_{(Z,\gamma),(Z',\gamma')} := \gamma'_Z. \tag{8.15}$$

PROOF. This is a direct verification and is left as an exercise. □

EXERCISE 8.5.2. Recall from Exercise 7.13.5 that there is a canonical tensor equivalence $\mathcal{Z}(\mathcal{C}^{\mathrm{op}}) \cong \mathcal{Z}(\mathcal{C})$. Show that it gives a *braided* equivalence

$$\mathcal{Z}(\mathcal{C}^{\mathrm{op}}) \cong \mathcal{Z}(\mathcal{C})^{\mathrm{rev}}. \tag{8.16}$$

Assume from now on that $\mathcal{C}$ is a finite tensor category.

PROPOSITION 8.5.3. *Let $\mathcal{C}$ be a finite tensor category and let $\mathcal{M}$ be an indecomposable exact $\mathcal{C}$-module category. Then $\mathcal{Z}(\mathcal{C}^*_{\mathcal{M}})$ and $\mathcal{Z}(\mathcal{C})^{rev}$ are equivalent as braided tensor categories, where $\mathcal{Z}(\mathcal{C})^{rev}$ denotes the reverse category, see Definition* 8.1.4.

PROOF. Let A be an algebra in $\mathcal{C}$ such that $\mathcal{M} \cong \mathsf{Mod}_{\mathcal{C}}(A)$, see Theorem 7.10.1. By Remark 7.16.3 we have a tensor equivalence

$$Z \mapsto Z \otimes A : \mathcal{Z}(\mathcal{C}) \xrightarrow{\sim} \mathcal{Z}(\mathsf{Bimod}_{\mathcal{C}}(A)).$$

It is immediate from construction in Remark 7.16.3 (see equation (7.51)) that this equivalence respects braiding. Since $\mathsf{Bimod}_{\mathcal{C}}(A) \cong (\mathcal{C}^*_{\mathcal{M}})^{\mathrm{op}}$ by Remarks 7.12.5 and 8.5.2 we have $\mathcal{Z}(\mathsf{Bimod}_{\mathcal{C}}(A)) \cong \mathcal{Z}(\mathcal{C}^*_{\mathcal{M}})^{\mathrm{rev}}$. □

EXAMPLE 8.5.4. (Center of $\mathsf{Vec}_{\mathbf{G}}$). Let us describe the center of the category of G-graded vector spaces, where G is a group. In the case of a finite group, we have already done this in Example 8.3.9 using the language of Hopf algebras.

By definition, an object in $\mathcal{Z}(\mathsf{Vec}_G)$ is a G-graded finite dimensional vector space $V = \oplus_{g \in G} V_g$ along with a collection of isomorphisms $\gamma_x : \delta_x \otimes V \xrightarrow{\sim} V \otimes \delta_x$ satisfying commutative diagram (7.41). They give rise to linear isomorphisms

$$u_{g,x} : V_{gxg^{-1}} \xrightarrow{\sim} V_x, \qquad g, x \in G.$$

Set $u_x := \oplus_{g \in G} u_{g,x} : V \xrightarrow{\sim} V$. It is straightforward to see that this collection of isomorphisms makes V a G-equivariant object (in the sense of Definition 2.7.2) with respect to the conjugation action of G on Vec_G.

Thus, objects of $\mathcal{Z}(\mathsf{Vec}_G)$ are identified with G-equivariant G-graded vector spaces. By Remark 4.15.8 simple objects of $\mathcal{Z}(\mathsf{Vec}_G)$ are in bijection with pairs (C, V) where C is a finite conjugacy class in G, and V is an irreducible finite dimensional representation of the centralizer of $g \in C$. [1]

---

[1]This implies that if G is a finitely generated infinite simple group (it is known that such groups exist), then $\mathcal{Z}(\mathsf{Vec}_G) = \mathsf{Vec}$. Indeed, the only finite conjugacy class in G is that of the identity, and the only irreducible finite dimensional representation of G is trivial (since any finitely generated linear group is residually finite and hence cannot be simple unless it is finite). This is somewhat counter-intuitive, as $\mathcal{Z}(\mathsf{Vec}_G)$ turns out to be "smaller" than Vec_G.

In particular, if G is finite then $\mathcal{Z}(\mathsf{Vec}_G)$ is a fusion category. By Proposition 4.15.9, the Frobenius-Perron dimension of the object corresponding to a pair (C, V) is $|C| \dim_{\Bbbk}(V)$. This implies that

$$\mathsf{FPdim}(\mathcal{Z}(\mathsf{Vec}_G)) = |G|^2. \tag{8.17}$$

Note that this can also be seen as a special case of Theorem 7.16.6.

In the special situation when $G = A$ is a finite abelian group, the fusion category $\mathcal{Z}(\mathsf{Vec}_A)$ is pointed and is equivalent to $\mathcal{C}(A \oplus \widehat{A}, q)$, where $\widehat{A} := \mathsf{Hom}(A, \Bbbk^\times)$ is the group of linear characters of A, and $(A \oplus \widehat{A}, q)$ is a metric group with respect to the canonical hyperbolic quadratic form $q((a, \chi)) := \chi(a)$, cf. Section 8.4.

EXERCISE 8.5.5. Let G be an affine (pro)algebraic group, and $\mathcal{C}$ be the category of finite dimensional $O(G)$-modules. Show that $\mathcal{Z}(\mathcal{C})$ is the category of G-equivariant finite dimensional $O(G)$-modules (so objects of $\mathcal{Z}(\mathcal{C})$ are supported at the set of elements of G which centralize the connected component of the identity $G_0 \subset G$).

EXAMPLE 8.5.6. (The center of $\mathsf{Rep}(H)$.) It is shown in Proposition 8.3.8 that the center of $\mathsf{Rep}(H)$ for a finite dimensional Hopf algebra H is equivalent as a braided tensor category to $\mathsf{Rep}(D(H))$, the representation category of the quantum double of H.

Similarly, if K is any Hopf algebra, and $\mathcal{C} = K\text{−}\mathsf{comod}$, then the center $\mathcal{Z}(\mathcal{C})$ is naturally equivalent as a braided tensor category to the category of Yetter-Drinfeld modules $YD(K)$ (see Section 7.15), with braiding defined by the formula

$$c_{X,Y} = \sigma \circ R_{XY},$$

where $R_{XY} : X \otimes Y \to X \otimes Y$ is given by the formula

$$R_{XY}(x \otimes y) = \eta_{Y,23}(\tau_X(x) \otimes y),$$

where $\eta_Y : K \otimes Y \to Y$ is the action, and $\tau_X : X \to K \otimes X$ is the coaction.

EXERCISE 8.5.7. Assume that $q \in \Bbbk^\times$ is not a root of unity, and let the Hopf algebra A be generated by $g^{\pm 1}, x$ with relations $gx = q^2 xg$ and coproduct $\Delta(g) = g \otimes g$, $\Delta(x) = x \otimes g + 1 \otimes x$. Show that $\mathcal{Z}(A\text{−}\mathsf{comod}) = \mathcal{C} \boxtimes \mathsf{Vec}_{\mathbb{Z}}$, where $\mathcal{C}$ is the category of representations of type I of the quantum group $U_q(\mathfrak{sl}_2)$. Calculate the braiding in $\mathcal{Z}(A\text{–}\mathsf{comod})$ in terms of this presentation. (*Hint:* use that $\mathcal{Z}(A\text{–}\mathsf{comod})$ is the category of Drinfeld-Yetter modules for A).

8.6. Factorizable braided tensor categories

Let $\mathcal{C}$ be a braided tensor category with braiding

$$c_{X,Y} : X \otimes Y \xrightarrow{\sim} Y \otimes X, \quad X, Y \in \mathcal{C}.$$

Let $\mathcal{C}^{\text{rev}}$ be the reverse braided tensor category, see Definition 8.1.4. The following proposition is straightforward.

PROPOSITION 8.6.1. *Then the assignment $X \mapsto (X, c_{-,X})$ extends to a braided tensor functor $\mathcal{C} \to \mathcal{Z}(\mathcal{C})$. Similarly, the assignment $X \mapsto (X, c_{X,-}^{-1})$ extends to a braided tensor functor $\mathcal{C}^{rev} \to \mathcal{Z}(\mathcal{C})$. Both functors are fully faithful. They combine together into a single braided tensor functor*

$$G : \mathcal{C} \boxtimes \mathcal{C}^{rev} \to \mathcal{Z}(\mathcal{C}). \tag{8.18}$$

DEFINITION 8.6.2. A braided tensor category $\mathcal{C}$ is called *factorizable* if the functor (8.18) is an equivalence.

PROPOSITION 8.6.3. *Let $\mathcal{C}$ be a tensor category. Then $\mathcal{Z}(\mathcal{C})$ is factorizable.*

PROOF. Recall from Proposition 7.13.8 that $\mathcal{Z}(\mathcal{C})$ is dual to $\mathcal{C}\boxtimes\mathcal{C}^{\text{op}}$ with respect to the module category $\mathcal{C}$. By Theorem 8.5.3 there is a braided tensor equivalence

$$\mathcal{Z}(\mathcal{C}\boxtimes\mathcal{C}^{\text{op}}) \cong \mathcal{Z}(\mathcal{C})\boxtimes\mathcal{Z}(\mathcal{C})^{\text{rev}} \xrightarrow{\sim} \mathcal{Z}(\mathcal{Z}(\mathcal{C})). \tag{8.19}$$

Namely, both categories are identified with the category of $(\mathcal{Z}(\mathcal{C})\boxtimes\mathcal{C}\boxtimes\mathcal{C}^{\text{op}})$-module endofunctors of $\mathcal{C}$. Under this identification the embeddings

$$\begin{aligned} \mathcal{Z}(\mathcal{C}) &= \mathcal{Z}(\mathcal{C})\boxtimes\mathsf{Vec} \hookrightarrow \mathcal{Z}(\mathcal{C})\boxtimes\mathcal{Z}(\mathcal{C})^{\text{rev}} \cong \mathcal{Z}(\mathcal{Z}(\mathcal{C})) \quad \text{and} \\ \mathcal{Z}(\mathcal{C})^{\text{rev}} &= \mathsf{Vec}\boxtimes\mathcal{Z}(\mathcal{C})^{\text{rev}} \hookrightarrow \mathcal{Z}(\mathcal{C})\boxtimes\mathcal{Z}(\mathcal{C})^{\text{rev}} \cong \mathcal{Z}(\mathcal{Z}(\mathcal{C})) \end{aligned}$$

assign to an object $Z \in \mathcal{Z}(\mathcal{C})$ the functor $(Z \otimes -) : \mathcal{C} \to \mathcal{C}$. The $\mathcal{Z}(\mathcal{C})$-module structure on this functor is given, respectively, by the braiding of $\mathcal{Z}(\mathcal{C})$ and its opposite. Thus, in this case functor (8.18) coincides with equivalence (8.19). □

EXERCISE 8.6.4. (i) Let (H, R) be a finite dimensional quasitriangular Hopf algebra. Show that the braided category $\mathsf{Rep}(H)$ is factorizable if and only if H is factorizable, i.e., $R^{21}R$ is a non-degenerate tensor in $H \otimes H$.

(ii) Show that any quasitriangular Hopf algebra H contains a unique factorizable Hopf subalgebra $\overline{H}$ with the same R-matrix (consider the Hopf subalgebra generated by the left or right component of $R^{21}R$).

(iii) Show that any factorizable Hopf algebra A is a quotient of the quantum double $D(A_+)$ of a finite dimensional Hopf algebra A_+, with the R-matrix being the image of the R-matrix of $D(A_+)$ (consider the Hopf algebra A_+ spanned by the first component of R). Thus, for any quasitriangular Hopf algebra H, we have a canonical quasitriangular Hopf algebra homomorphism $\psi : D(\overline{H}_+) \to H$, whose image is $\overline{H}$.

8.7. Module categories over braided tensor categories

Here we discuss a categorical analog of the fact that a left module over a commutative ring R is automatically an R-bimodule.

Let $\mathcal{C}$ be a braided tensor category with braiding c and let $\mathcal{M}$ be a $\mathcal{C}$-module category, see Definition 7.1.1, with the module associativity constraint

$$m_{X,Y,M} : (X \otimes Y) \otimes M \xrightarrow{\sim} X \otimes (Y \otimes M), \quad X, Y \in \mathcal{C},\ M \in \mathcal{M}.$$

Let $\mathcal{C}^*_{\mathcal{M}}$ be the tensor category dual to $\mathcal{C}$ with respect to $\mathcal{M}$ (i.e., the category of $\mathcal{C}$-module endofunctors of $\mathcal{M}$, see Definition 7.12.2). Observe that for every $X \in \mathcal{C}$ the action of X on $\mathcal{M}$ gives rise to a $\mathcal{C}$-module endofunctor of $\mathcal{M}$ in two ways (using the braiding and its opposite). We thus have assignments

$$H^{\pm} : X \mapsto ((X \otimes -),\, s^{\pm}_X) : \mathcal{C} \to \mathcal{C}^*_{\mathcal{M}}, \tag{8.20}$$

where the module functor structures $s^{\pm}$ are defined by

$$\begin{aligned} (s^+_X)_{Y,M} : H^+(X)(Y \otimes M) = X \otimes (Y \otimes M) &\xrightarrow{m^{-1}_{X,Y,M}} (X \otimes Y) \otimes M \xrightarrow{c_{X,Y}\otimes \mathsf{id}_M} \\ (Y \otimes X) \otimes M &\xrightarrow{m_{Y,X,M}} Y \otimes (X \otimes M) = Y \otimes H^+(X)(M), \end{aligned}$$

$$(s_X^-)_{Y,M} : H^-(X)(Y \otimes M) = X \otimes (Y \otimes M) \xrightarrow{m_{X,Y,M}^{-1}} (X \otimes Y) \otimes M \xrightarrow{c_{Y,X}^{-1} \otimes \mathrm{id}_M}$$
$$(Y \otimes X) \otimes M \xrightarrow{m_{Y,X,M}} Y \otimes (X \otimes M) = Y \otimes H^-(X)(M),$$

for all $X, Y \in \mathcal{C}$ and $M \in \mathcal{M}$.

PROPOSITION 8.7.1. *The assignments* (8.20) *define tensor functors*

$$H^+ : \mathcal{C} \to \mathcal{C}^*_{\mathcal{M}} \quad \textit{and} \quad H^- : \mathcal{C}^{\mathrm{op}} \to \mathcal{C}^*_{\mathcal{M}}. \tag{8.21}$$

PROOF. This is a straightforward verification using axioms (8.1) and (8.2) in the definition of a braided tensor category. □

REMARK 8.7.2. It follows from Proposition 8.7.1 that for a braided tensor category $\mathcal{C}$ a $\mathcal{C}$-module category is automatically a $\mathcal{C}$-bimodule category.

EXERCISE 8.7.3. Let (H, R) be a quasitriangular Hopf algebra.

(i) Show that evaluation on the second component of R defines a Hopf algebra homomorphism $R_+ : H^* \to H$ $(R_+(f) := (f \otimes \mathrm{id})(R))$, and evaluation on the first component of R^{-1} defines a Hopf algebra homomorphism $R_- : H^* \to H^{\mathrm{cop}}$ $(R_-(f) := (\mathrm{id} \otimes f)(R^{-1}))$.

(ii) Show that the image of R_+ is the Hopf algebra $\overline{H}_+$ defined in Exercise 8.6.4, and the image of R_- is isomorphic to $\overline{H}_+^{*\mathrm{cop}}$.

(iii) Let $\mathcal{M} = \mathsf{Vec}$ be the module category over $\mathcal{C} = \mathsf{Rep}(H)$ associated to the forgetful functor on $\mathsf{Rep}(H)$ (so that $\mathcal{C}^*_{\mathcal{M}} = \mathsf{Rep}(H^*)$). Show that the functors $H^{\pm}$ are induced by the Hopf algebra homomorphisms $R_{\pm}$.

8.8. Commutative algebras and central functors

We defined the notion of an (associative) algebra in a tensor category in Section 7.8. In general, the notion of commutativity for such algebras does not make sense, since a "permutation of factors" is not defined. However, this notion exists in the setting of *braided* tensor categories.

DEFINITION 8.8.1. Let $\mathcal{C}$ be a braided tensor category. An algebra A in $\mathcal{C}$ is said to be *commutative* if the following diagram commutes:

$$\begin{array}{ccc} A \otimes A & \xrightarrow{c_{A,A}} & A \otimes A \\ & {\scriptstyle m}\searrow \quad \swarrow{\scriptstyle m} & \\ & A, & \end{array} \tag{8.22}$$

where $m : A \otimes A \to A$ denotes the multiplication of A.

EXERCISE 8.8.2. (i) Show that an algebra is commutative in $\mathcal{C}$ if and only if it is commutative in $\mathcal{C}^{\mathrm{rev}}$.

(ii) Let (H, R) be a quasitriangular Hopf algebra, and A be an H-module algebra. Show that A is commutative in the category of H-modules if and only if $ab = m(R^{21}(b \otimes a))$, $a, b \in A$, where $m : A \otimes A \to A$ is the multiplication map. So, when regarded as a usual algebra, A is not necessarily commutative; one says that A is *braided-commutative*.

(iii) Let X be an object of a braided multitensor category $\mathcal{C}$. Define the *quantum symmetric algebra* S_qX to be the quotient of the tensor algebra TX by the ideal generated by the image of $c_{X,X} - \mathrm{id}_{X \otimes X}$. Show that S_qX is a commutative $\mathbb{Z}_+$-graded algebra in $\mathcal{C}$. [2] (The homogeneous components $S_q^N X$ of $S_q X$ are called the *quantum symmetric powers* of X). Show that $S_q X$ has the following universal property: any commutative algebra A in $\mathcal{C}$ generated by X (i.e., a quotient of TX) is a quotient of $S_q X$.

(iv) Let A, B be associative algebras in a braided category $\mathcal{C}$. Define a multiplication on $A \otimes B$ by the formula $m_{A\otimes B} = (m_A \otimes m_B) \circ (\mathsf{id}_A \otimes c_{A,B}^{-1} \otimes \mathsf{id}_B)$. Show that this makes $A \otimes B$ into an associative algebra.[3]

(v) Let $\mathcal{C}$ be the category $\mathsf{Vec}_{\mathbb{Z}}$ of $\mathbb{Z}$-graded vector spaces, with simple objects X_i, $i \in \mathbb{Z}$ (1-dimensional space in degree i). Let ζ be a nonzero scalar, and put a braiding on $\mathcal{C}$ by $c_{X_i,X_j} = \sigma\zeta^{ij}$, where σ is the permutation of components. Compute the quantum symmetric algebra $S_q X_m$ (show that $S_q^N X_m = 0$ for $N \geq 2$ unless $\zeta^{m^2} = 1$, and is 1-dimensional otherwise). Compute the algebra $S_q(X_m \oplus X_k)$, and show that it is not isomorphic to $S_q X_m \otimes S_q X_k$ (and does not map onto it) unless $\zeta^{2mk} = 1$. This shows that for commutative algebras A, B in a braided category $\mathcal{C}$, the tensor product $A \otimes B$ may be noncommutative.

(vi) Specialize (v) to the symmetric category case $\zeta = -1$, $\mathrm{char}\,\Bbbk \neq 2$ (this is the case of $\mathbb{Z}$-graded supervector spaces). In this case, show that for any object X, $S_q X = SX_+ \otimes \wedge X_-$, where X_+, X_- are the even and odd degree parts of X, respectively. Show also that in this case one always has $S_q(X \oplus Y) = S_q(X) \otimes S_q(Y)$ (note that at the same time, for the ordinary tensor product of algebras, $\wedge(X \oplus Y) \neq \wedge X \otimes \wedge Y$ for vector spaces X and Y).

(vii) Let $\mathcal{C} = \mathsf{Vec}_{\mathbb{Z}} \boxtimes \mathcal{C}_0$, where $\mathcal{C}_0$ is the category of type I representations of $U_q(\mathfrak{sl}_2)$. Put the braiding on $\mathcal{C}_0$ as in Exercise 8.3.13, and the braiding defined in (v) on $\mathsf{Vec}_{\mathbb{Z}}$, with $\zeta = q^{-1/2}$; this puts the tensor product braiding on $\mathcal{C}$. Let $V_1 \in \mathcal{C}_0$ be the 2-dimensional irreducible representation of $U_q(\mathfrak{sl}_2)$, i.e., the q-analog of the vector representation. Let $X = X_1 \boxtimes V_1$. Show that the quantum symmetric algebra $S_q X$, when regarded as an ordinary algebra, is isomorphic to the quantum polynomial algebra with generators x, y and the defining relation $yx = qxy$. Show that $S_q^N V = X_N \boxtimes V_N$, where V_N is the $N+1$-dimensional irreducible representation in $\mathcal{C}_0$.

(viii) (this is a higher rank generalization of (vii)) Let $V = \Bbbk^n$ be the n-dimensional defining representation of quantum GL_n (an irreducible $O_q(GL_n)$-comodule), with standard basis x_i, $i = 1, ..., n$. The (suitably normalized) R-matrix of V is given by $R(x_i \otimes x_i) = x_i \otimes x_i$, $R(x_i \otimes x_j) = q^{-1}(x_j \otimes x_i)$ for $i < j$, and $R(x_i \otimes x_j) = q^{-1}(x_j \otimes x_i) + (1 - q^{-2})(x_i \otimes x_j)$ for $i > j$. Show that $S_q V$, when regarded as an ordinary algebra, is isomorphic to the quantum polynomial algebra with generators $x_1, ..., x_n$ and defining relations $x_j x_i = q x_i x_j$ for $i > j$. Deduce that the dimension of $S_q^N V$ is $\binom{N+n-1}{n-1}$, and thus $S_q V$ is a flat deformation of the usual polynomial algebra SV (which is recovered when $q = 1$).

[2] In fact, strictly speaking, $S_q X$ lies in the ind-completion of $\mathcal{C}$ rather than $\mathcal{C}$ itself (as all of its homogeneous components may be nonzero). However, for simplicity we will abuse terminology, and refer to algebras in the ind-completion of $\mathcal{C}$ as algebras in $\mathcal{C}$.

[3] Note that in a non-braided category, we do not have a notion of the tensor product of algebras, and that the tensor products of algebras in $\mathcal{C}$ and $\mathcal{C}^{\mathrm{rev}}$ are different from each other.

Let A be a commutative algebra in a braided tensor category $\mathcal{C}$. Let $\mathsf{Mod}_{\mathcal{C}}(A)$ denote the category of right A-modules in $\mathcal{C}$.

EXERCISE 8.8.3. Let M be a right A-module in $\mathcal{C}$ with the structure morphism $p : M \otimes A \to M$. Show that each of the following morphisms

$$A \otimes M \xrightarrow{c_{A,M}} M \otimes A \xrightarrow{p} M \quad \text{and} \tag{8.23}$$

$$A \otimes M \xrightarrow{c_{M,A}^{-1}} M \otimes A \xrightarrow{p} M, \tag{8.24}$$

defines an A-bimodule structure on M.

Let A be an exact algebra in $\mathcal{C}$ (see Definition 7.8.20). Then the category $\mathsf{Bimod}_{\mathcal{C}}(A)$ is a multitensor category. Let

$$F_{\pm} : M \mapsto M_{\pm} : \mathsf{Mod}_{\mathcal{C}}(A) \to \mathsf{Bimod}_{\mathcal{C}}(A) \tag{8.25}$$

denote the functors constructed using (8.23) and (8.24), respectively. Each of these functors is a full embedding and, thus, makes $\mathsf{Mod}_{\mathcal{C}}(A)$ into a multitensor category with tensor product $\otimes_A$ (see Definition 7.8.21) by identifying it with a tensor subcategory of $\mathsf{Bimod}_{\mathcal{C}}(A)$.

EXERCISE 8.8.4. Show that the multitensor category structures on $\mathsf{Mod}_{\mathcal{C}}(A)$ constructed using functors $F_{\pm}$ are opposite to each other in the sense of Definition 2.1.5.

Let $\mathcal{M}$ denote $\mathsf{Mod}_{\mathcal{C}}(A)$ viewed as a $\mathcal{C}$-module category, see Proposition 7.8.10. The dual category $\mathcal{C}^*_{\mathcal{M}}$ is identified with $\mathsf{Bimod}_{\mathcal{C}}(A)^{\mathrm{op}}$, see Remark 7.12.5.

EXERCISE 8.8.5. Show that tensor functors $H^+ : \mathcal{C} \to \mathcal{C}^*_{\mathcal{M}}$ and $H^- : \mathcal{C}^{\mathrm{op}} \to \mathcal{C}^*_{\mathcal{M}}$ from Proposition 8.7.1 are identified with the composition functors

$$\mathcal{C} \xrightarrow{F_A} \mathsf{Mod}_{\mathcal{C}}(A) \xrightarrow{F_{\pm}} \mathsf{Bimod}_{\mathcal{C}}(A),$$

where $F_A(X) := X \otimes A$ is the free A-module functor.

Let $\mathcal{C}$ be a finite braided tensor category and let $\mathcal{A}$ be a finite tensor category.

DEFINITION 8.8.6. Let $F : \mathcal{C} \to \mathcal{A}$ be a tensor functor. A structure of a *central functor* on F is a braided tensor functor $F' : \mathcal{C} \to \mathcal{Z}(\mathcal{A})$ together with an isomorphism of F with the composition of the forgetful functor $\mathcal{Z}(\mathcal{A}) \to \mathcal{A}$ with F'.

REMARK 8.8.7.
(i) The forgetful functor $\mathcal{Z}(\mathcal{A}) \to \mathcal{A}$ itself is a central functor with $F' = \mathrm{id}_{\mathcal{Z}(\mathcal{A})}$.
(ii) Any braided tensor functor $F : \mathcal{C}_1 \to \mathcal{C}_2$ between braided tensor categories $\mathcal{C}_1$ and $\mathcal{C}_2$ is a central functor via the embedding $\mathcal{C}_2 \hookrightarrow \mathcal{Z}(\mathcal{C}_2)$.

PROPOSITION 8.8.8. *Let $F : \mathcal{C} \to \mathcal{A}$ be a central functor. Let $I : \mathcal{A} \to \mathcal{C}$ be the right adjoint functor of F. Then the object $A = I(\mathbf{1})$ has a canonical structure of commutative algebra in $\mathcal{C}$.*

PROOF. The algebra structure of A is defined in Section 7.9, see Example 7.9.10.

To prove commutativity, note that the multiplication $m : A \otimes A \to A$ is the image of a certain morphism $\tilde{m} \in \mathsf{Hom}_{\mathcal{A}}(F(A \otimes A), \mathbf{1})$ under the isomorphism

$$\mathsf{Hom}_{\mathcal{A}}(F(A \otimes A), \mathbf{1}) \cong \mathsf{Hom}_{\mathcal{C}}(A \otimes A,\, A). \tag{8.26}$$

Let c denote the braiding in $\mathcal{C}$ and let $\tilde{c}$ denote the braiding in $\mathcal{Z}(\mathcal{A})$, see (8.15). Under the above isomorphism (8.26) the opposite multiplication $m \circ c_{A,A}$ corresponds to $\tilde{m} \circ F(c_{A,A}) \in \mathsf{Hom}_{\mathcal{A}}(F(A \otimes A), \mathbf{1})$. The equality $\tilde{m} = \tilde{m} \circ F(c_{A,A})$ follows from commutativity of the following diagram, where $F' : \mathcal{C} \to \mathcal{Z}(\mathcal{A})$ is the central structure for the functor F:

$$\begin{CD}
F'(A \otimes A) @>{\sim}>> F'(A) \otimes F'(A) @>{l \otimes l}>> \mathbf{1} \otimes \mathbf{1} @>{\sim}>> \mathbf{1} \\
@V{F'(c_{A,A})}VV @V{\tilde{c}_{F'(A),F'(A)}}VV @V{\tilde{c}_{\mathbf{1},\mathbf{1}}}VV @V{\mathsf{id}_{\mathbf{1}}}VV \\
F'(A \otimes A) @>{\sim}>> F'(A) \otimes F'(A) @>{l \otimes l}>> \mathbf{1} \otimes \mathbf{1} @>{\sim}>> \mathbf{1}
\end{CD}$$

Here $l \in \mathsf{Hom}_{\mathcal{C}}(F(A), \mathbf{1})$ is the image of id_A under $\mathsf{Hom}_{\mathcal{C}}(A, A) \cong \mathsf{Hom}_{\mathcal{A}}(F(A), \mathbf{1})$. The left square commutes since F' is a braided functor, and the right one since $\tilde{c}_{\mathbf{1},\mathbf{1}} = \mathsf{id}_{\mathbf{1}}$. That the middle square commutes is a consequence of the naturality of the permutation morphism (7.40) in the definition of the center. □

EXAMPLE 8.8.9. (i) Let $\mathcal{C} = \mathsf{Rep}(G)$ and $F : \mathcal{C} \to \mathsf{Vec}$ be the forgetful braided tensor functor. Then the commutative algebra A from Proposition 8.8.8 is the algebra $\mathsf{Fun}(G, \Bbbk)$ of $\Bbbk$-valued functions on G. This algebra is called the *regular algebra* of $\mathsf{Rep}(G)$, cf. Definition 9.9.19.

(ii) Let Vec_G^{ω} be the fusion category of finite dimensional G-graded vector spaces with the associativity constraint defined by $\omega \in Z^3(G, \Bbbk^{\times})$, see Example 2.6.2. Let $\mathcal{C} = \mathcal{Z}(\mathsf{Vec}_G^{\omega})$ and $F : \mathcal{C} \to \mathsf{Vec}_G^{\omega}$ be the forgetful functor. Then the commutative algebra A from Proposition 8.8.8 is the regular algebra of $\mathsf{Rep}(G) \subset \mathcal{C}$.

(iii) Let $\mathcal{C} = \mathcal{Z}(\mathsf{Rep}(G)) \cong \mathcal{Z}(\mathsf{Vec}_G)$ and $F : \mathcal{C} \to \mathsf{Rep}(G)$ be the forgetful functor. Then the commutative algebra A from Proposition 8.8.8 is the group algebra of G considered as an algebra in $\mathcal{C}$. Notice that in this case the algebra $F(A)$ in the symmetric tensor category $\mathsf{Rep}(G)$ is non-commutative unless G is commutative.

We have the following construction, which is converse to Proposition 8.8.8. Namely, let A be a commutative exact algebra in $\mathcal{C}$. Suppose also that $\mathsf{Hom}_{\mathcal{C}}(\mathbf{1}, A) \cong \Bbbk$, so that the $\mathcal{C}$-module category $\mathsf{Mod}_{\mathcal{C}}(A)$ is indecomposable. Then the category $\mathsf{Bimod}_{\mathcal{C}}(A)$ is a finite tensor category and, hence, the category $\mathsf{Mod}_{\mathcal{C}}(A)$ inherits the finite tensor category structure via the embedding F_- from (8.25).

PROPOSITION 8.8.10. *The functor*

$$F_A : \mathcal{C} \to \mathsf{Mod}_{\mathcal{C}}(A) : X \mapsto X \otimes A$$

has a canonical structure of a central functor.

PROOF. We have $F_A(X) = X \otimes A$, and, hence, $F_A(X) \otimes_A Y = X \otimes Y$. Similarly, $Y \otimes_A F_A(X) = Y \otimes X$. These two objects are isomorphic via the braiding of $\mathcal{C}$ (using commutativity of A, one can check that the braiding gives an isomorphism of A-modules) and, hence, F_A lifts to a braided tensor functor

$$F'_A : \mathcal{C} \to \mathcal{Z}(\mathsf{Mod}_{\mathcal{C}}(A)) \tag{8.27}$$

whose composition with the forgetful functor $\mathcal{Z}(\mathsf{Mod}_{\mathcal{C}}(A)) \to \mathsf{Mod}_{\mathcal{C}}(A)$ equals F_A. □

8.9. The Drinfeld morphism

In this Section we assume that all monoidal categories are strict. Equivalently, we suppress all associativity and unit constraints.

LEMMA 8.9.1. *Let $\mathcal{C}$ be a braided monoidal category with braiding c, let $X, Y \in \mathcal{C}$, and let X^* be the left dual of X. We have*

$$c_{X^*,Y} = (\mathsf{ev}_X \otimes \mathsf{id}_{Y \otimes X^*}) \circ (\mathsf{id}_{X^*} \otimes c^{-1}_{X,Y} \otimes \mathsf{id}_{X^*}) \circ (\mathsf{id}_{X^* \otimes Y} \otimes \mathsf{coev}_X), \tag{8.28}$$

$$c^{-1}_{Y,X^*} = (\mathsf{ev}_X \otimes \mathsf{id}_{Y \otimes X^*}) \circ (\mathsf{id}_{X^*} \otimes c_{Y,X} \otimes \mathsf{id}_{X^*}) \circ (\mathsf{id}_{X^* \otimes Y} \otimes \mathsf{coev}_X). \tag{8.29}$$

PROOF. Consider the following diagram (where we suppress the identity morphisms, as usual):

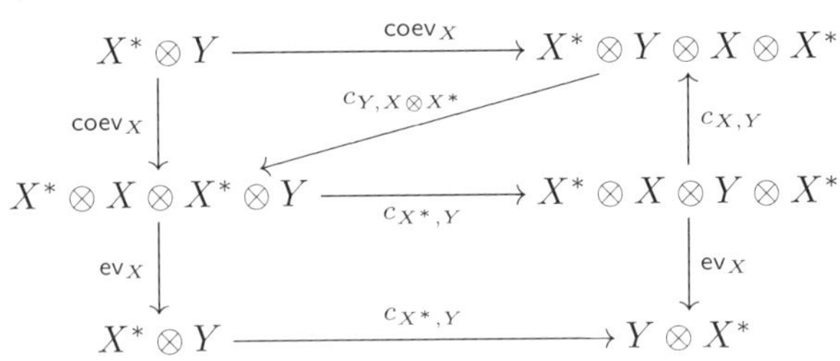

The upper triangle commutes by the naturality of the braiding. The lower triangle commutes by the hexagon axiom (8.1). The rectangle at the bottom commutes since $\otimes$ is a bifunctor. Hence, the diagram commutes. Comparing the compositions along the perimeter, we get the first equality. The second equality is obtained by replacing c by the opposite braiding. □

EXERCISE 8.9.2. Show that if X^*, Y^* are left duals of objects X, Y in a braided monoidal category $\mathcal{C}$ then $(c_{X,Y})^* = c_{X^*,Y^*}$.

Let $\mathcal{C}$ be a braided monoidal category. Let us define a natural transformation $u_X : X \to X^{**}$ as the composition

$$X \xrightarrow{\mathsf{id}_X \otimes \mathsf{coev}_{X^*}} X \otimes X^* \otimes X^{**} \xrightarrow{c_{X,X^*} \otimes \mathsf{id}_{X^{**}}} X^* \otimes X \otimes X^{**} \xrightarrow{\mathsf{ev}_X \otimes \mathsf{id}_{X^{**}}} X^{**}. \tag{8.30}$$

PROPOSITION 8.9.3. *We have*

$$u_X \otimes u_Y = u_{X \otimes Y} \circ c_{Y,X} \circ c_{X,Y} \tag{8.31}$$

for all $X, Y \in \mathcal{C}$.

PROOF. Consider the following diagram (to keep its size reasonable we omit tensor product signs):

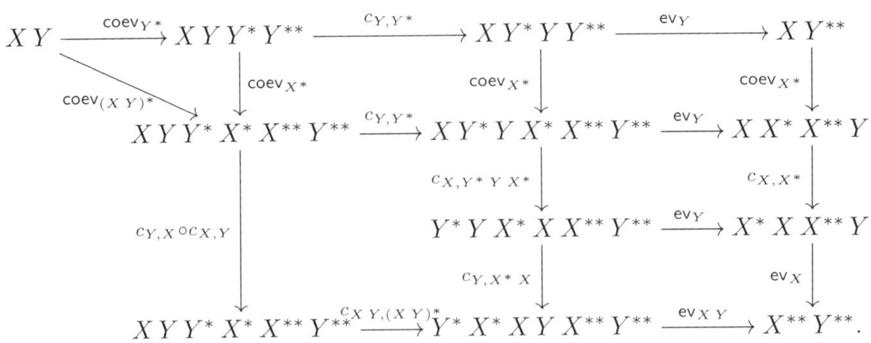

It is straightforward to see that it commutes. E.g., the big rectangle involving braidings commutes by the braided coherence theorem (see Exercise 8.2.7), and the two squares in the right column commute by the naturality of the braiding. Using definition (8.30), we see that the two compositions along the perimeter are precisely the two sides of (8.31). □

DEFINITION 8.9.4. The morphism u is called the *Drinfeld morphism.*

Note that (8.30) is not a morphism of tensor functors unless the category $\mathcal{C}$ is symmetric.

Let (H, R) be a quasitriangular Hopf algebra, and $R = \sum_i a_i \otimes b_i$.

PROPOSITION 8.9.5. *(Drinfeld) (i)* $R^{-1} = \sum_i S(a_i) \otimes b_i = \sum_i a_i \otimes S^{-1}(b_i)$*;*

(ii) The Drinfeld morphism u in the category of H-modules is defined by the Drinfeld element $u = \sum_i S(b_i)a_i$*;*

(iii) $u^{-1} = \sum_i b_i S^2(a_i)$*;*

(iv) $S^2(x) = uxu^{-1}$ *for* $x \in H$*;*

(v) $\Delta(u) = (u \otimes u)(R^{21}R)^{-1} = (R^{21}R)^{-1}(u \otimes u)$*;*

(vi) $z := uS(u)$ *is a central element and* $g := u^{-1}S(u)$ *is a grouplike element of H such that* $g^{-1}xg = S^4(x)$, $x \in H$.

PROOF. These properties follow immediately from the categorical properties proved above. Namely, (8.28) and (8.29) yield (i); parts (ii) and (iii) follow from the definition of u; part (iv) expresses the fact that u is a morphism $V \to V^{**}$; (v) follows from (8.31); and (vi) follows from (iv) and (v). We leave it to the reader to fill in the details. □

EXERCISE 8.9.6. (i) Let $H = \Bbbk G$ be the group algebra of a finite group G ($\mathrm{char}\,\Bbbk = 0$). Calculate the element $u \in D(H)$ explicitly, and show that it is central, and $g = 1$. Compute the values of u on the irreducible $D(H)$-modules.

(ii) Let $\mathcal{C}$ be the category of type I representations of $U_q(\mathfrak{sl}_2)$ (q is not a root of unity). Show that the Drinfeld morphism $u_X : X \to X^{**}$ is given by the formula $u_X = K\theta_X$, where $\theta_X : X \to X$ is a morphism which acts by the scalar $q^{-N(N+2)/2}$ on the $N+1$-dimensional simple object V_N. Deduce that $g = K^{-2}$.

Now, let H be a finite dimensional Hopf algebra. Since $g \in D(H)$ is a grouplike element, we have $g = a \otimes \alpha^{-1}$, where $a \in H$ and $\alpha \in H^*$ are grouplike elements. These elements are called the *distinguished grouplike elements* of H and $H^{*\mathrm{cop}}$, respectively.

COROLLARY 8.9.7. *(Radford's S^4 formula,* [**Ra2**]*). For any $x \in H$ one has* $S^4(x) = \alpha \circ (a^{-1}xa) \circ \alpha^{-1}$*, where $\alpha\circ$ is the action of α on H which is dual to its action on H^*.*

PROOF. This follows immediately from Proposition 8.9.5(vi) and the multiplication law in $D(H)$. □

EXERCISE 8.9.8. Check that g defines the distinguished character of $D(H)^*$. In other words, if λ is a left integral of $D(H)$ then $\lambda(x_1)x_2 = \lambda(x)g$. Deduce that α defines the distinguished character of H. Thus, our terminology is compatible to the one introduced in Chapter 6.

COROLLARY 8.9.9. *(Radford,* [**Ra2**]*). The antipode of a finite dimensional Hopf algebra H has order dividing* $4 \dim H$.

REMARK 8.9.10. In all known examples, the order of S actually divides $2\dim H$, but it is not known if this is true in general.

PROOF. By Corollary 6.2.7, the orders of a and α divide $\dim(H)$. Hence, the result follows from Corollary 8.9.7 (as the conjugations by a and by α commute). □

COROLLARY 8.9.11. *(Larson and Radford, [**LaR2**]) A semisimple Hopf algebra H over a field $\Bbbk$ of characteristic zero is cosemisimple.*

PROOF. We may assume that the ground field $\Bbbk$ is finitely generated, and then can embed $\Bbbk$ into the field of complex numbers. Thus, we may assume without loss of generality that $\Bbbk = \mathbb{C}$.

Let us compute the trace $\mathsf{Tr}|_H(S^2)$ of the squared antipode on H. To this end, let us write H as $\oplus_i \mathsf{End}(V_i)$, where V_i are the irreducible H-modules. The representations V_i which are nontrivially permuted by S^2 do not contribute to $\mathsf{Tr}|_H(S^2)$. So let us take i such that S^2 leaves V_i fixed, and hence acts on $\mathsf{End}(V_i)$. By Radford's theorem (Corollary 8.9.9), the squared antipode S^2 has a finite order on H. Hence, S^2 is induces an automorphism of $\mathsf{End}(V_i)$ of finite order. This automorphism must be inner (being an automorphism of the matrix algebra), and is implemented by an element $g_i \in GL(V_i)$, which can be chosen to have finite order. Let $\lambda_1, ..., \lambda_{d_i}$ be the eigenvalues of g_i ($d_i = \dim V_i$). They are roots of unity since g_i has finite order. We have $\mathsf{Tr}(S^2|_{\mathsf{End}(V_i)}) = \sum_{k,l} \lambda_k \lambda_l^{-1}$. But since λ_l are roots of unity, this equals $\sum_{k,l} \lambda_k \overline{\lambda}_l = |\sum_k \lambda_k|^2$. Thus, this trace is nonnegative. Moreover, if V_i is the trivial representation then the trace is 1. Thus, $\mathsf{Tr}|_H(S^2) \geq 1$. In particular, it is nonzero. By Corollary 7.18.10, this implies that H is cosemisimple, as desired. □

8.10. Ribbon monoidal categories

DEFINITION 8.10.1. A *twist* (or a *balancing transformation*) on a braided rigid monoidal category $\mathcal{C}$ is an element $\theta \in \mathsf{Aut}(\mathsf{id}_{\mathcal{C}})$ such that

$$\theta_{X\otimes Y} = (\theta_X \otimes \theta_Y) \circ c_{Y,X} \circ c_{X,Y} \tag{8.32}$$

for all $X, Y \in \mathcal{C}$. A twist is called *a ribbon structure* if $(\theta_X)^* = \theta_{X^*}$. A *ribbon* tensor category is a braided rigid monoidal category equipped with a ribbon structure.

EXAMPLE 8.10.2. As follows from Exercise 8.9.6(i), $\theta = u$ is a ribbon structure on the category of representations of the quantum double of a finite group G. Also, by Exercise 8.9.6(ii), the category $\mathcal{C}$ of type I representations of $U_q(\mathfrak{sl}_2)$ has a ribbon structure given by $\theta|_{V_N} = q^{-N(N+1)/2}$.

REMARK 8.10.3. The terminology in Definition 8.10.1 is justified as follows. Recall the braided monoidal category $\mathcal{T}$ of tangles introduced in Examples 2.3.14 and 8.2.4. This category has a generalization – the category $\mathcal{FT}$ of *framed* tangles in which intervals and circles are replaced by oriented "ribbons", which have to be glued preserving the orientation. This category has a natural twist, namely a straight ribbon with a double twist. Thus, it is a ribbon category.

Moreover, for any ribbon tensor category $\mathcal{C}$ and any object $X \in \mathcal{C}$, there is a natural uniquely determined monoidal functor $F : \mathcal{FT} \to \mathcal{C}$ which sends the generating object of $\mathcal{FT}$ to X and preserves the ribbon structure. Thus, the category of framed tangles is the "universal ribbon category".

Now note that $\mathrm{End}_{\mathcal{FT}}(\mathbf{1})$ is the set of isotopy classes of framed links. Therefore, the functor F in particular defines an invariant of framed links with values in $\mathsf{End}_{\mathcal{C}}(\mathbf{1}) = \Bbbk$. This invariant $RT_{\mathcal{C},X}(L)$ is called the *Reshetikhin-Turaev* invariant of L, and was introduced in [**RT1**]. In particular, if $\mathcal{C}$ is the representation category of the quantum group $U_q(\mathfrak{sl}_2)$ and $X = V_1$ is the 2-dimensional standard representation, then $RT_{\mathcal{C},X}(L)$ is essentially the Jones polynomial of L. More generally, if $X = V_n$ is the $n+1$-dimensional irreducible representation, then $RT_{\mathcal{C},X}(L)$ is the colored Jones polynomial (with all components of the link colored by V). Still more generally, one may color different components of the link by different objects of $\mathcal{C}$. Also, by renormalization, from these invariants it is easy to obtain invariants of ordinary (unframed) links. For more details about this, see the textbooks [**Kas, Tu4**].

REMARK 8.10.4. The notion of a ribbon structure can be understood as a "noncommutative" generalization of the notion of a quadratic form. Indeed, let G be a finite abelian group and $b : G \times G \to \Bbbk^\times$ be a bilinear form. As explained in Section 8.4, it defines a braiding on $\mathcal{C} = \mathsf{Vec}_G$. The corresponding quadratic form defines a ribbon structure on $\mathcal{C}$:

$$\theta_{\delta_x} = b(x, x)\, \mathsf{id}_{\delta_x}, \qquad x \in G.$$

From now on let $\mathcal{C}$ be a braided tensor category over a field $\Bbbk$.

LEMMA 8.10.5. *For any nonzero simple object X the composition*

$$f := \mathsf{ev}_X \circ c_{X,X^*} \circ \mathsf{coev}_X \in \mathsf{End}_{\mathcal{C}}(\mathbf{1}) \tag{8.33}$$

is nonzero.

PROOF. Since X is simple, the corresponding composition

$$\mathsf{End}_{\mathcal{C}}(\mathbf{1}) \to \mathsf{Hom}_{\mathcal{C}}(\mathbf{1}, X \otimes X^*) = \mathsf{End}_{\mathcal{C}}(X) \to \mathsf{End}_{\mathcal{C}}(\mathbf{1})$$

consists of nonzero maps between 1-dimensional spaces and hence is nonzero. So $f \neq 0$. □

PROPOSITION 8.10.6. *The natural transformation $u_X : X \to X^{**}$ defined in* (8.30) *is an isomorphism.*

PROOF. It suffices to prove this when X is simple. Define a natural transformation $v_X : X^{**} \to X$ by

$$X^{**} \xrightarrow{\mathsf{coev}_X \otimes \mathsf{id}_{X^{**}}} X \otimes X^* \otimes X^{**} \xrightarrow{\mathsf{id}_X \otimes c^{-1}_{X^{**},X^*}} X \otimes X^{**} \otimes X^* \xrightarrow{\mathsf{id}_X \otimes \mathsf{ev}_{X^*}} X. \tag{8.34}$$

Using the naturality of v, we compute

$$\begin{aligned} v_X \circ u_X &= (\mathsf{ev}_X \otimes \mathsf{id}_{X^{**}}) \circ (c_{X,X^*} \otimes \mathsf{id}_{X^{**}}) \circ (\mathsf{id}_{X \otimes X^*} \otimes v_X) \circ (\mathsf{id}_X \otimes \mathsf{coev}_{X^*}) \\ &= f' \otimes \mathsf{id}_X \otimes f, \end{aligned}$$

where f is the same as in (8.33) and $f' = \mathsf{ev}_{X^*} \circ c^{-1}_{X^*,X^{**}} \circ \mathsf{coev}_{X^*} \in \mathsf{End}_{\mathcal{C}}(\mathbf{1})$. By Lemma 8.10.5 f and f' are nonzero scalars and so $v_X \circ u_X$ is invertible. □

In view of Proposition 8.10.6, any natural isomorphism $\psi_X : X \simeq X^{**}$ in a braided tensor category $\mathcal{C}$ can be written as

$$\psi_X = u_X \theta_X, \quad \theta \in \mathsf{Aut}(\mathsf{id}_{\mathcal{C}}). \tag{8.35}$$

It follows from (8.31) that ψ is a tensor isomorphism (i.e., a pivotal structure on $\mathcal{C}$) if and only if θ is a twist.

The next Proposition relates the canonical natural isomorphism $u_X : X \xrightarrow{\sim} X^{**}$ with the natural isomorphism (7.64) from Section 7.19.

THEOREM 8.10.7. *Let $\mathcal{C}$ be a braided finite tensor category. Let D be the distinguished object of $\mathcal{C}$, let $\delta_X : X^{**} \to D \otimes {}^{**}X \otimes D^*$ be the tensor isomorphism constructed in Theorem 7.19.1, and let $u_X : X \to X^{**}$ be the isomorphism (8.30). Then there is an equality of natural isomorphisms*

$$(\mathsf{ev}_{D^*} \boxtimes \mathsf{id}_{^{**}X}) \circ (\mathsf{id}_D \otimes c_{^{**}X, D^*}) \circ \delta_X = u_{^{**}X}^{-1} \circ u_{^{*}X}^{*}, \tag{8.36}$$

where c denotes the braiding of $\mathcal{C}$.

PROOF. Let $\rho_X : (X \boxtimes \mathbf{1}) \otimes A \cong (\mathbf{1} \boxtimes X) \otimes A$ be as in Theorem 7.19.1. Recall that by definition,

$$(\delta_X \boxtimes \mathsf{id}_{\mathbf{1}}) \otimes \mathsf{id}_A = \rho_{D \otimes {}^{**}X \otimes D^*}^{-1} \circ (D^* \boxtimes \mathbf{1}) \otimes \rho_{D \otimes X \otimes D^*}^{**} \otimes (D \boxtimes \mathbf{1}). \tag{8.37}$$

Here and in the rest of this proof we identify $D \otimes D^*$ and $D^* \otimes D$ with $\mathbf{1}$ and suppress the evaluation and coevaluation maps for D and D^*. Let $\Sigma = c \boxtimes c$ denote the braiding on $\mathcal{C} \boxtimes \mathcal{C}^{\mathrm{op}}$. Define a natural isomorphism

$$\iota_X : (X \boxtimes \mathbf{1}) \otimes A \to (X^{**} \boxtimes \mathbf{1}) \otimes A \tag{8.38}$$

as the following composition:

$$\begin{aligned} \iota_X : (X \boxtimes \mathbf{1}) \otimes A \quad & \xrightarrow{\rho_X} \quad (\mathbf{1} \boxtimes X) \otimes A \xrightarrow{\Sigma_{\mathbf{1}\boxtimes X, A}} A \otimes (\mathbf{1} \boxtimes X) \\ & \xrightarrow{(D^* \boxtimes \mathbf{1}) \otimes \rho_{X^*}^*} A \otimes (X^{**} \boxtimes \mathbf{1}) \xrightarrow{\Sigma_{A, X^{**}\boxtimes \mathbf{1}}} (X^{**} \boxtimes \mathbf{1}) \otimes A. \end{aligned}$$

Observe that in terms of Hom spaces the isomorphism ι_X is given as the following sequence of natural isomorphisms:

$$\begin{aligned} \mathsf{Hom}_{\mathcal{C}\boxtimes\mathcal{C}^{\mathrm{op}}}(V_1 \boxtimes V_2, (X \boxtimes \mathbf{1}) \otimes A) & \cong \mathsf{Hom}_{\mathcal{C}\boxtimes\mathcal{C}^{\mathrm{op}}}(X^* \otimes V_1 \otimes V_2, \mathbf{1}) \\ & \cong \mathsf{Hom}_{\mathcal{C}\boxtimes\mathcal{C}^{\mathrm{op}}}(V_1 \otimes X^* \otimes V_2, \mathbf{1}) \\ & \cong \mathsf{Hom}_{\mathcal{C}\boxtimes\mathcal{C}^{\mathrm{op}}}(V_1 \otimes V_2 \otimes X^*, \mathbf{1}) \\ & \cong \mathsf{Hom}_{\mathcal{C}\boxtimes\mathcal{C}^{\mathrm{op}}}(V_1 \boxtimes V_2, (X^{**} \boxtimes \mathbf{1}) \otimes A) \end{aligned}$$

for all objects $V_1, V_2 \in \mathcal{C}$, where the two isomorphisms in the middle come from the braiding in $\mathcal{C}$. Namely, the composition of the first two arrows in (8.39) translates into the second isomorphism and the composition of the last two arrows in (8.39) translates into the third isomorphism. (One can easily check this using Remark 7.18.2 identifying $\mathcal{C} \boxtimes \mathcal{C}^{\mathrm{op}}$ with the category of right exact endofunctors of $\mathcal{C}$; the braiding isomorphisms in (8.39) are then identified with isomorphisms between left and right tensor multiplication by X).

On the other hand, the isomorphism $(u_X \boxtimes \mathsf{id}_{\mathbf{1}}) \otimes \mathsf{id}_A$ is given by

$$\begin{aligned} \mathsf{Hom}_{\mathcal{C}\boxtimes\mathcal{C}^{\mathrm{op}}}(V_1 \boxtimes V_2, (X \boxtimes \mathbf{1}) \otimes A) & \cong \mathsf{Hom}_{\mathcal{C}\boxtimes\mathcal{C}^{\mathrm{op}}}(X^* \otimes V_1 \otimes V_2, \mathbf{1}) \\ & \cong \mathsf{Hom}_{\mathcal{C}\boxtimes\mathcal{C}^{\mathrm{op}}}(V_1 \otimes V_2 \otimes X^*, \mathbf{1}) \\ & \cong \mathsf{Hom}_{\mathcal{C}\boxtimes\mathcal{C}^{\mathrm{op}}}(V_1 \boxtimes V_2, (X^{**} \boxtimes \mathbf{1}) \otimes A), \end{aligned}$$

therefore, the hexagon identity and Proposition 7.18.5(a) imply that

$$\iota_X = (u_X \boxtimes \mathsf{id}_{\mathbf{1}}) \otimes \mathsf{id}_A .$$

In particular, ι_X is an isomorphism in the category $\mathcal{H}$ of Hopf bimodules in $\mathcal{C}$, see Definition 7.18.1. The dual of ι_X in $\mathcal{H}$ can be found using isomorphism (7.59).

Using naturality of braiding, we compute:

$$\begin{aligned}
&\iota_{**X}^{-1} \circ \iota_{*X}^{*} = \\
&= \left(\rho_{**X}^{-1} \circ \Sigma_{1\boxtimes **X,A}^{-1} \circ (D^* \boxtimes \mathbf{1}) \otimes (\rho_{*X}^{*})^{-1} \circ \Sigma_{A,X\boxtimes\mathbf{1}}^{-1}\right) \\
&\quad \circ \Big(\Sigma_{A,X\boxtimes\mathbf{1}} \circ (D^* \boxtimes \mathbf{1}) \otimes \rho_{*X}^{*} \circ \Sigma_{\mathbf{1}\boxtimes **X,(\mathbf{1}\boxtimes D^*)\otimes A} \\
&\qquad \circ (D^* \boxtimes \mathbf{1}) \otimes \rho_{D\otimes X\otimes D^*}^{**} \otimes (D \boxtimes \mathbf{1})\Big) \\
&= \rho_{**X}^{-1} \circ (\mathbf{1} \boxtimes D) \otimes \Sigma_{\mathbf{1}\boxtimes **X,\mathbf{1}\boxtimes D^*} \otimes A \circ (D^* \boxtimes \mathbf{1}) \otimes \rho_{D\otimes X\otimes D^*}^{**} \otimes (D \boxtimes \mathbf{1}) \\
&= (c_{**X,D^*} \boxtimes \mathrm{id}_{\mathbf{1}}) \otimes A \circ \rho_{D\otimes **X\otimes D^*}^{-1} \circ (D^* \boxtimes \mathbf{1}) \otimes \rho_{D\otimes X\otimes D^*}^{**} \otimes (D \boxtimes \mathbf{1}).
\end{aligned}$$

Hence, comparing with (8.37) we get

$$\iota_{**X}^{-1} \circ \iota_{*X^*} = ((\mathrm{id}_D \otimes c_{**X,D^*}) \circ \delta_X \boxtimes \mathrm{id}_{\mathbf{1}}) \otimes \mathrm{id}_A .$$

On the other hand,

$$\iota_{**X}^{-1} \circ \iota_{*X}^{*} = (u_{**X}^{-1} \circ u_{*X}^{*} \boxtimes \mathrm{id}_{\mathbf{1}}) \otimes \mathrm{id}_A,$$

which implies the result. □

COROLLARY 8.10.8. *Let $\mathcal{C}$ be a finite braided tensor category with braiding c, and D be the distinguished invertible object of $\mathcal{C}$. Then for any $V \in \mathcal{C}$, we have $c_{V,D} \circ c_{D,V} = \mathrm{id}_{D\otimes V}$.*

PROOF. Observe that the right hand side of equation (8.36) does not change when the braiding c of $\mathcal{C}$ is replaced by its reverse c^{-1} (this is a consequence of the naturality of u and the identity $u_{c^{-1}} = (u_c^{-1})^*$). Hence, the left hand side of (8.36) also does not change. Since the definition of δ_X does not use braiding, we conclude that $c_{D,V} = c_{V,D}^{-1}$ for all objects V in $\mathcal{C}$, as claimed. □

REMARK 8.10.9. Here is another proof of Corollary 8.10.8. We have a canonical braided tensor functor $G : \mathcal{C}\boxtimes\mathcal{C}^{\mathrm{rev}} \to \mathcal{Z}(\mathcal{C})$. Equivalently, the functor $T : \mathcal{C}\boxtimes\mathcal{C}^{\mathrm{op}} \to \mathcal{C}$, $X \boxtimes Y \mapsto X \otimes Y$ has a natural structure of central tensor functor. Hence, by Proposition 8.8.8, the algebra $A := \underline{\mathrm{Hom}}_{\mathcal{C}\boxtimes\mathcal{C}^{\mathrm{op}}}(\mathbf{1}, \mathbf{1})$ is a commutative algebra in the category $\mathcal{C} \boxtimes \mathcal{C}^{\mathrm{rev}}$. Thus, A is a dyslectic A-module in the sense of Definition 3.12 in [**DaMNO**]. This implies that $A^* = (D \boxtimes \mathbf{1}) \otimes A$ is also dyslectic. Hence, by Lemma 3.15 in [**DaMNO**][4], $D \boxtimes \mathbf{1}$ centralizes A (i.e., the squared braiding on $(D\boxtimes\mathbf{1})\otimes A$ is the identity). Thus, $D\boxtimes\mathbf{1}$ centralizes $(\mathbf{1}\boxtimes P)\otimes A$ for any P; taking P projective and using Lemma 7.18.8, we see that D centralizes any projective object of $\mathcal{C}$. This implies the statement, since any object is a quotient of a projective one.

PROPOSITION 8.10.10. *If $\mathcal{C}$ is a factorizable finite tensor category then $\mathcal{C}$ is unimodular.*

PROOF. By Corollary 8.10.8, $c_{D,V} = c_{V,D}^{-1}$ for all $V \in \mathcal{C}$. Hence, D belongs to the intersection of images of $\mathcal{C}$ and $\mathcal{C}^{\mathrm{rev}}$ in $\mathcal{Z}(\mathcal{C})$, which is trivial since $\mathcal{C}$ is factorizable. Therefore $D = \mathbf{1}$. □

[4]This lemma is stated in [**DaMNO**] only for fusion categories, but its proof extends verbatim to general finite tensor categories.

REMARK 8.10.11. Proposition 8.10.10 appears as Proposition 4.5 in [**ENO1**], but the proof given there is not convincing, since it uses without proof the fact that the functor I commutes with duality (which is true but not obvious). Here we have given a different proof.

PROPOSITION 8.10.12. *Let $\mathcal{C}$ be a braided fusion category with a twist θ. The canonical pivotal structure $\psi = \theta \circ u$ is spherical if and only if θ is a ribbon structure.*

PROOF. The condition of $\theta_X^* = (\theta_X)^*$ translates to $\psi_{X^*} = (\psi_X \circ u_X^{-1})^* \circ u_{X^*}$ for all objects $X \in \mathcal{C}$. Using the naturality of u and Theorem 8.10.7 (recall that a fusion category is automatically unimodular, see Remark 6.5.9) we obtain

$$\psi_{X^*} = \psi_{X^{**}}^* \circ (u_{X^{**}}^{-1})^* \circ u_{X^*} = \psi_{X^{**}}^* \circ \delta_{X^{***}}^{-1}$$

for all objects X in $\mathcal{C}$. This is equivalent to the condition

$$\mathsf{Tr}(\psi_{X^*}) = \mathsf{Tr}(\psi_{X^{**}}^* \circ \delta_{X^{***}}^{-1}) \tag{8.39}$$

for every simple object X in $\mathcal{C}$. By Corollary 7.21.8 the last expression is equal to $\mathsf{Tr}(\psi_{X^{**}}^{**})$. Thus (8.39) is equivalent to $\mathsf{Tr}(\psi_{X^*}) = \mathsf{Tr}(\psi_{X^{**}}^{**}) = \mathsf{Tr}(\psi_{X^{**}})$, i.e., to ψ being spherical. □

Recall from Section 4.7 that in a spherical category there is a notion of a trace of an endomorphism $f \in \mathsf{End}_{\mathcal{C}}(X)$:

$$\mathsf{Tr}(f) : \mathbf{1} \xrightarrow{\mathsf{coev}_X} X \otimes X^* \xrightarrow{\psi_X \circ f \otimes \mathsf{id}_{X^*}} X^{**} \otimes X^* \xrightarrow{\mathsf{ev}_X} \mathbf{1}. \tag{8.40}$$

The dimension of $X \in \mathcal{C}$ is defined by $\dim(X) = \mathsf{Tr}(\mathsf{id}_X)$. Note that $\dim(X) \neq 0$ for every simple X.

The trace (8.40) is related to the left and right traces Tr^L and Tr^R from (4.8) and (4.9) by $\mathsf{Tr}(f) = \mathsf{Tr}^L(\psi_X f) = \mathsf{Tr}^R(f \psi_X^{-1})$.

REMARK 8.10.13. Note a difference between the Frobenius-Perron dimensions and the dimensions defined by formula (8.40). The former take values in $\mathbb{R}$, while the latter take values in $\Bbbk$. So these dimensions are not equal in general.

PROPOSITION 8.10.14. *Let $\mathcal{C}$ be a ribbon tensor category. For any object $X \in \mathcal{C}$ the composition*

$$\mathbf{1} \xrightarrow{\mathsf{coev}_X} X \otimes X^* \xrightarrow{\theta_X \otimes \mathsf{id}_{X^*}} X \otimes X^* \xrightarrow{c_{X,X^*}} X^* \otimes X \xrightarrow{\mathsf{ev}_X} \mathbf{1} \tag{8.41}$$

equals $\dim(X)$.

PROOF. It suffices to prove it for simple X. Consider the following diagram

$$\begin{array}{ccccccc}
X\otimes X^* & \xrightarrow{\mathsf{coev}_{X^*}} & X\otimes X^*\otimes X^{**}\otimes X^* & \xrightarrow{c_{X,X^*}} & X^*\otimes X\otimes X^{**}\otimes X^* & \xrightarrow{\mathsf{ev}_X} & X^{**}\otimes X^* \\
\uparrow{\scriptstyle \mathsf{coev}_X} & & \downarrow{\scriptstyle \mathsf{ev}_{X^*}} & & \downarrow{\scriptstyle \mathsf{ev}_{X^*}} & & \downarrow{\scriptstyle \mathsf{ev}_{X^*}} \\
\mathbf{1} & \xrightarrow{\mathsf{coev}_X} & X\otimes X^* & \xrightarrow{c_{X,X^*}} & X^*\otimes X & \xrightarrow{\mathsf{ev}_X} & \mathbf{1}.
\end{array}$$

It is easy to see that this diagram commutes. The composition through the top row is equal to

$$\mathsf{Tr}^L(u_X) = \theta_X^{-1}\,\mathsf{Tr}^L(\theta_X u_X) = \theta_X^{-1}\dim(X),$$

while the composition through the bottom row equals θ_X^{-1} times the right hand side of (8.41). □

EXERCISE 8.10.15. Let X be a simple object of a ribbon tensor category and $\theta_X \in \Bbbk^\times$ be the corresponding twist. Then

$$\theta_X^{-1} \dim(X) = \mathsf{Tr}(c_{X,X}^{-1}). \tag{8.42}$$

REMARK 8.10.16. By Exercise 4.7.16 ribbon structures on $\mathcal{C}$ form a torsor over the subgroup $\mathsf{Aut}_\otimes(\mathsf{id}_\mathcal{C})_2$ in $\mathsf{Aut}_\otimes(\mathsf{id}_\mathcal{C})$ of elements which have order ≤ 2 on simple objects. That is, if θ is a ribbon structure on $\mathcal{C}$ and $\psi \in \mathsf{Aut}_\otimes(\mathsf{id}_\mathcal{C})_2$ then $\theta \circ \psi$ is another ribbon structure, and every ribbon structure is of this form.

8.11. Ribbon Hopf algebras

Ribbon Hopf algebras are quasitriangular Hopf algebras which arise from ribbon categories. Namely, let H be a Hopf algebra, $\mathcal{C}$ be the category of H-modules, and assume that $\mathcal{C}$ is a ribbon monoidal category. Then H has the universal R-matrix $R \in H \otimes H$ coming from the braiding, and also the twist θ defines a central element $v \in H$. The axioms of a ribbon structure then translate into the following definition.

DEFINITION 8.11.1. A *ribbon Hopf algebra* is a triple (H, R, v), where (H, R) is a quasitriangular Hopf algebra, and $v \in H$ is an invertible central element satisfying

$$\Delta(v) = (v \otimes v)(R_{21}R)^{-1} \quad \text{and} \quad v = S(v).$$

Thus, we have the following proposition.

PROPOSITION 8.11.2. *Let (H, R, v) be a ribbon Hopf algebra. Then* Rep(H) *has a canonical structure of a ribbon category, where the balancing transformation θ is given by the action of v. Furthermore, there is a bijective correspondence between isomorphism classes of ribbon structures on a quasitriangular Hopf algebra (H, R) and equivalence classes of ribbon structures on the braided tensor category* Rep(H) *(where the braiding is the one given by R).*

PROOF. The proof is similar to the proof of Proposition 8.3.8. □

EXAMPLE 8.11.3. (i) Any triangular Hopf algebra has a ribbon structure with $v = 1$ as the ribbon element.

(ii) The quantum double $D(\Bbbk G)$ of the group algebra of a finite group G (in any characteristic) is a ribbon Hopf algebra with $v = u$ as the ribbon element.

(iii) Any semisimple cosemisimple quasitriangular Hopf algebra (H, R) has a ribbon structure with the ribbon element $v = u$ (we will prove later that in such a Hopf algebra $S^2 = \mathsf{id}$ and hence u is central).

EXERCISE 8.11.4. Show that $u_q(\mathfrak{sl}_2)$ (for q a root of unity of odd order) is a ribbon Hopf algebra.

8.12. Characterization of Morita equivalence

It follows from Proposition 8.5.3 that the centers of Morita equivalent finite tensor categories are braided equivalent. It turns out that the converse of this is also true, i.e., the center is a complete invariant of the categorical Morita equivalence class, see Theorem 8.12.3 below.

Let $\mathcal{C}$ be a finite tensor category and let $F_\mathcal{C} : \mathcal{Z}(\mathcal{C}) \to \mathcal{C}$ and $I_\mathcal{C} : \mathcal{C} \to \mathcal{Z}(\mathcal{C})$ denote the forgetful functor and its right adjoint. We can view $\mathcal{C}$ as an exact $\mathcal{Z}(\mathcal{C})$-module category via the functor $F_\mathcal{C}$.

LEMMA 8.12.1. *There is a natural isomorphism $I_\mathcal{C}(X) \cong \underline{\mathsf{Hom}}_{\mathcal{Z}(\mathcal{C})}(\mathbf{1}, X)$.*

PROOF. This is immediate from the definition (7.21) of internal Hom. Indeed, the action of $Z \in \mathcal{Z}(\mathcal{C})$ on $\mathcal{C}$ satisfies $Z \otimes \mathbf{1} = F(Z)$ and, hence,

$$\begin{aligned}\mathsf{Hom}_{\mathcal{Z}(\mathcal{C})}(Z, I(X)) &\cong \mathsf{Hom}_{\mathcal{C}}(F(Z), X)\\ &\cong \mathsf{Hom}_{\mathcal{C}}(Z \otimes \mathbf{1}, X)\\ &\cong \mathsf{Hom}_{\mathcal{Z}(\mathcal{C})}(Z, \underline{\mathsf{Hom}}_{\mathcal{Z}(\mathcal{C})}(\mathbf{1}, X))\end{aligned}$$

and the statement follows by the Yoneda Lemma. □

LEMMA 8.12.2. (i) *The object $A := I_{\mathcal{C}}(\mathbf{1})$ has a natural structure of commutative algebra in $\mathcal{Z}(\mathcal{C})$; moreover for any $X \in \mathcal{C}$ the object $I_{\mathcal{C}}(X)$ has a natural structure of a right A-module.*
(ii) *The functor $I_{\mathcal{C}}$ induces an equivalence of tensor categories $\mathcal{C} \simeq \mathsf{Mod}_{\mathcal{Z}(\mathcal{C})}(A)$.*

PROOF. That A is a commutative algebra in $\mathcal{Z}(\mathcal{C})$ follows from Proposition 8.8.8. By Theorem 7.10.1 for any $X \in \mathcal{C}$ the object $I_{\mathcal{C}}(X) = \underline{\mathsf{Hom}}(\mathbf{1}, X)$ is naturally a right A-module, and the functor

$$I_{\mathcal{C}} = \underline{\mathsf{Hom}}(\mathbf{1}, -) : \mathcal{C} \to \mathsf{Mod}_{\mathcal{Z}(\mathcal{C})}(A) \tag{8.43}$$

is an equivalence of categories.

It remains to prove that the functor $I_{\mathcal{C}}$ has a structure of a tensor functor.

For any $X \in \mathcal{C}$ let $\mu_X : F_{\mathcal{C}}(I_{\mathcal{C}}(X)) \to X$ be the image of id under the canonical isomorphism $\mathsf{Hom}_{\mathcal{Z}(\mathcal{C})}(I_{\mathcal{C}}(X), I_{\mathcal{C}}(X)) \simeq \mathsf{Hom}_{\mathcal{C}}(F_{\mathcal{C}}(I_{\mathcal{C}}(X)), X)$ (so we have $\mu_{\mathbf{1}} = \mu$ in the notation used above) and for $X, Y \in \mathcal{C}$ let $\mu_{X,Y} : I_{\mathcal{C}}(X) \otimes I_{\mathcal{C}}(Y) \to I_{\mathcal{C}}(X \otimes Y)$ be the image of $\mu_X \otimes \mu_Y$ under the canonical isomorphism

$$\mathsf{Hom}(F_{\mathcal{C}}(I_{\mathcal{C}}(X)) \otimes F_{\mathcal{C}}(I_{\mathcal{C}}(Y)), X \otimes Y) \simeq \mathsf{Hom}(I_{\mathcal{C}}(X) \otimes I_{\mathcal{C}}(Y), I_{\mathcal{C}}(X \otimes Y)).$$

(In the notation above $\mu_{\mathbf{1},\mathbf{1}} = m$ is the multiplication morphism on $A = I_{\mathcal{C}}(\mathbf{1})$ and $\mu_{X,\mathbf{1}}$ is the morphism making $I_{\mathcal{C}}(X)$ into a right A-module). It is straightforward to verify that $\mu_{X,Y}$ satisfies all the axioms of a tensor functor except for being an isomorphism. In particular, the morphism $\mu_{X,\mathbf{1}}$ makes $I_{\mathcal{C}}(X)$ into a right A-module; moreover, $\mu_{\mathbf{1},X}$ and $\mu_{X,\mathbf{1}}$ make $I_{\mathcal{C}}(X)$ into an A-bimodule. Next, $\mu_{\mathbf{1},X}$ is given by the composition

$$A \otimes I_{\mathcal{C}}(X) \xrightarrow{c^{-1}_{I_{\mathcal{C}}(X),A}} I_{\mathcal{C}}(X) \otimes A \xrightarrow{\mu_{X,\mathbf{1}}} I_{\mathcal{C}}(X),$$

so the structure of $I_{\mathcal{C}}(X)$ as an A-bimodule is the same as the structure used in the definition of the tensor structure on $\mathsf{Mod}_{\mathcal{Z}(\mathcal{C})}(A)$.

It is immediate to check that $\mu_{X,Y}$ admits a canonical factorization

$$I_{\mathcal{C}}(X) \otimes I_{\mathcal{C}}(Y) \to I_{\mathcal{C}}(X) \otimes_A I_{\mathcal{C}}(Y) \xrightarrow{\tilde{\mu}_{X,Y}} I_{\mathcal{C}}(X \otimes Y)$$

and that $\tilde{\mu}_{X,Y}$ satisfies all the axioms of a tensor functor with the possible exception of being an isomorphism. Finally, one verifies that for $X = F_{\mathcal{C}}(Z)$ with $Z \in \mathcal{Z}(\mathcal{C})$ we have $I_{\mathcal{C}}(X) \cong Z \otimes A$ (as A-modules), and this isomorphism identifies $\tilde{\mu}_{X,Y}$ with the canonical isomorphism $I_{\mathcal{C}}(X) \otimes_A I_{\mathcal{C}}(Y) = Z \otimes I_{\mathcal{C}}(Y) \simeq I_{\mathcal{C}}(F_{\mathcal{C}}(Z) \otimes Y)$ (the last equality is a special case of Lemma 7.9.4). Since the functor $F_{\mathcal{C}}$ is surjective (see Corollary 7.13.11), we get that $\tilde{\mu}_{X,Y}$ is always an isomorphism. Thus the isomorphisms $\tilde{\mu}_{X,Y}$ define a tensor structure on the functor $I_{\mathcal{C}}$, and the Lemma is proved. □

THEOREM 8.12.3. *Two finite tensor categories $\mathcal{C}$ and $\mathcal{D}$ are Morita equivalent if and only if $\mathcal{Z}(\mathcal{C})$ and $\mathcal{Z}(\mathcal{D})$ are equivalent as braided tensor categories.*

PROOF. The "only if" direction is established in Proposition 8.5.3.

Let $\mathcal{C}$, $\mathcal{D}$ be finite tensor categories such that there is a braided tensor equivalence $a : \mathcal{Z}(\mathcal{C}) \xrightarrow{\sim} \mathcal{Z}(\mathcal{D})$. Since $I_{\mathcal{D}}(\mathbf{1})$ is a commutative algebra in $\mathcal{Z}(\mathcal{D})$ and a preserves commutativity of algebras, $L := a^{-1}(I_{\mathcal{D}}(\mathbf{1}))$ is a commutative algebra in $\mathcal{Z}(\mathcal{C})$. Furthermore,

$$\mathcal{D} \cong \mathsf{Mod}_{\mathcal{Z}(\mathcal{C})}(L) \tag{8.44}$$

as a tensor category by Lemma 8.12.2.

We claim that the category $\mathsf{Mod}_{\mathcal{C}}(L)$ of L-modules in $\mathcal{C}$ is exact.[5] To prove this, observe the following equivalence of $\mathcal{C}$-module categories:

$$\mathsf{Fun}_{\mathcal{Z}(\mathcal{C})}(\mathsf{Mod}_{\mathcal{Z}(\mathcal{C})}(L),\, \mathcal{C}) \cong \mathsf{Mod}_{\mathcal{C}}(L).$$

Indeed, both categories are equivalent to the category of $(L - I_{\mathcal{C}}(\mathbf{1}))$-bimodules in $\mathcal{Z}(\mathcal{C})$.

By Theorem 7.12.16, the category $\mathsf{Fun}_{\mathcal{Z}(\mathcal{C})}(\mathsf{Mod}_{\mathcal{Z}(\mathcal{C})}(L),\, \mathcal{C})$ is an exact module category over $\mathsf{Fun}_{\mathcal{Z}(\mathcal{C})}(\mathcal{C},\, \mathcal{C})$. In particular, it is exact as a $\mathcal{C}$-module category (since $\mathsf{Fun}_{\mathcal{Z}(\mathcal{C})}(\mathcal{C},\, \mathcal{C}) \cong \mathcal{C} \boxtimes \mathcal{C}^{\mathrm{op}}$).

Note that the algebra L is indecomposable in $\mathcal{Z}(\mathcal{C})$ but L might be decomposable as an algebra in $\mathcal{C}$, i.e., the category $\mathsf{Bimod}_{\mathcal{C}}(L)$ is a multitensor category. It has a decomposition

$$\mathsf{Bimod}_{\mathcal{C}}(L) = \bigoplus_{i,j \in J} \mathsf{Bimod}_{\mathcal{C}}(L)_{ij},$$

where J is a finite set and each $\mathsf{Bimod}_{\mathcal{C}}(L)_{ii}$ is a finite tensor category. Let

$$L = \bigoplus_{i \in J} L_i,$$

be the decomposition of L such that $\mathsf{Bimod}_{\mathcal{C}}(L_i) \simeq \mathsf{Bimod}_{\mathcal{C}}(L)_{ii}$. Here L_i, $i \in J$, are indecomposable algebras in $\mathcal{C}$ such that the multiplication of L is zero on $L_i \otimes L_j$, $i \neq j$ (e.g., if $\mathcal{C} = \mathsf{Rep}(G)$ then $L = \mathsf{Fun}(G, k)$ with the adjoint action of G and J is the set of conjugacy classes of G). Note that $\mathcal{C}$-module categories $\mathsf{Mod}_{\mathcal{C}}(L_i)$, $i \in J$ are exact.

We would like to show that for each $i \in J$

$$\mathsf{Mod}_{\mathcal{Z}(\mathcal{C})}(L) \cong \mathsf{Bimod}_{\mathcal{C}}(L_i). \tag{8.45}$$

In view of (8.44) this will mean that $\mathcal{D}$ is dual to $\mathcal{C}$ with respect to the $\mathcal{C}$-module category $\mathsf{Mod}_{\mathcal{C}}(L_i)$ for any $i \in J$.

Consider the following commutative diagram of tensor functors:

$$\begin{array}{ccc}
\mathcal{Z}(\mathcal{C}) & \xrightarrow{\quad Z \mapsto Z \otimes L_i \quad} & \mathcal{Z}(\mathsf{Bimod}_{\mathcal{C}}(L_i)) \\
\Big\downarrow {\scriptstyle Z \mapsto Z \otimes L} & & \Big\downarrow {\scriptstyle F_{\mathsf{Bimod}_{\mathcal{C}}(L_i)}} \\
\mathsf{Mod}_{\mathcal{Z}(\mathcal{C})}(L) \xrightarrow{F_{\mathcal{C}}} \bigoplus_{i \in J} \mathsf{Bimod}_{\mathcal{C}}(L_i) & \subset \mathsf{Bimod}_{\mathcal{C}}(L) & \xrightarrow{\pi_i} \mathsf{Bimod}_{\mathcal{C}}(L_i).
\end{array}$$

---

[5]Here and below we abuse notation and write L both for an object of $\mathcal{Z}(\mathcal{C})$ and for its forgetful image in $\mathcal{C}$.

Here π_i is a projection from $\mathsf{Bimod}_{\mathcal{C}}(L) = \oplus_{ij} \mathsf{Bimod}_{\mathcal{C}}(L_i, L_j)$ to its (i, i) component. We have $\pi_i(X \otimes L) = X \otimes L_i$ for all $X \in \mathcal{C}$. The top arrow is the equivalence (7.52) and the forgetful functor $\mathcal{Z}(\mathsf{Bimod}_{\mathcal{C}}(L_i)) \to \mathsf{Bimod}_{\mathcal{C}}(L_i)$ (the right down arrow) is surjective. Hence, the composition $F_i := \pi_i F_{\mathcal{C}}$ of the functors in the bottom row is surjective. But F_i is a tensor functor between fusion categories of equal Frobenius-Perron dimension (thanks to Theorem 7.16.6 and Corollary 7.16.7) and hence it is an equivalence by Proposition 6.3.4. □

8.13. The S-matrix of a pre-modular category

From now on we assume that $\Bbbk$ is an algebraically closed field of characteristic 0.

DEFINITION 8.13.1. A *pre-modular* category is a ribbon fusion category (equivalently, a braided fusion category equipped with a spherical structure).

Let $\mathcal{C}$ be a pre-modular category with a spherical structure ψ. Let $\mathcal{O}(\mathcal{C})$ denote the set of (isomorphism classes of) simple objects of $\mathcal{C}$. For all $X, Y, Z \in \mathcal{O}(\mathcal{C})$ let N_{XY}^Z denote the multiplicity of Z in $X \otimes Y$.

We will identify the corresponding twist $\theta \in \mathsf{Aut}(\mathsf{id}_{\mathcal{C}})$ with a collection of scalars $\theta_X \in \Bbbk^\times$, $X \in \mathcal{O}(\mathcal{C})$. Let Tr and d denote the trace and dimension corresponding to ψ.

DEFINITION 8.13.2. Let $\mathcal{C}$ be a pre-modular category. The *S-matrix* of $\mathcal{C}$ is defined by

$$S := (s_{XY})_{X,Y \in \mathcal{O}(\mathcal{C})} \quad \text{where} \quad s_{XY} = \mathsf{Tr}(c_{Y,X} c_{X,Y}). \tag{8.46}$$

REMARK 8.13.3.
(i) The S-matrix of $\mathcal{C}$ is a symmetric n-by-n matrix where $n = |\mathcal{O}(\mathcal{C})|$ is the number of simple objects of $\mathcal{C}$. It satisfies $s_{X^*Y^*} = s_{XY}$ for all $X, Y \in \mathcal{O}(\mathcal{C})$. Also $s_{X\mathbf{1}} = s_{\mathbf{1}X} = \dim(X)$.
(ii) The S-matrix of $\mathcal{C}$ depends on the choice of the isomorphism $\psi : X \xrightarrow{\sim} X^{**}$ (which is chosen to be a spherical structure in (8.46)). A canonical alternative to this is to use u from (8.30) instead of ψ in (8.46). This would result in replacing s_{XY} by $\theta_X^{-1}\theta_Y^{-1}s_{XY}$.

DEFINITION 8.13.4. A pre-modular category $\mathcal{C}$ is said to be *modular* if its S-matrix is non-degenerate.

EXAMPLE 8.13.5. Let G be a finite abelian group. Let $q : G \to \Bbbk^\times$ be a quadratic form on G and let $b : G \times G \to \Bbbk^\times$ be the associated symmetric bilinear form. The S-matrix of the corresponding pointed pre-modular category $\mathcal{C}(G, q)$ (see Section 8.4) is $\{b(g, h)\}_{g,h \in G}$. Thus, $\mathcal{C}(G, q)$ is modular if and only if q is non-degenerate.

EXAMPLE 8.13.6. Let G be a finite group. Then $\mathcal{Z}(\mathsf{Vec}_G)$ is a pre-modular fusion category (we will see in Corollary 8.20.14 that it is, in fact, modular). Recall from Example 8.5.4 that simple objects of $\mathcal{Z}(\mathsf{Vec}_G)$ are parametrized by pairs (C, V), where C is a conjugacy class in G and V is an irreducible representation of the centralizer $C_G(a)$ of $a \in G$.

The S-matrix and twist of $\mathcal{Z}(\mathsf{Vec}_G)$ are given as follows:

$$
\begin{aligned}
s_{(C,V),(C',V')} &= \frac{|G|}{|C_G(a)||C_G(a')|} \sum_{g \in G(a,\,a')} \mathsf{Tr}_V(ga'g^{-1})\, \mathsf{Tr}_{V'}(g^{-1}ag), \\
\theta_{(C,V)} &= \frac{\mathsf{Tr}_V(a)}{\dim_{\Bbbk}(V)},
\end{aligned} \tag{8.47}
$$

where $a \in C$, $a' \in C'$, and $G(a,\, a') = \{g \in G \mid aga'g^{-1} = ga'g^{-1}a\}$.

EXERCISE 8.13.7. Prove formulas (8.47).

Let $\mathcal{C}$ be a pre-modular category.

PROPOSITION 8.13.8. *We have*

$$
s_{XY} = \theta_X^{-1}\theta_Y^{-1} \sum_{Z \in \mathcal{O}(\mathcal{C})} N_{XY}^Z \theta_Z \dim(Z). \tag{8.48}
$$

for all $X, Y \in \mathcal{O}(\mathcal{C})$.

PROOF. Apply Tr to both sides of formula (8.32). The right hand side is equal to $\theta_X\theta_Y s_{XY}$ while the left hand side is equal to

$$
\begin{aligned}
\mathsf{Tr}(\theta_{X\otimes Y}) &= \sum_{Z \in \mathcal{O}(\mathcal{C})} N_{XY}^Z\, \mathsf{Tr}(\theta_Z) \\
&= \sum_{Z \in \mathcal{O}(\mathcal{C})} N_{XY}^Z \theta_Z \dim(Z),
\end{aligned}
$$

where we used additivity of Tr. □

REMARK 8.13.9. Relation (8.48) between the twist and the S-matrix of a pre-modular category generalizes relation (8.9) between the quadratic form and the associated bilinear form. Indeed, for $\mathcal{C} = \mathcal{C}(G,\, q)$ equation (8.48) reduces to (8.9).

Assume $\mathcal{C}$ is strict. For any morphism $f : X \otimes Y \to X \otimes Y$ we can define the morphisms

$$
\mathsf{id}_X \otimes \mathsf{Tr}(f) : X \xrightarrow{\mathrm{coev}_Y} X \otimes Y \otimes Y^* \xrightarrow{(\mathsf{id}_X \otimes \psi_Y)(f \otimes \mathsf{id}_{Y^*})} X \otimes Y^{**} \otimes Y^* \xrightarrow{\mathrm{ev}_{Y^*}} X,
$$

$$
\mathsf{Tr} \otimes \mathsf{id}_Y(f) : Y \xrightarrow{\mathrm{coev}_{X^*}} X^* \otimes X^{**} \otimes Y \xrightarrow{(id_{X^*} \otimes f)(\psi_X^{-1} \otimes \mathsf{id}_Y)} X^* \otimes X \otimes Y \xrightarrow{\mathrm{ev}_X} Y.
$$

We thus can talk about "applying trace to factors of morphisms between tensor products". Note that $\mathsf{Tr}(\mathsf{Tr} \otimes \mathsf{id}_Y)(f) = \mathsf{Tr}(\mathsf{id}_X \otimes \mathsf{Tr})(f) = \mathsf{Tr}(f)$.

PROPOSITION 8.13.10. *The elements of the S-matrix satisfy the following relation:*

$$
s_{XY}s_{XZ} = \dim(X) \sum_{W \in \mathcal{O}(\mathcal{C})} N_{YZ}^W s_{XW}, \quad X, Y, Z \in \mathcal{O}(\mathcal{C}). \tag{8.49}
$$

PROOF. Consider the equality

$$
(c_{Y,X} \otimes \mathsf{id}_Z) \circ (\mathsf{id}_Y \otimes c_{Z,X}c_{X,Z}) \circ (c_{X,Y} \otimes \mathsf{id}_Z) = c_{Y\otimes Z,X} \circ c_{X,Y\otimes Z}, \qquad X,\, Y,\, Z \in \mathcal{O}(\mathcal{C}). \tag{8.50}
$$

We will compute the trace of both sides of (8.50). Applying trace to the third factor on the left hand side, we get $\dim(X)^{-1} s_{XZ} c_{Y,X} c_{X,Y}$. Thus, the trace of the left hand side of (8.50) is $\dim(X)^{-1} s_{XY} s_{XZ}$.

We use additivity of trace to compute the trace of the right hand side of (8.50):

$$\begin{aligned}\mathsf{Tr}(c_{Y\otimes Z,X}c_{X,Y\otimes Z}) &= \sum_{W\in\mathcal{O}(\mathcal{C})} N_{YZ}^{W}\,\mathsf{Tr}(c_{W,X}\circ c_{X,W}) \\ &= \sum_{W\in\mathcal{O}(\mathcal{C})} N_{YZ}^{W} s_{XW},\end{aligned}$$

which immediately implies the result. □

Thus, we have the following proposition.

PROPOSITION 8.13.11. *(i) For any fixed $X \in \mathcal{O}(\mathcal{C})$ the map*

$$h_X : Y \mapsto \frac{s_{XY}}{\dim(X)}, \quad Y \in \mathcal{O}(\mathcal{C}) \tag{8.51}$$

defines a homomorphism $K_0(\mathcal{C}) \to \Bbbk$. That is, simple objects of $\mathcal{C}$ give rise to characters of the Grothendieck ring $K_0(\mathcal{C})$.

(ii) The numbers $\frac{s_{XY}}{\dim(X)}$ are algebraic integers.

PROOF. Statement (i) follows from formula (8.49).

Statement (ii) follows since by (8.49) $h_X(Y)$ for various X are eigenvalues of the integer matrix $N_Y = \{N_{YZ}^X\})$. □

Note that in the case $X = \mathbf{1}$, this proposition says that the assignment $Y \mapsto \dim(Y)$ is a character of the Grothendieck ring $K_0(\mathcal{C})$, and its values are algebraic integers. This is also clear directly from the definition of $\dim(Y)$.

REMARK 8.13.12. Formula (8.49) and its proof remain valid in the case when the category $\mathcal{C}$ is a ribbon tensor category which is not necessarily finite or semisimple. In this case, it is possible that $\dim(X) = 0$; then formula (8.49) implies (by taking $Y = Z$) that $s_{XY} = 0$ for all Y. If $\dim(X) \neq 0$, then Proposition 8.13.11(i) holds for X with the same proof, and part (ii) holds for finite $\mathcal{C}$.

In fact, in the nonsemisimple case this construction gives us more: the formula $Y \mapsto (\mathsf{id}\otimes\mathsf{Tr})(c_{Y,-}\circ c_{-,Y})$ defines a homomorphism $h : K_0(\mathcal{C}) \to \mathrm{End}(\mathrm{Id}_{\mathcal{C}})$ ($Y \mapsto h(Y)$). The value $h_X(Y)$ of $h(Y)$ on a simple object X satisfies $\dim(X)h_X(Y) = s_{XY}$, including the case when both $\dim(X)$ and s_{XY} are zero. If $\mathcal{C}$ is finite, then $h_X(Y)$ is an algebraic integer (as the value of a character of $K_0(\mathcal{C})$).

EXERCISE 8.13.13. Let (H, R, v) be a ribbon Hopf algebra, and $\gamma = uv^{-1}$ be the corresponding grouplike element. For each $Y \in \mathsf{Rep}(H)$, let

$$C_Y = (\mathsf{id}\otimes\mathsf{Tr}_Y)(R^{21}R(1\otimes\gamma)).$$

(i) Show that C_Y is central, and the assignment $Y \mapsto C_Y$ is a homomorphism $K_0(\mathcal{C}) \to Z(H)$, where $Z(H)$ is the center of H. (This map is called the Drinfeld-Reshetikhin map, [**Dr5, Res**]). Show that $C_Y|_X = h_X(Y)$.

(ii) Calculate explicitly (in terms of the generators E, F, K) the *Casimir element* $C_V \in U_q(\mathfrak{sl}_2)$, where $V = V_1$ is the 2-dimensional irreducible representation, and explain how it degenerates into the usual Casimir element of $\mathfrak{sl}_2$ when $q \to 1$. Calculate the value of C_V on the irreducible $N+1$-dimensional representation V_N. Compute $C_{V_N}|_{V_M}$. Show that if q is not a root of unity then the center of $U_q(\mathfrak{sl}_2)$ is the algebra of polynomials in C_V, and express C_{V_N} as a polynomial of C_V.

8.14. Modular categories

Recall that a pre-modular category $\mathcal{C}$ is modular if its S-matrix is non-degenerate.

LEMMA 8.14.1. *Let A be a fusion ring with $\mathbb{Z}_+$-basis B, and let χ_1, χ_2 be distinct characters $A \to \Bbbk$. Then*

$$\sum_{X \in B} \chi_1(X)\chi_2(X^*) = 0. \tag{8.52}$$

PROOF. Let z denote the left hand side. Set $P := \sum_{X \in B} \chi_1(X)X^* \in A \otimes_{\mathbb{Z}} \Bbbk$. Clearly, $P \neq 0$. From equation (3.3) we see that that $YP = \chi_1(Y)P$ for all $Y \in A$. Applying χ_2 to this equation, we get $\chi_2(Y)z = \chi_1(Y)z$. Since $\chi_1 \neq \chi_2$, we must have $z = 0$. □

The categorical dimension of a fusion category $\mathcal{C}$ was defined in (7.71). When $\mathcal{C}$ is pre-modular, we have

$$\dim(\mathcal{C}) = \sum_{X \in \mathcal{O}(\mathcal{C})} \dim(X)^2, \tag{8.53}$$

where $\dim(x)$ is the dimension associated to the spherical structure of $\mathcal{C}$.

Let $E = \{E_{XY}\}_{X,Y \in \mathcal{O}(\mathcal{C})}$ be the square matrix such that $E_{XY} = 1$ if $X = Y^*$ and $E_{XY} = 0$ otherwise.

PROPOSITION 8.14.2. *Let $\mathcal{C}$ be a modular category and S be its S-matrix. Then $S^2 = \dim(\mathcal{C})E$. In other words, $S^{-1} = \{\dim(\mathcal{C})^{-1}s_{XY^*}\}$.*

PROOF. Since $\mathcal{C}$ is modular, the equality $h_Y = h_Z$ for $Y, Z \in \mathcal{O}(\mathcal{C})$ holds if and only if $Y = Z$, where $h_Y : K_0(\mathcal{C}) \to \Bbbk$ are the characters defined in (8.51).

Suppose $Y \neq Z$. By Lemma 8.14.1, we have

$$\sum_{X \in \mathcal{O}(\mathcal{C})} s_{XY}s_{XZ^*} = \sum_{X \in \mathcal{O}(\mathcal{C})} s_{XY}s_{X^*Z} = \dim(Y) \cdot \dim(Z) \sum_{X \in \mathcal{O}(\mathcal{C})} h_Y(X)h_Z(X^*) = 0.$$

It remains to check that $\sum_{X \in \mathcal{O}(\mathcal{C})} s_{XY}s_{XY^*} = \dim(\mathcal{C})$ for all $Y \in \mathcal{O}(\mathcal{C})$. We compute

$$\begin{aligned} \sum_{X \in \mathcal{O}(\mathcal{C})} s_{XY}s_{XY^*} &= \sum_{X \in \mathcal{O}(\mathcal{C})} \sum_{W \in \mathcal{O}(\mathcal{C})} \dim(X)s_{XW}N_{YY^*}^W \\ &= \dim(\mathcal{C})N_{YY^*}^{\mathbf{1}} = \dim(\mathcal{C}). \end{aligned}$$

Here the first equality is (8.49). The second equality is a consequence of Lemma 8.14.1, since

$$\sum_{X \in \mathcal{O}(\mathcal{C})} \dim(X)s_{XW} = \dim(W) \sum_{X \in \mathcal{O}(\mathcal{C})} \dim(X^*)h_W(X),$$

and the latter expression is equal to $\dim(\mathcal{C})$ if $W = \mathbf{1}$ and 0 otherwise. □

REMARK 8.14.3. Proposition 8.14.2 provides an alternative proof of Theorem 7.21.12 (saying that $\dim(\mathcal{C}) \neq 0$) for modular categories.

COROLLARY 8.14.4. ***(Verlinde formula).*** *Let $\mathcal{C}$ be a modular category. For all objects $Y, Z, W \in \mathcal{O}(\mathcal{C})$ we have*

$$\sum_{X \in \mathcal{O}(\mathcal{C})} \frac{s_{XY}s_{XZ}s_{XW^*}}{\dim(X)} = \dim(\mathcal{C})N_{YZ}^W. \tag{8.54}$$

Thus, the S-matrix determines the fusion rules of $\mathcal{C}$ (see Section 4.5).

PROOF. This follows immediately from Proposition 8.14.2 and formula (8.49). □

For any $Z \in \mathcal{O}(\mathcal{C})$ define the following square matrices:

$$D^Z := \left(\delta_{XY}\frac{s_{XZ}}{\dim(X)}\right)_{X,Y\in\mathcal{O}(\mathcal{C})} \qquad \text{and} \qquad N^Z = \left(N_{XY}^Z\right)_{X,Y\in\mathcal{O}(\mathcal{C})}.$$

COROLLARY 8.14.5. *Let $\mathcal{C}$ be a modular category with the S-matrix S. Then $D^Z = S^{-1}N^Z S$ for all $Z \in \mathcal{O}(\mathcal{C})$, i.e., conjugation by the S-matrix diagonalizes the fusion rules of $\mathcal{C}$.*

PROPOSITION 8.14.6. *Let $\mathcal{C}$ be a modular category and let $X \in \mathcal{O}(\mathcal{C})$. Then $\frac{\dim(\mathcal{C})}{\dim(X)^2}$ is an algebraic integer.*

PROOF. We compute, using Proposition 8.14.2:

$$\frac{\dim(\mathcal{C})}{\dim(X)^2} = \sum_{Y\in\mathcal{O}(\mathcal{C})}\frac{s_{XY}}{\dim(X)}\frac{s_{XY^*}}{\dim(X)} = \sum_{Y\in\mathcal{O}(\mathcal{C})} h_Y(X)h_{Y^*}(X), \tag{8.55}$$

where h_Y, $Y \in \mathcal{O}(\mathcal{C})$, are characters of $K_0(\mathcal{C})$ defined in (8.51). Since $h_Y(X)$ is an eigenvalue of the integer matrix N^X, it is an algebraic integer. Hence, the right hand side of (8.55) is an algebraic integer. □

Next, we discuss Galois properties of the S-matrix of a modular fusion category $\mathcal{C}$ over $\Bbbk = \mathbb{C}$. The following remarkable result due to de Boere, Goeree, Coste, and Gannon states that the entries of the S-matrix of $\mathcal{C}$ lie in a cyclotomic field, see [**deBG, CosG**].

THEOREM 8.14.7. *There exists a root of unity $\xi \in \Bbbk$ such that $s_{XY} \in \mathbb{Q}(\xi)$.*

PROOF. By Proposition 8.13.11(i) and Corollary 8.14.5, any homomorphism $K_O(\mathcal{C}) \to \Bbbk$ is of the form $Y \mapsto s_{XY}/\dim(X)$ for some well defined $X \in O(\mathcal{C})$. Hence for any automorphism g of $\Bbbk$ one has $g(s_{XY}/\dim(X)) = s_{g(X)Y}/\dim(g(X))$ for a well defined action of g on $\mathcal{O}(\mathcal{C})$.

Applying the automorphism g to equation (8.55) we compute:

$$\begin{aligned} g\left(\frac{\dim(\mathcal{C})}{\dim(X)^2}\right) &= g\left(\sum_{Y\in\mathcal{O}(\mathcal{C})}\frac{s_{XY}}{\dim(X)}\frac{s_{X^*Y}}{\dim(X)}\right)\\ &= \sum_{Y\in\mathcal{O}(\mathcal{C})}\frac{s_{g(X)Y}}{\dim(g(X))}\frac{s_{g(X^*)Y}}{\dim(g(X))}\\ &= \frac{\delta_{g(X)^*,g(X^*)}\dim(\mathcal{C})}{\dim(g(X))^2}.\end{aligned}$$

It follows that $g(X^*) = g(X)^*$ and that $g(\frac{\dim(X)^2}{\dim(\mathcal{C})}) = \frac{\dim(g(X))^2}{\dim(\mathcal{C})}$ for all $X \in \mathcal{O}(\mathcal{C})$. Hence,

$$\begin{aligned} g\left(\frac{s_{XY}^2}{\dim(\mathcal{C})}\right) &= g\left(\left(\frac{s_{XY}}{\dim(Y)}\right)^2\frac{\dim(Y)^2}{\dim(\mathcal{C})}\right)\\ &= \left(\frac{s_{Xg(Y)}}{\dim(g(Y))}\right)^2\frac{\dim(g(Y))^2}{\dim(\mathcal{C})} = \frac{s_{Xg(Y)}^2}{\dim(\mathcal{C})}.\end{aligned}$$

Let $\overline{s}_{XY} := s_{XY}/\sqrt{\dim(\mathcal{C})}$, $X, Y \in \mathcal{O}(\mathcal{C})$, be the normalized entries of the S-matrix. The extension of $\mathbb{Q}$ generated by all entries s_{XY}, $X, Y \in \mathcal{O}(\mathcal{C})$ is contained in the extension $L/\mathbb{Q}$ generated by $\overline{s}_{XY}$, $X, Y \in \mathcal{O}(\mathcal{C})$ (since $\overline{s}_{\mathbf{11}} = 1/\sqrt{\dim(\mathcal{C})}$). The above computation shows that $g(\overline{s}_{XY}) = \pm\overline{s}_{Xg(Y)}$. Moreover the sign $\epsilon_g(X) = \pm 1$ such that $g(\dim(X)) = \epsilon_g(X)\dim(g(X))$ is well defined since $\dim(X) \neq 0$, and

$$g(\overline{s}_{XY}) = g\left(\frac{\overline{s}_{XY}}{\dim(Y)}\dim(Y)\right) = \epsilon_g(Y)\overline{s}_{Xg(Y)} = \epsilon_g(X)\overline{s}_{g(X)Y}.$$

In particular, $L/\mathbb{Q}$ is finite and normal, that is, a Galois extension.

Now let f be another automorphism of $\Bbbk$. We have

$$gf(\overline{s}_{XY}) = g(\epsilon_f(Y)\overline{s}_{Xf(Y)}) = \epsilon_g(X)\epsilon_f(Y)\overline{s}_{g(X)f(Y)}$$

and

$$fg(\overline{s}_{XY}) = h(\epsilon_g(X)\overline{s}_{g(X)Y}) = \epsilon_f(Y)\epsilon_g(X)\overline{s}_{g(X)f(Y)} = gh(\overline{s}_{XY}).$$

Thus, the Galois group of $L/\mathbb{Q}$ is abelian. Now the Kronecker-Weber theorem, stating that any finite abelian extension of $\mathbb{Q}$ is contained in a cyclotomic field (see, e.g., [**Cas**]) implies the result. □

8.15. Gauss sums and the central charge

The *Gauss sums* of a pre-metric group (G, q) are defined by the classical formula

$$\tau^{\pm}(G, q) = \sum_{a\in G} q(a)^{\pm 1}. \tag{8.56}$$

Clearly, if (G, q) is the orthogonal direct sum of (G_1, q_1) and (G_2, q_2) then

$$\tau^{\pm}(G, q) = \tau^{\pm}(G_1, q_1)\tau^{\pm}(G_2, q_2).$$

Let $\mathcal{C}$ be a pre-modular category.

DEFINITION 8.15.1. The *Gauss sums* of $\mathcal{C}$ are defined by

$$\tau^{\pm}(\mathcal{C}) = \sum_{X\in\mathcal{O}(\mathcal{C})} \theta_X^{\pm 1}\dim(X)^2. \tag{8.57}$$

EXAMPLE 8.15.2. Let $\mathcal{C}$ be the pointed pre-modular category $\mathcal{C}(G, q)$ from Section 8.4. Then $\mathcal{O}(\mathcal{C}) = G$, $\dim(X) = 1$ for all $X \in \mathcal{O}(\mathcal{C})$, and $\theta_X = q(X)$. So $\tau^{\pm}(\mathcal{C})$ equals the classical Gauss sum $\tau^{\pm}(G, q)$ defined by (8.56).

LEMMA 8.15.3. *Let $\mathcal{C}$ be a pre-modular category. Then for any $Y \in \mathcal{O}(\mathcal{C})$ we have:*

$$\sum_{X\in\mathcal{O}(\mathcal{C})} \theta_X\dim(X)s_{XY} = \dim(Y)\theta_Y^{-1}\tau^{+}(\mathcal{C}). \tag{8.58}$$

PROOF. Using formula (8.48) we compute

$$\begin{aligned}\sum_{X\in\mathcal{O}(\mathcal{C})} \theta_X\dim(X)s_{XY} &= \theta_Y^{-1}\sum_{X\in\mathcal{O}(\mathcal{C})}\dim(X)\sum_{Z\in\mathcal{O}(\mathcal{C})} N_{XY}^{Z}\theta_Z\dim(Z)\\ &= \theta_Y^{-1}\sum_{Z\in\mathcal{O}(\mathcal{C})}\left(\sum_{X\in\mathcal{O}(\mathcal{C})} N_{ZY^*}^{X}\dim(X)\right)\theta_Z\dim(Z)\\ &= \theta_Y^{-1}\dim(Y)\sum_{Z\in\mathcal{O}(\mathcal{C})}\theta_Z\dim(Z)^2 = \dim(Y)\theta_Y^{-1}\tau^{+}(\mathcal{C}),\end{aligned}$$

as required. □

Below we summarize some basic properties of twists and Gauss sums.

PROPOSITION 8.15.4. *For a modular category $\mathcal{C}$ we have*

$$\tau^+(\mathcal{C})\tau^-(\mathcal{C}) = \dim(\mathcal{C}). \tag{8.59}$$

PROOF. Multiplying formula (8.58) in Lemma 8.15.3 by $\dim(Y)$ and taking the sum over $Y \in \mathcal{O}(\mathcal{C})$, we have:

$$\begin{aligned} \tau^+(\mathcal{C})\tau^-(\mathcal{C}) &= \sum_{X\in\mathcal{O}(\mathcal{C})} \left(\sum_{Y\in\mathcal{O}(\mathcal{C})} s_{XY} \dim(Y) \right) \theta_X \dim(X) \\ &= \sum_{Y\in\mathcal{O}(\mathcal{C})} \dim(Y)^2 = \dim(\mathcal{C}). \end{aligned}$$

□

By Theorem 7.21.12, $\dim(\mathcal{C})$ is a totally positive element of the subfield $\mathbb{k}_{alg} \subset \mathbb{k}$ of algebraic numbers. If we fix an embedding $\mathbb{k}_{alg} \hookrightarrow \mathbb{C}$, we can talk about the *multiplicative central charge*

$$\xi(\mathcal{C}) = \frac{\tau^+(\mathcal{C})}{\sqrt{\dim(\mathcal{C})}}, \tag{8.60}$$

where $\sqrt{\dim(\mathcal{C})}$ is the *positive* square root. From Proposition 8.15.4 we see that in a modular category one has $\xi(\mathcal{C})^2 = \tau^+(\mathcal{C})/\tau^-(\mathcal{C})$.

We have another version of Lemma 8.15.3.

COROLLARY 8.15.5. *Let $\mathcal{C}$ be a modular category. Then for any $Y \in \mathcal{O}(\mathcal{C})$ we have:*

$$\sum_{X\in\mathcal{O}(\mathcal{C})} \theta_X^{-1} \dim(X) s_{XY} = \dim(Y)\theta_Y \tau^-(\mathcal{C}). \tag{8.61}$$

PROOF. Multiply formula (8.58) by s_{Y^*Z}, $Z \in \mathcal{O}(\mathcal{C})$ and take sum over $Y \in \mathcal{O}(\mathcal{C})$:

$$\begin{aligned} \tau^+(\mathcal{C}) \sum_{Y\in\mathcal{O}(\mathcal{C})} \theta_Y^{-1} \dim(Y) s_{Y^*Z} &= \sum_{X\in\mathcal{O}(\mathcal{C})} \theta_X \dim(X) \sum_{Y\in\mathcal{O}(\mathcal{C})} s_{XY} s_{Y^*Z} \\ &= \dim(\mathcal{C})\, \theta_Z \dim(Z), \end{aligned}$$

where we used Proposition 8.14.2. So the result follows from (8.59). □

8.16. Representation of the modular group

The *modular group* is, by definition, the group $\Gamma := SL_2(\mathbb{Z})$ of 2×2 matrices with integer entries and determinant 1. This group plays a very important role in number theory. The quotient of Γ by its center $\langle \left(\begin{smallmatrix} -1 & 0 \\ 0 & -1 \end{smallmatrix}\right) \rangle$ is the group $PSL_2(\mathbb{Z})$ of fractional linear transformations of the Riemann sphere $\widehat{\mathbb{C}} := \mathbb{C} \cup \{\infty\}$:

$$\begin{pmatrix} a & b \\ c & d \end{pmatrix} (z) := \frac{az+b}{cz+d}, \qquad z \in \widehat{\mathbb{C}},$$

where a, b, c, d are integers such that $ad - bc = 1$. It is known that Γ is generated by two matrices

$$\mathfrak{s} := \begin{pmatrix} 0 & -1 \\ 1 & 0 \end{pmatrix} \quad \text{and} \quad \mathfrak{t} := \begin{pmatrix} 1 & 1 \\ 0 & 1 \end{pmatrix}. \tag{8.62}$$

These matrices correspond to transformations $z \mapsto -1/z$ and $z \mapsto z+1$. The modular group can be described abstractly in terms of generators and relations as

$$\Gamma = \langle \mathfrak{s}, \mathfrak{t} \mid (\mathfrak{st})^3 = \mathfrak{s}^2,\ \mathfrak{s}^4 = 1 \rangle, \tag{8.63}$$

and $\Gamma/\{\pm 1\} = \langle \mathfrak{s}, \mathfrak{t} \mid (\mathfrak{st})^3 = \mathfrak{s}^2 = 1 \rangle$.

It turns out that a modular category gives rise to a projective representation of Γ. This justifies the terminology. We explain this below. Let $T = (T_{XY})_{X,Y \in \mathcal{O}(\mathcal{C})}$ be the diagonal matrix with entries $T_{XY} = \delta_{X,Y}\theta_X^{-1}$.

THEOREM 8.16.1. *Let $\mathcal{C}$ be a modular category. We have $(ST)^3 = \tau^-(\mathcal{C})S^2$ and $S^4 = \dim(\mathcal{C})^2\,\mathsf{id}$, where id is the identity matrix. Hence, the assignments*

$$\mathfrak{s} \mapsto \frac{1}{\sqrt{\dim(\mathcal{C})}} S \quad \textit{and} \quad \mathfrak{t} \mapsto T \tag{8.64}$$

define a projective representation of Γ.

PROOF. We need to verify the relations from (8.63). By Proposition 8.14.2 we have the identity $S^4 = \dim(\mathcal{C})^2\,\mathsf{id}$, which corresponds to the second relation of (8.63). Let us prove the first relation. The XY-entry of the matrix $T^{-1}ST^{-1}$ is $\theta_X\theta_Y s_{XY}$. Let us compute the XY-entry of the matrix STS using equality (8.49) and Corollary 8.15.5:

$$\begin{aligned}
(STS)_{XY} &= \sum_{V \in \mathcal{O}(\mathcal{C})} s_{XV}\theta_V^{-1}s_{VY} \\
&= \sum_{W \in \mathcal{O}(\mathcal{C})} \left(\sum_{V \in \mathcal{O}(\mathcal{C})} \theta_V^{-1}\dim(V)s_{VW} \right) N_{XY}^W \\
&= \tau^-(\mathcal{C}) \sum_{W \in \mathcal{O}(\mathcal{C})} \dim(W)\theta_W N_{XY}^W \\
&= \tau^-(\mathcal{C})\theta_X\theta_Y s_{XY}.
\end{aligned}$$

Hence, $(ST)^3 = \tau^-(\mathcal{C})S^2$, as required. □

REMARK 8.16.2. The images of $\mathfrak{s}$ and $\mathfrak{t}$ under (8.64) satisfy the relations

$$\mathfrak{s}^4 = \mathsf{id}, \tag{8.65}$$

$$(\mathfrak{st})^3 = \xi(\mathcal{C})^{-1}\mathfrak{s}^2. \tag{8.66}$$

It is clear that one can normalize $\mathfrak{t}$ to make the above projective representation linear. But this only complicates things in applications and so we prefer not to do this.

REMARK 8.16.3. A theorem of Ng and Schauenburg [**NgS2**] says that the representation of the modular group constructed in Theorem 8.16.1 kills a congruence subgroup, i.e., factors through $SL_2(\mathbb{Z}/N\mathbb{Z})$ for some N which can be found explicitly. For modular categories coming from a rational conformal field theory, this result appeared earlier in physics literature, see [**Ban**].

REMARK 8.16.4. Theorem 8.16.1 is related to the fact that the modular group is the mapping class group (MCG) of the torus, and generalizes to MCG of higher genus surfaces (see [**BakK**]).

8.17. Modular data

The notion of a modular datum is a "numerical" counterpart of the notion of a modular tensor category, similarly to how the notion of a based ring is a numerical counterpart of the notion of a fusion category. It comprises many of the important preppies of modular categories (such as the Verlinde formula, $SL_2(\mathbb{Z})$-representation, etc.). This notion is important because modular categories are usually classified according to their modular data. Here we give an introduction to the basic theory of modular data. We loosely follow [**Gann3**] (see also [**Lus4**]).[6]

Let $\Bbbk$ be an algebraically closed field of characteristic zero.

DEFINITION 8.17.1. A modular datum M over $\Bbbk$ is a finite set I with a distinguished element $0 \in I$ and functions $S : I \times I \to \Bbbk$ (called the S-matrix) and $T : I \to \Bbbk$ (called the twist function), satisfying the following conditions:

(1) S is symmetric, $d_i := S_{0i} \neq 0\ \forall i \in I$, and $S_{00} = 1$;
(2) The functions $X_i : I \to \Bbbk$ defined by $X_i(j) := \frac{S_{ij}}{S_{0j}}$ span inside $\mathsf{Fun}(I, k)$ a based ring R_M (in which they are the distinguished basis);
(3) $d_i = d_{i^*}$, where $*$ is the involution coming from (Z);
(4) If we regard S and T as linear operators on $\mathsf{Fun}(I, \Bbbk)$ (by setting $(Sf)(i) = \sum S_{ij} f(j)$, $(Tf)(i) = T(i)f(i)$), then they define a projective representation of the modular group $SL_2(\mathbb{Z})$. That is, there exists $\tau \in \Bbbk^\times$ such that $(ST)^3 = \tau S^2$, and this operator commutes with T. Also, $T(0) = 1$.

The numbers d_i are called the quantum dimensions of M.

In the previous sections, we have proved the following proposition.

PROPOSITION 8.17.2. *If $\mathcal{C}$ is a modular tensor category, then its set of simple objects I, the S-matrix S and the twist function $T = \theta^{-1}$ form a modular datum.*

DEFINITION 8.17.3. A *categorification* of a modular datum M is a categorification $\mathcal{C}$ of the based ring R_M in the sense of Definition 4.10.1 which is a modular category with the S-matrix S and twists T_i^{-1}, $i \in I$.

THEOREM 8.17.4. *Fix a modular datum.*

(i) *Let $D = \sum_i d_i^2$. Then $D \neq 0$, and one has $S^2 = DE$, where $E_{ij} = \delta_{ij^*}$*
(ii) *The numbers S_{ij}/d_j (in particular, d_i) and D/d_i^2 are algebraic integers.*
(iii) *d_i are totally real and $D - 1$ is totally non-negative. The matrix $S/\sqrt{D}$ (well defined up to sign) is totally unitary.*

PROOF. (i) The proof is the same as the proof for modular categories given in the previous sections. Namely, by (2), S is invertible. So we have $\delta_j = S_{0j} \sum_i (S^{-1})_{ji} X_i$. Thus, for $j \neq m$ we have

$$\sum_{i,r} (S^{-1})_{ji} (S^{-1})_{mr} X_i X_r = 0.$$

[6]We note that our definition of a modular datum is slightly less restrictive than the definition of a modular datum in [**Gann3**] (where positivity of dimensions is assumed), and the definition of a fusion datum in [**Lus4**] (where there is an additional involution on the set of labels).

In particular, applying τ to this equality and using the symmetry of S, we get

$$\sum_i (S^{-1})_{ji}(S^{-1}_{i^*m}) = 0.$$

Thus, $S^{-1}ES^{-1}$ is a diagonal matrix, hence so is SES.

Now, let N_{ij}^m be the structure constants of R_M. Then by definition

$$S_{ir}S_{jr} = \sum N_{ij}^m S_{mr}S_{0r}.$$

So since $S_{rj} = S_{jr}$ and $S_{0r} = S_{r0} = S_{r^*0}$, we get

$$S_{ir}S_{rj} = \sum N_{ij}^m S_{mr}S_{r^*0}.$$

Summing this over r, and using that SES is diagonal, we get

$$(S^2)_{ij} = \delta_{ij^*}D.$$

(ii) By definition, S_{ij}/d_j is an eigenvalue of the integer matrix N_i of multiplication by X_i in R_M, thus it is an algebraic integer. Now,

$$\frac{D}{d_i^2} = \sum_j \frac{S_{ij}}{d_i}\frac{S_{ji^*}}{d_i^*},$$

so it is also an algebraic integer.

(iii) $2d_i = d_i + d_{i^*}$ is an eigenvalue of the matrix $N_i + N_{i^*}$, which is a symmetric integer matrix. So d_i is totally real. Next, $(S_{ij} + S_{i^*j})/d_j$ is also an eigenvalue of $N_i + N_{i^*}$, so it is also totally real and hence S_{i^*j} is totally complex conjugate to S_{ij}. This together with (i) implies that under any complex embedding of $\mathbb{Q}(S_{ij})$, $SS^\dagger = D$, where $\dagger$ is the Hermitian adjoint to S, as desired. $\square$

8.18. The Anderson-Moore-Vafa property and Verlinde categories

8.18.1. The Anderson-Moore-Vafa property. Let $\mathcal{C}$ be a pre-modular tensor category with simple objects X_i. Consider the space

$$E_{ijm}^r = \mathsf{Hom}_{\mathcal{C}}(X_r, X_i \otimes X_j \otimes X_m).$$

Using the braiding, this space is identified with E_{jim}^r. The hexagon relation for the braiding implies that

$$\det(c_{ij,m}^2|_{E_{ijm}^r}) = \det(c_{jm}^2|_{E_{ijm}^r})\det(c_{im}^2|_{E_{jim}^r}).$$

Recall that c^2 acts on $\mathsf{Hom}(X_s, X_p \otimes X_q)$ by the scalar T_s/T_pT_q. Therefore, the last identity can be rewritten in the form

$$\prod_p (T_pT_m/T_r)^{N_{ij}^p N_{pm}^r} = \prod_p (T_iT_m/T_p)^{N_{im}^p N_{jp}^r} \prod_p (T_jT_m/T_p)^{N_{jm}^p N_{ip}^r}.$$

This implies the following result.

THEOREM 8.18.1. *(Anderson-Moore-Vafa) In a pre-modular category, one has*

$$(T_iT_jT_mT_r)^{N_{ijm}^r} = \prod_p T_p^{N_{ij}^p N_{pm}^r + N_{im}^p N_{jp}^r + N_{jm}^p N_{ip}^r},$$

where $N_{ijm}^r := \sum_p N_{ij}^p N_{pm}^r$.

COROLLARY 8.18.2. ([**AndeM, Vaf, E**])

(i) *In a pre-modular category,* $T^N = 1$ *for some* N *dividing* $D^{5/2}$.

(ii) *In a modular category,* $\tau/\sqrt{D}$ *is a root of unity of order dividing* $4D^{5/2}|I|$ *(where* $D := \dim \mathcal{C}$*).*

PROOF. (i) For notational convenience, let us write the operation in $\Bbbk^\times$ additively, and to avoid confusion denote the elements of this additive group corresponding to T_j by t_j (e.g., if $\Bbbk = \mathbb{C}$, we can realize $\Bbbk^\times$ as $\mathbb{C}/\mathbb{Z}$, and then $t_j = \log(T_j)/2\pi i$).

The Anderson-Moore-Vafa equation can be written as the following equation in $\mathbb{Z} \otimes_{\mathbb{Z}} \Bbbk^\times$:

$$N_{ijm}^r \otimes (t_i + t_j + t_m + t_r) = \sum_p (N_{ij}^p N_{pm}^r + N_{im}^p N_{jp}^r + N_{jm}^p N_{ip}^r) \otimes t_p.$$

Let R be the ring of integers of the number field generated by d_i. Multiplying the last equation by d_r, and taking the sum over r, we get the following equation in $R \otimes_{\mathbb{Z}} \Bbbk^\times$:

$$d_i d_j d_m \otimes (t_i + t_j + t_m) + \sum_r d_r N_{ijm}^r \otimes t_r = \sum_p (N_{ij}^p d_m d_p + N_{im}^p d_j d_p + N_{jm}^p d_i d_p) \otimes t_p.$$

Let us now multiply this equation by d_i and sum over i. Then we get

$$\sum_i d_i^2 d_j d_m \otimes t_i + D d_j d_m \otimes (t_j + t_m) + \sum_r d_j d_m d_r^2 \otimes t_r = \sum_p (2 d_j d_m d_p^2 + D N_{jm}^p d_p) \otimes t_p.$$

After cancellations we get

$$D d_j d_m \otimes (t_j + t_m) = D \sum_p N_{jm}^p d_p \otimes t_p.$$

Now multiply this by d_j and sum over j. We get

$$D^2 d_m \otimes t_m + \sum_j D d_m d_j^2 \otimes t_j = D d_m \sum_p d_p^2 \otimes t_p.$$

After cancellation we get

$$D^2 d_m \otimes t_m = 0,$$

and using that d_m^2 divides D, we find that T_m is a root of unity of order dividing $D^{5/2}$.

(ii) The second statement is deduced from the first one by taking the determinant of the equation $(ST)^3 = \tau S^2$ and using that $\det(S)^4 = D^{2|I|}$. □

COROLLARY 8.18.3. *The central charge* $\xi(\mathcal{C})$ *of a modular category* $\mathcal{C}$ *is a root of unity.*

PROOF. The equality $\mathfrak{s}^4 = \mathsf{id}$ implies that $\det(\mathfrak{s})$ is a 4th root of unity. Since $\det(\mathfrak{t}) = \Pi_{X \in \mathcal{O}(\mathcal{C})} \theta_X^{-1}$, it follows from Corollary 8.18.2(ii) that $\det(\mathfrak{t})$ is a root of unity. Taking the determinant of both sides of (8.66), we see that $\xi(\mathcal{C})$ must be a root of unity. □

EXAMPLE 8.18.4. As an example consider the Ising category, with simple objects $X_0 = \mathbf{1}$, $X_1 = \chi$ and $X_2 = X$, where $\chi \otimes \chi = \mathbf{1}$, $\chi \otimes X = X \otimes \chi = X$, and $X \otimes X = \mathbf{1} \oplus \chi$. Then we have

$$S = \begin{pmatrix} 1 & 1 & \sqrt{2} \\ 1 & 1 & -\sqrt{2} \\ \sqrt{2} & -\sqrt{2} & 0 \end{pmatrix}.$$

In this case, it is easy to check that we must put $T_1 = -1$, but the axioms of a modular datum do not impose any conditions on $\theta = T_2^{-1}$. On the other hand, the Anderson-Moore-Vafa identity for $i = j = m = r = 2$ gives $\theta^8 = -1$, i.e., θ must be a primitive 16-th root of unity. Thus the Anderson-Moore-Vafa identity does not follow from the axioms of a modular datum. On the other hand, if $\theta^8 = -1$, the corresponding modular datum is categorifiable. (There are eight non-degenerate braided categories with such a fusion ring, and sixteen modular ones, parametrized by choices of $\sqrt{2}$ and by θ; namely, each braided category has two modular structures, differing by the sign of θ).

8.18.2. Verlinde categories.

EXAMPLE 8.18.5. (Verlinde modular categories)

Let $\mathfrak{g}$ be a simple complex Lie algebra. For simplicity let us assume that it is simply laced (so $(\alpha, \alpha) = 2$ for roots). Let h be the Coxeter number of $\mathfrak{g}$, let $\alpha_{\max}$ be the maximal root of $\mathfrak{g}$, and let ρ be half the sum of positive roots of $\mathfrak{g}$.

Let $l > h$ be a positive integer, and let $q \in \mathbb{C}$ be such that the order of q^2 is l.

Andersen and Paradowski [**AndeP**] associated to the pair $(\mathfrak{g}, q)$ a remarkable modular category $\mathcal{C}(\mathfrak{g}, q)$, referred to as the *Verlinde modular category*, which is some "semisimple part" of the representation category of the associated Lusztig quantum group $U_q^L(\mathfrak{g})$, more precisely the quotient of the subcategory of tilting modules by the subcategory of negligible modules. This construction is rather complicated, see e.g., [**BakK**] for a more detailed summary. The simple objects of $\mathcal{C}(\mathfrak{g}, q)$ are V_λ, where λ are dominant weights for $\mathfrak{g}$ such that $(\lambda + \rho, \alpha_{\max}) < l$ (e.g., $\mathcal{C}(\mathfrak{sl}_2, e^{\pi i/8})$ has 7 simple objects). The central charge of $\mathcal{C}(\mathfrak{g}, q)$ equals $e^{2\pi i c/8}$, where $c = \frac{(l-h)\dim(\mathfrak{g})}{l}$.

An important result of M. Finkelberg [**Fi2**] asserts that the Verlinde modular category $\mathcal{C}(\mathfrak{g}, q)$ and the category of integrable modules of level $k := l - h$ over the corresponding affine Lie algebra $\hat{\mathfrak{g}}$ are equivalent as modular categories when $q = e^{\pi i/l}$.

Let us discuss in a bit more detail the case of $\mathfrak{g} = \mathfrak{sl}_2$ (without proofs). In this case, $h = 2$. Let $k = l - 2$, and call the corresponding category $\mathcal{C}_k(q)$. So we have simple objects $V_0 = \mathbf{1}, ..., V_k$, and the fusion ring is the truncated Verlinde ring defined in Example 4.10.6. For instance, $\mathcal{C}_0(q) = \mathsf{Vec}$, $\mathcal{C}_1(q) = \mathsf{Vec}_{\mathbb{Z}/2\mathbb{Z}}^{\omega}$ (where ω is the nontrivial 3-cocycle), $\mathcal{C}_2(q)$ is one of the Ising model categories considered above, etc.

The simple objects of $\mathcal{C}_k(q)$ are the irreducible representations of the Lusztig quantum group $u_q(\mathfrak{sl}_2)$ (i.e., simple comodules for the quantum function algebra $O_q(SL_2)$) with highest weights $0, ..., k$. The tensor product is the truncation of the usual product defined by

$$V_i \otimes_{\text{usual}} V_j = V_i \otimes V_j \oplus N_{ij}$$

(and thus $V_i \otimes V_j = V_i \otimes_{\text{usual}} V_j / N_{ij}$), where N_{ij} is a certain "negligible" part of $V_i \otimes_{\text{usual}} V_j$, consisting of the direct summands which have the property that the quantum trace of any endomorphism is zero (where the quantum trace of a morphism $f : V \to V$ is defined as usually for quantum SL_2, namely $\mathsf{Tr}_q(f) = \mathsf{Tr}(fK)$; so in particular, for the quantum dimensions of V_i, we have $\dim V_i = \mathsf{Tr}_{V_i}(K) = [i+1]_q$). For example, $V_1 \otimes_{\text{usual}} V_k = V_{k-1} \oplus V_{k+1}$, but V_{k+1} (called the Steinberg module) is an irreducible module of dimension $[k+2]_q = [l]_q = 0$, so it is negligible in the above sense and hence $V_1 \otimes V_k = V_{k-1}$. It can be checked that this truncation procedure is compatible with the associativity morphisms and braiding, and gives rise to a modular tensor category (as mentioned above, the twists $\mathcal{C}_k(q)$ are given by $\theta_m = q^{-m(m+2)/2}$, for some choice of $q^{1/2}$).

EXERCISE 8.18.6. (i) Let $\mathcal{D}_k$ be the pointed subcategory of $\mathcal{C}_k(q)$, with simple objects $\mathbf{1} = V_0$ and $X = V_k$ (recall that $V_k \otimes V_k = \mathbf{1}$). Show that the braiding $c_{X,X}$ is given by the scalar i^k (for one of the choices of $q^{1/2}$). Deduce that $\mathcal{D}_k$ is equivalent to $\mathsf{Rep}(\mathbb{Z}/2\mathbb{Z})$ as a fusion category if and only if k is even, and that if k is odd then the associativity is defined by a nontrivial 3-cocycle on $\mathbb{Z}/2\mathbb{Z}$. Show that $\mathcal{D}_k$ is symmetric if and only if k is even, and is equivalent to $\mathsf{Rep}(\mathbb{Z}/2\mathbb{Z})$ if k is divisible by 4, and to the category sVec of super-vector spaces if $k \equiv 2 \mod 4$.

(ii) Compute the S-matrix of $\mathcal{C}_k(q)$ (*Hint:* you may ignore the truncation, as the trace of c^2 on the negligible part by its definition is zero). Check that the resulting representation of $SL_2(\mathbb{Z})$ factors through $SL_2(\mathbb{Z}/N\mathbb{Z})$ for some N. What is N? (The representation you will get is called the Weil representation).

EXERCISE 8.18.7. Show that the objects V_0 and V_2 of $\mathcal{C}_3(q)$ (where q is a primitive 10th root of 1) span a fusion category that categorifies the Yang-Lee fusion ring with basis $1, X$ and $X^2 = X + 1$. We will denote this category YL_+ for $q = \exp(\pi i/5)$ and YL_- for $q = \exp(2\pi i/5)$.

EXERCISE 8.18.8. Show that $\mathcal{C}_k(q)$ is equivalent to $\mathcal{C}_k(q')$ as a fusion category if and only if $q' = q^{\pm 1}$.

Hint: Let $\phi : V_1 \to V_1^*$ be an isomorphism, where $V_1 \in \mathcal{C}_k(q)$ is the two-dimensional representation. Show that $\mathsf{Tr}^L(\phi \circ (\phi^*)^{-1})$ equals to $-q - q^{-1}$.

EXERCISE 8.18.9. Let $\mathcal{C}$ be a spherical tensor category with spherical structure $a_X : X \to X^{**}$. A morphism $f : X \to Y$ in $\mathcal{C}$ is called *negligible* if for any morphism $g : Y \to X$ one has $\mathsf{Tr}^L(a_X \circ g \circ f) = 0$.

(i) Show that if X, Y are indecomposable, and $f : X \to Y$ is not an isomorphism, then f is negligible.

Hint: Take $g : Y \to X$ and show that $g \circ f$ is not an isomorphism. Deduce that $g \circ f$ is nilpotent and conclude that $\mathsf{Tr}^L(a_X \circ g \circ f) = 0$.

(ii) Show that the composition of a negligible morphism with any morphism is negligible, and the tensor product of a negligible morphism with any morphism is negligible.

(iii) Use (ii) to conclude that the category $\overline{\mathcal{C}}$ with the same objects as $\mathcal{C}$, and $\mathsf{Hom}_{\overline{\mathcal{C}}}(X, Y) := \mathsf{Hom}(X, Y)/\mathsf{Hom}_{\mathrm{negligible}}(X, Y)$ (where $\mathsf{Hom}_{\mathrm{negligible}}(X, Y)$ is the subspace of negligible morphisms) is a semisimple spherical tensor category, whose simple objects are the indecomposables of $\mathcal{C}$ of nonzero dimension. Show that if $\mathcal{C}$ is braided or symmetric, so is $\overline{\mathcal{C}}$. Show that $\overline{\mathcal{C}}$ is a fusion category if $\mathcal{C}$ has finitely many indecomposables. Deduce that if $\mathcal{C} = \mathsf{Rep}_{\Bbbk}(G)$, where G is a finite group, and $\operatorname{char}\Bbbk = p > 0$, then $\overline{\mathcal{C}}$ is a symmetric fusion category if the p-Sylow subgroup of G is cyclic.

Hint: For the last question, use the result of [**Hi**] that $\Bbbk[G]$ has finite representation type if and only if the p-Sylow subgroup of G is cyclic.

(iv) Let $\mathcal{C}$ be the category of modules over the Hopf algebra $\Bbbk[x]$, where $\operatorname{char}\Bbbk = 0$, and $\Delta(x) = x \otimes 1 + 1 \otimes x$. Show that $\overline{\mathcal{C}}$ is equivalent to the category $\mathsf{Rep}(SL_2(\Bbbk))$ as a symmetric tensor category.

(v) Let $\operatorname{char}\Bbbk = p > 0$, and $\mathcal{C}$ be the symmetric category $\mathsf{Rep}_{\Bbbk}(\mathbb{Z}/p)$ (so it has p indecomposables of vector space dimensions $1, ..., p$). Show that $\overline{\mathcal{C}}$ is a symmetric fusion category with $p-1$ simple objects, which categories the Verlinde fusion ring Ver_{p-2} in characteristic p, see Example 4.10.6. [7]

(vi) Do part (v) in the case $\mathcal{C} = \mathsf{Rep}\,\Bbbk[x]/(x^p)$, where x is primitive. Does the resulting category $\overline{\mathcal{C}}$ have the same Grothendieck ring as in (v)?

(vii) Let $\mathrm{char}\Bbbk = 0$, and H be the Taft Hopf algebra of dimension n^2 (Example 5.5.6), and $\mathcal{C} = \mathsf{Rep}(H)$. Classify indecomposable objects in $\mathcal{C}$, and show that $\overline{\mathcal{C}}$ is a fusion category that categorifies the fusion ring $\mathbb{Z}[\mathbb{Z}/n] \otimes \mathrm{Ver}_{n-2}$. Explain the relation of $\overline{\mathcal{C}}$ to the Verlinde categories $\mathcal{C}_k(q)$ from Example 8.18.5.

(viii) Let $\mathcal{C}$ be a nonsemisimple tensor category and $\mathcal{C}_{\mathrm{stab}}$ be the *stable category* of $\mathcal{C}$, which by definition is the additive monoidal category that has the same objects as $\mathcal{C}$, and $\mathsf{Hom}_{\mathcal{C}_{\mathrm{stab}}}(X,Y) = \mathsf{Hom}(X,Y)/I(X,Y)$, where $I(X,Y)$ is the space of morphisms that factor through a projective object. Show that the natural functor $F : \mathcal{C} \to \overline{\mathcal{C}}$ factors through $\mathcal{C}_{\mathrm{stab}}$.

Hint: Use the method of the proof of Theorem 6.6.1.

(ix) Show that the functor F from (viii) is an additive monoidal functor (bilinear on morphisms). Is F exact on any side?

REMARK 8.18.10. The construction of the category $\overline{\mathcal{C}}$ is discussed in [**BarW**] and more recently in [**AAITV**].

8.19. A non-spherical generalization of the S-matrix

Let $\mathcal{C}$ be a braided (but not necessarily spherical) fusion category. Let us fix some notation. Given an object $X \in \mathcal{C}$, a natural (but not necessarily tensor) isomorphism $\psi_X : X \to X^{**}$, and a morphism $f : X \to X$, we define

$$\mathsf{Tr}_+(f) := \mathsf{Tr}^L(\psi_X \circ f) \quad \text{and} \quad \mathsf{Tr}_-(f) := \mathsf{Tr}^R(f \circ \psi_X^{-1}), \tag{8.67}$$

where the traces Tr^L and Tr^R were defined in (4.8) and (4.9). We also set

$$d_\pm(X) := \mathsf{Tr}_\pm(\mathsf{id}_X). \tag{8.68}$$

The numbers $\mathsf{Tr}_\pm(f)$ and $d_\pm(X)$, $X \in \mathcal{O}(\mathcal{C})$ depend on ψ. Namely, replacing ψ by $\psi \circ \alpha$, where $\alpha = \{\alpha_X\}_{X \in \mathcal{O}(\mathcal{C})}$ is an automorphism of the identity functor of $\mathcal{C}$ changes $d_\pm(X)$ to $\alpha_X^{\pm 1} d_\pm(X)$.

DEFINITION 8.19.1. Define the matrix $\tilde{S} = (\tilde{s}_{XY})_{X,Y \in \mathcal{O}(\mathcal{C})}$ by

$$\tilde{s}_{XY} := \frac{(\mathsf{Tr}_- \otimes \mathsf{Tr}_+)(c_{Y,X} \circ c_{X,Y})}{d_-(X) d_+(Y)}. \tag{8.69}$$

Note that if one replaces ψ by $\psi \circ \alpha$ as above then both the numerator and the denominator of (8.69) are multiplied by $\alpha_X^{-1}\alpha_Y$, so that $\tilde{s}_{XY}$ does not change. Thus, the matrix $\tilde{S}$ is an invariant of a braided fusion category $\mathcal{C}$ and does *not* depend on the choice of ψ (unlike the S-matrix of a pre-modular category, cf. Remark 8.13.3(ii)).

DEFINITION 8.19.2. A braided fusion category $\mathcal{C}$ is said to be *non-degenerate* if the corresponding matrix $\tilde{S}$ is non-degenerate.

[7] This is the simplest one of the categories discussed in Remark 9.9.34 below (namely, the one associated to SL_2).

For a pre-modular category $\mathcal{C}$ we have

$$\tilde{s}_{XY} = \frac{s_{XY}}{\dim(X)\dim(Y)},$$

i.e., the $\tilde{S}$-matrix can be thought of as a normalized S-matrix. Note that S and $\tilde{S}$ have the same rank.

REMARK 8.19.3. Clearly, a pre-modular category is modular if and only if it is non-degenerate in the sense of Definition 8.19.2.

We will need the following "non-spherical" analog of the Verlinde formula (8.49). Let $\mathcal{C}$ be a braided fusion category. Pick a natural isomorphism $\psi_X : X \xrightarrow{\sim} X^{**}$, $X \in \mathcal{O}(\mathcal{C})$. For all $Y, Z, W \in \mathcal{O}(\mathcal{C})$ consider the linear automorphism

$$t_{YZ}^W : \mathsf{Hom}(W, Y \otimes Z) \to \mathsf{Hom}(W, Y \otimes Z), \quad f \mapsto \psi_{Y\otimes Z}^{-1}(\psi_Y \otimes \psi_Z)f.$$

The automorphisms t_{YZ}^W measure the failure of ψ to be an isomorphism of *tensor* functors. Let T_{YZ}^W denote the trace of t_{YZ}^W.

REMARK 8.19.4. One can show that for a certain choice of the isomorphisms ψ_X one has $T_{YZ}^W \in \mathbb{Z}$ (we will not need this fact in the sequel). When ψ is a spherical structure (i.e., an isomorphism of tensor functors) we have $T_{YZ}^W = N_{YZ}^W$, where N_{YZ}^W denotes the multiplicity of W in $Y \otimes Z$.

The numbers T_{YZ}^W depend on the choice of ψ. Namely, if one replaces ψ by $\psi \circ \alpha$, where $\alpha = \{\alpha_X\}_{X\in\mathcal{O}(\mathcal{C})}$ is an automorphism of the identity functor of $\mathcal{C}$, then T_{YZ}^W gets multiplied by $\frac{\alpha_Y\alpha_Z}{\alpha_W}$. Therefore the numbers

$$\tilde{N}_{YZ}^W := \frac{d_+(W)}{d_+(Y)d_+(Z)} T_{YZ}^W, \qquad Y, Z, W \in \mathcal{O}(\mathcal{C}), \tag{8.70}$$

do not depend on the choice of ψ. Note that $\tilde{N}_{YZ}^W = 0$ when $N_{YZ}^W = 0$.

LEMMA 8.19.5. *For all $X, Y, Z \in \mathcal{O}(\mathcal{C})$ we have*

$$\tilde{s}_{XY}\tilde{s}_{XZ} = \sum_{W\in\mathcal{O}(\mathcal{C})} \tilde{N}_{YZ}^W \tilde{s}_{XW}. \tag{8.71}$$

PROOF. Consider the equality

(8.72)

$$(c_{Y,X} \otimes \mathsf{id}_Z) \circ (\mathsf{id}_Y \otimes c_{Z,X} \circ c_{X,Z}) \circ (c_{X,Y} \otimes \mathsf{id}_Z) = c_{Y\otimes Z,X} \circ c_{X,Y\otimes Z}, \qquad X \in \mathcal{O}(\mathcal{C}).$$

Let us apply $\mathsf{Tr}_+(\mathsf{Tr}_- \otimes \mathsf{id}_Y \otimes \mathsf{Tr}_+)$ to both sides of (8.72). The left hand side gives $d_-(X)d_+(Y)d_+(Z)\tilde{s}_{XY}\tilde{s}_{XZ}$. For $W \in \mathcal{O}(\mathcal{C})$ let $\pi_{Y\otimes Z}^{(W)}$ be the idempotent endomorphism of $Y \otimes Z$ corresponding to the projection on the maximal multiple of W contained in $Y \otimes Z$. Then the right hand side of (8.72) gives

$$d_-(X) \sum_{W\in\mathcal{O}(\mathcal{C})} \mathsf{Tr}_+(\mathsf{id}_Y \otimes \mathsf{Tr}_+)(\pi_{Y\otimes Z}^{(W)})\tilde{s}_{XW} = d_-(X) \sum_{W\in\mathcal{O}(\mathcal{C})} d_+(W)T_{Y,Z}^W \tilde{s}_{XW}.$$

Comparing the results, we obtain (8.71). $\square$

8.20. Centralizers and non-degeneracy

DEFINITION 8.20.1. Objects X and Y of a braided monoidal category $\mathcal{C}$ are said to *centralize* each other if

$$c_{Y,X} \circ c_{X,Y} = \mathrm{id}_{X\otimes Y}\,. \tag{8.73}$$

EXERCISE 8.20.2. Let A, B be commutative algebras in a braided category $\mathcal{C}$ which centralize each other. Show that $A \otimes B$ is a commutative algebra.

Let $\mathcal{C}$ be a fusion category. The *centralizer* $\mathcal{D}'$ of a fusion subcategory $\mathcal{D} \subset \mathcal{C}$ is defined to be the full subcategory of objects of $\mathcal{C}$ that centralize each object of $\mathcal{D}$. It is easy to see that $\mathcal{D}'$ is a fusion subcategory. Clearly, $\mathcal{D}$ is symmetric (see Definition 8.1.12) if and only if $\mathcal{D} \subset \mathcal{D}'$.

Note that if $X, Y \in \mathcal{O}(\mathcal{C})$ centralize each other then $\tilde{s}_{XY} = 1$ (the $\tilde{S}$-matrix of $\mathcal{C}$ is introduced in Definition 8.19.1). Proposition 8.20.5 below provides a characterization of the centralizer of a fusion subcategory of a braided fusion category in terms of its $\tilde{S}$-matrix.

Lemma 8.20.3 below was proved by Altschüler and Bruguières in [**AlB**] under the assumption that ψ is an isomorphism of tensor functors, i.e., a pivotal structure. Our proof is a slight modification of that in [**AlB**]. Let $\mathcal{O}(\mathcal{C})$ denote the set of simple objects of $\mathcal{C}$.

Recall that for any object X and morphism $f : X \to X$ we defined the traces $\mathrm{Tr}_{\pm}(f)$ and dimensions $d_{\pm}(X)$ in (8.67) and (8.68) (depending on a choice of an isomorphism $\psi_X : X \xrightarrow{\sim} X^{**}$).

LEMMA 8.20.3. *Let $\mathcal{C}$ be a fusion category (not necessarily pivotal) and let $\mathcal{D} \subset \mathcal{C}$ be a fusion subcategory. Let V be a fixed object of $\mathcal{C}$ and let $\beta_U : U \otimes V \xrightarrow{\sim} U \otimes V$, $U \in \mathcal{D}$ be a natural family of isomorphisms. Define*

$$H_U := \sum_{Y \in \mathcal{O}(\mathcal{D})} d_+(Y)(\mathrm{Tr}_- \otimes \mathrm{id}_{U\otimes V})\beta_{Y\otimes U}. \tag{8.74}$$

Then $H_U = \mathrm{id}_U \otimes H_{\mathbf{1}}$.

PROOF. For any object Y in $\mathcal{D}$ and any $X \in \mathcal{O}(\mathcal{D})$ let $\pi_Y^{(X)} : Y \to Y$ be the idempotent endomorphism of Y corresponding to the projection on the maximal multiple of X in Y. Define auxiliary morphisms $A_X^Y,\, B_X^Y : U \otimes V \to U \otimes V$ by

$$\begin{aligned} A_X^Y &:= (\mathrm{id}_U \otimes \mathrm{ev}_X \otimes \mathrm{id}_V)(\pi_{U\otimes X^*}^{(Y^*)} \otimes \beta_X)(\mathrm{id}_U \otimes \mathrm{coev}'_X \otimes \mathrm{id}_V),\\ B_X^Y &:= (\mathrm{ev}_Y \otimes \mathrm{id}_{U\otimes V})(\mathrm{id}_{Y^*} \otimes \beta_{Y\otimes U}(\pi_{Y\otimes U}^{(X)} \otimes \mathrm{id}_V))(\mathrm{coev}'_Y \otimes \mathrm{id}_{U\otimes V}). \end{aligned}$$

To prove the Lemma it suffices to check the identity

$$d_+(X)A_X^Y = d_+(Y)B_X^Y, \quad X, Y \in \mathcal{O}(\mathcal{D}), \tag{8.75}$$

since then the result follows by taking the sum of both sides of (8.75) over $X, Y \in \mathcal{O}(\mathcal{D})$.

If X, Z are objects in $\mathcal{C}$ then there is a non-degenerate pairing

$$\mathrm{Hom}(X,\, Z) \times \mathrm{Hom}(Z,\, X) : (f, g) \mapsto \mathrm{Tr}_+(fg). \tag{8.76}$$

For any basis $\{f_i\}_{i\in I}$ of $\mathrm{Hom}(X,\, Z)$ let $\{f^i\}_{i\in I}$ be the basis of $\mathrm{Hom}(Z,\, X)$ dual to $\{f_i\}_{i\in I}$ with respect to the pairing (8.76). In the case when X is simple, we have the following straightforward identities:

$$f^i f_j = d_+(X)^{-1}\delta_{ij}\,\mathsf{id}_X, \qquad \pi_Z^{(X)} = d_+(X)\sum_{i\in I} f_i f^i.$$

Let us take $Z = Y\otimes U$ above. Using the naturality of β, we obtain

$$\begin{aligned}\beta_{Y\otimes U}(\pi_{Y\otimes U}^{(X)}\otimes \mathsf{id}_V) &= d_+(X)\sum_{i\in I}\beta_{Y\otimes U}(f_i f^i\otimes \mathsf{id}_V)\\ &= d_+(X)\sum_{i\in I}(f_i\otimes \mathsf{id}_V)\beta_X(f^i\otimes \mathsf{id}_V).\end{aligned}$$

For every $i\in I$ consider morphisms

$$\begin{aligned} g_i &:= (\mathrm{ev}_Y\otimes \mathsf{id}_{U\otimes X^*})(\mathsf{id}_{Y^*}\otimes f_i\otimes \mathsf{id}_{X^*})(\mathsf{id}_{Y^*}\otimes \mathrm{coev}_X) : Y^*\to U\otimes X^*,\\ g^i &:= (\mathsf{id}_{Y^*}\otimes \mathrm{ev}'_X)(\mathsf{id}_{Y^*}\otimes f^i\otimes \mathsf{id}_{X^*})(\mathrm{coev}'_Y\otimes \mathsf{id}_{U\otimes X^*}) : U\otimes X^*\to Y^*.\end{aligned}$$

Using the axioms of evaluation and coevaluation, we obtain

$$\begin{aligned}(\mathsf{id}_U\otimes \mathrm{ev}_X)(g_i\otimes \mathsf{id}_X) &= (\mathrm{ev}_Y\otimes \mathsf{id}_U)(\mathsf{id}_{Y^*}\otimes f_i),\\ (g^i\otimes \mathsf{id}_X)(\mathsf{id}_U\otimes \mathrm{coev}'_X) &= (\mathsf{id}_{Y^*}\otimes f^i)(\mathrm{coev}'_Y\otimes \mathsf{id}_U).\end{aligned}$$

Using the definitions of $\mathsf{Tr}_\pm$ and the properties of the evaluation and coevaluation maps, one computes

$$\mathsf{Tr}_-(g^i g_j) = \frac{d_-(Y^*)}{d_+(Y)}\,\mathsf{Tr}_+(f^i f_j) = \frac{d_-(Y^*)}{d_+(Y)}\delta_{ij}.$$

Hence, $g^i g_j = d_+(Y)^{-1}\delta_{ij}\,\mathsf{id}_{Y^*}$ and, therefore, $\sum_{i\in I} g_i g^i = d_+(Y)^{-1}\pi_{U\otimes X^*}^{(Y^*)}$. Using the previously obtained identities, we compute

$$\begin{aligned} B_X^Y &= d_+(X)(\mathrm{ev}_Y\otimes \mathsf{id}_{U\otimes V})(\mathsf{id}_Y\otimes \sum_{i\in I}(f_i\otimes \mathsf{id}_V)\beta_X(f^i\otimes \mathsf{id}_V))(\mathrm{coev}'_Y\otimes \mathsf{id}_{U\otimes V})\\ &= d_+(X)(\mathsf{id}_U\otimes \mathrm{ev}_X\otimes \mathsf{id}_V)(\sum_{i\in I} g_i g^i\otimes \beta_X)(\mathsf{id}_U\otimes \mathrm{coev}'_X\otimes \mathsf{id}_V)\\ &= d_+(X)(\mathsf{id}_U\otimes \mathrm{ev}_X\otimes \mathsf{id}_V)(\dim(Y)^{-1}\pi_{U\otimes X^*}^{(Y^*)}\otimes \beta_X)(\mathsf{id}_U\otimes \mathrm{coev}'_X\otimes \mathsf{id}_V)\\ &= d_+(X)d_+(Y)^{-1}A_X^Y,\end{aligned}$$

and the result follows. □

REMARK 8.20.4. Lemma 8.20.3 can be thought of as a categorical analog of the Haar theorem. Indeed, for $\mathcal{C} = \mathsf{Vec}_G$, where G is a finite group, it says that the counting measure is invariant under translations.

Let $\mathcal{C}$ be a braided fusion category with braiding c. Let $\tilde{S} = \{\tilde{s}_{XY}\}_{X,Y\in\mathcal{O}(\mathcal{C})}$ be the $\tilde{S}$-matrix of $\mathcal{C}$, see Definition 8.19.1. Let $\mathcal{D}\subset\mathcal{C}$ be a fusion subcategory. The next Proposition was proved by D. Altschüler and A. Bruguières in [**AlB**] for pre-modular categories (see [**Mu3**] for the proof in the special case of pseudo-unitary braided fusion categories).

PROPOSITION 8.20.5. *Let $\mathcal{C}$ be a braided fusion category, let V be a simple object in $\mathcal{C}$, and let $\mathcal{D}\subset\mathcal{C}$ be a fusion subcategory. We have the following dichotomy:*

(i) *$V\in\mathcal{O}(\mathcal{D}')$ if and only if $\tilde{s}_{YV} = 1$ for all $Y\in\mathcal{O}(\mathcal{D})$,*

(ii) *$V\notin\mathcal{O}(\mathcal{D}')$ if and only if $\sum_{Y\in\mathcal{O}(\mathcal{D})}|Y|^2\tilde{s}_{YV} = 0$ and if and only if $\sum_{Y\in\mathcal{O}(\mathcal{D})}\tilde{s}_{VY}|Y|^2 = 0$.*

PROOF. Note that conditions in the statement of the Proposition are independent from the choice of a natural isomorphism $\psi_X : X \xrightarrow{\sim} X^{**}$.

Fix $V \in \mathcal{O}(\mathcal{C})$ and set $\beta_U = c_{V,U}c_{U,V}$ in Lemma 8.20.3. Then

$$\begin{aligned} H_U &= \sum_{Y\in\mathcal{O}(\mathcal{D})} d_+(Y)(\mathsf{Tr}_- \otimes \mathsf{id}_{U\otimes V})(c_{V,Y\otimes U}c_{Y\otimes U,V}) \\ &= \sum_{Y\in\mathcal{O}(\mathcal{D})} d_+(Y)(\mathsf{Tr}_- \otimes \mathsf{id}_{U\otimes V})(\mathsf{id}_Y \otimes c_{V,U})(c_{V,Y}\circ c_{Y,V}\otimes \mathsf{id}_U)(\mathsf{id}_Y \otimes c_{U,V}) \\ &= (\mathsf{id}_U \otimes H_{\mathbf{1}})c_{V,U}\circ c_{U,V}. \end{aligned}$$

We have $H_{\mathbf{1}} = \sum_{Y\in\mathcal{O}(\mathcal{D})} d_+(Y)(\mathsf{Tr}_- \otimes \mathsf{id}_V)(c_{V,Y}\circ c_{Y,V}) : V \to V$. Applying Tr_+, we obtain $H_{\mathbf{1}} = \sum_{Y\in\mathcal{O}(\mathcal{D})} |Y|^2\tilde{s}_{YV}\,\mathsf{id}_V$. Combining this with Lemma 8.20.3, we obtain

$$(8.77)\qquad \left(\sum_{Y\in\mathcal{O}(\mathcal{D})} |Y|^2\tilde{s}_{YV}\right)(c_{V,U}\circ c_{U,V} - \mathsf{id}_{U\otimes V}) = 0, \qquad \text{for all } U \in \mathcal{O}(\mathcal{D}).$$

Note that, when $\tilde{s}_{YV} = 1$ for all $Y \in \mathcal{O}(\mathcal{D})$, the first factor in (8.77) equals $\dim(\mathcal{D}) \neq 0$ and hence, $c_{V,U}\circ c_{U,V} = \mathsf{id}_{U\otimes V}$ for all $U \in \mathcal{O}(\mathcal{D})$, which proves (i). If the second factor is not zero for some $U \in \mathcal{O}(\mathcal{D})$ then the first factor must vanish, and the first part of (ii) follows. The second part of (ii) is proved similarly. □

EXAMPLE 8.20.6. Let (G, q) be a pre-metric group and $H \subset G$ a subgroup. If $\mathcal{C}$ is a pointed braided category corresponding to a pre-metric group (G, q) and $\mathcal{D}$ corresponds to a subgroup $H \subset G$ then $\mathcal{D}'$ corresponds to $H^\perp$.

Now we will formulate the main properties of centralizers. In the case of *pointed* braided categories (see Example 8.20.6) Theorems 8.20.7, 8.21.1, and 8.21.4 below amount to standard properties of orthogonal complements.

For a fusion subcategory $\mathcal{D}$ of a braided fusion category $\mathcal{C}$, let us consider the following submatrix of the $\tilde{S}$-matrix of $\mathcal{C}$:

$$(8.78)\qquad \tilde{S}_{\mathcal{D}} := \{\tilde{s}_{XY}\}_{X\in\mathcal{O}(\mathcal{D}),Y\in\mathcal{O}(\mathcal{C})}.$$

Note that $\tilde{S} = \tilde{S}_{\mathcal{C}}$.

Let $\mathcal{C}$ be a braided fusion category and let $\mathcal{D} \subset \mathcal{C}$ be a fusion subcategory. Since $\mathcal{C}$ is a $\mathcal{D}'$-module category, it decomposes into a direct sum of indecomposable $\mathcal{D}'$-module subcategories, called $\mathcal{D}'$-module components of $\mathcal{C}$, see Proposition 7.6.7 and Definition 7.6.8.

THEOREM 8.20.7. *The rank of $\tilde{S}_{\mathcal{D}}$ is equal to the number of $\mathcal{D}'$-module components of $\mathcal{C}$. In particular, $\mathcal{C}$ is non-degenerate if and only if $\mathcal{C}' = \mathsf{Vec}$.*

PROOF. Given a fusion subcategory $\mathcal{D} \subset \mathcal{C}$ let us define an inner product of functions $f, g : \mathcal{O}(\mathcal{D}) \to \Bbbk$ by

$$(8.79)\qquad (f, g)_{\mathcal{D}} := \sum_{X\in\mathcal{O}(\mathcal{D})} |X|^2 f(X)g(X^*).$$

For any $Y \in \mathcal{O}(\mathcal{C})$ define $h_Y : \mathcal{O}(\mathcal{D}) \to \Bbbk$ by $h_Y(X) = \tilde{s}_{XY}$.

The idea of the proof is to show that functions h_Y and h_Z are equal if and only if Y, Z belong to the same component and are orthogonal with respect to the inner product (8.79) if and only if they belong to different $\mathcal{D}'$-module components of $\mathcal{C}$, see Lemmas 8.20.8 and 8.20.9 below. This will imply that distinct columns of $\tilde{S}_{\mathcal{D}}$ are linearly independent and are parametrized by $\mathcal{D}'$-components of $\mathcal{C}$.

LEMMA 8.20.8. *Let $\mathcal{D} \subset \mathcal{C}$ be a fusion subcategory. Suppose $Y, Z \in \mathcal{O}(\mathcal{C})$ are in the same $\mathcal{D}'$-module component of $\mathcal{C}$. Then*

$$h_Y = h_Z.$$

PROOF. Let us choose W in $\mathcal{D}'$ so that Y is contained in $Z \otimes W$. Let $p^{(Y)}_{Z\otimes W} : Z \otimes W \to Y$ be a projection from $Z \otimes W$ onto Y and let $i^{(Y)}_{Z\otimes W} : Y \to Z \otimes W$ be an inclusion of Y into $Z \otimes W$ such that $p^{(Y)}_{Z\otimes W} i^{(Y)}_{Z\otimes W} = \mathsf{id}_Y$. For any $X \in \mathcal{O}(\mathcal{D})$ we compute

$$\begin{aligned}
\tilde{s}_{XY} &= d_-(X)^{-1} d_+(Y)^{-1} (\mathsf{Tr}_- \otimes \mathsf{Tr}_+)(c_{Y,X} \circ c_{X,Y}) \\
&= d_-(X)^{-1} d_+(Y)^{-1} (\mathsf{Tr}_- \otimes \mathsf{Tr}_+)((\mathsf{id}_X \otimes p^{(Y)}_{Z\otimes W})(c_{Z\otimes W,X} \circ c_{X,Z\otimes W})(\mathsf{id}_X \otimes i^{(Y)}_{Z\otimes W})) \\
&= d_-(X)^{-1} d_+(Y)^{-1} (\mathsf{Tr}_- \otimes \mathsf{Tr}_+)((\mathsf{id}_X \otimes p^{(Y)}_{Z\otimes W})(c_{Z,X} \circ c_{X,Z} \otimes \mathsf{id}_W)(\mathsf{id}_X \otimes i^{(Y)}_{Z\otimes W})) \\
&= d_-(X)^{-1} d_+(Y)^{-1} \left(d_-(Z)^{-1} (\mathsf{Tr}_- \otimes \mathsf{Tr}_+)(c_{Z,X} \circ c_{X,Z}) \right) \mathsf{Tr}_+(\mathsf{id}_Y) \\
&= \tilde{s}_{XZ},
\end{aligned}$$

as required. □

LEMMA 8.20.9. *For $Y, Z \in \mathcal{O}(\mathcal{C})$ we have $(h_Y, h_Z)_{\mathcal{D}} = 0$ if and only if Y and Z are in different $\mathcal{D}'$-module components of $\mathcal{C}$.*

PROOF. Fix $Y \in \mathcal{O}(\mathcal{C})$. If $Z \in \mathcal{O}(\mathcal{C})$ is such that Y and Z are in different $\mathcal{D}'$-components of $\mathcal{C}$, then

$$\begin{aligned}
(h_Y, h_Z)_{\mathcal{D}} &= \sum_{X \in \mathcal{O}(\mathcal{D})} |X|^2 \tilde{s}_{XY} \tilde{s}_{XZ^*} \\
&= \sum_{W \in \mathcal{O}(\mathcal{C})} \tilde{N}^W_{YZ^*} \left(\sum_{X \in \mathcal{O}(\mathcal{D})} |X|^2 \tilde{s}_{XW} \right) \\
&= \dim(\mathcal{C}) \sum_{W \in \mathcal{O}(\mathcal{D}')} \tilde{N}^W_{YZ^*} = 0,
\end{aligned}$$

where we used Lemma 8.19.5 and Proposition 8.20.5(ii).

Assume there is Z_0 in the $\mathcal{D}'$-module component of Y such that $(h_Y, h_{Z_0})_{\mathcal{D}} = 0$. Then by Lemma 8.20.8 we have $(h_Y, h_Z)_{\mathcal{D}} = 0$ for any simple Z from the $\mathcal{D}'$-component of Y. Thus we have seen that for any $Z \in \mathcal{O}(\mathcal{C})$

$$\sum_{X \in \mathcal{O}(\mathcal{D})} |X|^2 \tilde{s}_{XY} \tilde{s}_{XZ^*} = (h_Y, h_Z)_{\mathcal{D}} = 0.$$

Multiplying the last equation by $|Z|^2$, taking the sum over $Z \in \mathcal{O}(\mathcal{C})$, and using Proposition 8.20.5, we obtain

$$\begin{aligned}
0 &= \sum_{X \in \mathcal{O}(\mathcal{D})} |X|^2 \tilde{s}_{XY} \left(\sum_{Z \in \mathcal{O}(\mathcal{C})} \tilde{s}_{XZ} |Z|^2 \right) \\
&= \dim(\mathcal{C}) \sum_{X \in \mathcal{O}(\mathcal{D} \cap \mathcal{C}')} |X|^2 \tilde{s}_{XY} \\
&= \dim(\mathcal{C}) \dim(\mathcal{D} \cap \mathcal{C}'),
\end{aligned}$$

which is a contradiction, since the categorical dimension of a fusion category is nonzero by Theorem 7.21.12. □

Let us complete the proof of Theorem 8.20.7. It follows from Lemma 8.20.9 that the functions h_Y, where Y runs through the set of simple objects representing different $\mathcal{D}'$-module components of $\mathcal{C}$, form an orthogonal basis of the vector space generated by the columns of $\tilde{S}_{\mathcal{D}}$. □

COROLLARY 8.20.10. *Let $\mathcal{C}$ be a braided fusion category. A fusion subcategory $\mathcal{D} \subset \mathcal{C}$ is a non-degenerate braided category if and only if $\mathcal{D} \cap \mathcal{D}' = \mathsf{Vec}$.* □

COROLLARY 8.20.11. *Let $\mathcal{D}$ be a fusion subcategory of a non-degenerate braided fusion category $\mathcal{C}$. Then the number of $\mathcal{D}'$-components of $\mathcal{C}$ is equal to $|\mathcal{O}(\mathcal{D})|$.*

PROOF. Since $\mathcal{C}$ is non-degenerate, the rows of $\tilde{S}_D$ must be linearly independent. □

Now we collect various conditions equivalent to the non-degeneracy of a braided fusion category. Recall from Definition 8.6.2 that a braided tensor category is factorizable if the braided tensor functor $G : \mathcal{C} \boxtimes \mathcal{C}^{\text{rev}} \to \mathcal{Z}(\mathcal{C})$ defined in (8.18) is an equivalence.

PROPOSITION 8.20.12. *The following conditions are equivalent for a braided fusion category $\mathcal{C}$:*

(i) *$\mathcal{C}$ is non-degenerate (see Definition 8.19.2);*
(ii) *$\mathcal{C}' = \mathsf{Vec}$;*
(iii) *$\mathcal{C}$ is factorizable.*

PROOF. The equivalence (i)⇔(ii) is already proved in Theorem 8.20.7. So we just need to prove the equivalence (ii)⇔(iii). By Theorem 7.16.6 we have

$$\mathsf{FPdim}(\mathcal{Z}(\mathcal{C})) = \mathsf{FPdim}(\mathcal{C})^2 = \mathsf{FPdim}(\mathcal{C} \boxtimes \mathcal{C}^{\text{rev}}).$$

Hence, it suffices to prove that $\mathcal{C}' = \mathsf{Vec}$ if and only if G is fully faithful. The latter condition holds if and only if the images of $\mathcal{C}$ and $\mathcal{C}^{\text{rev}}$ in $\mathcal{Z}(\mathcal{C})$ intersect trivially. But their intersection is precisely $\mathcal{C}'$. □

We shall prove later in Theorem 9.3.2 that the center of a fusion category is a fusion category. Given this result, we can deduce the following.

COROLLARY 8.20.13. *Let $\mathcal{C}$ be a fusion category. Then its center $\mathcal{Z}(\mathcal{C})$ is non-degenerate.*

PROOF. Combine Propositions 8.6.3 and 8.20.12. □

COROLLARY 8.20.14. *The center of a spherical fusion category is modular.*

EXAMPLE 8.20.15. Let H be a complex semisimple Hopf algebra. It is known that the Drinfeld double $D(H)$ of H (see Section 8.3) is also semisimple (see Proposition 7.18.15). Therefore, $\mathsf{Rep}(D(H))$ is a modular category (e.g., $\mathsf{Rep}(D(G))$, where G is a finite group, is a modular category).

In conclusion of this section, let us prove a result about non-degenerate braided categories in an arbitrary characteristic.

PROPOSITION 8.20.16. *Let $\mathcal{C}$ be a non-degenerate braided fusion category over an algebraically closed field $\Bbbk$ of any characteristic. Then $\dim(\mathcal{C}) \neq 0$. Moreover, in this case we have*

$$\sum_X |X|^2 \tilde{s}_{XY} \tilde{s}_{X^*Z} = \dim(\mathcal{C}) \delta_{YZ},$$

where the sum is over the simple objects of X, and in particular (for $Z = \mathbf{1}$)

$$\sum_X |X|^2 \tilde{s}_{XY} = \dim(\mathcal{C}) \delta_{Y\mathbf{1}}$$

PROOF. Let $\mathcal{D} = \mathcal{C}$. Since $\mathcal{C}$ is non-degenerate, $\mathcal{D}' = \mathsf{Vec}$. Then, arguing as in the beginning of the proof of Lemma 8.20.9, we see that

$$\sum_X |X|^2 \tilde{s}_{XY} \tilde{s}_{X^*Z} = \dim(\mathcal{C}) \delta_{YZ}$$

(as $\tilde{N}_{YZ^*}^{\mathbf{1}} = \delta_{YZ}$). Since the matrix $(\tilde{s}_{XY})$ is non-degenerate, this implies that $\dim(\mathcal{C}) \neq 0$. The proposition is proved. □

Now we can derive the converse to the Larson-Radford theorem (Corollary 7.18.10).

COROLLARY 8.20.17. *Let H be a semisimple and cosemisimple Hopf algebra over an algebraically closed field $\Bbbk$ of any characteristic. Then $\mathsf{Tr}_H(S^2) \neq 0$.*

PROOF. Consider the Drinfeld double $D(H)$ of H. Clearly, $\mathsf{Tr}_{D(H)}(S^2) = \mathsf{Tr}_H(S^2)\,\mathsf{Tr}_{H^{*\mathrm{op}}}(S^2)$, so it suffices to show that $\mathsf{Tr}_{D(H)}(S^2) \neq 0$.

By Proposition 7.18.15, $D(H)$ is semisimple, so the category $\mathcal{C} = \mathsf{Rep}(D(H))$ is a fusion category. This category is braided, since it is the Drinfeld center of $\mathsf{Rep}(H)$. Also, it is clearly non-degenerate, since the square of the braiding is defined by the element $R^{21}R \in D(H) \otimes D(H)$, which is a non-degenerate tensor (i.e., defines a non-degenerate bilinear form on $D(H)^*$); this follows from the fact that $R \in H \otimes H^*$ is the canonical element.

According to Example 7.21.4, $\mathsf{Tr}_{D(H)}(S^2)$ is the categorical dimension $\dim(\mathcal{C})$ of the category $\mathcal{C} = \mathsf{Rep}(D(H))$. So it suffices to show that $\dim(\mathcal{C}) \neq 0$. But this follows from Proposition 8.20.16. □

REMARK 8.20.18. In fact, Larson and Radford show that $\mathsf{Tr}_H(S^2) = \varepsilon(I)\lambda(1)$, where λ is a left integral of H^* and I is a left or right integral of H^*, such that $\lambda(I) = 1$ (see Exercise 7.10.10 for a proof). This gives another proof of the fact that $\mathsf{Tr}_H(S^2) \neq 0$ if and only if H is semisimple and cosemisimple.

8.21. Dimensions of centralizers

THEOREM 8.21.1. *Let $\mathcal{C}$ be a braided fusion category and let $\mathcal{B}$, $\mathcal{D} \subset \mathcal{C}$ be fusion subcategories.*

(i) *We have*

$$\dim(\mathcal{B} \cap \mathcal{D}') \dim(\mathcal{D}) = \dim(\mathcal{D} \cap \mathcal{B}') \dim(\mathcal{B}), \tag{8.80}$$

in particular,

$$\dim(\mathcal{D}) \dim(\mathcal{D}') = \dim(\mathcal{C}) \dim(\mathcal{D} \cap \mathcal{C}'). \tag{8.81}$$

(ii) *If $\mathcal{D} \supset \mathcal{C}'$ then*

$$\mathcal{D}'' = \mathcal{D}. \tag{8.82}$$

PROOF. We use Proposition 8.20.5 to compute

$$\begin{aligned}\sum_{X\in\mathcal{O}(\mathcal{D})}\sum_{Z\in\mathcal{O}(\mathcal{B})}|X|^2\tilde{s}_{XZ}|Z|^2 &= \sum_{X\in\mathcal{O}(\mathcal{D})}|X|^2\left(\sum_{Z\in\mathcal{O}(\mathcal{B})}\tilde{s}_{XZ}|Z|^2\right)\\ &= \dim(\mathcal{B})\sum_{X\in\mathcal{O}(\mathcal{B}'\cap\mathcal{D})}|X|^2\\ &= \dim(\mathcal{B})\dim(\mathcal{B}'\cap\mathcal{D}).\end{aligned}$$

Changing the order of summation in the above computation, we see that the same expression also equals $\dim(\mathcal{D})\dim(\mathcal{B}\cap\mathcal{D}')$, proving (8.80). Taking $\mathcal{B}=\mathcal{C}$, we obtain (8.81). Finally, when $\mathcal{D}\supset\mathcal{C}'$, we have

$$\dim(\mathcal{D})\dim(\mathcal{D}')=\dim(\mathcal{C})\dim(\mathcal{C}')=\dim(\mathcal{D}')\dim(\mathcal{D}''),$$

whence $\dim(\mathcal{D})=\dim(\mathcal{D}'')$ and $\mathcal{D}=\mathcal{D}''$, proving (8.82). □

For a pair of fusion subcategories $\mathcal{A},\mathcal{B}\subset\mathcal{C}$ let $\mathcal{A}\vee\mathcal{B}$ denote the smallest fusion subcategory of $\mathcal{C}$ containing $\mathcal{A}$ and $\mathcal{B}$.

COROLLARY 8.21.2. *If $\mathcal{D}\subset\mathcal{C}$ is any fusion subcategory then $\mathcal{D}''=\mathcal{D}\vee\mathcal{C}'$.*

PROOF. Let $\widetilde{\mathcal{D}}=\mathcal{D}\vee\mathcal{C}'$. Then $\mathcal{D}'=\widetilde{\mathcal{D}}'$, so $\mathcal{D}''=\widetilde{\mathcal{D}}''=\widetilde{\mathcal{D}}$ by Theorem 8.21.1(ii). □

COROLLARY 8.21.3. *Let $\mathcal{C}$ be a braided fusion category and let $\mathcal{A},\mathcal{B}\subset\mathcal{C}$ be fusion subcategories. Then*

$$\dim(\mathcal{A})\dim(\mathcal{B})=\dim(\mathcal{A}\vee\mathcal{B})\dim(\mathcal{A}\cap\mathcal{B}). \tag{8.83}$$

PROOF. We can assume that $\mathcal{C}$ is non-degenerate, for otherwise we replace $\mathcal{C}$ by $\mathcal{Z}(\mathcal{C})$ which is non-degenerate by Corollary 8.20.13. Using equation (8.81) twice and equation (8.80) with $\mathcal{D}=\mathcal{A}'$, we have

$$\dim(\mathcal{A}\vee\mathcal{B})=\frac{\dim(\mathcal{C})}{\dim(\mathcal{A}'\cap\mathcal{B}')}=\frac{\dim(\mathcal{C})\dim(\mathcal{B})}{\dim(\mathcal{A}')\dim(\mathcal{A}\cap\mathcal{B})}=\frac{\dim(\mathcal{A})\dim(\mathcal{B})}{\dim(\mathcal{A}\cap\mathcal{B})}.$$

Note that the categorical dimensions are nonzero by Theorem 7.21.12. □

The following result is an easy consequence of Theorems 8.20.7 and 8.21.1(ii).

THEOREM 8.21.4. *Let $\mathcal{C}$ be a braided fusion category and $\mathcal{K}\subset\mathcal{C}$ a fusion subcategory. Suppose that the braided category $\mathcal{K}$ is non-degenerate. Then*

(i) *the natural braided tensor functor $\mathcal{K}\boxtimes\mathcal{K}'\to\mathcal{C}$ is an equivalence;*
(ii) *$\mathcal{K}'$ is non-degenerate if and only if $\mathcal{C}$ is.*

PROOF. By Theorem 8.20.7, since $\mathcal{K}$ is non-degenerate, $\mathcal{K}\cap\mathcal{K}'=\mathsf{Vec}$. Let $\mathcal{D}$ be the fusion subcategory of $\mathcal{C}$ generated by $\mathcal{K}$ and $\mathcal{K}'$. The tensor functor $\mathcal{K}\boxtimes\mathcal{K}'\to\mathcal{D}$ defined by $X\boxtimes Y\mapsto X\otimes Y$ is a braided equivalence.

By Corollary 8.21.2, $\mathcal{K}''$ is the smallest subcategory of $\mathcal{C}$ containing $\mathcal{K}$ and $\mathcal{C}'$. Therefore, $\mathcal{K}''\simeq\mathcal{K}\boxtimes\mathcal{C}'\subset\mathcal{K}\boxtimes\mathcal{K}'\simeq\mathcal{D}$ and $\mathcal{K}''\cap\mathcal{K}'=\mathcal{C}'$. In view of Theorem 8.20.7 this proves (ii). Applying Theorem 8.21.1(ii) to $\mathcal{D}$, we have $\mathcal{D}=\mathcal{D}''=(\mathcal{K}''\cap\mathcal{K}')'=\mathcal{C}''=\mathcal{C}$, proving (i). □

The next theorem is a modification of equation (8.81) of Theorem 8.21.1(i) with categorical dimensions replaced by Frobenius-Perron ones.

THEOREM 8.21.5. *Let $\mathcal{C}$ be a braided fusion category and let $\mathcal{D} \subset \mathcal{C}$ be a fusion subcategory. Then*

$$\mathsf{FPdim}(\mathcal{D})\,\mathsf{FPdim}(\mathcal{D}') = \mathsf{FPdim}(\mathcal{C})\,\mathsf{FPdim}(\mathcal{D} \cap \mathcal{C}'). \tag{8.84}$$

PROOF. We start with the following result. Let $\mathcal{C}$ be a fusion category and let $\mathcal{A}, \mathcal{B} \subset \mathcal{C}$ be fusion subcategories. Suppose that

$$X \otimes Y \simeq Y \otimes X \text{ for all } X \in \mathcal{O}(\mathcal{A}),\ Y \in \mathcal{O}(\mathcal{B}). \tag{8.85}$$

Property (8.85) implies that the simple objects of $\mathcal{A} \vee \mathcal{B}$ are the simple summands of $X \otimes Y$ with $X \in \mathcal{O}(\mathcal{A})$ and $Y \in \mathcal{O}(\mathcal{B})$.

LEMMA 8.21.6. *Let $\mathcal{C}$, $\mathcal{A}$, $\mathcal{B}$ be as above. Then*

$$\mathsf{FPdim}(\mathcal{A})\,\mathsf{FPdim}(\mathcal{B}) = \mathsf{FPdim}(\mathcal{A} \vee \mathcal{B})\,\mathsf{FPdim}(\mathcal{A} \cap \mathcal{B}). \tag{8.86}$$

PROOF. Recall from Definition 3.3.8 that the regular element of $K_0(\mathcal{C}) \otimes_{\mathbb{Z}} \mathbb{R}$ is

$$R_{\mathcal{C}} = \sum_{X \in \mathcal{O}(\mathcal{C})} \mathsf{FPdim}(X) X.$$

It is defined up to a scalar multiple by the property that $Y \otimes R_{\mathcal{C}} = \mathsf{FPdim}(Y) R_{\mathcal{C}}$ for all $Y \in \mathcal{O}(\mathcal{C})$.

One can see that

$$R_{\mathcal{A}} \otimes R_{\mathcal{B}} = a R_{\mathcal{A} \vee \mathcal{B}}, \tag{8.87}$$

for some positive a. Indeed, take $V = \oplus_{X \in \mathcal{O}(\mathcal{A}), Y \in \mathcal{O}(\mathcal{B})} X \otimes Y$. Then the multiplication matrix of V in $K_0(\mathcal{A} \vee \mathcal{B})$ has strictly positive entries and both sides of (8.87) are Frobenius-Perron eigenvectors of V. Hence, they differ by a positive scalar.

The scalar a equals the multiplicity of the unit object $\mathbf{1}$ in $R_{\mathcal{A}} \otimes R_{\mathcal{B}}$, which is the same as the multiplicity of $\mathbf{1}$ in $\sum_{Z \in \mathcal{O}(\mathcal{A} \cap \mathcal{B})} \mathsf{FPdim}(Z)^2 Z \otimes Z^*$. Hence, $a = \mathsf{FPdim}(\mathcal{A} \cap \mathcal{B})$. Taking the Frobenius-Perron dimensions of both sides of (8.87), we get the result. □

Now we will prove Theorem 8.21.5. For any fusion category $\mathcal{C}$ and any fusion subcategory $\mathcal{D} \subset \mathcal{C}$ let $\mathcal{Z}_{\mathcal{D}}(\mathcal{C})$ be the fusion category whose objects are pairs $(X, c_{X,-})$, where X is an object of $\mathcal{C}$ and $c_{X,-}$ is a natural family of isomorphisms $c_{X,V} : X \otimes V \simeq V \otimes X$ for all objects V in $\mathcal{D}$ satisfying the same compatibility conditions as in the definition of the center of $\mathcal{C}$ (so, in particular, $\mathcal{Z}_{\mathcal{C}}(\mathcal{C}) = \mathcal{Z}(\mathcal{C})$). Note that $\mathcal{Z}_{\mathcal{D}}(\mathcal{C})$ is dual to the fusion category $\mathcal{D} \boxtimes \mathcal{C}^{\text{op}}$ with respect to its module category $\mathcal{C}$, where $\mathcal{D}$ and $\mathcal{C}^{\text{op}}$ act on $\mathcal{C}$ via the left and right multiplication, respectively. So we have

$$\mathsf{FPdim}(\mathcal{Z}_{\mathcal{D}}(\mathcal{C})) = \mathsf{FPdim}(\mathcal{D} \boxtimes \mathcal{C}^{\text{op}}) = \mathsf{FPdim}(\mathcal{D})\,\mathsf{FPdim}(\mathcal{C}). \tag{8.88}$$

Also, the forgetful tensor functor $F : \mathcal{Z}(\mathcal{C}) \to \mathcal{Z}_{\mathcal{D}}(\mathcal{C})$ is surjective.

Now assume that $\mathcal{C}$ is a non-degenerate braided fusion category with braiding c. There are two embeddings $X \mapsto (X, c_{X,-})$ and $X \mapsto (X, c^{-1}_{-,X})$ of $\mathcal{C}$ into $\mathcal{Z}_{\mathcal{D}}(\mathcal{C})$. Their images will be denoted by $\mathcal{C}_+$ and $\mathcal{C}_-$. Clearly $\mathcal{C}_+ \cap \mathcal{C}_- = \mathcal{D}'$. On the other hand, $\mathcal{C}_+ = FG(\mathcal{C} \boxtimes \mathbf{1})$ and $\mathcal{C}_- = FG(\mathbf{1} \boxtimes \mathcal{C})$, where $G : \mathcal{C} \boxtimes \mathcal{C}^{\text{op}} \cong \mathcal{Z}(\mathcal{C})$ is the functor from Proposition 8.20.12. Since F is surjective, we see that $\mathcal{C}_+ \vee \mathcal{C}_- = \mathcal{Z}_{\mathcal{D}}(\mathcal{C})$. So applying Lemma 8.21.6 to $\mathcal{A} = \mathcal{C}_+$ and $\mathcal{B} = \mathcal{C}_-$ we obtain

$$\mathsf{FPdim}(\mathcal{Z}_{\mathcal{D}}(\mathcal{C}))\,\mathsf{FPdim}(\mathcal{D}') = \mathsf{FPdim}(\mathcal{C}_+)\,\mathsf{FPdim}(\mathcal{C}_-) = \mathsf{FPdim}(\mathcal{C})^2.$$

By (8.88), this means that

$$\mathsf{FPdim}(\mathcal{D})\,\mathsf{FPdim}(\mathcal{D}') = \mathsf{FPdim}(\mathcal{C}). \tag{8.89}$$

Now let $\mathcal{C}$ be any (possibly degenerate) braided fusion category. Let $\mathcal{D}^\wr, \mathcal{C}^\wr$ denote the centralizers of $\mathcal{D}, \mathcal{C}$ in the non-degenerate braided fusion category $\mathcal{Z}(\mathcal{C})$. Then the usual centralizer of $\mathcal{D}$ in $\mathcal{C}$ is $\mathcal{D}' = \mathcal{D}^\wr \cap \mathcal{C}$ and $\mathcal{D} \cap \mathcal{C}^\wr = \mathcal{D} \cap \mathcal{C}'$.

We can compute $\mathsf{FPdim}(\mathcal{D} \vee \mathcal{C}^\wr)$ in two different ways. First, by Lemma 8.21.6,

$$\mathsf{FPdim}(\mathcal{D} \vee \mathcal{C}^\wr) = \frac{\mathsf{FPdim}(\mathcal{D})\,\mathsf{FPdim}(\mathcal{C}^\wr)}{\mathsf{FPdim}(\mathcal{D} \cap \mathcal{C}^\wr)} = \frac{\mathsf{FPdim}(\mathcal{D})\,\mathsf{FPdim}(\mathcal{C})}{\mathsf{FPdim}(\mathcal{D} \cap \mathcal{C}')}, \tag{8.90}$$

(we have used the equality $\mathsf{FPdim}(\mathcal{C}^\wr) = \mathsf{FPdim}(\mathcal{C})$, which follows from (8.89)). Second, by (8.89) we have

$$\mathsf{FPdim}(\mathcal{D} \vee \mathcal{C}^\wr) = \frac{\mathsf{FPdim}(\mathcal{Z}(\mathcal{C}))}{\mathsf{FPdim}(\mathcal{D}^\wr \cap \mathcal{C})} = \frac{\mathsf{FPdim}(\mathcal{C})^2}{\mathsf{FPdim}(\mathcal{D}')} \tag{8.91}$$

Comparing (8.90) and (8.91) we get Theorem 8.21.5. □

COROLLARY 8.21.7. *Let $\mathcal{C}$ be a non-degenerate braided fusion category and let $\mathcal{D} \subset \mathcal{C}$ be a fusion subcategory. Then*

$$\mathsf{FPdim}(\mathcal{D})\,\mathsf{FPdim}(\mathcal{D}') = \mathsf{FPdim}(\mathcal{C}). \tag{8.92}$$

8.22. Projective centralizers

As before, let $\mathcal{C}$ be a braided fusion category.

LEMMA 8.22.1. *Let $X \in \mathcal{C}$. For each $\lambda \in \Bbbk^\times$ let $\mathcal{D}_\lambda$ be the full subcategory of objects Y such that $c_{Y,X}c_{X,Y} = \lambda \cdot \mathsf{id}_{X \otimes Y}$.*

(i) *If $Y \in \mathcal{D}_\lambda$ and $Z \in \mathcal{D}_\mu$ then $Y \otimes Z \in \mathcal{D}_{\lambda\mu}$;*
(ii) *If $Y \in \mathcal{D}_\lambda$ then $Y^* \in \mathcal{D}_{\lambda^{-1}}$;*
(iii) *The category*

$$\mathcal{D} = \bigoplus_{\lambda \in \Bbbk^\times} \mathcal{D}_\lambda \tag{8.93}$$

is a fusion subcategory of $\mathcal{C}$.

PROOF. Statement (i) is immediate from the hexagon axiom for the braiding. To prove statement (ii), observe that if $Y^* \in \mathcal{D}_\lambda$ (i.e., $c_{Y^*,X} = \lambda c^{-1}_{X,Y^*}$) then by (8.28)-(8.29), $c^{-1}_{Y,X} = \lambda c_{X,Y}$, i.e., $Y \in \mathcal{D}_{\lambda^{-1}}$. Finally, (iii) follows from (i) and (ii) combined. □

REMARK 8.22.2. According to (i), the decomposition (8.93) is a grading of $\mathcal{D}$.

DEFINITION 8.22.3. We say that simple objects $X, Y \in \mathcal{C}$ *projectively centralize* each other if $c_{Y,X} \circ c_{X,Y} = \lambda \cdot \mathsf{id}_{X \otimes Y}$ for some $\lambda \in \Bbbk^\times$. If X and Y are arbitrary objects of $\mathcal{C}$ we say that they projectively centralize each other if every simple component of X projectively centralizes every simple component of Y. We say that full subcategories $\mathcal{C}_1, \mathcal{C}_2 \subset \mathcal{C}$ projectively centralize each other if each object of $\mathcal{C}_1$ projectively centralizes each object of $\mathcal{C}_2$.

DEFINITION 8.22.4. The *projective centralizer of an object $X \in \mathcal{C}$* is the full subcategory of objects of $\mathcal{C}$ projectively centralizing X. The *projective centralizer of a full subcategory $\mathcal{D} \subset \mathcal{C}$* is the full subcategory of objects of $\mathcal{C}$ projectively centralizing each object of $\mathcal{D}$.

Lemma 8.22.1 implies that the projective centralizer of an object of $\mathcal{C}$ (or of a full subcategory $\mathcal{D} \subset \mathcal{C}$) is a fusion subcategory.

PROPOSITION 8.22.5. *For any simple objects $X, Y \in \mathcal{C}$ the following conditions are equivalent:*

(i) *X centralizes $Y \otimes Y^*$;*
(ii) *$X \otimes X^*$ centralizes Y;*
(iii) *X and Y projectively centralize each other.*

PROOF. It suffices to show that (i) and (iii) are equivalent. The implication (iii)⇒(i) follows from Lemma 8.22.1(i-ii).

To deduce (iii) from (i) note that (i) is equivalent to the identity

$$c_{Y,X}^{-1} \circ c_{X,Y}^{-1} \otimes \mathsf{id}_{Y^*} = \mathsf{id}_Y \otimes c_{Y^*,X} \circ c_{X,Y^*}.$$

Applying trace to the last factor we see that $c_{Y,X} \circ c_{X,Y}$ must be a scalar multiple of $\mathsf{id}_{X \otimes Y}$. □

Now we apply the above results to compute the centralizer of the adjoint subcategory introduced in Section 4.14. Recall that if $\mathcal{K}$ is a fusion subcategory of a braided category $\mathcal{C}$ then $\mathcal{K}_{ad}$ is the fusion subcategory of $\mathcal{K}$ generated by the objects $X \otimes X^*$, where X runs through the set of simple objects of $\mathcal{K}$.

The next proposition describes $(\mathcal{K}_{ad})'$. Recall that the commutator $\mathcal{K}^{co}$ was introduced in Definition 4.14.10.

PROPOSITION 8.22.6. *Let $\mathcal{K}$ be a fusion subcategory of a braided fusion category $\mathcal{C}$. Then $(\mathcal{K}_{ad})' = (\mathcal{K}')^{co}$. In fact, both $(\mathcal{K}_{ad})'$ and $(\mathcal{K}')^{co}$ are equal to the projective centralizer of $\mathcal{K}$.*

PROOF. By definition, the simple objects of $(\mathcal{K}_{ad})'$ are those $X \in \mathcal{O}(\mathcal{C})$ which centralize $Y \otimes Y^*$ for any $Y \in \mathcal{O}(\mathcal{K})$. Similarly, the simple objects of $(\mathcal{K}')^{co}$ are those $X \in \mathcal{O}(\mathcal{C})$ for which $X \otimes X^*$ centralizes any $Y \in \mathcal{O}(\mathcal{K})$. Now use Proposition 8.22.5. □

COROLLARY 8.22.7. *Let $\mathcal{K}$ be a fusion subcategory of a braided fusion category $\mathcal{C}$. Then $(\mathcal{K}')_{ad} \subset (\mathcal{K}^{co})'$. If $\mathcal{C}$ is non-degenerate then $(\mathcal{K}')_{ad} = (\mathcal{K}^{co})'$.*

PROOF. By Proposition 8.22.6, $((\mathcal{K}')_{ad})' = (\mathcal{K}'')^{co} \supset \mathcal{K}^{co}$, so $(\mathcal{K}')_{ad} \subset (\mathcal{K}^{co})'$.

If $\mathcal{C}$ is non-degenerate then $\mathcal{C}' = \mathsf{Vec}$, so by Theorem 8.21.1(ii), $\mathcal{D}'' = \mathcal{D}$ for any fusion subcategory $\mathcal{D} \subset \mathcal{C}$. Therefore in the non-degenerate case Proposition 8.22.6 immediately implies the equality $(\mathcal{K}')_{ad} = (\mathcal{K}^{co})'$. □

For any fusion category $\mathcal{C}$ let $\mathcal{C}_{pt} \subset \mathcal{C}$ be the maximal pointed subcategory of $\mathcal{C}$, i.e., the fusion subcategory generated by the invertible objects of $\mathcal{C}$.

COROLLARY 8.22.8. *If $\mathcal{C}$ is a non-degenerate braided fusion category then*

$$(\mathcal{C}_{ad})' = \mathcal{C}_{pt} \subset \mathcal{C} \quad \textit{and} \quad (\mathcal{C}_{pt})' = \mathcal{C}_{ad}. \tag{8.94}$$

PROOF. Apply Proposition 8.22.6 in the case $\mathcal{K} = \mathcal{C}$. Since $\mathcal{C}$ is non-degenerate $\mathcal{C}' = \mathsf{Vec}$, so we get $(\mathcal{C}_{ad})' = (\mathsf{Vec})^{co}$. Finally, $(\mathsf{Vec})^{co} = \mathcal{C}_{pt}$ (this is immediate from Definition 4.14.10). The second statement follows immediately from the first one by Theorem 8.21.1(ii). □

Here is a more explicit version of Corollary 8.22.8. An invertible object X of a braided fusion category $\mathcal{C}$ defines an automorphism $\alpha(X)$ of the identity functor given by the composition

$$Y \xrightarrow{\mathsf{id}_Y \otimes \mathrm{coev}_X} Y \otimes X \otimes X^* \xrightarrow{c_{X,Y} \circ c_{Y,X} \otimes \mathsf{id}_{X^*}} Y \otimes X \otimes X^* \xrightarrow{\mathsf{id}_Y \otimes \mathrm{coev}_X^{-1}} Y. \tag{8.95}$$

LEMMA 8.22.9. *Let X be an invertible object of a braided fusion category $\mathcal{C}$.*

(i) *$\alpha(X)$ is a tensor automorphism of the identity functor;*
(ii) *the map $\alpha : \mathcal{O}(\mathcal{C}_{pt}) \to \mathsf{Aut}_{\otimes}(\mathsf{id}_{\mathcal{C}})$ is a homomorphism of groups;*
(iii) *for a non-degenerate category $\mathcal{C}$ the homomorphism α is isomorphism.*

PROOF. (i) and (ii) are straightforward. For (iii) note that $X \in \mathcal{O}(\mathcal{C}_{pt})$ is in the kernel of α if and only if $X \in \mathcal{C}'$. Thus under assumptions of (iii) α is injective. Recall that by Proposition 4.14.3 (iii), we have $\mathsf{Aut}_{\otimes}(\mathsf{id}_{\mathcal{C}}) = \mathsf{Hom}(U_{\mathcal{C}}, \Bbbk^{\times})$ where $U_{\mathcal{C}}$ is the universal grading group. It is clear that for a braided category $\mathcal{C}$ the group $U_{\mathcal{C}}$ is commutative and hence $U_{\mathcal{C}} = \mathsf{Hom}(\mathsf{Aut}_{\otimes}(\mathsf{id}_{\mathcal{C}}), \Bbbk^{\times})$. The homomorphism $\alpha^{\vee} : U_{\mathcal{C}} = \mathsf{Hom}(\mathsf{Aut}_{\otimes}(\mathsf{id}_{\mathcal{C}}), \Bbbk^{\times}) \to \mathsf{Hom}(\mathcal{O}(\mathcal{C}_{pt}), \Bbbk^{\times})$ dual to α is surjective and hence defines a faithful grading of $\mathcal{C}$ by the group $\mathsf{Hom}(\mathcal{O}(\mathcal{C}_{pt}), \Bbbk^{\times})$, see Proposition 4.14.3. It is clear from the definitions that the trivial component of this grading is precisely $(\mathcal{C}_{pt})'$. By Lemma 8.22.8 we have $(\mathcal{C}_{pt})' = \mathcal{C}_{ad}$ and by Corollary 3.6.6 $\mathcal{C}_{ad}$ is the trivial component of the universal grading. Hence $\alpha^{\vee}$ and α are isomorphisms, and the lemma is proved. □

Proposition 4.14.3 provides a canonical isomorphism $\mathsf{Aut}_{\otimes}(\mathsf{id}_{\mathcal{C}}) = \mathsf{Hom}(U_{\mathcal{C}}, k^{\times})$, where $U_{\mathcal{C}}$ is the group of the universal grading. So we can consider the homomorphism $\alpha : \mathcal{O}(\mathcal{C}_{pt}) \to \mathsf{Aut}_{\otimes}(\mathsf{id}_{\mathcal{C}})$ as a pairing

$$\langle\, , \rangle : \mathcal{O}(\mathcal{C}_{pt}) \times U_{\mathcal{C}} \to \Bbbk^{\times}. \tag{8.96}$$

Now Lemma 8.22.9(iii) says that for a non-degenerate braided fusion category $\mathcal{C}$ this pairing is perfect.

8.23. De-equivariantization

The goal of this section is to describe the construction opposite to the equivariantization.

Let $\mathcal{C}$ be a fusion category. Recall from Section 4.15 that given an action of a finite group G on $\mathcal{C}$, one constructs a fusion category $\mathcal{C}^G$ of G-equivariant objects in $\mathcal{C}$ (see Definition 2.7.2). Furthermore, there is a canonical fully faithful tensor functor

$$\mathsf{Rep}(G) = \mathsf{Vec}^G \hookrightarrow \mathcal{C}^G. \tag{8.97}$$

PROPOSITION 8.23.1. *There is a canonical braided tensor functor*

$$F : \mathsf{Rep}(G) \to \mathcal{Z}(\mathcal{C}^G) \tag{8.98}$$

such that

(i) *embedding (8.97) is canonically isomorphic to the composition of (8.98) and the forgetful functor $\mathcal{Z}(\mathcal{C}^G) \to \mathcal{C}^G$,*
(ii) *the composition of embedding (8.97) with the forgetful functor $\mathcal{C}^G \to \mathcal{C}$ maps $\mathsf{Rep}(G)$ to Vec and has a canonical structure of a braided tensor functor.*

PROOF. We can interpret embedding (8.97) as follows. For every representation $\rho : G \mapsto GL(V)$ the object $V \otimes \mathbf{1}$ has a natural G-equivariant structure given by

$$u_g : T_g(V \otimes \mathbf{1}) = V \otimes \mathbf{1} \xrightarrow{\rho(g)\otimes \mathrm{id}_{\mathbf{1}}} V \otimes \mathbf{1},$$

where $g \mapsto T_g$, $g \in G$, denotes the action of G on $\mathcal{C}$. The unit constraint of $\mathcal{C}$ equips $V \otimes \mathbf{1}$ with a canonical structure of a central equivariant object, i.e., (8.97) lifts to a braided tensor functor $\mathsf{Rep}(G) \to \mathcal{Z}(\mathcal{C}^G)$. This proves (i). To prove (ii), note that the forgetful functor $\mathcal{Z}(\mathcal{C}^G) \to \mathcal{C}^G$ preserves the unit constraint and, hence, restricts to a *braided* tensor functor $\mathsf{Rep}(G) \to \mathsf{Vec}$. □

REMARK 8.23.2. We will see in Theorem 9.9.26(ii) that the braided tensor functor $\mathsf{Rep}(G) \to \mathsf{Vec}$ constructed in the proof of Proposition 8.23.1 is isomorphic to the forgetful functor.

THEOREM 8.23.3. *Let $\mathcal{D}$ be a fusion category and let G be a finite group. Suppose there is a braided tensor functor $\mathsf{Rep}(G) \to \mathcal{Z}(\mathcal{D})$ such that its composition with the forgetful functor $\mathcal{Z}(\mathcal{D}) \to \mathcal{D}$ is fully faithful. Then there is a fusion category $\mathcal{C}$ and an action of G on $\mathcal{C}$ such that $\mathcal{D} \cong \mathcal{C}^G$.*

PROOF. Let A be the image of algebra $\mathsf{Fun}(G, \Bbbk) \in \mathsf{Rep}(G)$ in $\mathcal{D}$. Let $\mathcal{C} := \mathsf{Mod}_{\mathcal{D}}(A)$ be the category of right A-modules in $\mathcal{D}$. It is a fusion category thanks to the full faithfulness condition. Given a right A-module (N, p), where $p : N \otimes A \to N$ is the module structure, we can view it as a left A-module via

$$A \otimes N \xrightarrow{\gamma_N} N \otimes A \xrightarrow{p} N,$$

where $\gamma_X : X \otimes A \xrightarrow{\sim} A \otimes X$ denotes the central structure of A. Furthermore, commutativity of A implies that N is an A-bimodule, see Definition 7.8.25. Hence, the tensor product of right A-modules is again a right A-module, cf. Definition 7.8.21 and Exercise 7.8.27. Thus, $\mathcal{C}$ is a fusion category (see Remark 8.23.5 below) with tensor product $\otimes_A$. The free module functor

$$\mathcal{D} \to \mathcal{C} : X \mapsto X \otimes A \tag{8.99}$$

is a surjective tensor functor. Furthermore, the action of G on $\mathsf{Fun}(G, \Bbbk)$ by right translations induces the action $g \mapsto T_g$ of G on $\mathcal{C}$.

We define a tensor functor $\mathcal{D} \to \mathcal{C}^G$ by assigning to $X \in \mathcal{D}$ the free module $X \otimes A$ with the G-equivariant structure coming from the action of G on A. The full faithfulness condition implies that this functor is injective. Since the Frobenius-Perron dimensions of $\mathcal{D}$ and $\mathcal{C}^G$ are equal (cf. Exercise 6.1.10(ii)) the result follows from Corollary 6.3.5. □

DEFINITION 8.23.4. The category $\mathcal{C}$ constructed in Theorem 8.23.3 is called the *de-equivariantization* of $\mathcal{D}$ and denoted $\mathcal{D}_G$.

REMARK 8.23.5. The term *de-equivariantization* in Definition 8.23.4 is justified by the fact that the assignments $\mathcal{C} \to \mathcal{C}^G$ and $\mathcal{D} \mapsto \mathcal{D}_G$ are mutual inverses. These assignments can be understood as 2-equivalences between certain 2-categories, see [**DrGNO2**, Section 4.2]. In particular, since $\mathcal{C}$ is a fusion category if and only if $\mathcal{C}^G$ is a fusion category and the functor $\mathsf{Rep}(G) = \mathsf{Vec}^G \to \mathcal{C}^G$ is fully faithful, we see that a de-equivariantization of a fusion category is a fusion category.

Next we briefly discuss equivariantization and de-equivariantization for braided fusion categories.

DEFINITION 8.23.6. Let $\mathcal{C}$ be a braided fusion category. We will say that an action $g \mapsto T_g$, $g \in G$, of G on $\mathcal{C}$ is *braided* if each T_g is a braided autoequivalence of $\mathcal{C}$.

Let $\mathcal{C}$ be a braided fusion category with a braided action of G. The braiding of $\mathcal{C}$ extends to $\mathcal{C}^G$, making the latter a braided fusion category such that the forgetful functor $\mathcal{C}^G \to \mathcal{C}$ is braided. Note that the image of the canonical embedding $\mathsf{Rep}(G) \hookrightarrow \mathcal{C}^G$ belongs to $(\mathcal{C}^G)'$.

Conversely, let us show that the de-equivariantization of a braided fusion category $\mathcal{D}$ by a subcategory $\mathsf{Rep}(G) \subset \mathcal{D}'$ is braided.

We start with an example.

EXAMPLE 8.23.7. Let (G, q) be a pre-metric group, see Section 8.4. Let H be a subgroup of G such that $q|_H = 1$ and $H \subset G^{\perp}$. Then there is a well-defined quadratic form $\tilde{q}$ on the quotient group G/H:

$$\tilde{q}(x + H) := q(x), \quad x \in G.$$

Thus, one can take the quotient of a pre-metric group G by an isotropic subgroup contained in $G^{\perp}$.

Here is a categorical construction analogous to Example 8.23.7.

PROPOSITION 8.23.8. *Let $\mathcal{D}$ be a braided fusion category with braiding c and let G be a finite group. Suppose there is a fully faithful braided tensor functor $\mathsf{Rep}(G) \to \mathcal{D}'$. Then $\mathcal{D}_G$ is a braided fusion category with braiding $\tilde{c} : M \otimes_A N \to N \otimes_A M$ such that the diagram*

$$\begin{array}{ccc} M \otimes N & \xrightarrow{\quad c_{M,N} \quad} & N \otimes M \\ \downarrow & & \downarrow \\ M \otimes_A N & \xrightarrow{\quad \tilde{c}_{M,N} \quad} & N \otimes_A M \end{array} \tag{8.100}$$

commutes. Furthermore, the forgetful tensor functor $\mathcal{D} \to \mathcal{D}_G$ is braided.

PROOF. We use the proof given by Pareigis in [**Pa**]. Let $A = \mathsf{Fun}(G, \Bbbk) \in \mathsf{Rep}(G)$ and let M, N be right A-modules with action maps $\rho_M : M \otimes A \to M$, $\rho_N : N \otimes A \to N$. We claim that the diagram

$$\begin{array}{ccccc} M \otimes A \otimes N & \overset{\rho_M \otimes \mathsf{id}_N}{\underset{\mathsf{id}_M \otimes \rho_N \circ c_{N,A}}{\rightrightarrows}} & M \otimes N & \longrightarrow & M \otimes_A N \\ \Big\downarrow{\scriptstyle \mathsf{id}_M \otimes c_{A,N}} & & & & \\ M \otimes N \otimes A & & \Big\downarrow{\scriptstyle c_{M,N}} & & \\ \Big\downarrow{\scriptstyle c_{M,N} \otimes A} & & & & \\ N \otimes A \otimes M & \overset{\mathsf{id}_N \otimes \rho_M \circ c_{A,M}}{\underset{\rho_N \otimes \mathsf{id}_M}{\rightrightarrows}} & N \otimes M & \longrightarrow & N \otimes_A M \end{array} \tag{8.101}$$

is commutative. That is, both squares in (8.101) commute. The lower square (i.e., the one containing the lower horizontal arrows) commutes since braiding is a natural isomorphism. The upper square in (8.101) is precisely the perimeter of the following diagram:

$$
\begin{array}{ccccc}
 & & M\otimes N\otimes A & & \\
 & \nearrow^{\mathrm{id}_M\otimes c_{A,N}} & & \searrow^{c_{M,N\otimes A}} & \\
M\otimes A\otimes N & \xrightarrow{c_{M\otimes A,N}} & N\otimes M\otimes A & \xrightarrow{\mathrm{id}_N\otimes c_{M,A}} & N\otimes A\otimes M \\
\downarrow{\scriptstyle \rho_M\otimes\mathrm{id}_N} & & \downarrow{\scriptstyle \mathrm{id}_N\otimes\rho_M} & & \downarrow{\scriptstyle \mathrm{id}_N\otimes c_{A,M}}\\
M\otimes N & \xrightarrow{c_{M,N}} & N\otimes M & \xleftarrow{\mathrm{id}_N\otimes\rho_M} & N\otimes M\otimes A.
\end{array} \tag{8.102}
$$

In diagram (8.102) the triangle on top commutes by the hexagon axiom (8.1), the left square commutes by naturality of braiding, and the right square commutes since $A \in \mathcal{D}'$. Hence, the upper square in (8.101) commutes. Hence, there is a canonical isomorphism of difference cokernels $\tilde{c}_{M,N} : M \otimes_A N \xrightarrow{\sim} N \otimes_A M$. It is clear that this isomorphism provides $\mathcal{D}_G$ with a braiding such that the statement of the Proposition follows. □

EXERCISE 8.23.9. Let (G, q) be a pre-metric group. Prove that there exists a finite abelian group $\tilde{G}$ with surjective homomorphism $\tilde{G} \to G$ such that the pullback of q to $\tilde{G}$ is of the form $B(g, g)$ for some bicharacter $B : \tilde{G} \times \tilde{G} \to \Bbbk^\times$.

EXAMPLE 8.23.10. Let $\mathcal{D}$ be a pointed braided fusion category with associated pre-metric group (G, q) (see Section 8.4). Assume that $H \subset G$ is as in Example 8.23.7. Then the full subcategory of $\mathcal{D}$ generated by simple objects with isomorphism classes in H is contained in $\mathcal{D}'$. Moreover, this subcategory is equivalent to $\mathsf{Rep}(H^*)$ where H^* is the dual group of H, see Theorem 9.9.22. Thus Proposition 8.23.8 applies. Observe that the category $\mathcal{D}_{H^*}$ is pointed with associated metric group $(G/H, \tilde{q})$. Note that this observation together with Exercises 8.23.9, 8.4.4 and 8.4.5 provides a proof of the surjectivity statement in Theorem 8.4.9.

We use the de-equivariantization construction to analyze centers of graded fusion categories.

PROPOSITION 8.23.11. *Let G be a finite group and let*

$$\mathcal{C} = \bigoplus_{g \in G} \mathcal{C}_g$$

be a fusion category faithfully graded by G. Then $\mathcal{Z}(\mathcal{C})$ contains a Tannakian subcategory $\mathcal{E} = \mathsf{Rep}(G)$ such that the de-equivariantization of $\mathcal{E}'$ by $\mathcal{E}$ in $\mathcal{Z}(\mathcal{C})$ is equivalent to $\mathcal{Z}(\mathcal{C}_1)$ as a braided fusion category.

PROOF. We construct a subcategory $\mathcal{E} \subset \mathcal{Z}(\mathcal{C})$ as follows. For any representation $\pi : G \to GL(V)$ of G consider an object Y_π in $\mathcal{Z}(\mathcal{C})$ where $Y_\pi = V \otimes \mathbf{1}$ as an object of $\mathcal{C}$ with the permutation isomorphism

$$c_{Y_\pi, X} := \pi(g) \otimes \mathsf{id}_X : Y_\pi \otimes X \cong X \otimes Y_\pi, \qquad \text{when } X \in \mathcal{C}_g, \tag{8.103}$$

where we identified $Y_\pi \otimes X$ and $X \otimes Y_\pi$ with $V \otimes X$. Let $\mathcal{E}$ be the fusion subcategory of $\mathcal{Z}(\mathcal{C})$ consisting of the objects Y_π, where π runs through all finite dimensional representations of G. Clearly, $\mathcal{E}$ is equivalent to $\mathsf{Rep}(G)$ with its standard braiding.

By construction, the forgetful functor maps $\mathcal{E}$ to Vec and $\mathcal{E}'$ consists of all objects in $\mathcal{Z}(\mathcal{C})$ whose forgetful image is in $\mathcal{C}_1$. Consider the surjective braided functor $H : \mathcal{E}' \to \mathcal{Z}(\mathcal{C}_1)$ obtained by restricting the braiding of $X \in \mathcal{E}'$ from $\mathcal{C}$ to $\mathcal{C}_1$. This functor H can be factored through the de-equivariantization functor $\mathcal{E}' \to (\mathcal{E}')_G$. Hence, there is a surjective tensor functor $\tilde{H} : (\mathcal{E}')_G \to \mathcal{Z}(\mathcal{C}_1)$.

Using Theorems 7.16.6 and 8.21.5, we conclude that categories $(\mathcal{E}')_G$ and $\mathcal{Z}(\mathcal{C}_1)$ both have Frobenius-Perron dimension $\mathsf{FPdim}(\mathcal{C}_1)^2$. By Proposition 6.3.4, $\tilde{H}$ is a braided equivalence. □

COROLLARY 8.23.12. *In the notation of Proposition* 8.23.11 *we have*

$$\dim(\mathcal{Z}(\mathcal{C})) = |G|^2 \dim(\mathcal{Z}(\mathcal{C}_1)).$$

PROOF. Let $\mathcal{E} = \mathsf{Rep}(G)$ be the canonical Tannakian subcategory of $\mathcal{Z}(\mathcal{C})$. Then $\dim(\mathcal{E}) = |G|$. Using Corollary 8.20.13 and Theorem 8.21.1 we obtain

$$\dim(\mathcal{E}') = \frac{\dim(\mathcal{Z}(\mathcal{C}))}{\dim(\mathcal{E})} = \frac{\dim(\mathcal{Z}(\mathcal{C}))}{|G|}.$$

Next, by Proposition 7.21.15

$$\dim((\mathcal{E}')_G) = \frac{\dim(\mathcal{E}')}{|G|} = \frac{\dim(\mathcal{Z}(\mathcal{C}))}{|G|^2}.$$

The result follows since $(\mathcal{E}')_G \cong \mathcal{Z}(\mathcal{C}_1)$ by Proposition 8.23.11. □

COROLLARY 8.23.13. *Let G be a finite group acting on a fusion category $\mathcal{C}$. Then*

$$\dim(\mathcal{Z}(\mathcal{C}^G)) = |G|^2 \dim(\mathcal{Z}(\mathcal{C})).$$

PROOF. We saw in Example 7.12.25 that $\mathcal{C}^G$ is categorically Morita equivalent to the graded fusion category $\mathcal{C} \rtimes G$ whose trivial component is $\mathcal{C}$. Therefore by Corollary 7.16.2, $\mathcal{Z}(\mathcal{C}^G) \cong \mathcal{Z}(\mathcal{C} \rtimes G)$, and the claim follows from Corollary 8.23.12. □

8.24. Braided G-crossed categories

Let G be a finite group. Kirillov Jr. [**Kir1**] and Müger [**Mu5**] found a description of all braided fusion categories $\mathcal{D}$ containing $\mathsf{Rep}(G)$. Namely, they showed that the datum of a braided fusion category $\mathcal{D}$ containing $\mathsf{Rep}(G)$ is equivalent to the datum of a *braided G-crossed category* $\mathcal{C}$, see Theorem 8.24.3.

DEFINITION 8.24.1. A *braided G-crossed fusion category* is a fusion category $\mathcal{C}$ equipped with the following structures:

(i) a (not necessarily faithful) grading $\mathcal{C} = \bigoplus_{g \in G} \mathcal{C}_g$,
(ii) an action $g \mapsto T_g$ of G on $\mathcal{C}$ such that $T_g(\mathcal{C}_h) \subset \mathcal{C}_{ghg^{-1}}$,
(iii) a natural collection of isomorphisms, called the *G-braiding*:

$$c_{X,Y} : X \otimes Y \simeq T_g(Y) \otimes X, \qquad X \in \mathcal{C}_g,\ g \in G \text{ and } Y \in \mathcal{C}. \tag{8.104}$$

Let $\gamma_{g,h} : T_g T_h \xrightarrow{\sim} T_{gh}$ denote the tensor structure of the functor $g \mapsto T_g$ and let μ_g denote the tensor structure of T_g.

The above structures are required to satisfy the following compatibility conditions:

(a) the diagram

(8.105)

$$\begin{CD}
T_g(X) \otimes T_g(Y) @>{c_{T_g(X),T_g(Y)}}>> T_{ghg^{-1}}(T_g(Y)) \otimes T_g(X) \\
@A{(\mu_g)^{-1}_{X,Y}}AA @VV{(\gamma_{ghg^{-1},g})_Y \otimes \mathsf{id}_{T_g(X)}}V \\
T_g(X \otimes Y) @. T_{gh}(Y) \otimes T_g(X) \\
@V{T_g(c_{X,Y})}VV @AA{(\gamma_{g,h})_Y \otimes \mathsf{id}_{T_g(X)}}A \\
T_g(T_h(Y) \otimes X) @>>{(\mu_g)^{-1}_{T_g(Y),X}}> T_g(T_h(Y)) \otimes T_g(X),
\end{CD}$$

commutes for all $g, h \in G$ and objects $X \in \mathcal{C}_h$, $Y \in \mathcal{C}$,

(b) the diagram

(8.106)

$$\begin{array}{ccccc}
 & & (X \otimes Y) \otimes Z & & \\
 & \swarrow^{\alpha_{X,Y,Z}} & & \searrow^{c_{X,Y} \otimes \mathsf{id}_Z} & \\
X \otimes (Y \otimes Z) & & & & (T_g(Y) \otimes X) \otimes Z \\
\downarrow{\scriptstyle c_{X,Y \otimes Z}} & & & & \downarrow{\scriptstyle \alpha_{T_g(Y),X,Z}} \\
T_g(Y \otimes Z) \otimes X & & & & T_g(Y) \otimes (X \otimes Z) \\
\downarrow{\scriptstyle (\mu_g)^{-1}_{Y,Z} \otimes \mathsf{id}_X} & & & & \downarrow{\scriptstyle \mathsf{id}_{T_g(Y)} \otimes c_{X,Z}} \\
(T_g(Y) \otimes T_g(Z)) \otimes X & & \xrightarrow{\alpha_{T_g(Y),T_g(Z),X}} & & T_g(Y) \otimes (T_g(Z) \otimes X)
\end{array}$$

commutes for all $g \in G$ and objects $X \in \mathcal{C}_g, Y, Z \in \mathcal{C}$, and

(c) the diagram

(8.107)

$$\begin{array}{ccccc}
 & & X \otimes (Y \otimes Z) & & \\
 & \nearrow^{\alpha_{X,Y,Z}} & & \searrow^{\mathsf{id}_X \otimes c_{Y,Z}} & \\
(X \otimes Y) \otimes Z & & & & X \otimes (T_h(Z) \otimes Y) \\
\uparrow{\scriptstyle c^{-1}_{X \otimes Y,Z}} & & & & \downarrow{\scriptstyle \alpha^{-1}_{X,T_h(Z),Y}} \\
T_{gh}(Z) \otimes (X \otimes Y) & & & & (X \otimes T_h(Z)) \otimes Y \\
\uparrow{\scriptstyle (\gamma_{g,h})_Z \otimes \mathsf{id}_{X \otimes Y}} & & & & \downarrow{\scriptstyle c_{X,T_h(Z)} \otimes \mathsf{id}_Y} \\
T_gT_h(Z) \otimes (X \otimes Y) & & \xrightarrow{\alpha^{-1}_{T_gT_h(Z),X,Y}} & & (T_gT_h(Z) \otimes X) \otimes Y.
\end{array}$$

commutes for all $g, h \in G$ and objects $X \in \mathcal{C}_g$, $Y \in \mathcal{C}_h$, $Z \in \mathcal{C}$.

REMARK 8.24.2. The trivial component $\mathcal{C}_1$ of a braided G-crossed fusion category $\mathcal{C}$ is a braided fusion category with the action of G by braided autoequivalences. This can be seen by taking $X, Y \in \mathcal{C}_1$ in diagrams (8.105) – (8.107).

THEOREM 8.24.3 ([**Kir1, Mu5**]). *The equivariantization and de-equivariantization constructions establish a bijection between the set of equivalence classes of braided G-crossed fusion categories and the set of equivalence classes of braided fusion categories containing* $\mathsf{Rep}(G)$ *as a symmetric fusion subcategory.*

We shall now sketch the proof of this theorem. An alternative approach is given in [**DrGNO1**].

Suppose $\mathcal{C}$ is a braided G-crossed fusion category. We define a braiding $\tilde{c}$ on its equivariantization $\mathcal{C}^G$ as follows.

Let $(X, \{u_g\}_{g\in G})$ and $(Y, \{v_g\}_{g\in G})$ be objects in $\mathcal{C}^G$. Let $X = \oplus_{g\in G} X_g$ be a decomposition of X with respect to the grading of $\mathcal{C}$. Define an isomorphism

(8.108)
$$\tilde{c}_{X,Y} : X\otimes Y = \bigoplus_{g\in G} X_g\otimes Y \xrightarrow{\oplus c_{X_g,Y}} \bigoplus_{g\in G} T_g(Y)\otimes X_g \xrightarrow{\oplus v_g\otimes \mathsf{id}_{X_g}} \bigoplus_{g\in G} Y\otimes X_g = Y\otimes X.$$

It follows from condition (a) of Definition 8.24.1 that $\tilde{c}_{X,Y}$ respects the equivariant structures, i.e., it is an isomorphism in $\mathcal{C}^G$. Its naturality is clear. The fact that $\tilde{c}$ is a braiding on $\mathcal{C}^G$ (i.e., the hexagon axioms) follows from the commutativity of diagrams (8.106) and (8.107). It is easy to check that $\tilde{c}$ restricts to the standard braiding on $\mathsf{Rep}(G) = \mathsf{Vec}^G \subset \mathcal{C}^G$. Hence, $\mathcal{C}^G$ contains a Tannakian subcategory $\mathsf{Rep}(G)$.

Conversely, let $\mathcal{C}$ be a braided fusion category with braiding c containing a Tannakian subcategory $\mathsf{Rep}(G)$. The restriction of the de-equivariantization functor F from (8.99) to $\mathsf{Rep}(G)$ is isomorphic to the fiber functor $\mathsf{Rep}(G) \to \mathsf{Vec}$. Hence for any object X in $\mathcal{C}_G$ and any object V in $\mathsf{Rep}(G)$ we have an automorphism of $F(V) \otimes X$ defined as the composition

$$F(V) \otimes X \xrightarrow{\sim} X \otimes F(V) \xrightarrow{\sim} F(V) \otimes X, \tag{8.109}$$

where the first isomorphism comes from the fact that $F(V) \in \mathsf{Vec}$ and the second one is a consequence of de-equivariantization being a central functor.

When X is simple, we have an isomorphism $\mathsf{Aut}_{\mathcal{C}}(F(V) \otimes X) \cong \mathsf{Aut}_{\mathsf{Vec}}(F(V))$, hence we obtain a tensor automorphism i_X of $F|_{\mathsf{Rep}(G)}$. Since $\mathsf{Aut}_{\otimes}(F|_{\mathsf{Rep}(G)}) \cong G$, we have an assignment $X \mapsto i_X \in G$. The hexagon axiom of the braiding implies that this assignment is multiplicative, i.e., that $i_Z = i_X i_Y$ for any simple object Z contained in $X \otimes Y$. Thus, it defines a G-grading on $\mathcal{C}$:

$$\mathcal{C} = \bigoplus_{g\in G} \mathcal{C}_g, \text{ where } \mathcal{O}(\mathcal{C}_g) = \{X \in \mathcal{O}(\mathcal{C}) \mid i_X = g\}. \tag{8.110}$$

It is straightforward to check that $i_{T_g(X)} = ghg^{-1}$ whenever $i_X = h$.

Finally, to construct a G-crossed braiding on $\mathcal{C}$, observe that $\mathcal{C}$ and $\mathcal{C}^{\text{rev}}$ are embedded into the crossed product category $\mathcal{C} \rtimes G = (\mathcal{C}^G)^*_{\mathcal{C}}$ as subcategories $\mathcal{C}_{\text{left}}$ and $\mathcal{C}_{\text{right}}$ consisting, respectively, of functors of left and right multiplications by objects of $\mathcal{C}$. Clearly, there is a natural family of isomorphisms

$$X \otimes Y \xrightarrow{\sim} Y \otimes X, \qquad X \in \mathcal{C}_{\text{left}},\ Y \in \mathcal{C}_{\text{right}}, \tag{8.111}$$

satisfying obvious compatibility conditions. Note that $\mathcal{C}_{\text{left}}$ is identified with the diagonal subcategory of $\mathcal{C} \rtimes G$ spanned by objects $X \boxtimes g$, $X \in \mathcal{C}_g$, $g \in G$, and $\mathcal{C}_{\text{right}}$ is identified with the trivial component subcategory $\mathcal{C} \boxtimes e$. Using (4.22), we conclude that isomorphisms (8.111) give rise to a G-crossed braiding on $\mathcal{C}$.

One can check that the two above constructions (from braided fusion categories containing $\mathsf{Rep}(G)$ to braided G-crossed categories and vice versa) are inverses of each other, see [**Kir1, Mu5, DrGNO2**] for details.

REMARK 8.24.4. Let $\mathcal{C} = \oplus_{g \in G} \mathcal{C}_g$ be a braided G-crossed fusion category. It was shown in [**DrGNO1**] that the braided category $\mathcal{C}^G$ is non-degenerate if and only if $\mathcal{C}_1$ is non-degenerate and the G-grading of $\mathcal{C}$ is faithful.

8.25. Braided Hopf algebras, Nichols algebras, pointed Hopf algebras

8.25.1. Braided bialgebras and Hopf algebras. Let $\mathcal{C}$ be a braided tensor category, and B be an algebra in $\mathcal{C}$. Then, as we have seen above (see Exercise 8.8.2(iv)), $B \otimes B$ is also an algebra in $\mathcal{C}$. Therefore, it makes sense to talk about B being a bialgebra or a Hopf algebra in $\mathcal{C}$ (namely, the usual definition makes sense). Moreover, if B is a bialgebra (respectively, Hopf algebra) in $\mathcal{C}$ then the category $B - \mathbf{mod}$ of B-modules in $\mathcal{C}$ is a tensor category (again, the usual definition of the tensor product makes sense).

In particular, if (H, R) is a quasitriangular Hopf algebra, then a bialgebra (respectively, Hopf algebra) B in the category $\mathbf{Rep}(H)$ is called an (H, R)-*braided bialgebra* (respectively, *braided Hopf algebra*). Applying the forgetful functor, we see that B is an algebra in the ordinary sense together with a coassociative coproduct $\overline{\Delta} : B \to B \otimes B$, but in general $\overline{\Delta}$ is not an algebra homomorphism from B to the usual tensor product $B \otimes B$. Rather, using the definition of the tensor product of algebras from Exercise 8.8.2(iv), we find that it is a twisted homomorphism in the following sense:

$$\overline{\Delta}(ab) = (m \otimes m)(\mathsf{id} \otimes R^{-1}\sigma \otimes \mathsf{id})(\overline{\Delta}(a) \otimes \overline{\Delta}(b)),$$

where $m : B \otimes B \to B$ is the multiplication and σ is the permutation of components (this differs from the usual Hopf algebra axiom by the additional factor R^{-1}).

EXERCISE 8.25.1. Let $R^{-1} = \sum \alpha_i \otimes \beta_i$. Show that if B is an (H, R)-braided bialgebra (respectively, Hopf algebra), then the smash product algebra $B\#H$ (see Exercise 7.8.32) has a natural bialgebra (respectively, Hopf algebra) structure (in the usual sense), with coproduct defined by the formula

$$\Delta(b) = \sum_i (b_1 \otimes \alpha_i) \otimes \beta_i(b_2), \; b \in B,$$

where $\overline{\Delta}(b) = b_1 \otimes b_2$ (Sweedler's notation).

In particular, all this applies to the case $H = D(K)$ (the quantum double of K), where K is a finite dimensional Hopf algebra. In this case, $R, R^{-1} \in K \otimes K^*$, so we see that the subalgebra $B\#K \subset B\#H$ is a Hopf subalgebra. The Hopf algebra $B\#K$ is called *the Radford biproduct* or *bosonization* of B (following Majid, [**Maj3**]), see [**Ra5, Ra6**] and references therein. Moreover, it is easy to see that the construction of bosonization makes sense in the more general case, when K is not necessarily finite dimensional, and B is an algebra in the category $YD(K)$ of Yetter-Drinfeld modules over K (with opposite braiding). We will use the terms "braided bialgebra" and "braided Hopf algebra" in this case as well.

EXAMPLE 8.25.2. $K = \Bbbk G$, where G is a group. Recall that in this case, a Yetter-Drinfeld module over K is just a G-graded G-module (where G acts on itself by conjugation). Thus, the algebra B in this case has a decomposition $B = \oplus_{g\in G} B_g$, and $R^{-1} = \sum_{g\in G} g^{-1} \otimes \delta_g$, so for $b \in B_g$ we have

$$\Delta(b) = \sum_{g\in G} \overline{\Delta}_g(b) \otimes (g^{-1} \otimes 1),$$

where $\overline{\Delta}_g(b)$ is the projection of $\overline{\Delta}(b)$ to $B \otimes B_g$.

An important theorem about braided Hopf algebras is the following *Radford's biproduct theorem* ([**Ra5, Ra6, AndrS**]).

THEOREM 8.25.3. *Let K be a Hopf algebra, and A be a Hopf algebra containing K. Suppose that $\pi : A \to K$ is a Hopf algebra homomorphism such that $\pi|_K = \mathsf{id}$. Let $p : A \to A$ be the linear map defined by $\pi(a) = a_1 S(\pi(a_2))$, where we use Sweedler's notation. Let $B = p(A)$. Then B has a natural structure of a braided Hopf algebra in $YD(K)$, and the multiplication map $B \otimes K \to A$ is an isomorphism of Hopf algebras $B\#K \to A$.*

Now let K be any Hopf algebra, and V a (possibly infinite-dimensional) Yetter-Drinfeld module over K. Then the tensor algebra $B := TV$ is a Hopf algebra in the category $YD(K)$, with coproduct given by the condition that $v \in V$ is primitive, i.e.,

$$\overline{\Delta}(v) = v \otimes 1 + 1 \otimes v, \ v \in V$$

(it is easy to see that this coproduct extends uniquely to higher degrees). In particular, in the setting of Example 8.25.2 (i.e., for $K = \Bbbk G$), we have $V = \oplus_{g\in G} V_g$, and for $v \in V_g$ we have

$$\Delta(v) = v \otimes 1 + g^{-1} \otimes v,$$

i.e., v is a $(g^{-1}, 1)$-skew-primitive element. So, the biproduct $TV\#\Bbbk G$ is a coradically graded pointed Hopf algebra generated by grouplike elements $g \in G$ and skew-primitive elements $v \in V_g$, $g \in G$.

In particular, if G is a finite abelian group of order coprime to char$\Bbbk$, then we can find a biproduct $\{x_i\}$ of V compatible to the grading, i.e. $x_i \in V_{g_i}$, $g_i \in G$. Then we have

$$gx_i = \chi_i(g) x_i g, \ \Delta(x_i) = x_i \otimes 1 + g_i^{-1} \otimes x_i,$$

where χ_i are characters of G.

EXERCISE 8.25.4. Show that any pointed Hopf algebra $\mathcal{H}$ generated by grouplike and skew-primitive elements is a quotient of $TV\#\Bbbk G$ for a suitable $V \in YD(\Bbbk G)$. Namely, $V = \oplus_{h\in G} \mathsf{Ext}^1_{\mathcal{H}-\mathrm{comod}}(h, 1)$. (Recall that by Proposition 1.9.12, $\mathsf{Ext}^1_{\mathcal{H}-\mathrm{comod}}(h, g) \cong \mathsf{Prim}_{g,h}(\mathcal{H})/\Bbbk(g-h)$).

In particular, Conjecture 5.11.7 would imply that any finite dimensional pointed Hopf algebra is a quotient $TV\#\Bbbk G$ for some finite group G and some finite dimensional $V \in YD(\Bbbk G)$.

It turns out that for coradically graded pointed Hopf algebras, one can say more: any such Hopf algebra is a biproduct of a group algebra $\Bbbk G$ with a braided Hopf algebra B. Namely we have the following proposition, which is a special case of Radford's biproduct theorem (see [**Ra5, Ra6**]).

PROPOSITION 8.25.5. *Let $\mathcal{A}$ be a pointed Hopf algebra, and $A = \mathrm{gr}\mathcal{A}$. Then $A \cong B\#\Bbbk G$ is a Radford biproduct, where $G = \mathbf{G}(\mathcal{A}) = \mathbf{G}(A)$ is the group of grouplike elements of $\mathcal{A}$ and A, and B is a Hopf algebra in the category of Yetter-Drinfeld modules over $\Bbbk G$.*

PROOF. We have a Hopf algebra projection $\pi : A \to \Bbbk G$ such that $\pi|_G = \mathsf{id}$. This is the setting in which the Radford biproduct theorem applies. Namely, let $p : A \to A$ be the linear map defined by the formula $p(a) = a_1 S(\pi(a_2))$ (using Sweedler's notation). Then by Radford's theorem (Theorem 8.25.3), $B := \mathsf{Im}(p)$ is a subalgebra of A which has a structure of a braided algebra in $YD(\Bbbk G)$, and the multiplication map $B \otimes \Bbbk G \to A$ defines an isomorphism of Hopf algebras $B\#\Bbbk G \to A$. □

EXERCISE 8.25.6. Fill in the details in the proof of Proposition 8.25.5 (i.e., verify Radford's theorem in this special case).

EXERCISE 8.25.7. (i) Let A be the canonical algebra of a braided tensor category $\mathcal{C}$, see Definition 7.9.12, and let $T : \mathcal{C} \boxtimes \mathcal{C}^{\mathrm{op}} \cong \mathcal{C} \boxtimes \mathcal{C} \to \mathcal{C}$ be the tensor functor defined by tensor multiplication. Show that $T(A)$ has a natural structure of a Hopf algebra in $\mathcal{C}$.

(ii) In (i), let $\mathcal{C} = \mathsf{Rep}\, H$, where H is a finite dimensional quasitriangular Hopf algebra. Show that $T(A)$ is naturally isomorphic to H^* as a vector space, and describe the algebra and coalgebra structures on it in terms of this isomorphism. The algebra H^* with these operations is called the *braided dual* of H, see [**Maj2**].

(iii) Compute $T(A)$ in the case when $\mathcal{C}$ is the representation category (of type I) for the quantum group $U_q(\mathfrak{g})$, where q is not a root of 1. (You should get the corresponding quantum function algebra $O_q(G)$ with a modified multiplication (see Section 5.8); for $\mathfrak{g} = \mathfrak{gl}_n$ this is the so-called *reflection equation algebra*).

REMARK 8.25.8. The algebra A is commutative in the braided category $\mathcal{C} \boxtimes \mathcal{C}^{\mathrm{op}}$, but in general, the tensor functor T does not preserve braiding, and the algebra $T(A)$ is not commutative (in the sense of Definition 8.8.1). It is, however, commutative as a braided Hopf algebra in the sense of [**Maj4**], Definition 2.3, see also [**Maj3**]. Note that this definition makes sense only for Hopf algebras and not for algebras.

8.25.2. Classification of finite dimensional pointed Hopf algebras and Nichols algebras. Proposition 8.25.5 plays a central role in attacking the problem of classification of finite dimensional pointed Hopf algebras. This problem is still open, but a lot of progress has been made in recent years, and a powerful theory has been developed. Solving this problem amounts to two steps.

Step 1. Classification of coradically graded finite dimensional pointed Hopf algebras A, i.e., those for which $A \cong \mathrm{gr}(A)$.

By Proposition 8.25.5, in this case one has $A \cong B\#\Bbbk G$, where G is a finite group, so the problem reduces to the classification of finite dimensional $\mathbb{Z}_+$-graded braided Hopf algebras B in $YD(\Bbbk G)$ such that $B[0] = \Bbbk$, and the coradical filtration of B is induced by the grading.

Step 2. Given B, classification of *liftings* of $A := B\#\Bbbk G$, i.e., Hopf algebras $\mathcal{A}$ such that $\mathrm{gr}(\mathcal{A}) \cong A$.

This is essentially a deformation theory problem.

Let us focus on Step 1 (classification of graded algebras B). Assume that A is generated by grouplike and skew-primitive elements (as conjectured in Conjecture

5.11.7). In this case, B is generated by primitive elements, all sitting in degree ≤ 1, and the space of primitive elements of degree 1 is $V = B[1]$. This implies that $B = TV/I$, where $I \subset \bigoplus_{m\ge 2} V^{\otimes m}$ is a graded Hopf ideal of TV (of finite codimension).

This leads us to the fundamental notion of a *Nichols algebra*[8], going back to the paper [**Nic**]. The theory of Nichols algebras is a vast subject, and we will limit ourselves to a very brief review of this theory.

To define Nichols algebras, let us return to considering the Hopf algebra TV in the category $YD(K)$ for any Hopf algebra K.

PROPOSITION 8.25.9. *There exists the largest graded Hopf ideal*

$$I \subset \bigoplus_{m\ge 2} V^{\otimes m} \subset TV,$$

i.e. a graded ideal such that such that $\overline{\Delta}(I) \subset TV \otimes I + I \otimes TV$.

PROOF. Take the sum of all ideals I with this property. □

REMARK 8.25.10. Note that it is possible that $I = 0$ (see examples below).

DEFINITION 8.25.11. The Nichols algebra $\mathfrak{B}(V)$ is the Hopf algebra TV/I in the category $YD(K)$.

In other words, the Nichols algebra $\mathcal{B}(V)$ is the smallest possible algebra B as above (i.e., generated in degree 1) with $B[1] = V$. In particular, if a given B is finite dimensional and $B[1] = V$, then $\mathcal{B}(V)$ is finite dimensional.

Moreover, if G is abelian and $K = \Bbbk G$, it is known (see [**Ang2**]) that in this case $B = \mathcal{B}(V)$ (i.e. any finite dimensional Hopf quotient of TV is necessarily the smallest one); this is, in fact, equivalent to Conjecture 5.11.7 in the abelian case (proved in [**Ang2**]) by taking dual Hopf algebras (check it!). This shows that classification of finite dimensional Nichols algebras in $YD(\Bbbk G)$ is the main part of Step 1 (the classification of possible algebras B).

EXERCISE 8.25.12. Let $G = \mathbb{Z}/n\mathbb{Z}$ with generator g, and $q \in \Bbbk^\times$ be such that $q^n = 1$. Let V be the 1-dimensional Yetter-Drinfeld module for $\Bbbk G$ defined by $V = V_{g^m}$, $0 \le m \le n-1$, and $gv = qv$.

(i) Compute the algebra $\mathcal{B}(V)$. Show that it is finite dimensional if and only if $q^m \ne 1$; and in this case $\mathcal{B}(V) = \Bbbk[x]/(x^r)$, where r is the order of q^m.

(ii) Show that if q^m is a primitive n-th root of unity, then $\mathcal{B}(V)\#\Bbbk G$ is isomorphic to the Taft algebra of dimension n^2 (Example 5.5.6).

(iii) Show that the Taft algebra does not admit nontrivial liftings (i.e., any lifting of the Taft algebra is isomorphic to the Taft algebra itself as a Hopf algebra).

(iv) Assume that $m = 1$ and $q \ne 1$, but the order of q is $r < n$. Show that there exists a unique nontrivial lifting $H(q,n)$ of $\mathcal{B}(V)\#\Bbbk G$, generated by g, x with relations

$$gx = qxg,\ g^n = 1,\ x^r = 1 - g^{-r},$$

and coproduct

$$\Delta(x) = x \otimes 1 + g^{-1} \otimes x.$$

(the so-called *generalized Taft algebra*).

(v) Describe simple $H(q,n)$-comodules, and show that $H(q,n)$ is not pointed.

[8]Nichols algebras should not be confused with Nichols Hopf algebras of dimension 2^{n+1} considered in Example 5.5.8.

EXERCISE 8.25.13. Let $n > 1$ be odd, $G = \mathbb{Z}/n\mathbb{Z}$ with generator K, and $q \in \Bbbk^\times$ be a primitive n-th root of 1. Let V be two-dimensional with basis e and f and $K \cdot e = q^2 e$, $K \cdot f = q^{-2} f$, and $V = V_K$.

(i) Compute the Nichols algebra $\mathcal{B}(V)$, and show that it has dimension n^2 and is generated by e, f with defining relations $e^n = 0$, $f^n = 0$, $ef = q^2 fe$. Show that the corresponding biproduct $\mathcal{B}(V)\#\Bbbk G$ is isomorphic to $\mathrm{gr}(u_q(\mathfrak{sl}_2))$.

(ii) Show that $\mathcal{B}(V)\#\Bbbk G$ has a unique nontrivial lifting, namely, $u_q(\mathfrak{sl}_2)$.

EXERCISE 8.25.14. Generalize Exercises 8.25.12 and 8.25.13 to the case when $G = \mathbb{Z}$ and $q \in \Bbbk^\times$ is any number.

EXERCISE 8.25.15. Let $G = \mathbb{Z}/2\mathbb{Z} = < g >$, and $V = V_{-1}$, with $g|_V = -\mathrm{id}$. Show that $\mathcal{B}(V) = \wedge V$, and $\mathcal{B}(V)\#\Bbbk G$ is the Nichols Hopf algebra of dimension 2^{n+1}, where $n = \dim V$ (see Example 5.5.8).

In general, there are rather few pairs (G, V) for which the algebra $\mathcal{B}(V)$ is finite dimensional. To illustrate it, we give the following example.

EXAMPLE 8.25.16. Let A be a symmetric generalized Cartan matrix of size r by r. This means that $a_{ii} = 2$, $a_{ij} = a_{ji} \in \mathbb{Z}_{\le 0}$ for $i \neq j$, and $a_{ij} = 0$ iff $a_{ji} = 0$. Let $n > 1$ be an odd integer, and $q \in \Bbbk^\times$ be a primitive root of unity of order n. Let $G = (\mathbb{Z}/n\mathbb{Z})^r$, with generators K_i, $i = 1, ..., r$, such that $K_i^n = 1$, and let V have basis $e_1, ..., e_r$ with $K_i \cdot e_j = q^{a_{ij}} e_j$, and $e_j \in V_{K_j}$. Then it is known that $\mathcal{B}(V)$ is finite dimensional if and only if A is positive definite, i.e., is a Cartan matrix of a simply laced semisimple Lie algebra. In this case, $\mathcal{B}(V)\#\Bbbk G = u_q(\mathfrak{b}_+)$, the positive Borel subalgebra of the small quantum group $u_q(\mathfrak{g})$ (i.e., the subalgebra generated by K_i and $E_i := K_i^{-1} e_i$). In general, $\mathcal{B}(V)\#\Bbbk G = u_q(\mathfrak{b}_+)$ is the positive Borel subalgebra of the small quantum group attached to the Kac-Moody algebra $\mathfrak{g} = \mathfrak{g}(A)$. This example generalizes in a straightforward way to symmetrizable generalized Cartan matrices, so that finite dimensional Nichols algebras correspond to all semisimple Lie algebras.

Example 8.25.16 is in a certain sense prototypical. Namely, a complete classification of finite dimensional Nichols algebras with abelian group G is now known ([**He**]), and this classification is, in essence, a quantum version of the classification of finite dimensional simple Lie algebras and superalgebras, including positive (in particular, small) characteristic. In particular, a fundamental role in this classification is played by the notions of a Weyl groupoid and an arithmetic root system, due to Heckenberger, which are suitable generalizations of the notions of the Weyl group and root system of a simple Lie algebra. Also, the order of the roots of unity involved is a quantum analog of the characteristic, but things are more complicated since several different orders may be involved at the same time. In particular, there are many exceptional cases related to roots of unity of small order.

The lifting problem for abelian G is also expected to be completely solved in the near future. It has been solved in many cases, e.g. if $|G|$ is coprime to $2, 3, 5, 7$.

The case of nonabelian G is more complicated, and only partial results are available. There seems to be rather few non-abelian groups G generated by g such that $V_g \neq 0$, for which $\mathcal{B}(V)$ is finite dimensional. Very recent results on this topic can be found in [**HecV**].

Infinite dimensional Nichols algebras are less well understood. They should be viewed as quantum analogs of Kac-Moody Lie algebras and superalgebras in positive characteristic.

An exhaustive review on recent results on finite dimensional Nichols algebras and pointed Hopf algebras (as of March 2014) can be found in Andruskiewitsch's ICM-2014 talk, [**Andr**]. In particular, on Step 2 and specifically on construction of liftings as cocycle deformations, see [**Andr**], 3.8.

8.26. Bibliographical notes

8.1. Braided monoidal categories and functors between them were introduced by Joyal and Street in [**JoyS1, JoyS3, JoyS5**]. The notion of symmetric monoidal category appeared earlier in the works of Mac Lane [**Mac1**] and Bénabou [**Ben2**]. The Yang-Baxter equation has its origins in statistical mechanics and was studied in relation with quantum groups in [**Dr3**] and in relation with link invariants in [**Tu1**].

8.2. Relation between braided categories and braid groups (motivating the name "braiding") is discussed in [**JoyS5**].

8.3. Quasi-triangular Hopf algebras were introduced by Drinfeld in [**Dr3, Dr5**]. The notion of quantum double is also due to Drinfeld [**Dr3**]. Example 8.3.7 is due to Radford [**Ra3**]. The construction of R-matrices for quantum groups appeared in [**Dr3**]. A discussion of the cactus group and its connection to coboundary categories is contained in [**HenK**].

8.4. The equivalence between pre-metric groups and pointed braided fusion categories is due to Joyal and Street [**JoyS5**]. The proof in the text follows [**JoyS5**, Section 3]. An alternative approach can be found in [**DrGNO2**, Appendix D].

8.5. The center construction is due to Drinfeld (unpublished) and appears in the work of Majid [**Maj1**] and Joyal and Street [**JoyS2**]. Proposition 8.5.3 is proved in [**EtO1**]. Interpretation of the representation category of the quantum double of a Hopf algebra H as the center of $\mathsf{Rep}(H)$ can be found in Kassel's book [**Kas**].

8.6. The notion of a factorizable Hopf algebra was introduced by Reshetikhin and Semenov-Tian-Shansky in [**RS**]. They also proved that the quantum double of a finite-dimensional Hopf algebra is factorizable. The categorical notion of factorizability was introduced in [**ENO1**], where factorizability of $\mathcal{Z}(\mathcal{C})$ was also proved. Proposition 8.10.10 generalizes the corresponding result of Radford for Hopf algebras [**Ra4**] and the one of Bulacu and Torrecillas [**BuT**] for quasi-Hopf algebras.

8.7. Tensor functors (8.20) were considered in [**BocEK3**] under the name of *α-induction*.

8.8. The correspondence between central functors and commutative algebras in braided tensor categories was discussed in [**DaMNO**]. Quantum symmetric algebras were first considered in [**FaRT**].

8.9. See [**Dr5**] for Hopf algebra version of the Drinfeld isomorphism and textbooks [**BakK**, Section 2.2] and [**Kas**, Chapter XIV] for the categorical version.

8.10. The definition of a twist in a braided category is given by Shum [**Sh**]. Ribbon categories were called *balanced* in [**JoyS5**]. Theorem 8.10.7 is proved in [**ENO1**].

8.11. Ribbon Hopf algebras were introduced by Reshetikhin and Turaev [**RT1**].

8.12. The results of this section are taken from [**ENO3**] and [**DaN**].

8.13. The notion of S-matrix is due to Turaev [**Tu3**]. Pre-modular categories appeared in the work of Bruguiéres [**Bru**]. The S-matrix (8.47) was first considered by Lusztig in [**Lus4**] under the name of "exotic Fourier transform". It also appeared in the work [**DiVVV**] in connection with orbifold conformal field theories.

8.14. The notion of a modular category was introduced by Turaev [**Tu3**] in connection with the study of invariants of 3-manifolds. The notion of a *modular Hopf algebra*, i.e., a semisimple Hopf algebra whose representation category is modular was introduced by Reshetikhin and Turaev on [**RT2**]. Formula (8.54) was conjectured by Verlinde [**Verl**] in the context of conformal field theory and proved by Moore and Seiberg [**MooS2**]. Proposition 8.14.6 was proved in [**EtG10**]. A discussion of modular categories and their applications can be found in textbooks [**Tu2**] and [**BakK**]. Proof of Theorem 8.14.7 is taken from [**ENO2**, Appendix].

8.15. The results of this Section are taken from [**Tu2**, Section II.1] and [**BakK**, Section 3.1]. Our proofs follow [**DrGNO2**, Section 6].

8.16. The projective representation (8.64) of $SL_2(\mathbb{Z})$ is described in [**MooS2**] in the setting of 2-dimensional conformal field theory and in [**Tu2**, Section 3.9] in the categorical setting.

8.17. Our exposition of the theory of modular data follows [**Gann3**] and [**Lus4**]. Verlinde fusion rings are considered in [**Verl**], which is why we refer to their categorifications as Verlinde categories.

8.18. Corollaries 8.18.2 and 8.18.3 were established by Anderson and Moore [**AndeM**] and Vafa [**Vaf**]. Our exposition follows [**E**].

8.19. This Section follows [**DrGNO2**, Sections 2.8.1 and 3.4.2].

8.20. The fundamental notions of the centralizer of a set of objects in a braided category and non-degeneracy were introduced by Müger in [**Mu3**]. Results of this section were first obtained in [**Mu3**] in the setting of modular categories and generalized to arbitrary braided fusion categories in [**DrGNO2**, Section 3].

8.21. Properties of centralizers considered in this section were established in [**Mu3**] for modular categories and in [**DrGNO2**, Section 3] for braided fusion categories.

8.22. Projective centralizers were introduced and studied in [**DrGNO2**, Section 3.3].

8.23. The idea of de-equivariantization is due to Bruguiéres [**Bru**] and Müger [**Mu1**]. For the most part, our exposition follows [**DrGNO2**, Section 4]. Proposition 8.23.11 is proved in [**ENO3**]. Theorem 8.23.3 is similar to Theorem 4.1 in [**Kir1**].

8.24. The notion of a braided G-crossed category is due to Turaev [**Tu5, Tu6**]. These categories were extensively studied in [**Kir1, Mu5, DrGNO2**].

8.25. A discussion of braided bialgebras and Hopf algebras can be found in [**Ra5**]. For a review of pointed Hopf algebras and Nichols algebras see [**Ra5, AndrS**], and for more recent results [**Andr**].

8.27. Other results

8.27.1. Categorical Witt equivalence and completely anisotropic braided fusion categories. Non-degenerate braided fusion categories $\mathcal{C}_1$ and $\mathcal{C}_2$ are *Witt equivalent* if there exists a braided equivalence $\mathcal{C}_1 \boxtimes \mathcal{Z}(\mathcal{A}_1) \simeq \mathcal{C}_2 \boxtimes \mathcal{Z}(\mathcal{A}_2)$, where $\mathcal{A}_1$, $\mathcal{A}_2$ are fusion categories [**DaMNO**].

The *categorical Witt group* $\mathcal{W}$ is the group of Witt equivalence classes of non-degenerate braided fusion categories. The class of Drinfeld centers is the trivial element of $\mathcal{W}$. The classical counterpart of $\mathcal{W}$ is the group $\mathcal{W}_{cl}$ of quadratic forms (over $\mathbb{k}$) on finite Abelian groups (this group is well known, see, e.g., [**Schar**]). Since any quadratic form on a finite Abelian group gives rise to a pointed braided fusion category, $\mathcal{W}_{cl}$ is a subgroup of $\mathcal{W}$. In the classical case Drinfeld centers are precisely Abelian groups with hyperbolic quadratic forms.

By definition, an *étale* algebra in $\mathcal{C}$ is a commutative separable algebra in $\mathcal{C}$. A braided fusion category $\mathcal{C}$ is *completely anisotropic* if there are no non-trivial étale algebras in $\mathcal{C}$. It was shown in [**DaMNO**] that every element of $\mathcal{W}$ is represented by a unique completely anisotropic category. One can pass from a braided fusion category $\mathcal{C}$ to its completely anisotropic representative by taking a certain subcategory $\mathcal{C}_A^0$ of the category of A-modules in $\mathcal{C}$, where A is a maximal étale algebra in $\mathcal{C}$. This is analogous to passing from a quadratic form on a group B to the one on $C^{\perp}/C$, where $C \subset B$ is an isotropic subgroup.

It was proved in [**DaNO**] that a completely anisotropic braided fusion category admits a unique decomposition into the tensor product of its pointed part and *simple* anisotropic categories (i.e., categories without proper fusion subcategories).

All examples of completely anisotropic categories known at present (apart from those coming from quadratic forms) come from fusion categories $\mathcal{C}(\mathfrak{g}, \ell)$ associated to affine Lie algebras.

Examples of étale algebras in $\mathcal{C}(\mathfrak{g}, \ell)$ come from the theory of conformal inclusions of affine Lie algebras [**KacVW, ScheW**] and coset models [**GodKO**]. They lead to non-trivial relations among the Witt classes of categories $\mathcal{C}(\mathfrak{g}, \ell)$. Note also that any étale algebra in $\mathcal{C}(\mathfrak{g}, \ell)$ gives rise to a modular invariant, i.e., to a matrix with integral entries invariant under (i.e., commuting with) the action of the modular group associated to $\mathcal{C}(\mathfrak{g}, \ell)$.

The categorical Witt group $\mathcal{W}$ provides a natural framework for classification of conformal field theories. There is a common belief among physicists that all rational conformal field theories come from lattice and WZW models via coset and orbifold (and perhaps chiral extension) constructions (see [**MooS1**]). A corresponding statement for braided fusion categories means that $\mathcal{W}$ is generated by classes of categories $\mathcal{C}(\mathfrak{g}, \ell)$. While this statement might be out of reach at the moment, any progress in the study of the Witt group $\mathcal{W}$ will have important consequences for classification of braided fusion categories and conformal field theories.

It was shown in [**DaNO**] that there is a canonical homomorphism $S : \mathcal{W} \to s\mathcal{W}$, where $s\mathcal{W}$ is the "super analogue" of the Witt group (consisting of classes of braided fusion categories over sVec). The homomorphism S sends the Witt class of non-degenerate braided fusion category $\mathcal{C}$ to that of $\mathcal{C} \boxtimes \mathsf{sVec}$. The structure of $s\mathcal{W}$ was also described in [**DaNO**]. Namely,

$$s\mathcal{W} = s\mathcal{W}_{cl} \oplus s\mathcal{W}_2 \oplus s\mathcal{W}_{\infty}, \tag{8.112}$$

where $s\mathcal{W}_{cl}$ is the classical part, $s\mathcal{W}_2$ is an elementary Abelian 2-group, and $s\mathcal{W}_{\infty}$ is a free Abelian group. The kernel of S is isomorphic to $\mathbb{Z}/16\mathbb{Z}$ and is generated by Ising categories. In particular, the group $\mathcal{W}$ is 2-primary, i.e., has no odd torsion.

8.27.2. Brauer-Picard groups as categorical orthogonal groups. The Brauer-Picard group $\mathsf{BrPic}(\mathcal{C})$ of a finite tensor category $\mathcal{C}$ is the group of equivalence classes of exact invertible $\mathcal{C}$-bimodule categories with respect to the tensor product $\boxtimes_{\mathcal{C}}$ defined by a universal property. It was proved in [**ENO4**] that when $\mathcal{C}$ is a fusion category, $\mathsf{BrPic}(\mathcal{C})$ is isomorphic to the group $\mathsf{Aut}_{\otimes}^{br}(\mathcal{Z}(\mathcal{C}))$ of braided autoequivalences of $\mathcal{Z}(\mathcal{C})$. This result was generalized to finite tensor categories in [**DaN**]. E.g., for $\mathcal{C} = \mathsf{Vec}_A$, the category of vector spaces graded by a finite Abelian group A, one has $\mathsf{BrPic}(\mathsf{Vec}_A) \cong O(A \oplus A^*)$, where A^* is the group of characters of

A and $O(A \oplus A^*)$ is the group of automorphisms of $A \oplus A^*$ preserving its canonical quadratic form. Thus, the Brauer-Picard group of a tensor category can be viewed as a categorical analogue of the orthogonal group.

If $\mathcal{C}$ is braided then one-sided $\mathcal{C}$-module categories can be viewed as $\mathcal{C}$-bimodule categories (similarly to how modules over a commutative ring can be viewed as bimodules). In particular, one can talk about invertible $\mathcal{C}$-module categories. Equivalence classes of such categories form a subgroup in $\mathsf{BrPic}(\mathcal{C})$, called the *Picard group* of $\mathcal{C}$ and denoted $\mathsf{Pic}(\mathcal{C})$. The image of $\mathsf{Pic}(\mathcal{C})$ under the isomorphism $\mathsf{BrPic}(\mathcal{C}) \cong \mathsf{Aut}^{br}_{\otimes}(\mathcal{Z}(\mathcal{C}))$ consists of equivalence classes of autoequivalences of $\mathcal{Z}(\mathcal{C})$ trivializable on $\mathcal{C}$, where $\mathcal{C}$ is viewed as a subcategory of $\mathcal{Z}(\mathcal{C})$ via the embedding $X \mapsto (X, c_{X,-})$ [**DaN**].

One has $\mathsf{BrPic}(\mathcal{C}) \cong \mathsf{Pic}(\mathcal{Z}(\mathcal{C}))$ for any finite tensor category $\mathcal{C}$ [**ENO4, DaN**].

More generally, two finite tensor categories are categorically Morita equivalent if and only if their centers are equivalent as braided tensor categories [**ENO3, DaN**].

In practice it is much easier to work with the group $\mathsf{Aut}^{br}_{\otimes}(\mathcal{Z}(\mathcal{C}))$ than with $\mathsf{BrPic}(\mathcal{C})$, since the multiplication of the latter is defined by an abstract universal property while for the former it is simply the composition of functors. In addition, $\mathsf{Aut}^{br}_{\otimes}(\mathcal{Z}(\mathcal{C}))$ can be viewed as a generalization of the classical orthogonal group which brings important geometric insights.

The group $\mathsf{BrPic}(\mathcal{C})$ has a natural action on the set of $\mathcal{C}$-module categories. In view of the above result it identifies with the action of $\mathsf{Aut}^{br}_{\otimes}(\mathcal{Z}(\mathcal{C}))$ on the set of Lagrangian étale algebras in $\mathcal{Z}(\mathcal{C})$, i.e., on the "quantum Lagrangian Grassmannian". This action was studied in [**NikR**] and [**BonN**] where Brauer-Picard groups of symmetric tensor categories were computed.

8.27.3. Core of a braided fusion category. The *core* of a braided fusion category $\mathcal{C}$ was introduced in [**DrGNO2**]. It is defined as the fiber category $\mathcal{E}' \boxtimes_{\mathcal{E}} \mathsf{Vec}$, where $\mathcal{E}$ is a maximal Tannakian subcategory of $\mathcal{E}$. The core does not depend on the choice of $\mathcal{E}$ and is *weakly anisotropic*, i.e., has no Tannakian subcategories stable under the group of braided autoequivalences. The core separates the part of a braided fusion category that does not come from finite groups.

It was explained in [**DrGNO1, DrGNO2**] that any braided fusion category $\mathcal{C}$ is obtained from a weakly anisotropic category (namely, the core of $\mathcal{C}$) using finite groups (via the equivariantization and extension procedures). Thus, the classification of braided fusion categories reduces to that of weakly anisotropic ones.

Let $\mathcal{C}$ be a weakly anisotropic fusion category, let $\mathcal{C}_{pt}$ be its maximal pointed fusion subcategory, and for any $\mathcal{D} \subset \mathcal{C}$ let $\mathcal{D}'$ denote the centralizer of $\mathcal{D}$. Then $\mathcal{C}$ has one of the following types:

(1) *ordinary*, if $\mathcal{C}_{pt} \cap (\mathcal{C}_{pt})' = \mathcal{C}' = \mathsf{Vec}$;
(2) *super*, if $\mathcal{C}_{pt} \cap (\mathcal{C}_{pt})' = \mathcal{C}' = \mathsf{sVec}$;
(3) *Ising*, if $\mathcal{C}' = \mathsf{Vec}$ and $\mathcal{C}_{pt} \cap (\mathcal{C}_{pt})' = \mathsf{sVec}$.

The smallest categories of the last type are the so-called *Ising categories*, i.e., non-pointed categories of dimension 4 (see, e.g., [**DrGNO2, MooS1**]).

A weakly anisotropic fusion category $\mathcal{C}$ of ordinary type (respectively, of super type) is said to be *simple* if it is not pointed and has no proper fusion subcategories except Vec (respectively, except Vec and sVec). Examples of such simple categories can be constructed using quantum groups at roots of unity.

It was proved in [**DrGNO2**] and [**DaNO**] that a weakly anisotropic braided fusion category of ordinary or super type admits a unique decomposition into the tensor product of its pointed part and simple categories. Here by "tensor product" we mean the usual Deligne tensor product $\boxtimes$ for ordinary type and the "super" tensor product over sVec for the super type.

8.27.4. G-extensions of tensor categories, homotopy theory, and topological field theory. In [**ENO4**] the *Brauer-Picard groupoid* $\underline{\underline{\underline{\mathsf{BrPic}}}}$ of fusion categories was introduced. By definition, this is a 3-groupoid, whose objects are fusion categories, 1-morphisms from $\mathcal{C}$ to $\mathcal{D}$ are invertible $(\mathcal{C}, \mathcal{D})$-bimodule categories (with the composition given by tensor product of bimodule categories), 2-morphisms are equivalences of such bimodule categories, and 3-morphisms are isomorphisms of such equivalences. This 3-groupoid can be truncated in the usual way to a 2-groupoid $\underline{\underline{\mathsf{BrPic}}}$ and further to a 1-groupoid (i.e., an ordinary groupoid) BrPic; the group of automorphisms of $\mathcal{C}$ in this groupoid is $\mathsf{BrPic}(\mathcal{C})$. It was shown in [**ENO4**, Theorem 1.1] that the 2-functor of taking the center of a fusion category is a fully faithful embedding of $\underline{\underline{\mathsf{BrPic}}}$ into the 2-groupoid $\underline{\underline{\mathsf{Aut}}}_{\otimes}^{br}$ of braided equivalences (this generalizes the above mentioned group isomorphism $\mathsf{BrPic}(\mathcal{C}) \cong \mathsf{Aut}_{\otimes}^{br}(\mathcal{Z}(\mathcal{C}))$). The objects of $\underline{\underline{\mathsf{Aut}}}_{\otimes}^{br}$ are braided fusion categories, 1-morphisms are braided equivalences, and 2-morphisms are isomorphisms of such equivalences.

Let G be a finite group. It turns out that Brauer-Picard groupoid is the right tool to describe G-extensions (see Section 4.14) of fusion categories. To do this, recall that to the 3-groupoid $\underline{\underline{\underline{\mathsf{BrPic}}}}$ one can attach its classifying space $B\underline{\underline{\underline{\mathsf{BrPic}}}}$, defined up to homotopy equivalence. This space falls into connected components, labeled by categorical Morita equivalence classes of fusion categories. Each connected component $B\underline{\underline{\underline{\mathsf{BrPic}}}}(\mathcal{C})$ corresponding to a fusion category $\mathcal{C}$ is a 3-type, i.e., it has three nontrivial homotopy groups: its fundamental group π_1 is $\mathsf{BrPic}(\mathcal{C}) \left(= \mathsf{Aut}_{\otimes}^{br}(\mathcal{Z}(\mathcal{C})\right)$, π_2 is the group of isomorphism classes of invertible objects of $\mathcal{Z}(\mathcal{C})$, and $\pi_3 = \Bbbk^{\times}$ (the multiplicative group of the ground field). It was proved in [**ENO4**] that G-extensions of $\mathcal{C}$ are parametrized by maps of classifying spaces $BG \to B\underline{\underline{\underline{\mathsf{BrPic}}}}(\mathcal{C})$. Thus, to classify G-extensions of $\mathcal{C}$, one needs to classify the homotopy classes of such maps, which can be done using the classical obstruction theory. Namely, by [**ENO4**, Theorem 1.3] G-extensions of $\mathcal{C}$ are parameterized by pairs (T, α), where $T : \mathrm{Cat}(G) \to \underline{\underline{\mathsf{BrPic}}}$ is a monoidal functor (equivalently, T is an action of G by braided autoequivalences of $\mathcal{Z}(\mathcal{C})$, see Definition 2.7.1) such that a certain obstruction $O(T) \in H^4(G, \Bbbk^{\times})$ vanishes, and α is an element of a certain torsor over $H^3(G, \Bbbk^{\times})$.

Note that $\underline{\underline{\underline{\mathsf{BrPic}}}}$ is a part of a 3-category whose objects, morphisms, 2-morphisms, and 3-morphisms are, respectively, finite tensor categories, bimodule categories, bimodule functors, and bimodule natural transformations. In [**DouSS**] Douglas, Schommer-Pries, and Snyder investigated the relationship between this 3-category and local topological field theories. In particular they showed that fusion categories are *fully dualizable* objects in this 3-category in the sense of Lurie [**Lur**], i.e., they give rise to 3-dimensional local field theories. In general, finite tensor categories are 2-dualizable and so give rise to 2-dimensional framed field theories (they produce representations of mapping class groups of closed surfaces). The approach of [**DouSS**] provides an important topological insight into the structure of finite tensor categories. In particular, there is a purely topological construction of

Radford's tensor isomorphism between the identity functor and the conjugate of the quadruple duality functor (cf. Section 7.19).

8.27.5. Categorification of the representation ring and the Verlinde ring of SL_n. According to Example 7.22.6, if G is a simple complex algebraic group then there exists a unique formal deformation of the tensor category $\mathsf{Rep}(G)$, which is the category of representations of the corresponding quantum group. This category can be realized as the category of comodules over the quantum function algebra $O_q(G)$, where $q = e^{\hbar}$, and $\hbar$ is a formal parameter; it does not change under $q \mapsto q^{-1}$, or $\hbar \mapsto -\hbar$ (as the actual deformation parameter is $\hbar^2$).

However, it is interesting to ask a "non-perturbative" version of this question: what are the possible categorifications of the representation ring of G over $\mathbb{C}$? Obviously, if q is not a root of unity, then one such categorification is the category of comodules over $O_q(G)$. Moreover, since this category is graded by $Z(G)^\vee$ (the character group of the center of G), it can be twisted by an element $\omega \in H^3(Z(G)^\vee, \mathbb{C}^*)$ (by multiplying the associativity morphism by values of a cocycle representing ω).

If $G = SL_n$, it is shown by Kazhdan and Wenzl in [**KazW**] that these are the only examples. A generalization of this result to orthogonal and symplectic groups (in the case of braided categories) is obtained in Tuba and Wenzl [**TubW**]. Categorifications in the limiting case $n \to \infty$ were discussed by Davydov in [**Da1**].

A similar question can be asked for the Verlinde algebra, i.e., the fusion ring of the Verlinde fusion category attached to the quantum group at a root of unity. In this case, it was also shown in [**KazW**] that for $G = SL_n$, the Verlinde fusion categories are the only categorifications, up to twisting, and it is explained in [**TubW**] how similar results can be obtained in the braided case for the orthogonal and symplectic groups.

8.27.6. Vertex operator algebras. The notion of vertex operator algebra (VOA) was introduced by Borcherds in [**Bo**] as a mathematically rigorous (and in fact, purely algebraic) version of the holomorphic part of the OPE (operator product expansion) algebra in 2-dimensional conformal field theory (CFT); see the textbook [**DiMS**] by Di Francesco, Mathieu, and Senechal for the physical side of the story. Namely, a VOA V and its category of modules is the structure that encodes "genus zero" conformal blocks of a CFT (if V is conformal, i.e., contains a Virasoro Lie algebra) then one can extend the construction to conformal blocks on higher genus Riemann surfaces, which gives the full structure of the holomorphic part of the theory). In particular, under some conditions on V, conformal blocks on $\mathbb{CP}^1$ with three marked points (viewed as two inputs and one output) define a tensor product functor on the category $V -$ mod, and the structure of conformal blocks with four marked points (viewed as three inputs and one output) determines an associativity constraint on this category, giving it a structure of a tensor category. This latter construction is no longer purely algebraic but rather complex-analytic: it requires solving differential equations arising from variation of the cross-ratio of the four points. Also, moving the marked points around each other and considering the corresponding monodromy gives rise to a braiding on this tensor category.

The theory of VOA is by now a vast subject. Formally, a VOA is a vector space V (infinite dimensional in all interesting cases) together with a product $V \otimes V \to V((z))$ (usually encoded as a map $V \to \mathsf{End}(V)((z, z^{-1})))$, a unit (vacuum vector) $\mathbf{1} \in V$, and an operator $T : V \to V$ (the shift) satisfying certain axioms (locality, unit axiom, shift axiom). We will not give the precise axioms here, but will just mention that the idea is to capture the physical notion of local operators $a(z)$ "living" at a point z of a complex plane. Such operators commute with each other: $a(z)b(w) = b(w)a(z)$ (which is called locality), but the catch is that they can be composed only when $z \neq w$, and the product $a(z)b(w)$ has a pole at $z = w$. This catch is the source of all the nontrivial structure, making VOA, despite their "commutativity", an object from the world of noncommutative algebra; in fact, the locality axiom is a curious hybrid of commutativity, associativity, and Jacobi identity from classical algebra. The shift operator and the shift axiom encode the idea that the operator product expansion in field theory is invariant with respect to translations $z \mapsto z + c$. There is also an important notion of a superVOA, which is, essentially, a VOA in the category of supervector spaces. Super VOA are needed for a rigorous approach to conformal field theory which involves fermions.

Some of the most important examples of VOA include free bosonic and fermionic VOAs, lattice VOAs, affine Kac-Moody VOA (corresponding to the Wess-Zumino-Witten model), Virasoro VOA, W-algebras, etc. These and other examples are discussed in [**FrLM**], [**KacV**], [**FreB**].

Here, we will not give a serious discussion of VOA, referring the reader to many good sources, such as the classical textbook of Frenkel, Lepowsky and Meurman [**FrLM**] and more recent books by Kac [**KacV**] and Frenkel and Ben-Zvi [**FreB**], as well as the book by Beilinson and Drinfeld on chiral algebras [**BeD**], which offers a geometric approach to VOA, based on the theory of D-modules (namely, chiral algebras live on algebraic curves, and VOA correspond to the case of formal disk). Rather, we will limit ourselves to a short discussion of the connection of VOAs with the theory of tensor categories.

The fact that a (super) VOA satisfying appropriate conditions gives rise to a tensor category was realized by mathematical physicists already in 1980s; see, e.g., [**MooS1**] and references therein. Namely, they realized that a rational VOA (i.e., one with a finite semisimple category of modules) gives rise to a braided (in fact, modular) fusion category. However, a rigorous general construction of the tensor category structure and braiding is nontrivial and did not appear in the literature until much later. By now, it is known that VOA satisfying suitable finiteness conditions (such as C_2-cofiniteness), give rise to both braided fusion categories (in the case of rational CFT) and braided finite tensor categories (in the case of logarithmic CFT).[9] This is done in a series of papers by Huang, Lepowsky and Zhang; see [**HuL**] for a review of these works and additional references. See the papers by Miyamoto [**Miy1, Miy2**], which also treat tensor products of VOA modules (in the case of logarithmic CFT).

A more geometric approach to this problem, which is more high-tech but might be potentially simpler, is outlined in [**BeD**]. Namely, the chiral homology construction of [**BeD**] provides a collection of local systems on configuration spaces of genus zero curves with n marked points, with some factorization structure. Then one can obtain a braided tensor category using the Riemann-Hilbert correspondence, see

[9] We note that a very nontrivial VOA can produce the trivial category of modules, i.e. the category Vec. Such a VOA is called *holomorphic*. A basic example of a holomorphic VOA is the basic representation of the affine Lie algebra of type E_8. Another example is the VOA discussed in [**FrLM**], whose automorphism group is the Monster simple group.

[**BakK**], Chapter 6. However, this approach has not yet been implemented in the literature.

For a comprehensive review of logarithmic CFT from the physical viewpoint and various examples and references on this subject, see the paper of Creutzig and Ridout [**CreR**].

We note that in many important examples of VOA, the existence of a braided tensor category structure on the category of modules was known rigorously already in the early 1990s. Notably, for the Kac-Moody VOA at generic and negative rational level, this was done in the series of papers by Kazhdan and Lusztig [**KazL1, KazL2, KazL3, KazL4, KazL5**], and at positive integer level by Finkelberg [**Fi2, Fi1**] [10]; see also the paper by Tsuchiya, Ueno, and Yamada [**TUY**], Chapter 7 of Bakalov-Kirillov [**BakK**], and a more recent review by Loojenga [**Loo**]. In this case, it was shown by Kazhdan and Lusztig that the corresponding braided category is equivalent to the appropriate category of representations of the corresponding quantum group. For the case of the Virasoro algebra, see the paper by Beilinson, Feigin, and Mazur [**BeiFM**].

However, in spite of so many impressive results, the relationship between VOA and braided tensor categories remains to be understood better, even in the semisimple case, corresponding to rational CFT. For example, it is not clear at all which examples of modular categories arise from VOA.

8.27.7. Drinfeld associators and the Drinfeld-Kohno theorem. Interesting examples of manifestly non-strict monoidal categories (i.e., ones with a nontrivial associator) were first studied by Drinfeld in [**Dr4, Dr5**]. Namely, let $\mathfrak{g}$ be a quasitriangular Lie quasi-bialgebra (over a field $\Bbbk$ of characteristic zero), i.e., a Lie algebra $\mathfrak{g}$ equipped with an element $t \in (S^2\mathfrak{g})^{\mathfrak{g}}$. In [**Dr4, Dr5**] Drinfeld used deformation theory to show that this structure can be quantized, i.e., there exists a *Drinfeld associator* $\Phi \in (U(\mathfrak{g}^{\otimes 3}))^{\mathfrak{g}_{\rm ad}}[[\hbar]]$ of the form

$$\Phi = 1 + \frac{\hbar^2}{24}[t_{12}, t_{23}] + O(\hbar^3)$$

which satisfies the pentagon relation. This associator makes the category $\mathcal{C}$ of topologically free $\mathfrak{g}$-modules over $k[[\hbar]]$ into a monoidal category, where it serves as the associativity morphism. Moreover, this category is braided, with braiding $c = \exp(\hbar t/2)$. Finally, he showed that Φ is uniquely determined by these properties up to twisting, and moreover both Φ and the twists are determined by universal formulas in terms of t.

The Hopf algebra $U(\mathfrak{g})$ with the usual coproduct, $R = \exp(\hbar t/2)$, and the Drinfeld associator Φ form a structure called a *quasitriangular quasi-Hopf algebra*, defined by Drinfeld in [**Dr4, Dr5**]. This is simply a quasi-Hopf algebra whose category of modules is braided (so the axioms of this structure can be guessed from reconstruction theory, and one can prove a reconstruction theorem in the style of Chapter 5).

If $k = \mathbb{C}$, then there is an explicit construction of Φ, which is based on complex analysis. Namely, one considers the Knizhnik-Zamolodchikov equation

$$\frac{dF}{dz} = \frac{\hbar}{2\pi i}\left(\frac{t_{12}}{z} + \frac{t_{23}}{z-1}\right)F, \tag{8.113}$$

[10] See the erratum to [**Fi2**] which shows that this requires Verlinde's formula, proved by Faltings.

where $F = F(z)$ takes values in the free algebra generated by t_{12} and t_{23}. Then Φ can be taken to be the suitably defined monodromy of this equation from $z = 0$ to $z = 1$. This associator is called the Knizhnik-Zamolodchikov associator and denoted by Φ_{KZ}.

If $\mathfrak{g}$ is a finite dimensional simple Lie algebra over $\mathbb{C}$ and t is the Casimir tensor, then it is not hard to see that the (strict) braided monoidal category of representations of the quantum group $U_q(\mathfrak{g})$ over $\mathbb{C}[[\hbar]]$, where $q = \exp(\hbar/2)$, is of the above form (this follows from the fact that $U_q(\mathfrak{g})$ is isomorphic to $U(\mathfrak{g})[[\hbar]]$ as an algebra, as $U(\mathfrak{g})$ has no nontrivial first order deformations by Whitehead's theorem). In particular, it is equivalent to the Drinfeld category defined using Φ_{KZ}. From this, Drinfeld deduced a fundamental result called the Drinfeld-Kohno theorem: the monodromy of the KZ equations is given, up to an isomorphism, by the R-matrices of the quantum group $U_q(\mathfrak{g})$.

Let us explain the Drinfeld-Kohno theorem in a bit more detail. Let V be a finite dimensional representation of $\mathfrak{g}$. Then we may consider the system of differential equations

$$\frac{\partial F}{\partial z_\ell} = \frac{\hbar}{2\pi i} \sum_{j \neq \ell} \frac{t_{\ell j}}{z_\ell - z_j} F,$$

where $F(z_1, ..., z_n) \in V^{\otimes n}$ is an analytic function in $z_1, ..., z_n$ in some region, and t_{ij} is the element t acting in the i-th and j-th copies of F. These equations are called the Knizhnik-Zamolodchikov equations (note that equation (8.113) is obtained from this system for $n = 3$, $i = 2$, $z_1 = 0, z_2 = z, z_3 = 1$). It is easy to see that the KZ equations are equivariant with respect to the symmetric group S_n, so they define a flat connection on the configuration space $\mathcal{C}_n$, consisting of subsets of $\mathbb{C}$ of size n. Thus, we have a monodromy representation of this connection, which is a representation ρ of the fundamental group $\pi_1(\mathcal{C}_n) = B_n$ on $V^{\otimes n}$, and which is independent, up to an isomorphism, on the choice of the base point in $\mathcal{C}_n$.

The Drinfeld-Kohno theorem gives a description of ρ in terms of the quantum group $U_q(\mathfrak{g})$, where $q = \exp(\hbar/2)$. Namely, let V_q be the deformation of V to a $U_q(\mathfrak{g})$-module. Then, since the category of representations of $U_q(\mathfrak{g})$ is a braided tensor category, $V_q^{\otimes n}$ carries a natural action ρ_q of the braid group B_n. The Drinfeld-Kohno theorem says that ρ is isomorphic to ρ_q.

Drinfeld's proof of this theorem works for formal $\hbar$, and hence, by using standard "abstract nonsense", for Weil generic $\hbar$, i.e., for all $\hbar$ outside of an unspecified countable set. It is in fact known that the result holds for $\hbar \notin \pi i\mathbb{Q}$ and extends to Kac-Moody algebras (this follows, for instance, from the paper [**KazL3**], or from the results of the book [**Var**] or [**BezFS**] on integral formulas for solutions of the KZ equations). Another proof that works for Kac-Moody algebras and formal (or Weil generic) $\hbar$ and arbitrary highest weight modules follows from quantization theory of Lie bialgebras (see [**EtKa3**]). This approach also allowed Geer to prove the Drinfeld-Kohno theorem for Lie superalgebras, [**Gee**].

In [**Dr4**], Drinfeld also gave a non-braided version of the above quantization result. Namely, define a *coboundary Lie quasibialgebra* to be a Lie algebra $\mathfrak{g}$ together with an element $\phi \in (\wedge^3\mathfrak{g})^{\mathfrak{g}}$. Then Drinfeld proved that there exists an element $\Phi \in (U(\mathfrak{g})^{\otimes 3})^{\mathfrak{g}_{\mathrm{ad}}}[[\varepsilon]]$ of the form $\Phi = 1+\varepsilon\phi+O(\varepsilon^2)$ which satisfies the pentagon identity, and therefore defines a structure of a tensor category on the category $\mathcal{C}$ defined above. Moreover, this category is a coboundary category, with the coboundary structure defined by the usual classical flip. Finally, if $\varepsilon = \hbar^2$ and $\phi = \frac{1}{24}[t^{12}, t^{23}]$, we recover the previous setting.

8.27.8. Finiteness of braid group images. Let $\mathcal{C}$ be a braided fusion category. We say that $\mathcal{C}$ has finite braid group images if for any $X \in \mathcal{C}$ and $n \geq 2$, the action of the braid group B_n on $X^{\otimes n}$ factors through a finite group. If $\mathcal{C}$ comes from a rational conformal field theory, then this property is equivalent to saying that (genus 0) correlation functions of the corresponding conformal field theory are algebraic. It is an interesting question which braided fusion categories have finite braid group images (see [**NaR**]). This question is, in particular, motivated by quantum computation. It is conjectured in [**NaR**] that a braided fusion category has finite braid group images if and only if it is weakly integral. This is known in many cases. In particular, this is true for group-theoretical categories, see [**ERW**].

8.27.9. The modular functor and the Reshetikhin-Turaev invariants. Let $\mathcal{C}$ be a ribbon category, and $X \in \mathcal{C}$. We have explained in Remark 8.10.3 how to attach to any framed link L its Reshetikhin-Turaev invariant $RT_{\mathcal{C},X}(L)$. In particular, if $\mathcal{C}$ is the representation category of quantum SL_2, and $X = V_n$ is the $n+1$-dimensional irreducible representation, then we obtain the collored Jones polynomial $J_n(L, q)$.

It turns out that, as was shown in [**RT2**], if $\mathcal{C}$ is a modular category (say, over $\mathbb{C}$) then this construction can be strengthened to give invariants of closed oriented 3-manifolds. Let us briefly describe the main idea of the construction of these invariants.

For simplicity, assume that $\mathcal{C}$ is a modular category of multiplicative central charge 1 (one can extend to the general case by considering appropriate central extensions of mapping class groups). By the Lickorish-Wallace theorem, any closed connected oriented 3-manifold may be obtained by surgery of the 3-sphere S^3 along a framed link L. However, different links may define the same 3-manifold. In terms of the link projection, the condition for two framed links to define the same 3-manifold is that they are related by a sequence of usual Reidemeister moves (which do not really change the link) and also so-called Kirby moves. Thus, a framed link invariant defines a closed oriented 3-manifold invariant if and only if it is invariant under Kirby moves. The Reshetikhin-Turaev invariant $RT_{\mathcal{C},X}(L)$ does not satisfy this condition, in general; however, if $\mathcal{C}$ is a modular category, then Reshetikhin and Turaev showed in [**RT2**] that the (suitably normalized) sum

$$RT_{\mathcal{C}}(L) = \sum_X \dim(X) RT_{\mathcal{C},X}(L)$$

(i.e., $RT_{\mathcal{C}} = RT_{\mathcal{C},R_{\mathcal{C}}}$, where $R_{\mathcal{C}}$ is the virtual "regular object" of $\mathcal{C}$) is actually invariant under Kirby moves and thus defines a 3-manifold invariant. In particular, taking $\mathcal{C}$ to be the Verlinde category $\mathcal{C}_k(q)$ from Example 8.18.5, where q is a root of unity of order $2(k+2)$, we see that upon suitable normalization, $\sum_{m=0}^{k}[m+1]_q J_m(L,q)$ is a closed oriented 3-manifold invariant (the invariant of a disconnected manifold is the product of the invariants of its connected components).

Another method of constructing the same invariants is via the *modular functor*, described in detail in [**BakK**]. The modular functor $\tau = \tau_{\mathcal{C}}$ is attached to any modular tensor category $\mathcal{C}$ (again, for simplicity, of multiplicative central charge 1), and attaches to every closed oriented surface S a finite dimensional vector space $\tau(S)$ (in fact, it also attaches data to surfaces with punctures). Moreover, if M is an oriented 3-manifold with boundary $\partial M = S_1 - S_2$ (i.e., a cobordism between S_1 and S_2), then we have a linear operator $\tau(M) : \tau(S_1) \to \tau(S_2)$, and concatenation of cobordisms corresponds to composition of operators. In this sense, τ is a functor from the category of cobordisms between closed oriented 2-manifolds to the category of vector spaces. Furthermore, this functor is monoidal with respect to the symmetric monoidal structure on cobordisms defined by disjoint union. Such a data is called a 3-dimensional *topological quantum field theory* (TQFT); thus, any modular category gives rise to a 3-dimensional TQFT.

Note that the mapping class group $\mathsf{Aut}(S)$ acts on the space $\tau(S)$, since every orientation preserving automorphism g of S in particular defines a cobordism M_g from S to S. The collection of data $\tau(S)$ and $\tau(M_g)$ is called the *Modular functor* attached to $\mathcal{C}$; this is part of the information of the 3-dimensional TQFT attached to $\mathcal{C}$.

Note that a closed oriented 3-manifold M may be viewed as a cobordism of an empty surface to itself. Since $\tau(\emptyset) = \mathbb{C}$, this means that $\tau(M)$ is a complex number, which is a closed oriented 3-manifold invariant. This is in fact the same Reshetikhin-Turaev invariant as the one described above, defined using surgery.

There is an explicit way to calculate the invariant $\tau(M)$ for a closed manifold M using a *Heegaard splitting*. Namely, recall that any 3-manifold admits a Heegard splitting by a closed oriented surface S into a union of two handlebodies B, B^* bounded by S. Now, let B be a handlebody with $\partial B = S$. Then B may be viewed as a cobordism from $\emptyset$ to S, so $\tau(B)$ is a vector in $\tau(S)$. Similarly, if B^* is the same handlebody with opposite orientation (so that $\partial B^* = -S$), then $\tau(B^*) \in \tau(S)^*$. So if a 3-manifold M is obtained by gluing B and B^* using $g \in \mathsf{Aut}(S)$ (provided by a Heegaard splitting), then $\tau(M) = (\tau(B^*), \tau(g)\tau(B))$.

Let us briefly describe the construction of the space $\tau(S)$. To construct $\tau(S)$, first consider the case when S is a genus zero surface with boundary: $\partial S = \sum_i \alpha_i - \sum_j \beta_j$, and assume that to each component γ of ∂S we have attached a simple object X_γ of $\mathcal{C}$. Then we set $\tau(S) = \mathsf{Hom}_{\mathcal{C}}(\otimes_i X_{\alpha_i}, \otimes_j X_{\beta_j})$ (note that at least up to a non-canonical isomorphism, this is independent on the order of boundary components, as $\mathcal{C}$ is braided). Then for a closed S, cut it into genus zero surfaces with boundary S_k, and define $\tau(S)$ by the formula $\tau(S) := \oplus_L \otimes_k \tau(S_k^{(L)})$, where L runs over all labelings of all the cuts by simple objects of $\mathcal{C}$. One of the main results is that $\tau(S)$ defined in this way is independent on the way of cutting. For example, if S is a torus, then we can cut it into a sphere with two holes (a pipe) by one cut along a curve γ. Thus, $\tau(S) = \oplus_X \mathsf{Hom}_{\mathcal{C}}(X, X) = \oplus_X \mathsf{Hom}_{\mathcal{C}}(\mathbf{1}, X \otimes X^*)$, where X runs over simple objects of $\mathcal{C}$. The action of the mapping class group $\mathsf{Aut}(S) = SL_2(\mathbb{Z})$ on this space is exactly the $SL_2(\mathbb{Z})$ representation attached to the modular category $\mathcal{C}$ (defined by the S-matrix and the twists).

To define the vector $\tau(B)$, let g be the genus of S, and fix a cutting of S by g cuts along the A-cycles $\gamma_1, ..., \gamma_g$. We will then obtain a sphere with $2g$ holes, so $\tau(S) = \oplus_{X_1,...,X_g} \mathsf{Hom}_{\mathcal{C}}(\mathbf{1}, X_1 \otimes X_1^* \otimes ... \otimes X_g \otimes X_g^*$. In particular, one direct summand in this sum corresponds to the case $X_1 = ... = X_g = \mathbf{1}$; this gives an inclusion of $\mathbb{C}$ into $\tau(S)$, and the vector $\tau(B)$ is just the image of 1.

An interesting special case is $\mathcal{C} = \mathcal{Z}(\mathcal{D})$, where $\mathcal{D}$ is a spherical fusion category (so the multiplicative central charge of $\mathcal{C}$ is 1 by the result of Müger [**Mu4**]). Turaev and Virelizier showed in [**TurVire**] that in this case the Reshetikhin-Turaev invariants attached to $\mathcal{C}$ coincide with the Turaev-Viro invariants corresponding to

$\mathcal{D}$ introduced in [**TurViro**], see [**BalK**] and references therein. Moreover, in this case the invariants come from a fully extended topological quantum field theory, see [**Lur**] and [**DouSS**].

For more details on the basics of 3-dimensional topological quantum field theories and the corresponding link and 3-manifold invariants, see the book [**Tu4**].

8.27.10. Kernel of the modular representation and higher Frobenius-Schur indicators. Recall from Section 8.16 that given a modular tensor category, one constructs a modular representation of the modular group $SL_2(\mathbb{Z})$. It was conjectured that this representation has the following *congruence subgroup property*: its kernel contains a congruence subgroup, that is, the kernel of the reduction homomorphism $SL_2(\mathbb{Z}) \to SL_2(\mathbb{Z}/N\mathbb{Z})$ for some positive integer N. In particular, the image of the modular representation is finite. A physical argument in favor of this conjecture was given by Bantay [**Ban**] in the context of conformal field theory. It was shown by Xu [**X2**] that it is possible to make Bantay's arguments completely rigorous in the operator algebras approach to conformal field theory. However since it is not known whether an arbitrary modular category can be realized via a conformal field theory, the conjecture remained open.

Then Sommerhauser and Zhu [**SoZ**] introduced a new approach based on the theory of higher *Frobenius-Schur indicators*, see e.g. [**KaSZ**]. These indicators encode the information on the action of the cyclic group $\mathbb{Z}/n\mathbb{Z}$ on the vector space $\mathsf{Hom}(\mathbf{1}, X^{\otimes n})$ for a simple object X. For instance, in the case $n = 2$ the Frobenius-Schur indicator takes value 0 if the object X is not self-dual and the value ± 1 otherwise; the sign here expresses the distinction between the objects which are "orthogonal" or "symplectic" in a suitable sense. Sommerhauser and Zhu [**SoZ**] introduced some further generalizations of Frobenius-Schur indicators and gave a proof of the congruence subgroup property for the Drinfeld doubles of semisimple Hopf algebras. The theory of higher Frobenius-Schur indicators was extended by Ng and Schauenburg to the case of (spherical) fusion categories in [**NgS1**]. Furthermore, a complete proof of the congruence subgroup property was given in [**NgS2**].

These ideas play a significant role in a recent proof [**BrNRW**] of important Wang's conjecture: there are just finitely many modular tensor categories of a given rank (i.e., the number of simple objects). The classification of modular categories of rank ≤ 5 is currently known, see [**BrNRW**].

8.27.11. Modular invariants. Let $\mathcal{C}$ be a modular tensor category realized as a representation category of a suitable vertex operator algebra V, see Section 8.27.6. From a physical point of view, the algebra V encodes the data of a chiral conformal field theory. In order to pass to the full conformal field theory, one needs to specify a *bulk algebra* which is a Lagrangian étale algebra (see 8.27.2) in the modular tensor category $\mathcal{C} \boxtimes \mathcal{C}^{\text{rev}}$, see e.g. [**FuRS1**]. The class of such an algebra in the Grothendieck ring of $\mathcal{C} \boxtimes \mathcal{C}^{\text{rev}}$ can be considered as a matrix with non-negative integer entries where the columns and rows are labeled by the simple objects of $\mathcal{C}$. It is known (see [**BocEK1, FuRS1**]) that this matrix commutes with the modular representation associated with $\mathcal{C}$, see Section 8.16. In other words, this matrix is a *modular invariant*, i.e., a matrix with nonnegative integer entries commuting with the modular representation and such that the diagonal entry corresponding to the unit object of $\mathcal{C}$ equals 1.

The knowledge of the modular invariant is equivalent to the knowledge of the *partition function* of the corresponding conformal field theory which is an important quantity from the point of view of physics. This motivates an interest in search and classification of modular invariants for various vertex algebras V. Many results in this direction are available in the literature, see e.g. [**DiMS**, Chapter 17] and references therein. Modular invariants for $\mathcal{C} = \mathcal{Z}(\mathsf{Vec}_G)$ were classified by Davydov in [**Da5**]. A famous result of Cappelli, Itzykson, Zuber and Kato [**CapIZ, Kat**] states that the classification of modular invariants for the categories associated with sl_2 (see Example 8.18.5) follows the same pattern as Dynkin diagrams of ADE type. In a subsequent work, Gannon gave a similar classification of modular invariants in the case of sl_3, see [**Gann1**]. Conversely, given a modular invariant, one can ask whether it comes from a Lagrangian algebra. There are examples when such an algebra does not exist (see e.g. [**ScheY, Vers**]) and when such an algebra is not unique, see e.g. [**BocE**]. Group-theoretical examples of non-unique Lagrangian algebras behind the Cardy modular invariant were classified in [**Da6**].

8.27.12. Modular categories and character sheaves. Modular categories play a role in the theory of *character sheaves*. Namely let G be an algebraic group over a finite field $\mathbb{F}_q$. For any $n \in \mathbb{Z}_{>0}$ the group $G(\mathbb{F}_{q^n})$ of $\mathbb{F}_{q^n}$-points of G is finite. One observes that for some groups G complex characters of the finite groups $G(\mathbb{F}_{q^n})$ vary nicely with the value of n; for example in the case $G = GL_m$ the degrees of the irreducible complex characters are given by some polynomials (independent of q and n but depending on m) evaluated at q^n. Lusztig suggested that such phenomena are explained by the existence of a geometric theory of characters (making sense over arbitrary field) where the complex characters are replaced by certain perverse ℓ-adic sheaves on the group G. He developed such a theory in the case of reductive groups G, see [**Lus1**]. Later, Boyarchenko and Drinfeld developed a similar theory in the case of unipotent groups G following an idea by Lusztig, see [**BoyD**].

An important question in such a theory is a classification of irreducible character sheaves. It turns out that isomorphism classes of such sheaves naturally split into a disjoint union of finite subsets of size which is independent of n (and, sometimes, of q). Moreover, the elements of such subsets are naturally labeled by irreducible objects of some modular categories; furthermore, the transition matrix between the *characteristic functions* of character sheaves (where the values of the characteristic functions are given by the traces of the action of the Frobenius endomorphism on the stalks) and the complex characters coincides with the S-matrix of the corresponding modular category. This is explained by the existence of a tensor operation on the character sheaves (*convolution* or its truncated version) such that the suitable categories of character sheaves acquire a structure of a modular tensor category, see [**BezFO, BoyD, Lus6**].

Modular categories that appear in the theory of character sheaves are group-theoretical (see Section 9.7). However it is expected that there exists a generalization of this theory where the Weyl group of G is replaced by an arbitrary Coxeter group, see [**Lus6**]. In particular, the dihedral Coxeter groups should produce modular categories closely related with the Verlinde categories (see Example 4.10.6) and the Coxeter group of type H_4 should give rise to a new exotic example of modular tensor category, see [**Mal**].

8.27.13. Modular categories and anyons. It is well known that the statistical properties of point-like objects (or quasi-particles) in physics are described either by the Bose-Einstein (for bosons) or by the Fermi-Dirac (for fermions) statistics. However for quasi-particles in two dimensions there is a theoretical possibility of more general statistics described via representations of the braid groups (this is a reflection of the fact that the fundamental group of the configuration space of points in 2 dimensions is the braid group while in 3 and higher dimensions it is isomorphic to the symmetric group). Such quasi-particles are called *anyons* and there are experimental indications that they do exist, see e.g. [**Wi, Ster**]. It is expected that the statistics of anyons is described by modular tensor categories, and we refer the reader to [**Kit2, KitK, Kon**] for an account of the theory. For example pointed modular categories correspond to *abelian anyons* and such quasi-particles were detected in quasi-particle systems related to the fractional quantum Hall effect. The non-abelian anyons should correspond to non-pointed modular categories and are expected to exist in some states of the fractional quantum Hall effect. However the experimental evidence for the existence of non-abelian anyons is not yet conclusive. On the theoretical side Kitaev and Levin-Wen constructed explicit hamiltonians describing statistical physics models with anyonic behaviour [**Kit1, LeW**]. The non-abelian anyons play a significant role in topological quantum computing, which is one of the approaches to actual implementation of quantum computers, see e.g. [**NSSFS, KiSV, WangZ, KoKR**].

8.27.14. Pointed tensor categories. An interesting problem is to classify pointed finite tensor categories, i.e. those whose simple objects are invertible. This problem is equivalent to classifying finite dimensional basic quasi-Hopf algebras up to twisting (recall that "basic" means that every simple module is 1-dimensional). Isomorphism classes of simple objects in such a category form a finite group G, and it is interesting to classify the corresponding categories for any given G.

In the case of Hopf algebras, this problem (i.e., the problem of classification of pointed Hopf algebras) has been thoroughly studied in the works of Angiono, Andruskiewitsch, Heckenberger, Schneider, and others, see Section 8.25. Let us discuss the general (quasi-Hopf) case, in which less is known.

Let us assume that char$\Bbbk = 0$. In this case, we have seen in Theorem 4.4.1 that if $G = 1$ then $\mathcal{C} = \mathsf{Vec}$. In the case of $G = \mathbb{Z}/p\mathbb{Z}$, where p is a prime, the classification was obtained in [**EtG8**]; besides semisimple examples and the well known pointed Hopf algebras whose grouplike elements form a cyclic group of prime order ([**AndrS**]) , this classification involves some basic quasi-Hopf algebras which are not Hopf algebras, introduced in [**EtG7, EtG6**], [**Gel1**]. These examples and classification were extended to cyclic groups G of composite order coprime to $2, 3, 5, 7$ in [**Ang1**].

CHAPTER 9

Fusion categories

In this Chapter the ground field $\Bbbk$ is assumed to be algebraically closed of characteristic 0 unless otherwise specified. Section 9.16 is an exception from this rule.

9.1. Ocneanu rigidity (absence of deformations)

Here we discuss vanishing of the Davydov-Yetter cohomology and absence of deformations for multifusion categories.

We start with the following algebraic warm-up. Let A be a Frobenius algebra in Vec (see Definition 7.20.3). This means that A has a coassociative comultiplication $\Delta : A \to A \otimes_{\Bbbk} A$ that is an A-bimodule homomorphism, i.e.,

$$\Delta(xy) = (x \otimes 1)\Delta(y) = \Delta(x)(1 \otimes y)$$

for all $x, y \in A$. Note that $m \circ \Delta : A \to A$ is a homomorphism of bimodules; in particular, the element $u := m \circ \Delta(1)$ is central in A. Recall that we say that A is *separable* if the composition $m \circ \Delta : A \to A$, where $m : A \otimes A \to A$ is the multiplication of A, is an isomorphism. This is equivalent to u being invertible. In this case the map $x \mapsto \Delta(u^{-1}x)$ is a right inverse of m.

Consider the Hochschild complex $C^\bullet(A)$ of A with $C^n(A) = \mathsf{Hom}_{\Bbbk}(A^{\otimes n}, A)$ for $n \geq 0$ (we set $A^{\otimes 0} = \Bbbk$) and the differential map $d_n : C^n(A) \to C^{n+1}(A)$ given by

$$\begin{aligned} d_n f(x_1, \dots, x_{n+1}) &= x_1 f(x_2, \dots, x_{n+1}) + \sum_{i=1}^{n} (-1)^i f(x_1, \dots, x_i x_{i+1}, \dots, x_{n+1}) \\ &\quad + (-1)^{n+1} f(x_1, \dots, x_n) x_{n+1} \end{aligned}$$

for $f \in C^n(A)$.

Let $H^n(A)$ denote the n-th cohomology of $C^\bullet(A)$, called the n-th Hochschild cohomology of A (it is commonly denoted $HH^n(A, A)$). The following result is well known.

PROPOSITION 9.1.1. *For a separable algebra A one has $H^n(A) = 0$ for $n > 0$.*

PROOF. Let $f \in C^n(A)$, $n \geq 1$ be such that $d_n f = 0$. Define $\tilde{f} \in C^{n-1}$ by

$$\tilde{f}(x_1, \dots, x_{n-1}) := u^{-1} 1_1 f(1_2, x_1, \dots x_{n-1}). \tag{9.1}$$

For $f \in C^1(A)$ we define $\tilde{f} \in C^0(A)$ by $\tilde{f}(1) = u^{-1} 1_1 f(1_2)$, where $\Delta(1) = 1_1 \otimes 1_2$ (Sweedler's notation). We compute

$$\begin{aligned} & u\, d_n \tilde{f}(x_1, \dots, x_n) = \\ &= x_1 1_1 f(1_2, x_2, \dots, x_n) - 1_1 f(1_2, x_1 x_2, \dots, x_n) + \cdots \\ &\quad \cdots + (-1)^{n-1} 1_1 f(1_2, x_1, \dots, x_{n-1} x_n) + (-1)^n 1_1 f(1_2, x_1, \dots, x_{n-1}) x_n. \end{aligned}$$

The condition $d_n f = 0$ yields

$$\begin{aligned}(-1)^{n-1} 1_1 f(1_2, x_1, \dots, x_{n-1}x_n) = \\ = \ & 1_1 1_2 f(x_1, \dots, x_n) - 1_1 f(1_2 x_1, \dots, x_n) + \cdots \\ & \cdots + (-1)^{n-1} 1_1 f(1_2, x_1, \dots, x_{n-2}x_{n-1}, x_n) \\ & +(-1)^{n+1} 1_1 f(1_2, x_1, \dots, x_{n-1}) x_n.\end{aligned}$$

Combining these two formulas and using that $x_1 1_1 \otimes 1_2 = 1_1 \otimes 1_2 x_1$ we obtain, after cancellations, that $d_n \tilde{f} = f$, i.e., each n-cocycle is a coboundary. Thus, $H^n(A) = 0$. □

REMARK 9.1.2. Another standard proof of this result is as follows. For any algebra A, one has $H^n(A) = \mathrm{Ext}^n_{A-\mathbf{bimod}}(A, A)$. But if A is separable then it is semisimple, so the category of A-modules and A-bimodules is semisimple. Hence the functor Ext^n vanishes in this category for all $n > 0$.

Now let $\mathcal{C}$ be a multifusion category, and let $A := \underline{\mathrm{Hom}}_{\mathcal{C} \boxtimes \mathcal{C}^{\mathrm{op}}}(\mathbf{1}, \mathbf{1})$ be the canonical Frobenius algebra associated to $\mathcal{C}$, see Definition 7.9.12 and Section 7.20. Recall the Hochschild complex $C^\bullet(A)$ of A from Section 7.22; by Proposition 7.22.7, this complex is isomorphic to the Davydov-Yetter complex of $\mathcal{C}$.

THEOREM 9.1.3. *Let $\Bbbk$ be an algebraically closed field of characteristic 0 and let $\mathcal{C}$ be a multifusion category over $\Bbbk$. Then $H^n(\mathcal{C}) = 0$ for all $n > 0$.*

PROOF. By Proposition 7.22.7, the statement is equivalent to $H^n(A) = 0$ for all $n > 0$. To prove the latter we use the same argument as in the proof of Proposition 9.1.1. Namely, for an n-cocycle $f \in C^n(A)$ we define $\tilde{f} \in C^{n-1}(A)$ as follows. Let $u = m \circ \Delta \circ e : \mathbf{1} \boxtimes \mathbf{1} \to A$, where $e : \mathbf{1} \boxtimes \mathbf{1} \to A$ is the unit of A, and let $u^{-1} : \mathbf{1} \boxtimes \mathbf{1} \to A$ be the convolution inverse of u, i.e., the morphism such that $m(u \otimes u^{-1}) = e$ (this inverse exists by Corollary 7.21.19). Set $\tilde{f}$ to be the composition

$$A^{\otimes(n-1)} \cong (\mathbf{1} \boxtimes \mathbf{1}) \otimes A^{\otimes(n-1)} \xrightarrow{\Delta \circ u^{-1} \otimes \mathrm{id}} A^{\otimes(n+1)} \xrightarrow{\mathrm{id} \otimes f} A \otimes A \xrightarrow{m} A. \tag{9.2}$$

The morphism (9.2) is a categorical analog of the map (9.1). The same computation as in the proof of Proposition 9.1.1 (with linear maps replaced by morphisms in $\mathcal{C} \boxtimes \mathcal{C}^{\mathrm{op}}$) shows that $d_n \tilde{f} = f$, i.e., $H^n(A) = 0$. □

THEOREM 9.1.4. *A multifusion category does not admit nontrivial deformations. In particular, the number of multifusion categories (up to equivalence) with a given Grothendieck ring is finite.*

PROOF. Let $\mathcal{C}$ be a multifusion category. By Theorem 9.1.3, $H^i(\mathcal{C}) = 0$, $i > 0$, in particular $H^3(\mathcal{C}) = 0$.

Now, consider the set X of all admissible associativity constraints for a multifusion category with the same Grothendieck ring as $\mathcal{C}$. It is an affine algebraic variety. This variety is acted upon by the affine algebraic group G of twists (i.e., gauge transformations). For $x \in X$, let $\mathcal{C}_x$ denote the multifusion category with associativity constraint x. For any point $x \in X$ we have the morphism of varieties $f_x : G \to X$, $g \mapsto g \cdot x$, with differential $(df_x)_1 : \mathrm{Lie}(G) \to T_x X$, where $T_x X$ is the tangent space to X at x. It is straightforward to derive from the definition of an associativity constraint that $T_x X$ equals the space of 3-cocycles $Z^3(\mathcal{C}_x)$ and $(df_x)_1(\mathrm{Lie}(G))$ equals the space of 3-coboundaries $B^3(\mathcal{C}_x)$ (the details are left as an exercise). Therefore, the quotient $T_xX/(df_x)_1(\mathrm{Lie}(G))$ is equal to $H^3(\mathcal{C}_x) = 0$, i.e., $(df_x)_1$ is surjective. This implies by a standard argument from algebraic geometry that all orbits Gx are open in X, and hence that there are finitely many orbits. We are done. □

We also have the following theorem.

THEOREM 9.1.5. *A tensor functor between multifusion categories does not have nontrivial deformations. In particular, the number of such functors (up to equivalence) for fixed source and target categories is finite.*

PROOF. The proof of Theorem 9.1.5 is analogous to the proof of Theorem 9.1.4. Namely, the first part of the result follows directly from Theorem 9.1.3 for $i = 2$. The second part is a consequence of the obvious combinatorial fact that there are only finitely many homomorphisms between the Grothendieck rings of the categories under consideration. □

COROLLARY 9.1.6. *(i) A module category $\mathcal{M}$ over a multi-fusion category $\mathcal{C}$ does not admit nontrivial deformations. In particular, the number of equivalence classes of such module categories with a given number of simple objects is finite.*

(ii) The number of classes of indecomposable module categories over a given multi-fusion category $\mathcal{C}$ is finite.

PROOF. (i) It suffices to choose a semisimple $\mathbb{k}$-algebra R with number of blocks equal to the number of simple objects in $\mathcal{M}$, and apply Theorem 9.1.5 to the functor $F : \mathcal{C} \to R-$bimod associated with $\mathcal{M}$.

(ii) This follows from (i), as there is a bound for the number of simple objects in any such module category. □

In particular, Corollary 9.1.6 implies that a given fusion category has finitely many fiber functors. This is also shown in [**IK**] in the special case of categories coming from subfactors (see Theorem 2.4 in [**IK**]).

COROLLARY 9.1.7. *([**Stef**]) The number of isomorphism classes of semisimple Hopf algebras of a given dimension d over an algebraically closed field of characteristic zero is finite.*

PROOF. It is clear that there are finitely many fusion rings of Frobenius-Perron dimension d. Each of them has finitely many categorifications by Theorem 9.1.4. Finally, each of these categorifications has finitely many fiber functors by Theorem 9.1.5. So the result follows from Theorem 5.3.12. □

COROLLARY 9.1.8. *Any multifusion category, any tensor functor between such categories, and any semisimple Hopf algebra is defined over an algebraic number field.*

PROOF. This follows in a standard way from Theorem 9.1.4 and Theorem 9.1.5. Namely, any orbit Gx from the proof of Theorem 9.1.4 is an irreducible component of a variety defined over $\mathbb{Q}$; thus it is defined over $\overline{\mathbb{Q}}$ and has a point with coordinates in $\overline{\mathbb{Q}}$. □

9.2. Induction to the center

Let $\mathcal{C}$ be a fusion category. Let $\mathcal{Z}(\mathcal{C})$ be the center of $\mathcal{C}$. Let $I : \mathcal{C} \to \mathcal{Z}(\mathcal{C})$ be the right adjoint of the forgetful functor $F : \mathcal{Z}(\mathcal{C}) \to \mathcal{C}$ (7.43).

LEMMA 9.2.1. *Consider $\mathcal{C}$ as a $\mathcal{C} \boxtimes \mathcal{C}^{\text{op}}$-module category. Then for all objects V, W of $\mathcal{C}$ we have*

$$\underline{\mathsf{Hom}}_{\mathcal{C}\boxtimes\mathcal{C}^{\text{op}}}(V, W) \cong \bigoplus_{X \in \mathcal{O}(\mathcal{C})} X \boxtimes ({}^*V \otimes {}^*X \otimes W). \tag{9.3}$$

PROOF. Let Y_1, Y_2 be objects of $\mathcal{C}$. We compute

$$\begin{aligned}
\mathsf{Hom}_{\mathcal{C}}(Y_1 \otimes V \otimes Y_2, W) &\cong \mathsf{Hom}_{\mathcal{C}}(Y_1, W \otimes Y_2^* \otimes V^*) \\
&\cong \bigoplus_{X \in \mathcal{O}(\mathcal{C})} \mathsf{Hom}_{\mathcal{C}}(Y_1, X) \otimes_{\Bbbk} \mathsf{Hom}_{\mathcal{C}}(X, W \otimes Y_2^* \otimes V^*) \\
&\cong \bigoplus_{X \in \mathcal{O}(\mathcal{C})} \mathsf{Hom}_{\mathcal{C}}(Y_1, X) \otimes_{\Bbbk} \mathsf{Hom}_{\mathcal{C}}(Y_2, {}^*V \otimes {}^*X \otimes W) \\
&\cong \mathsf{Hom}_{\mathcal{C}\boxtimes\mathcal{C}^{\text{op}}}(Y_1 \boxtimes Y_2, \bigoplus_{X \in \mathcal{O}(\mathcal{C})} X \boxtimes {}^*V \otimes {}^*X \otimes W).
\end{aligned}$$

Using definition (7.21) of the internal Hom, we obtain the isomorphism (9.3). □

PROPOSITION 9.2.2. *Let $F : \mathcal{Z}(\mathcal{C}) \to \mathcal{C}$ and $I : \mathcal{C} \to \mathcal{Z}(\mathcal{C})$ be the forgetful functor and induction. For any object Y in $\mathcal{C}$ we have*

$$FI(Y) \cong \bigoplus_{X \in \mathcal{O}(\mathcal{C})} X \otimes Y \otimes X^*. \tag{9.4}$$

PROOF. Observe that $I(Y) = \underline{\mathsf{Hom}}_{\mathcal{Z}(\mathcal{C})}(\mathbf{1}, Y)$ (see [**Bal**], Theorem 2.3, for an explicit formula for $\gamma_X : X \otimes I(Y) \to I(Y) \otimes X$). The result follows from combining Proposition 7.12.28 (applied to the pair of dual categories $\mathcal{C} \boxtimes \mathcal{C}^{\text{op}}$ and $\mathcal{Z}(\mathcal{C})$) and Lemma 9.2.1. □

REMARK 9.2.3. The object $I(\mathbf{1})$ has a canonical structure of an algebra in $\mathcal{Z}(\mathcal{C})$. Its forgetful image

$$FI(\mathbf{1}) \cong \bigoplus_{X \in \mathcal{O}(\mathcal{C})} X \otimes X^* \tag{9.5}$$

is the direct sum of the algebras $X \otimes X^*$ from Example 7.8.4 (cf. Proposition 7.16.5). This follows from the explicit description of the multiplication of A in Example 7.9.14.

LEMMA 9.2.4. *The central structure on the object $I(\mathbf{1})$ (i.e., a natural isomorphism $\gamma_Y : Y \otimes I(\mathbf{1}) \xrightarrow{\sim} I(\mathbf{1}) \otimes Y$, $Y \in \mathcal{C}$) is the image under $\otimes$ of isomorphism (7.55):*

$$\rho_Y : Y \otimes \left(\bigoplus_{X \in \mathcal{O}(\mathcal{C})} {}^*X \boxtimes X \right) \xrightarrow{\sim} \left(\bigoplus_{X \in \mathcal{O}(\mathcal{C})} {}^*X \boxtimes X \right) \otimes Y \tag{9.6}$$

from Proposition 7.18.5.

PROOF. This is a consequence of Remark 7.12.29. It says that when we view $\mathcal{C}$ as a $\mathcal{C} \boxtimes \mathcal{C}^{\mathrm{op}}$-module category (so that the dual category is $\mathcal{Z}(\mathcal{C})$) then the central structure on $I(\mathbf{1}) = \underline{\mathsf{Hom}}_{\mathcal{Z}(\mathcal{C})}(\mathbf{1}, \mathbf{1})$ comes from the isomorphism

$$(Y \boxtimes \mathbf{1}) \otimes \underline{\mathsf{Hom}}_{\mathcal{C}\boxtimes\mathcal{C}^{\mathrm{op}}}(\mathbf{1},\ \mathbf{1}) \cong \underline{\mathsf{Hom}}_{\mathcal{C}\boxtimes\mathcal{C}^{\mathrm{op}}}(\mathbf{1}, Y)$$

which is precisely (7.55). □

EXERCISE 9.2.5. Let $\mathcal{C}$ be a fusion category. Prove that invertible subobjects of $I(\mathbf{1})$ form a group isomorphic to the group $\mathsf{Aut}_{\otimes}(\mathsf{id}_{\mathcal{C}})$ of tensor automorphisms of the identity functor of $\mathcal{C}$.

9.3. Duality for fusion categories

By Example 7.5.4, a $\mathcal{C}$-module category $\mathcal{M}$ is exact if and only if it is semisimple. Our goal in this section is to prove that in this case the dual category $\mathcal{C}^*_{\mathcal{M}}$ is also a fusion category.

LEMMA 9.3.1. *For any natural isomorphism $\phi_X : X \xrightarrow{\sim} X^{**}$ the canonical isomorphism*

$$\Phi := \bigoplus_{X\in\mathcal{O}(\mathcal{C})} ({}^*\phi_X)^{-1} \otimes \phi_X : \bigoplus_{X\in\mathcal{O}(\mathcal{C})} {}^*X \otimes X \xrightarrow{\sim} \bigoplus_{X\in\mathcal{O}(\mathcal{C})} X^* \otimes X^{**}. \tag{9.7}$$

is an isomorphism between $I(\mathbf{1})$ and $I(\mathbf{1})^{**}$ in $\mathcal{Z}(\mathcal{C})$.

PROOF. Note that the isomorphism

$$\bigoplus_{X\in\mathcal{O}(\mathcal{C})} ({}^*\phi_X)^{-1} \boxtimes \phi_X : \bigoplus_{X\in\mathcal{O}(\mathcal{C})} {}^*X \boxtimes X \xrightarrow{\sim} \bigoplus_{X\in\mathcal{O}(\mathcal{C})} X^* \boxtimes X^{**}$$

commutes with isomorphism (9.6), and so the result follows from Lemma 9.2.4. □

THEOREM 9.3.2. *The center of a fusion category is a fusion category.*

PROOF. Let $\mathcal{C}$ be a fusion category. By Theorem 7.16.1 its center $\mathcal{Z}(\mathcal{C})$ is a finite tensor category. Let $I : \mathcal{C} \to \mathcal{Z}(\mathcal{C})$ be the right adjoint of the forgetful functor $F : \mathcal{Z}(\mathcal{C}) \to \mathcal{C}$. We claim that $I(\mathbf{1})$ is a projective object in $\mathcal{Z}(\mathcal{C})$. Indeed, $\mathsf{Hom}_{\mathcal{Z}(\mathcal{C})}(-, I(\mathbf{1})) \cong \mathsf{Hom}_{\mathcal{C}}(F(-), \mathbf{1})$. The forgetful functor $F : \mathcal{Z}(\mathcal{C}) \to \mathcal{C}$ is clearly exact. Since $\mathcal{C}$ is semisimple, $\mathbf{1}$ is injective, and so the functor $\mathsf{Hom}_{\mathcal{Z}(\mathcal{C})}(-, I(\mathbf{1}))$ is exact. Thus, $I(\mathbf{1})$ is injective. Hence, it is projective by Proposition 6.1.3.

Lemma 9.3.1 provides a canonical isomorphism $\Phi : I(\mathbf{1}) \xrightarrow{\sim} I(\mathbf{1})^{**}$, see (9.7). Using Proposition 4.7.3 we compute its trace:

$$\mathsf{Tr}^L(\Phi) = \bigoplus_{X\in\mathcal{O}(\mathcal{C})} \mathsf{Tr}^L(({}^*\phi_X)^{-1})\, \mathsf{Tr}^L(\phi_X) = \bigoplus_{X\in\mathcal{O}(\mathcal{C})} |X|^2 = \dim(\mathcal{C}).$$

Hence, $\mathsf{Tr}^L(\Phi) \neq 0$ by Theorem 7.21.12. It follows that the unit object $\mathbf{1}$ of $\mathcal{Z}(\mathcal{C})$ is a direct summand of the projective object $I(\mathbf{1}) \otimes I(\mathbf{1})^*$. Hence, $\mathbf{1}$ is projective and so $\mathcal{Z}(\mathcal{C})$ is semisimple by Corollary 4.2.13. □

COROLLARY 9.3.3. *Let $\mathcal{C}$ be a fusion category and let $\mathcal{M}$ be a semisimple $\mathcal{C}$-module category. Then the dual tensor category $\mathcal{C}^*_{\mathcal{M}}$ is multifusion.*

PROOF. It suffices to prove the statement in the case when $\mathcal{M}$ is indecomposable. By Theorem 9.3.2 the center $\mathcal{Z}(\mathcal{C})$ is a fusion category. Hence, $\mathcal{Z}(\mathcal{C}^*_{\mathcal{M}})$ is a fusion category by Theorem 7.16.1. The forgetful functor $F : \mathcal{Z}(\mathcal{C}^*_{\mathcal{M}}) \to \mathcal{C}^*_{\mathcal{M}}$ is surjective by Corollary 7.13.11. By Theorem 6.1.16 the object $\mathbf{1} \in \mathcal{C}^*_{\mathcal{M}}$ is projective and so $\mathcal{C}^*_{\mathcal{M}}$ is semisimple by Corollary 4.2.13. □

Recall that for a finite tensor category $\mathcal{C}$ one has $\mathsf{FPdim}(\mathcal{Z}(\mathcal{C})) = \mathsf{FPdim}(\mathcal{C})^2$, see Theorem 7.16.6. For pivotal fusion categories a similar formula holds for categorical dimensions (7.71).

PROPOSITION 9.3.4. *In any fusion category* $\mathcal{C}$ *we have*

$$\dim(\mathcal{Z}(\mathcal{C})) = \dim(\mathcal{C})^2. \tag{9.8}$$

PROOF. First, let us prove the statement when $\mathcal{C}$ is pivotal. Observe that in this case $\mathcal{Z}(\mathcal{C})$ inherits the pivotal structure from $\mathcal{C}$ (see Exercise 7.13.6). Let d denote the dimension function determined by the pivotal structure on $\mathcal{C}$. The forgetful functor $F : \mathcal{Z}(\mathcal{C}) \to \mathcal{C}$ preserves dimensions, therefore, applying it to the virtual object $\sum_{Z\in\mathcal{O}(\mathcal{Z}(\mathcal{C}))} d(Z)Z \in K(\mathcal{Z}(\mathcal{C})) \otimes_{\mathbb{Z}} \Bbbk$, we obtain

$$\begin{aligned} F(\sum_{Z\in\mathcal{O}(\mathcal{Z}(\mathcal{C}))} d(Z)Z) &= \sum_{Z\in\mathcal{O}(\mathcal{Z}(\mathcal{C}))}\ \sum_{X\in\mathcal{O}(\mathcal{C})} d(Z)[F(Z):X]X \\ &= \sum_{X\in\mathcal{O}(\mathcal{C})}\ \sum_{Z\in\mathcal{O}(\mathcal{Z}(\mathcal{C}))} d(Z)[I(X):Z]X \\ &= \sum_{X\in\mathcal{O}(\mathcal{C})} d(I(X))X = \dim(\mathcal{C}) \sum_{X\in\mathcal{O}(\mathcal{C})} d(X)X. \end{aligned}$$

Here $I : \mathcal{C} \to \mathcal{Z}(\mathcal{C})$ denotes the induction functor and $d(I(X)) = \dim(\mathcal{C})d(X)$ by Proposition 9.2.2. Taking dimensions of both sides of the equality

$$F(\sum_{Z\in\mathcal{O}(\mathcal{Z}(\mathcal{C}))} d(Z)Z) = \dim(\mathcal{C}) \sum_{X\in\mathcal{O}(\mathcal{C})} d(X)X,$$

we obtain (9.8).

Next, consider an arbitrary fusion category $\mathcal{C}$. Let $\tilde{\mathcal{C}}$ be its pivotalization, see Definition 7.21.9. We have $\dim(\tilde{\mathcal{C}}) = 2\dim(\mathcal{C})$. Since $\tilde{\mathcal{C}}$ is a $\mathbb{Z}/2\mathbb{Z}$-equivariantization of $\mathcal{C}$, we can use Corollary 8.23.13 to compute

$$\dim(\mathcal{Z}(\mathcal{C})) = \frac{1}{4}\dim(\mathcal{Z}(\tilde{\mathcal{C}})) = \frac{1}{4}\dim(\tilde{\mathcal{C}})^2 = \dim(\mathcal{C})^2,$$

as required. □

COROLLARY 9.3.5. *Let* $\mathcal{C}$ *be a ribbon fusion category. For each simple object* $X \in \mathcal{C}$ *the number* $\frac{\dim(\mathcal{C})}{\dim(X)}$ *is an algebraic integer.*

PROOF. By Proposition 9.3.4 we have $\dim(\mathcal{Z}(\mathcal{C})) = \dim(\mathcal{C})^2$. Since $\mathcal{C}$ is a fusion subcategory of the modular category $\mathcal{Z}(\mathcal{C})$ (see Proposition 8.6.1 and Corollary 8.20.14) the result follows from Proposition 8.14.6. □

We recover the classical result of Frobenius in the representation theory of finite groups.

COROLLARY 9.3.6. *Let* G *be a finite group and let* V *be an irreducible representation of* G. *Then* $\dim_{\Bbbk}(V)$ *divides* $|G|$.

THEOREM 9.3.7. *Let $\mathcal{C}$ be a spherical fusion category. Then* $\dim(\mathcal{C})$ *can be written as a sum of its divisors in the ring of algebraic integers, namely,*

$$\dim(\mathcal{C}) = 1 + \sum_{Z \in \mathcal{O}(\mathcal{Z}(\mathcal{C})),\, Z \not\cong \mathbf{1}} [F(Z) : \mathbf{1}] \dim(Z). \tag{9.9}$$

PROOF. This result is immediately obtained by computing the dimension of $I(\mathbf{1})$ in two ways. On the one hand, the dimension is equal to $\dim(\mathcal{C})$ by (9.5). On the other hand, $I(\mathbf{1}) = \oplus_{Z \in \mathcal{O}(\mathcal{Z}(\mathcal{C}))} [F(Z) : \mathbf{1}]Z$, hence $\dim(I(\mathbf{1}))$ equals to the right hand side of (9.9). The divisibility condition follows from Proposition 8.14.6. □

EXAMPLE 9.3.8. Let $\mathcal{C} = \mathsf{Rep}(G)$, where G is a finite group. In this case $I(\mathbf{1})$ is the group algebra of G viewed as a representation of G under the conjugation action. So the simple subobjects of $I(\mathbf{1})$ in $\mathcal{Z}(\mathsf{Rep}(G))$ are conjugacy class sums of G and (9.9) is precisely the class equation of G.

PROPOSITION 9.3.9. *Let $\mathcal{C}$ be a fusion category and let $\mathcal{M}$ be an indecomposable $\mathcal{C}$-module category. Then*

$$\dim(\mathcal{C}) = \dim(\mathcal{C}^*_{\mathcal{M}}).$$

PROOF. By Corollary 9.3.3 the category $\mathcal{C}^*_{\mathcal{M}}$ is fusion. By Corollary 7.16.2 $\mathcal{Z}(\mathcal{C}) \cong \mathcal{Z}(\mathcal{C}^*_{\mathcal{M}})$ and so

$$\dim(\mathcal{C})^2 = \dim(\mathcal{Z}(\mathcal{C})) = \dim(\mathcal{Z}(\mathcal{C}^*_{\mathcal{M}})) = \dim(\mathcal{C}^*_{\mathcal{M}})^2.$$

Since the categorical dimension of a fusion category is totally positive (Proposition 7.21.14) the result follows. □

Next, we discuss cyclotomicity of dimension functions. As before, let $\mathcal{C}$ be a fusion category and let $\mathcal{Z}(\mathcal{C})$ be its center. The forgetful functor $\mathcal{Z}(\mathcal{C}) \to \mathcal{C}$ induces a homomorphism of the Grothendieck rings $F : \mathsf{Gr}(\mathcal{Z}(\mathcal{C})) \to \mathsf{Gr}(\mathcal{C})$. It is clear that the image of this map is contained in the center $Z(\mathsf{Gr}(\mathcal{C}))$ of the ring $\mathsf{Gr}(\mathcal{C})$.

LEMMA 9.3.10. *The map* $\mathsf{Gr}(\mathcal{Z}(\mathcal{C})) \otimes \mathbb{Q} \to Z(\mathsf{Gr}(\mathcal{C})) \otimes \mathbb{Q}$ *is surjective.*

PROOF. Let $I : \mathcal{C} \to \mathcal{Z}(\mathcal{C})$ denote the induction functor, see Section 9.2. We will denote the induced map $\mathsf{Gr}(\mathcal{C}) \to \mathsf{Gr}(\mathcal{Z}(\mathcal{C}))$ by the same letter. By Proposition 9.2.2 we have $F(I(X)) = \sum_{Y \in \mathcal{O}(\mathcal{C})} YXY^*$ for any $X \in \mathsf{Gr}(\mathcal{C})$. In particular, for $X \in Z(\mathsf{Gr}(\mathcal{C}))$ we have $F(I(X)) = X \sum_{Y \in \mathcal{O}(\mathcal{C})} YY^*$. Note that the operator of multiplication by the element $\sum_{Y \in \mathcal{O}(\mathcal{C})} YY^* \in Z(\mathsf{Gr}(\mathcal{C}))$ is a self-adjoint positive definite operator $\mathsf{Gr}(\mathcal{C}) \to \mathsf{Gr}(\mathcal{C})$ (with respect to the usual scalar product defined by $(X, Y) = \delta_{X,Y}$ for $X, Y \in \mathcal{O}\mathcal{C})$) and, hence, invertible. The lemma is proved. □

THEOREM 9.3.11. *Let $\mathcal{C}$ be a fusion category and let L be an irreducible representation of* $\mathsf{Gr}(\mathcal{C})$ *over* $\Bbbk$. *There exists a root of unity ξ such that for any object* $X \in \mathsf{Gr}(\mathcal{C})$ *one has* $\mathsf{Tr}(X, L) \in \mathbb{Z}[\xi]$.

PROOF. Recall that the pivotalization of $\mathcal{C}$ is a spherical fusion category $\tilde{\mathcal{C}}$ endowed with a surjective tensor functor $\tilde{\mathcal{C}} \to \mathcal{C}$ which maps simple objects of $\tilde{\mathcal{C}}$ to simple objects of $\mathcal{C}$, see Definition 7.21.9 and Corollary 7.21.10. Hence the induced map $\mathsf{Gr}(\tilde{\mathcal{C}}) \to \mathsf{Gr}(\mathcal{C})$ is surjective, and pullback of any irreducible representation of $\mathsf{Gr}(\mathcal{C})$ is an irreducible representation of $\mathsf{Gr}(\tilde{\mathcal{C}})$. Therefore if Theorem 9.3.11 holds for $\tilde{\mathcal{C}}$ then it also holds for $\mathcal{C}$. Thus in the rest of the proof we can and will assume that the category $\mathcal{C}$ is spherical. In particular the center $\mathcal{Z}(\mathcal{C})$ is modular, see Corollary 8.20.14

By Exercise 9.3.12 below the element $e_L = \sum_{Y \in \mathcal{O}(\mathcal{C})} \mathsf{Tr}(Y, L)Y^*$ is proportional to a primitive central idempotent in $\mathsf{Gr}(\mathcal{C}) \otimes \Bbbk$. By Lemma 9.3.10, there exists a primitive idempotent $\tilde{e}_L \in \mathsf{Gr}(\mathcal{Z}(\mathcal{C})) \otimes \Bbbk$ such that e_L is proportional to $F(\tilde{e}_L)$. It follows from Corollary 8.14.5 that any idempotent of $\mathsf{Gr}(\mathcal{Z}(\mathcal{C})) \otimes \Bbbk$ can be expressed in terms of the S-matrix of the category $\mathcal{Z}(\mathcal{C})$. By Theorem 8.14.7 there exists a root of unity ξ such that the entries of the S-matrix lie in $\mathbb{Q}(\xi)$ and thus $\tilde{e}_L \in \mathsf{Gr}(\mathcal{Z}(\mathcal{C})) \otimes \mathbb{Q}(\xi)$. Hence e_L is proportional to some element of $\mathsf{Gr}(\mathcal{C}) \otimes \mathbb{Q}(\xi)$. Now the coefficient of $\mathbf{1} \in \mathsf{Gr}(\mathcal{C})$ in e_L equals to $\dim(L) \in \mathbb{Q}(\xi)$ and is nonzero. Hence $e_L \in \mathsf{Gr}(\mathcal{C}) \otimes \mathbb{Q}(\xi)$ and $\mathsf{Tr}(Y, L) \in \mathbb{Q}(\xi)$. Obviously the number $\mathsf{Tr}(Y, L)$ is an algebraic integer, and the ring of integers in $\mathbb{Q}(\xi)$ is $\mathbb{Z}[\xi]$. The theorem is proved. □

EXERCISE 9.3.12. Let A be a based ring of finite rank with basis $\{b_i\}$ and let L be an irreducible representation of A over an algebraically closed field $\Bbbk$ of characteristic zero. Show that the element $e_L = \sum_i \mathsf{Tr}(b_i, L)b_i^* \in A \otimes \Bbbk$ is central. Moreover, it acts by zero on any irreducible representation of A which is not isomorphic to L. In other words, e_L is proportional to a primitive central idempotent of $A \otimes \Bbbk$.

Hint: the ring $A \otimes \Bbbk$ is semisimple by Corollary 3.7.7.

COROLLARY 9.3.13. *Any irreducible representation of* $\mathsf{Gr}(\mathcal{C})$ *is defined over some cyclotomic field. In particular for any homomorphism* $\phi : \mathsf{Gr}(\mathcal{C}) \to \mathbb{C}$ *and any object* $X \in C$ *we have* $\phi(X) \in \mathbb{Q}[\xi]$ *for some root of unity* ξ.

PROOF. Let $\mathbb{Q}^{ab}$ be the field of all cyclotomic numbers. We already proved that $\mathsf{Gr}(\mathcal{C}) \otimes \mathbb{Q}^{ab}$ decomposes into direct sum of simple algebras (such a decomposition is the same as the decomposition of $\mathbf{1} \in \mathsf{Gr}(\mathcal{C})$ into the sum of primitive central idempotents). Now it is well known that the Brauer group of $\mathbb{Q}^{ab}$ is trivial (see [**Se**, p.162]). The result follows. □

COROLLARY 9.3.14. *Let* $\mathcal{C}$ *be a fusion category. There exists a root of unity* ξ *such that for any object* X *of* $\mathcal{C}$ *one has* $\mathsf{FPdim}(X) \in \mathbb{Z}[\xi]$.

9.4. Pseudo-unitary fusion categories

PROPOSITION 9.4.1. *Let* $\mathcal{C}$ *be a fusion category over* $\mathbb{C}$. *For every* $X \in \mathcal{O}(\mathcal{C})$ *one has*

$$|X|^2 \leq \mathsf{FPdim}(X)^2 \tag{9.10}$$

and, hence, $\dim(\mathcal{C}) \leq \mathsf{FPdim}(\mathcal{C})$.

PROOF. This is [**ENO2**, Proposition 8.21]. It suffices to prove this statement when $\mathcal{C}$ has a pivotal structure. Indeed, one can always pass to the pivotalization $\tilde{\mathcal{C}}$ of $\mathcal{C}$ from Section 7.21 (since the forgetful functor $\tilde{\mathcal{C}} \to \mathcal{C}$ preserves the squared norms of simple objects and Frobenius-Perron dimensions). As we have seen in the proof of Theorem 7.21.12, in the pivotal situation for any simple X its squared norm $|X|^2$ is an eigenvalue of the non-negative integral matrix $N_X N_X^*$. By the Perron-Frobenius Theorem 3.2.1 we have $|X|^2 \leq \mathsf{FPdim}(X \otimes X^*) = \mathsf{FPdim}(X)^2$. □

PROPOSITION 9.4.2. *For any fusion category* $\mathcal{C}$ *over* $\mathbb{C}$, *the ratio* $\frac{\dim(\mathcal{C})}{\mathsf{FPdim}(\mathcal{C})}$ *is an algebraic integer* ≤ 1.

PROOF. That the ratio is ≤ 1 follows from inequality (9.10).

First, note that it suffices to prove the algebraic integrality statement in the case when $\mathcal{C}$ is spherical. Indeed, let $\tilde{\mathcal{C}}$ be the pivotalization of $\mathcal{C}$, see Definition 7.21.9. By Corollary 7.21.10 $\tilde{\mathcal{C}}$ is spherical and

$$\frac{\dim(\tilde{\mathcal{C}})}{\mathsf{FPdim}(\tilde{\mathcal{C}})} = \frac{2\dim(\mathcal{C})}{2\,\mathsf{FPdim}(\mathcal{C})} = \frac{\dim(\mathcal{C})}{\mathsf{FPdim}(\mathcal{C})}.$$

Next, we can assume that $\mathcal{C}$ is modular. This is because we can replace $\mathcal{C}$ by its center $\mathcal{Z}(\mathcal{C})$ since

$$\frac{\dim(\mathcal{Z}(\mathcal{C}))}{\mathsf{FPdim}(\mathcal{Z}(\mathcal{C}))} = \left(\frac{\dim(\mathcal{C})}{\mathsf{FPdim}(\mathcal{C})}\right)^2,$$

thanks to Theorem 7.16.6 and Proposition 9.3.4.

Let $S = \{s_{XY}\}$ denote the S-matrix of $\mathcal{C}$, see Section 8.13. The Frobenius-Perron dimension is a homomorphism from $K_0(\mathcal{C})$ to $\mathbb{C}$, hence, it must be of the form (8.51). Thus, there exists a canonical simple object $X \in \mathcal{C}$ such that $\mathsf{FPdim}(Z) = \frac{s_{ZX}}{d_X}$ for all simple objects Z in $\mathcal{C}$. Therefore,

$$\mathsf{FPdim}(\mathcal{C}) = \sum_Z \mathsf{FPdim}(Z)^2 = \sum_Z \frac{s_{ZX}}{d_X}\frac{s_{Z^*X}}{d_X} = \frac{\dim(\mathcal{C})}{d_X^2}.$$

Thus, $\frac{\dim(\mathcal{C})}{\mathsf{FPdim}(\mathcal{C})} = d_X^2$. The latter is an algebraic integer since $d : K_0(\mathcal{C}) \to \mathbb{C}$ is a homomorphism. $\square$

REMARK 9.4.3. It follows from (9.10) that if $\dim(\mathcal{C}) = \mathsf{FPdim}(\mathcal{C})$ then $|X|^2 = \mathsf{FPdim}(X)^2$ for all objects $X \in \mathcal{O}(\mathcal{C})$.

DEFINITION 9.4.4. A fusion category $\mathcal{C}$ over $\mathbb{C}$ is called *pseudo-unitary* if its categorical and Frobenius-Perron dimensions are equal, i.e., $\dim(\mathcal{C}) = \mathsf{FPdim}(\mathcal{C})$.

We see that $\mathcal{C}$ is pseudo-unitary if and only if $|X|^2 = \mathsf{FPdim}(X)^2$ for all simple $X \in \mathcal{C}$.

EXAMPLE 9.4.5. Let G be a finite group. Then $\mathsf{Rep}(G)$ is a pseudo-unitary category.

EXERCISE 9.4.6. (i) Let $\mathcal{C}_k(q)$ be the Verlinde categories from Example 8.18.5. Show that in these categories $|V_m|^2 = [m]_q^2$.

(ii) Using the above formula, determine for which q the category $\mathcal{C}_k(q)$ is pseudo-unitary.

(iii) In particular, show that among the two non-equivalent Yang-Lee categories YL_+, YL_- (see Exercise 8.18.7), one is pseudo-unitary and the other is not. Compute explicitly $|X|^2$ in both of these categories.

(iv) Show that Deligne's tensor product $\mathcal{C} := YL_+ \boxtimes YL_-$ of the two non-equivalent Yang-Lee categories has integer categorical dimension $\dim(\mathcal{C})$, but non-integer Frobenius-Perron dimension $\mathsf{FPdim}(\mathcal{C})$. What are these dimensions?

REMARK 9.4.7. If $\Bbbk = \mathbb{C}$, then one can introduce the notion of a Hermitian fusion category; it is a fusion category with nondegenerate Hermitian forms on morphism spaces between tensor products, subject to certain axioms. A Hermitian category is called unitary if these forms are positive definite (see **[Gal]**, **[Kir2]**, **[Tu4]** for more details). For instance, $\mathsf{Rep}(G)$ for a finite group G is a unitary category (since all irreducible complex representations of G have positive definite

Hermitian inner products); also, fusion categories coming from subfactors are unitary by construction.

It can be shown that any unitary fusion category is pseudounitary, which justifies the term "pseudounitary category" (the advantage of this notion is that it does not require using any Hermitian forms).

We note that we do not know an example of a fusion category over $\mathbb{C}$ which does not admit a Hermitian structure, or a pseudo-unitary fusion category which does not admit a unitary structure.

9.5. Canonical spherical structure

Let $\mathcal{C}$ be a fusion category over $\mathbb{C}$. By Corollary 7.19.3, in any fusion category there is a natural tensor isomorphism $g_X : X \xrightarrow{\sim} X^{****}$. Let $a_X : X \xrightarrow{\sim} X^{**}$ be a natural (not necessarily tensor) isomorphism such that $a_{X^{**}} \circ a_X = g_X$ (i.e., a_X is a square root of g_X). For all $X, Y, V \in \mathcal{O}(\mathcal{C})$ let

$$b_{XY}^V : \mathsf{Hom}_{\mathcal{C}}(V, X \otimes Y) \xrightarrow{\sim} \mathsf{Hom}_{\mathcal{C}}(V^{**}, X^{**} \otimes Y^{**}) \tag{9.11}$$

be a linear isomorphism such that

$$a_X \otimes a_Y = \bigoplus_{V \in \mathcal{O}(\mathcal{C})} b_{XY}^V \otimes a_V .$$

Note that the source and target of (9.11) are canonically isomorphic so that we can regard b_{XY}^V as an automorphism of $\mathsf{Hom}_{\mathcal{C}}(V, X \otimes Y)$. The natural isomorphism a_X is tensor (i.e., is a pivotal structure) if and only if $b_{XY}^V = \mathsf{id}$ for all $X, Y, V \in \mathcal{O}(\mathcal{C})$. Since a_X is a square root of a tensor isomorphism g_X, we see that $(b_{XY}^V)^2 = \mathsf{id}$.

The integers

$$N_{XY}^V = \dim_{\mathbb{C}} \mathsf{Hom}_{\mathcal{C}}(V, X \otimes Y) \qquad \text{and} \qquad T_{XY}^V = \mathsf{Tr}(b_{XY}^V),$$

where Tr denotes the trace of a linear transformation, satisfy inequality

$$|T_{XY}^V| \leq N_{XY}^V . \tag{9.12}$$

The equality $T_{XY}^V = N_{XY}^V$ occurs if and only if $b_{XY}^V = \mathsf{id}$, i.e., if and only if a_X is a pivotal structure.

For any $X \in \mathcal{O}(\mathcal{C})$ let $d_X = \mathsf{Tr}(a_X)$. Then

$$d_X d_Y = \sum_{V \in \mathcal{O}(\mathcal{C})} T_{XY}^V d_V .$$

Furthermore, $|X|^2 = |d_X|^2$ for every $X \in \mathcal{O}(\mathcal{C})$.

PROPOSITION 9.5.1. *A pseudo-unitary fusion category admits a unique spherical structure $a_X : X \xrightarrow{\sim} X^{**}$ with respect to which $d_X = \mathsf{FPdim}(X)$ for every simple object X.*

PROOF. Let $\mathcal{C}$ be a pseudo-unitary fusion category. Let $g_X : X \xrightarrow{\sim} X^{****}$ be a tensor isomorphism and $a_X : X \xrightarrow{\sim} X^{**}$ be its square root as before. The idea of the proof is to twist g_X by an appropriate tensor automorphism of the identity endofunctor of $\mathcal{C}$ in such a way that the dimensions corresponding to the square root of the resulting isomorphism become positive real numbers.

We have $|d_X| = \mathsf{FPdim}(X)$ for any simple object X, therefore, using (9.12), we obtain:

$$\begin{aligned}\mathsf{FPdim}(X)\,\mathsf{FPdim}(Y) &= |d_X d_Y| = |\sum_{V\in\mathcal{O}(\mathcal{C})} T_{XY}^V d_V| \\ &\leq \sum_{V\in\mathcal{O}(\mathcal{C})} N_{XY}^V\,\mathsf{FPdim}(V) = \mathsf{FPdim}(X)\,\mathsf{FPdim}(Y),\end{aligned}$$

for all $X, Y \in \mathcal{O}(\mathcal{C})$. Hence, the inequality in the above chain is an equality, i.e., $T_{XY}^V = \pm N_{XY}^V$ and the ratio $\frac{d_X d_Y}{d_V}$ is a real number whenever $N_{XY}^V \neq 0$. Thus, $\frac{d_X^2 d_Y^2}{d_V^2}$ is a positive number whenever V is contained in $X \otimes Y$.

The latter property is equivalent to $\sigma_X := \frac{|d_X|^2}{d_X^2}\,\mathsf{id}_X$ being a tensor automorphism of the identity endofunctor of $\mathcal{C}$. Let us replace g_X by $g_X \circ \sigma_X$ (so it is still a tensor isomorphism $X \xrightarrow{\sim} X^{****}$). The square root of the latter is $a_X \circ \tau_X$, where $\tau_X = \frac{|d_X|}{d_X}\,\mathsf{id}_X$. The dimensions corresponding to $a_X \circ \tau_X$ are now such that $d_X = |d_X|$, i.e., are positive real numbers (and $\sigma_X = \mathsf{id}_X$). This forces $T_{XY}^V = N_{XY}^V$. Thus, $a_X \circ \tau_X$ is a spherical structure on $\mathcal{C}$. □

9.6. Integral and weakly integral fusion categories

DEFINITION 9.6.1. A fusion category $\mathcal{C}$ is called *weakly integral* if $\mathsf{FPdim}(\mathcal{C}) \in \mathbb{Z}$. It is called *integral* if $\mathsf{FPdim}(X) \in \mathbb{Z}$ for any object X of $\mathcal{C}$.

EXERCISE 9.6.2. Let $\mathcal{C}$ be a spherical fusion category such that dimensions of all its objects are integers. Prove that $\mathcal{C}$ is integral, and that $\dim(X) = \pm\,\mathsf{FPdim}(X)$ for any simple object X of $\mathcal{C}$. *Hint:* Use argument in the proof of Proposition 9.4.2 to show that $\frac{\dim(\mathcal{C})}{\mathsf{FPdim}(\mathcal{C})}$ is an integer since it equals dimension of some object in $\mathcal{C}$.

Clearly, $\mathcal{C}$ is integral (respectively, weakly integral) if $K_0(\mathcal{C})$ is integral (respectively, weakly integral) in the sense of Definition 3.5.5.

EXERCISE 9.6.3. Show that the categories $\mathcal{C}_2(q)$, where q is a primitive 8th root of unity (the Ising categories, see Example 8.18.5) are weakly integral, but not integral. Show that $\mathcal{C}_k(q)$ are not weakly integral for any $k > 2$.

REMARK 9.6.4. By Proposition 6.1.14, the most general example of an integral fusion category is the representation category of a finite dimensional semisimple quasi-Hopf algebra.

A basic example of a weakly integral fusion category which is not integral is the *Tambara-Yamagami category* (see [**TaY**]), which categorifies the Tambara-Yamagami fusion ring, see Example 4.10.5. The Ising category $\mathcal{C}_2(q)$ is the simplest special case of this class of fusion categories.

PROPOSITION 9.6.5. *Let $\mathcal{C}$ be a weakly integral fusion category defined over $\mathbb{C}$. Then $\mathcal{C}$ is pseudo-unitary.*

PROOF. It is shown in Proposition 9.4.2 that the ratio $\frac{\dim(\mathcal{C})}{\mathsf{FPdim}(\mathcal{C})}$ is an algebraic integer ≤ 1. Let $D := \dim(\mathcal{C})$, let $D_1 = D, D_2, \ldots, D_N$ be algebraic conjugates of D, and let $g_1, \ldots, g_N$ be the elements of the Galois group $Gal(\overline{\mathbb{Q}}/\mathbb{Q})$ such that $D_i = g_i(D)$. Applying Proposition 9.4.2 to the category $g_i(\mathcal{C})$ (obtained from $\mathcal{C}$ by applying g_i to its structural constants, which can be chosen to be algebraic numbers by Corollary 9.1.8) we see that $\frac{\dim(g_i(\mathcal{C}))}{\mathsf{FPdim}(\mathcal{C})} = \frac{D_i}{\mathsf{FPdim}(\mathcal{C})}$ is an algebraic integer ≤ 1. Therefore,

$$\prod_{i=1}^{N} \frac{D_i}{\mathsf{FPdim}(\mathcal{C})}$$

is an algebraic integer ≤ 1. But this product is a rational number. So it must be equal to 1, and so all the factors are equal to 1. Thus, $\dim(\mathcal{C}) = \mathsf{FPdim}(\mathcal{C})$, as desired. □

COROLLARY 9.6.6. *A weakly integral fusion category $\mathcal{C}$ admits a unique spherical structure $a_X : X \xrightarrow{\sim} X^{**}$ with respect to which $d_X = \mathsf{FPdim}(X)$ for every simple object X in $\mathcal{C}$.*

PROOF. This follows from Propositions 9.5.1 and 9.6.5. □

COROLLARY 9.6.7. *(Larson and Radford, [**LaR1**]). Let H be a semisimple Hopf algebra over a field $\Bbbk$ of characteristic zero. Then $S^2 = \mathsf{id}$ on H.*

PROOF. The category $\mathsf{Rep}(H)$ is an integral fusion category, in which Frobenius-Perron dimensions are the usual vector space dimensions. So by Corollary 9.6.6, $\mathsf{Rep}(H)$ has a unique spherical structure in which categorical dimensions coincide with the vector space dimensions. This spherical structure is a grouplike element $u \in H$ such that $uxu^{-1} = S^2(x)$, and for every simple H-module V, $\mathsf{Tr}_V(u) = \dim V$. Clearly, u has finite order (as it generates a group algebra inside H), so its eigenvalues on V are roots of unity. But if $\lambda_1, ..., \lambda_m$ are roots of unity, and $\lambda_1 + ... + \lambda_m = m$ then $\lambda_j = 1$ for all j. Thus, $u = 1$, and thus $S^2 = \mathsf{id}$. □

REMARK 9.6.8. This gives another proof that a semisimple Hopf algebra H over a field of characteristic zero is cosemisimple. Indeed, $S^2 = \mathsf{id}$ on H, so $\mathsf{Tr}(S^2) \neq 0$, and thus H is cosemisimple.

PROPOSITION 9.6.9. *Let $\mathcal{C}$ be a weakly integral fusion category.*

(i) *For any $X \in \mathcal{O}(\mathcal{C})$ we have $\mathsf{FPdim}(X) = \sqrt{n_X}$ for some integer n_X.*
(ii) *The map $\deg : \mathcal{O}(\mathcal{C}) \to \mathbb{Q}_{>0}^{\times}/(\mathbb{Q}_{>0}^{\times})^2$ that takes $X \in \mathcal{O}(\mathcal{C})$ to the image of $\mathsf{FPdim}(X)^2$ in $\mathbb{Q}_{>0}^{\times}/(\mathbb{Q}_{>0}^{\times})^2$ is a grading of $\mathcal{C}$.*

PROOF. This is a direct consequence of the corresponding result about fusion rings (Proposition 3.5.7). □

COROLLARY 9.6.10. *Let $\mathcal{C}$ be a fusion category such that $\mathsf{FPdim}(\mathcal{C})$ is an odd integer. Then $\mathcal{C}$ is integral.*

PROOF. Follows from Corollary 3.5.8. □

The next proposition shows that the class of (weakly) integral fusion categories is closed under categorical Morita equivalence.

PROPOSITION 9.6.11. *Let $\mathcal{C}$ be an integral (respectively, weakly integral) fusion category and let $\mathcal{M}$ be an indecomposable $\mathcal{C}$-module category. Then the dual fusion category $\mathcal{C}^*_{\mathcal{M}}$ is integral (respectively, weakly integral).*

PROOF. When $\mathcal{C}$ is weakly integral, the statement follows from Corollary 7.16.7. Suppose $\mathcal{C}$ is integral. Then $\mathcal{Z}(\mathcal{C})$ is integral since the forgetful functor $\mathcal{Z}(\mathcal{C}) \to \mathcal{C}$ preserves Frobenius-Perron dimensions.

By Corollary 7.16.2, $\mathcal{Z}(\mathcal{C}_{\mathcal{M}}^*) \cong \mathcal{Z}(\mathcal{C})$, so $\mathcal{Z}(\mathcal{C}_{\mathcal{M}}^*)$ is integral. The forgetful functor $\mathcal{Z}(\mathcal{C}_{\mathcal{M}}^*) \to \mathcal{C}_{\mathcal{M}}^*$ is surjective by Corollary 7.13.11. Hence, every object X of $\mathcal{C}_{\mathcal{M}}^*$ is a subobject of an object of integral Frobenius-Perron dimension, and $\mathsf{FPdim}(X) \in \mathbb{Z}$ by Lemma 3.5.6. □

EXERCISE 9.6.12. Let $\mathcal{C}_{ad}$ be the adjoint subcategory of a weakly integral fusion category $\mathcal{C}$, see Definition 4.14.5. Prove that $\mathcal{C}_{ad}$ is integral.

EXERCISE 9.6.13. Let $\mathcal{C}$ be an arbitrary fusion category. Show that the full abelian subcategory of $\mathcal{C}$ consisting of objects of integral Frobenius-Perron dimension is a fusion subcategory of $\mathcal{C}$.

9.7. Group-theoretical fusion categories

In this Section we use categorical Morita equivalence (see Definition 7.12.17) to introduce a basic class of fusion categories important for classification purposes. Two additional classes will be introduced in Section 9.8.

Recall from Definition 5.11.1 that a fusion category $\mathcal{C}$ is pointed if every simple object of $\mathcal{C}$ is invertible. The most general example of a pointed fusion category is the category Vec_G^ω of finite dimensional vector spaces graded by a finite group G with the associativity constraint twisted by the 3-cocycle $\omega \in Z^3(G, \Bbbk^\times)$, see Example 2.3.8.

DEFINITION 9.7.1. A fusion category is called *group-theoretical* if it is categorically Morita equivalent to a pointed fusion category.

Recall that indecomposable Vec_G-module categories were classified in Example 7.4.10 (this is the case when $\omega = 1$). It is easy to generalize this description to the general case.

EXAMPLE 9.7.2. Equivalence classes of indecomposable right Vec_G^ω-module categories correspond to pairs (L, ψ), where L is a subgroup of G such that $\omega|_{L\times L\times L}$ is cohomologically trivial and $\psi \in C^2(L, \Bbbk^\times)$ is a 2-cochain satisfying

$$d_2\psi = \omega|_{L\times L\times L}. \tag{9.13}$$

The corresponding Vec_G^ω-module category is constructed as follows. Given a pair (L, ψ) as above, define an algebra

$$A(L,\psi) = \bigoplus_{a\in L} \delta_a \tag{9.14}$$

in Vec_G^ω with the multiplication

$$\bigoplus_{a,b\in L} \psi(a, b)\, \mathsf{id}_{\delta_{ab}} : A(L, \psi) \otimes A(L, \psi) \to A(L, \psi). \tag{9.15}$$

Let $\mathcal{M}(L, \psi)$ denote the category of left $A(L, \psi)$-modules in Vec_G^ω.

To see that any Vec_G^ω-module category $\mathcal{M}$ is equivalent to some $\mathcal{M}(L, \psi)$, note that the set of simple objects of $\mathcal{M}$ is a homogeneous G-space and so is identified with G/L for some subgroup L of G, and that the module associativity constraint of $\mathcal{M}$ gives rise to a 2-cochain $\psi \in C^2(L, \Bbbk^\times)$ satisfying (9.13).

REMARK 9.7.3. Two Vec_G^ω-module categories $\mathcal{M}(L, \psi)$ and $\mathcal{M}(L', \psi')$ are equivalent if and only if there is $g \in G$ such that $L' = gLg^{-1}$ and ψ' is cohomologous to ψ^g in $H^2(L', \Bbbk^\times)$, where $\psi^g(x, y) := \psi(gxg^{-1}, gyg^{-1})$ for all $x, y \in L$. Here we abuse notation and identify ψ and ψ' with cocycles representing them.

Thus, a group-theoretical fusion category is equivalent to $(\mathrm{Vec}_G^\omega)^*_{\mathcal{M}(L,\psi)}$, where $\mathcal{M}(L,\psi)$ is the Vec_G^ω-module category from Example 9.7.2.

EXAMPLE 9.7.4. The category $(\mathrm{Vec}_G^\omega)^*_{\mathcal{M}(L,\psi)}$ can be described quite explicitly in terms of finite groups and their cohomology as the category of $A(L,\psi)$-bimodules in Vec_G^ω, where $A(L,\psi)$ is the algebra introduced in (9.14).

For example, simple objects of $(\mathrm{Vec}_G^\omega)^*_{\mathcal{M}(L,\psi)}$ can be described as follows.

For any $g \in G$ the group $L^g := L \cap gLg^{-1}$ has a well-defined 2-cocycle

$$\begin{aligned}\psi^g(h,h') : &= \psi(h,h')\psi(g^{-1}h'^{-1}g,\, g^{-1}h^{-1}g)\omega(hh'g,\, g^{-1}h'^{-1}g,\, g^{-1}h^{-1}g)^{-1} \\ &\quad \times \omega(h,\, h',\, g)\omega(h,\, h'g,\, g^{-1}h'^{-1}g), \qquad h, h' \in L^g.\end{aligned}$$

One can check that irreducible $A(L,\psi)$-bimodules in Vec_G^ω are parametrized by pairs (Z,π), where Z is a double L-coset in G and π is an irreducible projective representation of L^g with the Schur multiplier ψ^g, $g \in Z$.

EXERCISE 9.7.5. (i) Let G, L be finite groups such that G acts on L by automorphisms. Let $\Bbbk^L$ be the Hopf algebra of functions on L, and $\Bbbk G$ be the group algebra of G. Let $H = \Bbbk^L \# \Bbbk G$ be the semidirect product Hopf algebra, i.e., H is generated by the Hopf subalgebra $\Bbbk^L$ and the group of grouplike elements G modulo commutation relations $fg = gf^g$ for $f \in \Bbbk^L$, $g \in G$, where $f^g(x) := f(gx)$. Show that H is a semisimple Hopf algebra (for $\mathrm{char}\Bbbk = 0$), and that $\mathrm{Rep}(H)$ is the G-equivariantization of Vec_L. Show that $\mathrm{Rep}(H)$ is a group-theoretical fusion category.

(ii) More generally, let $K = GL$ be an exact factorization of a finite group K, i.e., G and L are subgroups of K such that the multiplication map $m : G \times L \to K$ is bijective. In this case, L acts on the left on $K/L = G$ (as a set). Similarly, G acts on $L = G\backslash K$ on the right. So we can define a Hopf algebra $H = \Bbbk^L \# \Bbbk G$ similarly to (i). Namely, H has basis $b_{x,g} := \delta_x \otimes g$ for $x \in L$ and $g \in G$, with multiplication law

$$b_{x,g} b_{y,h} = \delta_{x^g,y} b_{x,gh}$$

and coproduct

$$\Delta(b_{x,g}) = \sum_{y,z \in L : yz = x} b_{y,{}^z g} \otimes b_{z,g}.$$

Show that H is a well defined semisimple Hopf algebra (it was introduced in works of G. I. Kac, and is called a Kac algebra), and $\mathrm{Rep}(H)$ is a group-theoretical fusion category.

REMARK 9.7.6. The class of group-theoretical fusion categories is not closed under equivariantizations. In particular, the class of Hopf algebras with group-theoretical representation categories is not closed under Hopf algebra extensions.

The smallest example of a semisimple Hopf algebra whose representation category is not group-theoretical has dimension 36 [**Nik2**]. It is obtained as a $\mathbb{Z}/2\mathbb{Z}$-extension of the Kac-Paljutkin Hopf algebra [**KacGP**] of dimension 18. Several families of non-group-theoretical Hopf algebras obtained by equivariantization of Tambara-Yamagami categories (see Example 4.10.5) were obtained in [**GeNN, Nik2**].

REMARK 9.7.7. In view of Proposition 9.6.11, group-theoretical categories are integral.

DEFINITION 9.7.8. A multifusion category $\mathcal{D}$ is called a *quotient* of a fusion category $\mathcal{C}$ if there is a surjective tensor functor $F : \mathcal{C} \to \mathcal{D}$, cf. Definition 1.8.3.

Components of multifusion categories were studied in Section 4.3.

PROPOSITION 9.7.9. (i) *A fusion subcategory of a group-theoretical category is group-theoretical.*
(ii) *A component in a quotient category of a group-theoretical category is group-theoretical.*

PROOF. (i) Let $\mathcal{C} \subset \mathcal{D}$ be fusion categories and let $\mathcal{D}$ be group-theoretical. Let $\mathcal{M}$ be an indecomposable module category over $\mathcal{D}$ such that $\mathcal{D}^*_{\mathcal{M}} \simeq \mathrm{Vec}_G^{\omega}$. Then by Theorem 7.17.4, $\mathcal{C}^*_{\mathcal{M}}$ is a quotient of Vec_G^{ω}.

Let I label the component categories of $\mathcal{C}^*_{\mathcal{M}}$. It is clear that for each $g \in G$ and $i \in I$ there exists a unique $g(i) \in I$ such that the functor of tensor multiplication by δ_g is an equivalence $\mathcal{C}_{ir} \simeq \mathcal{C}_{g(i)r}$ for each $r \in I$. Let us denote this functor by g. So we have an action of G on I. Let $j \in I$ and H be the stabilizer of j in G. For any $g \in H$, denote by $\bar{g}_{jj}$ the projection of g to the fusion category $(\mathcal{C}^*_{\mathcal{M}})_{jj}$. Then the assignment $g \to \bar{g}_{jj}$ is a surjective tensor functor $\mathrm{Vec}_H^{\omega} \to (\mathcal{C}^*_{\mathcal{M}})_{jj}$. This functor must map invertible objects to invertible objects. Hence, all simple objects of $\mathcal{C}^*_{jj}$ are invertible, so $\mathcal{C}^*_{jj}$ is pointed, and hence $\mathcal{C}$ is group-theoretical (as it is dual to $\mathcal{C}^*_{jj}$ with respect to a module category).

(ii) Let $\mathcal{D}$ be group-theoretical, $F : \mathcal{D} \to \mathcal{C}$ be surjective ($\mathcal{C}$ is indecomposable), and $\mathcal{C}_{ii}$ a component of $\mathcal{C}$. We need to show that $\mathcal{C}_{ii}$ is group-theoretical. Let $\mathcal{M}$ be an indecomposable module category over $\mathcal{C}$. It suffices to show that $\mathcal{C}^*_{\mathcal{M}}$ is group-theoretical, as $\mathcal{C}^*_{\mathcal{M}}$ is dual to $\mathcal{C}_{ii}$ with respect to the i-th part of $\mathcal{M}$. But by Theorem 7.17.4, $\mathcal{C}^*_{\mathcal{M}}$ is embedded into $\mathcal{D}^*_{\mathcal{M}}$, so it suffices to know that $\mathcal{D}^*_{\mathcal{M}}$ is group-theoretical, which follows by duality from the fact that $\mathcal{D}$ is group-theoretical. □

9.8. Weakly group-theoretical fusion categories

The notions of a nilpotent and cyclically nilpotent fusion category were introduced in Definition 4.14.14 (see also Remark 4.14.16).

DEFINITION 9.8.1. A fusion category is *weakly group-theoretical* if it is categorically Morita equivalent to a nilpotent fusion category. A fusion category is *solvable* if it is categorically Morita equivalent to a cyclically nilpotent fusion category.

REMARK 9.8.2. Since the Frobenius-Perron dimension of a fusion category is invariant under categorical Morita equivalence, we have $\mathrm{FPdim}(\mathcal{A}) \in \mathbb{Z}$ for every weakly group-theoretical fusion category $\mathcal{A}$.

Let G be a finite group. Recall from Section 4.14 that a fusion category $\mathcal{C}$ is called a *G-extension* of a fusion category $\mathcal{D}$ if there is a G-grading of $\mathcal{C}$:

$$\mathcal{C} = \bigoplus_{g \in G} \mathcal{C}_g,$$

such that $\mathcal{C}_1 \cong \mathcal{D}$.

LEMMA 9.8.3. *Let G be a finite group, let $\mathcal{A}$ be a G-extension of a fusion category $\mathcal{A}_0$, and let $\mathcal{B}_0$ be a fusion category Morita equivalent to $\mathcal{A}_0$. There exists a G-extension $\mathcal{B}$ of $\mathcal{B}_0$ which is Morita equivalent to $\mathcal{A}$.*

PROOF. Let A be an algebra in $\mathcal{A}_0$ such that $\mathcal{B}_0$ is equivalent to the category of A-bimodules in $\mathcal{A}_0$. Let $\mathcal{B}$ be the category of A-bimodules in $\mathcal{A}$ (we can view A as an algebra in $\mathcal{A}$ since $\mathcal{A}_0 \subset \mathcal{A}$). Then $\mathcal{B}$ inherits the G-grading, thanks to A being in the trivial component of the G-graded fusion category $\mathcal{A}$. By construction, $\mathcal{B}$ is categorically Morita equivalent to $\mathcal{A}$. □

PROPOSITION 9.8.4. *The class of weakly group-theoretical fusion categories is closed under taking graded extensions, equivariantizations, categorically Morita equivalent categories, tensor products, centers, subcategories and component categories of quotient categories.*

PROOF. The invariance under taking Morita equivalent categories and tensor products is obvious. The invariance under taking extensions follows from Lemma 9.8.3, and the invariance under equivariantizations follows from Example 7.12.25. The invariance under taking the center then follows from Morita invariance, as $\mathcal{Z}(\mathcal{C})$ is categorically Morita equivalent to $\mathcal{C} \boxtimes \mathcal{C}^{\mathrm{op}}$. The rest of the proof is similar to the proof of Proposition 9.7.9. To prove the invariance under taking subcategories, let $\mathcal{C}$ be a weakly group-theoretical category, and $\mathcal{D} \subset \mathcal{C}$ a fusion subcategory. Let $\mathcal{M}$ be an indecomposable $\mathcal{C}$-module category such that $\mathcal{C}^*_{\mathcal{M}}$ is nilpotent. Then every component category of $\mathcal{D}^*_{\mathcal{M}}$ is nilpotent, since it is easy to see that every component category in a quotient of a nilpotent category is nilpotent. The case of a component in a quotient category reduces to the case of a subcategory by taking duals. □

Here is a list of properties of solvable categories.

PROPOSITION 9.8.5. (i) *The class of solvable categories is closed under taking extensions and equivariantizations by solvable groups, Morita equivalent categories, tensor products, subcategories and component categories of quotient categories.*

(ii) *The categories* $\mathsf{Vec}_{G,\omega}$ *and* $\mathsf{Rep}(G)$ *are solvable if and only if* G *is a solvable group.*

(iii) *A solvable fusion category* $\mathcal{A} \neq \mathsf{Vec}$ *contains a nontrivial invertible object.*

PROOF. (i) As in the proof of Proposition 9.8.4, everything follows from the easy fact that a component category in a quotient of a cyclically nilpotent category is cyclically nilpotent.

(ii) One direction is obvious, since if G is solvable, Vec_G^{ω} is cyclically nilpotent. Since $\mathsf{Rep}(G)$ is Morita equivalent to Vec_G, it is also solvable by (i).

To prove the converse implication, it suffices to show that if $\mathsf{Rep}(G)$ is solvable then so is G. Indeed, $\mathcal{Z}(\mathsf{Vec}_G^{\omega})$ contains $\mathsf{Rep}(G)$ as a fusion subcategory, so the solvability of Vec_G^{ω} implies the solvability of $\mathsf{Rep}(G)$ by (i). We have two possibilities: either $\mathsf{Rep}(G)$ is an H-extension or $\mathsf{Rep}(G) = \mathcal{C}^H$ for some fusion category $\mathcal{C}$, where H is a cyclic group of prime order. In the former situation G must have a non-trivial center Z and we can pass to the fusion subcategory $\mathsf{Rep}(G/Z) \subset \mathsf{Rep}(G)$ which is again solvable by (i). In the latter situation $\mathsf{Rep}(G)$ contains a fusion subcategory of prime order therefore, G contains a normal subgroup G_1 of prime index and we can pass to the solvable quotient category $\mathsf{Rep}(G_1)$. So the required statement follows by induction.

(iii) The proof is by induction on the dimension of $\mathcal{C}$. The base of induction is clear, and only the induction step needs to be justified. If $\mathcal{C}$ is an extension of a smaller solvable category $\mathcal{D}$, then either $\mathcal{D} \neq \mathsf{Vec}$ and the statement follows from the induction assumption, or $\mathcal{D} = \mathsf{Vec}$ and $\mathcal{C}$ is pointed, so the statement is obvious. On the other hand, if $\mathcal{C}$ is a $\mathbb{Z}/p\mathbb{Z}$-equivariantization of a smaller solvable category $\mathcal{D}$, then $\mathsf{Rep}(\mathbb{Z}/p\mathbb{Z})$ sits inside $\mathcal{C}$, so we are done. □

The following question is one of the main open problems in the theory of finite dimensional Hopf algebras.

QUESTION 9.8.6. Is the representation category of any semisimple Hopf algebra weakly group-theoretical?

The same question may be asked about quasi-Hopf algebras.

We will show below that the answer is "yes" for categories of dimension $p^a q^b$, where p and q are primes, and a, b are non-negative integers.

9.9. Symmetric and Tannakian fusion categories

Recall from Definition 8.1.12 that a braided fusion category $\mathcal{C}$ is *symmetric* if $c_{Y,X} \circ c_{X,Y} = \mathsf{id}_{X \otimes Y}$ for all objects $X, Y \in \mathcal{C}$; in this case the braiding c is called *symmetric*.

EXAMPLE 9.9.1.
(1) Let $\mathcal{C}$ be a pointed braided category corresponding to a pre-metric group (Γ, q) (see Section 8.4). Such $\mathcal{C}$ is symmetric if and only if the bicharacter b associated to q is trivial. Equivalently, $q : \Gamma \to \Bbbk^\times$ is a character such that $q(x) = \pm 1$ for any $x \in \Gamma$.
(2) The category $\mathsf{Rep}(G)$ of representations of a finite group G equipped with its standard symmetric braiding $\sigma_{XY}(x \otimes y) := y \otimes x$ is an example of a symmetric fusion category.
(3) Let G be a finite group and let $z \in G$ be a central element such that $z^2 = 1$. Then the fusion category $\mathsf{Rep}(G)$ has a braiding σ' defined as follows:

$$\sigma'_{XY}(x \otimes y) = (-1)^{mn} y \otimes x \text{ if } x \in X,\ y \in Y,\ zx = (-1)^m x,\ zy = (-1)^n y.$$

The fusion category $\mathsf{Rep}(G)$ equipped with this braiding will be denoted by $\mathsf{Rep}(G, z)$.

REMARK 9.9.2. Equivalently, $\mathsf{Rep}(G, z)$ can be described as a full subcategory of the category of super-representations of G; namely, $\mathsf{Rep}(G, z)$ consists of those super-representations V on which z acts by the *parity automorphism* (i.e., $zv = v$ if $v \in V$ is even and $zv = -v$ if $v \in V$ is odd).

EXAMPLE 9.9.3. Let $G = \mathbb{Z}/2\mathbb{Z}$ and z be the nontrivial element of G. Then $\mathsf{Rep}(G, z)$ is the category of super-vector spaces sVec.

The main result of this section is Theorem 9.9.26 (due to Deligne) which states among other things that any symmetric fusion category over an algebraically closed field of characteristic zero is equivalent to one from Example 9.9.1 (3) above.

EXERCISE 9.9.4. Prove that the categories from Example 9.9.1 (1) are of the form $\mathsf{Rep}(G, z)$ for suitable G and z.

Recall that for any object V of a braided tensor category $\mathcal{C}$ we have an action of the braid group B_n on the tensor power $V^{\otimes n}$, see Remark 8.2.5. In the case of a symmetric tensor category $\mathcal{C}$ this action factors through the symmetric group S_n which is the quotient of B_n by the relations $\sigma_i^2 = 1$.

DEFINITION 9.9.5. The *symmetric nth power* $S^n(V)$ (or S^nV) of V is the maximal quotient of $V^{\otimes n}$ on which the action of S_n is trivial. Similarly, the *exterior nth power* $\wedge^n(V)$ (or $\wedge^nV$) of V is the maximal quotient of $V^{\otimes n}$ on which the action of S_n factors through the sign representation.

REMARK 9.9.6. In other words, S^nV is the degree n part of the *symmetric algebra* SV, which is the quotient of the tensor algebra TV by the ideal generated by the image of $\mathrm{id}_{V\otimes V} - \sigma$, where $\sigma : V\otimes V \to V\otimes V$ is the permutation. Similarly, $\wedge^nV$ is the degree n part of the *exterior algebra* $\wedge V$, which is the quotient of the tensor algebra TV by the ideal generated by the image of $\mathrm{id}_{V\otimes V} + \sigma$.

EXERCISE 9.9.7. Note that in a symmetric tensor category $\mathcal{C}$, one can define any linear algebraic structures that we encounter in ordinary algebra (i.e., modules over any linear operad). The following exercise studies this notion for two important kinds of structures – Lie algebras and Hopf algebras.

(i) A Lie algebra in a symmetric tensor category $\mathcal{C}$ is an (ind-)object L together with a bracket operation $\beta : L\otimes L \to L$ which satisfies the skew-symmetry and the Jacobi identity:

$$\beta(\mathrm{id} + (12)) = 0, \quad \beta\circ(\beta\otimes \mathrm{id})\circ(\mathrm{id} + (123) + (132)) = 0,$$

where (123) and (132) are the nontrivial cyclic permutations in S_3.

(i) Show that any associative algebra in $\mathcal{C}$ has a natural structure of a Lie algebra (with the bracket being the commutator). In particular, if V is any object of $\mathcal{C}$ then $\underline{\mathrm{End}}(V) := V\otimes V^*$ is a Lie algebra. This Lie algebra is denoted $\mathfrak{gl}(V)$.

(ii) Suppose that V is equipped with a symmetric isomorphism $B : V \to V^*$, i.e., such that $B^* = B$ (note that in a symmetric rigid monoidal category we have a canonical identification $V \cong V^{**}$). Show that $\wedge^2V \subset V\otimes V \cong V\otimes V^*$ has a natural Lie algebra structure, obtained by restricting the bracket in $V\otimes V^*$; this Lie algebra is called the orthogonal Lie algebra of V and denoted $\mathfrak{o}(V)$. Similarly, if V is equipped with a skew-symmetric isomorphism $B : V \to V^*$ (i.e., $B^* = -B$), show that $S^2V \subset V\otimes V \cong V\otimes V^*$ has a natural Lie algebra structure, obtained by restricting the bracket in $V\otimes V^*$; this Lie algebra is called the symplectic Lie algebra of V and denoted $\mathfrak{sp}(V)$.

(iii) In (i), let $\mathcal{C} = \mathsf{sVec}$, and let $F : \mathsf{sVec} \to \mathsf{Vec}$ be the forgetful functor. Let $F(L) = L_+ \oplus L_-$, where L_+ is the the image of the even part of L and L_- is the image of the odd part of L under F. Since F is not braided, $F(L)$ is, in general, not a Lie algebra; however, it carries a bracket operation $[,] = F(\beta)$. Write down explicitly the properties of $[,]$ obtained from the skew-symmetry and Jacobi identity for β. You will obtain the definition of a *Lie superalgebra*. Thus, a Lie superalgebra is just the image of a Lie algebra in sVec under the forgetful functor to Vec.

(iv) In (ii), assume that $F(V) = V_+ \oplus V_-$, where $\dim V_+ = m$ and $\dim V_- = n$. Describe explicitly (in linear-algebraic terms) the Lie superalgebras $F(\mathfrak{gl}(V))$ (called $\mathfrak{gl}(m|n)$), $F(\mathfrak{o}(V))$ (called $\mathfrak{osp}(m|n)$), and $F(\mathfrak{sp}(V))$ (called $\mathfrak{osp}(n|m)$), and the operations on them.

(v) Show that if L is a Lie algebra in $\mathcal{C}$ then the category of L-modules in $\mathcal{C}$ is a symmetric tensor category.

(vi) Define a bialgebra and a Hopf algebra in a symmetric tensor category $\mathcal{C}$, and show that the categories of modules and comodules over a Hopf algebra H in $\mathcal{C}$ are tensor categories.

(vii) Let L be a Lie algebra in a symmetric tensor category $\mathcal{C}$. Define the associative algebra $U(L)$ in $\mathcal{C}$ to be the quotient of the tensor algebra TL of L by the ideal generated by the image of the morphism $\mathsf{id}_{L\otimes L} - \sigma - \beta : L \otimes L \to TL$, where σ is the permutation of components. Show that $U(L)$ is a Hopf algebra in $\mathcal{C}$, and that the category of $U(L)$-modules is equivalent to the category of L-modules as a tensor category.

(viii) Prove the categorical PBW theorem: the associated graded algebra $\mathrm{gr}U(L)$ under the natural filtration is naturally isomorphic to the symmetric algebra $SL = \oplus_{n\geq 0} S^n L$.

Hint: Use the Campbell-Hausdorff formula to define a noncommutative $*$-product on SL which defines the algebra $U(L)$.

(ix) An affine scheme X in a symmetric tensor category $\mathcal{C}$ is, by definition, the same thing as its algebra of regular functions $O(X)$, which is a commutative algebra in $\mathcal{C}$; but morphisms between affine schemes go in the opposite direction (i.e., the category of affine schemes in $\mathcal{C}$ is the dual category to the category of commutative algebras in $\mathcal{C}$). An affine group scheme is an affine scheme G together with a Hopf algebra structure on the commutative algebra $O(G)$ (i.e., as in the classical case, an affine group scheme in $\mathcal{C}$ is a group object in the category of affine schemes in $\mathcal{C}$). Show that if G is an affine group scheme in $\mathcal{C}$ then the category $\mathsf{Rep}(G)$ of $O(G)$-comodules in $\mathcal{C}$ is a symmetric tensor category.

(x) Let $V \in \mathcal{C}$. Define the scheme $G = GL(V)$ as follows: $O(G)$ is the quotient of $S(V^* \otimes V \oplus V^* \otimes V)$ by the ideal generated by the images of the morphisms $\delta - \mathrm{ev}_V$ and $\delta^{op} - \mathrm{ev}_V$, where $\delta : V^* \otimes V \to (V^* \otimes V) \otimes (V^* \otimes V)$ is given by $\delta = \mathsf{id} \otimes \mathrm{coev}_V \otimes \mathsf{id}$, and $\delta^{op} = P_{12,34} \circ \delta$ (where $P_{12,34}$ is the permutation of the components 12 and 34). This mimicks in the categorical setting defining the usual $GL(V)$ as the set of pairs (X, Y) of endomorphisms of V such that $XY = YX = \mathsf{id}$. Show that G is an affine group scheme in $\mathcal{C}$.

(xi) Show that in (x), if $\mathcal{C} = \mathsf{Rep}(K)$ for an algebraic group K, then the symmetric tensor category of representations of $GL(V)$ in $\mathcal{C}$ is equivalent to $\mathsf{Rep}(GL(V) \rtimes K)$.

(xii) In the setting of (ii), define the groups schemes $O(V)$, $Sp(V)$ (note that in general we cannot define $SO(V)$, since we do not have a notion of determinant).

By Proposition 8.10.12 a symmetric fusion category $\mathcal{C}$ has a spherical structure such that the corresponding twist (see Definition 8.10.1) equals the identity. In general, dimensions defined by this spherical structure are not positive: it is easy to see that if $\mathcal{C} = \mathsf{Rep}(G, z)$ then the categorical dimension of $X \in \mathcal{C}$ equals the *super*-dimension of the super-vector space X (i.e., the trace of the operator z acting on the vector space X), so the dimension of a simple object can be a negative integer.

EXERCISE 9.9.8. Let $\mathcal{C}$ be as in Example 9.9.1 (1). Then dimension of a simple object with isomorphism class $x \in \Gamma$ equals $q(x)$. (*Hint:* Use Exercise 8.10.15).

EXERCISE 9.9.9. (i) Let $\mathcal{C}$ be a symmetric fusion category over a field $\Bbbk$ of characteristic zero. For $\alpha \in \Bbbk$, let $\binom{\alpha}{n} := \frac{\alpha(\alpha-1)\cdots(\alpha-n+1)}{n!}$. Let $V \in \mathcal{C}$ be an object of dimension α. Prove that

$$\dim S^n(V) = \binom{\alpha+n-1}{n}, \quad \dim \wedge^n(V) = \binom{\alpha}{n}.$$

Hint: S^nV and $\wedge^nV$ are the images of the symmetrizer and skew-symmetrizer in $k[S_n]$, respectively.

(ii) Let $\mathcal{D}$ be a symmetric tensor category over a field $\Bbbk$ of characteristic p. Show that the dimension of any object $X \in \mathcal{D}$ lies in $\mathbb{F}_p \subset \Bbbk$.

Hint. Let $a = 1 - s \in \Bbbk S_p$, where $s \in S_p$ is a cyclic permutation. Show that $a^p = 0$ and deduce that $\mathsf{Tr}_{X^{\otimes p}}(a) = 0$. On the other hand, show that $\mathsf{Tr}_{X^{\otimes p}}(a) = (\dim X)^p - \dim X$.

Recall from Corollary 4.7.13 that the dimension of an object of a pivotal fusion category is an algebraic integer.

EXERCISE 9.9.10. ([**AEG**, Section 7]). Let $\alpha \in \Bbbk$ be such that $\binom{\alpha+k-1}{k}$ and $\binom{\alpha}{k}$ are algebraic integers for all $k \in \mathbb{N}$. Prove that $\alpha \in \mathbb{Z}$.

Thus we see that dimensions of all objects of a symmetric fusion category are integers.

COROLLARY 9.9.11. *A symmetric fusion category is integral. In particular its Frobenius-Perron dimension is an integer.*

PROOF. This follows from Exercise 9.6.2. □

We will say that a symmetric fusion category is *positive* if all dimensions are non-negative integers. Thus in a positive symmetric fusion category we have by Exercise 9.6.2 that $\dim(X) = \mathsf{FPdim}(X)$ for any $X \in \mathcal{C}$.

EXAMPLE 9.9.12. A pointed braided category $\mathcal{C}$ with associated pre-metric group (Γ, q) (see Section 8.4) is positive if and only if $q = 1$, see Exercise 9.9.8.

COROLLARY 9.9.13. *Let $\mathcal{C}$ be a positive symmetric fusion category and let $X \in \mathcal{C}$. Assume that* $\dim(X) = d \in \mathbb{Z}_{\geq 0}$. *Then* $\wedge^{d+1}(X) = 0$.

PROOF. By Exercise 9.9.9(i), we have $\mathsf{FPdim}(\wedge^{d+1}(X)) = \dim(\wedge^{d+1}(X)) = 0$. □

COROLLARY 9.9.14. *Let $\mathcal{C}$ be a symmetric fusion category over $\mathbb{C}$. Then there is a unique (possibly non-faithful) $\mathbb{Z}/2\mathbb{Z}$-grading $\mathcal{C} = \mathcal{C}_1 \oplus \mathcal{C}_{-1}$ such that $\mathcal{C}_1$ is positive and for any $0 \neq X \in \mathcal{C}_{-1}$ we have* $\dim(X) < 0$.

PROOF. Note that $\mathcal{C}$ has two spherical structures: the one described earlier in this section and the one provided by Proposition 9.5.1. By Remark 8.10.16 they differ by a tensor automorphism of the identity endofunctor of $\mathcal{C}$ which gives rise to the required $\mathbb{Z}/2\mathbb{Z}$-grading. □

Let $\mathcal{C} = \mathcal{C}_1 \oplus \mathcal{C}_{-1}$ be an arbitrary $\mathbb{Z}/2\mathbb{Z}$-graded symmetric fusion category with braiding c. For $X \in \mathcal{C}_{(-1)^m}$ and $Y \in \mathcal{C}_{(-1)^n}$ we consider the modified morphism

$$c^{mod}_{X,Y} := (-1)^{mn} c_{X,Y}. \tag{9.16}$$

EXERCISE 9.9.15. (1) Prove that morphism (9.16) extended by linearity to $\mathcal{C}$ is again a symmetric braiding on $\mathcal{C}$. We will denote the category $\mathcal{C}$ with modified braiding c^{mod} by $\mathcal{C}^{mod}$.
(2) We have $(\mathcal{C}^{mod})^{mod} = \mathcal{C}$.
(3) Let $\dim^{mod}$ be the dimension function of the category $\mathcal{C}^{mod}$. Then for $X \in \mathcal{C}_{(-1)^m}$ we have $\dim^{mod}(X) = (-1)^m \dim(X)$.

Thus this construction applied to a $\mathbb{Z}/2\mathbb{Z}$-grading from Corollary 9.9.14 produces a positive category $\mathcal{C}^{mod}$. Hence any symmetric fusion category can be obtained from a $\mathbb{Z}/2\mathbb{Z}$-graded positive symmetric fusion category by the modification of braiding as above.

DEFINITION 9.9.16. Let $\mathcal{C}$ be a symmetric fusion category. A braided tensor functor $F : \mathcal{C} \to \mathsf{Vec}$ is called a *symmetric fiber functor*. We say that $\mathcal{C}$ is *Tannakian* if it admits a symmetric fiber functor.

REMARK 9.9.17. The above fiber functor is defined in the setting of *symmetric* fusion categories and should not be confused with the more general (not necessarily braided) fiber functor considered in Section 5.1. In this section we will consider only symmetric fiber functors, and will drop the adjective "symmetric" for brevity (which is usually done in the literature).

EXERCISE 9.9.18. Prove that a Tannakian fusion category is positive.

Given a fiber functor $F : \mathcal{C} \to \mathsf{Vec}$, let $I : \mathsf{Vec} \to \mathcal{C}$ be its right adjoint and let $A_F = I(\mathbf{1})$. By Proposition 8.8.8, $A_F \in \mathcal{C}$ is a commutative algebra. Moreover, the category $\mathsf{Mod}_{\mathcal{C}}(A_F)$ is tensor equivalent to the category Vec via the functor $M \mapsto \mathsf{Hom}_{\mathcal{C}}(\mathbf{1}, M)$. In other words, any right (or left) A_F-module is isomorphic to a direct sum of several copies of A_F.

DEFINITION 9.9.19. The algebra A_F constructed above is called the *regular algebra* of $(\mathcal{C}, F)$.

EXERCISE 9.9.20. Prove that the tensor functor F is canonically isomorphic to the tensor functor

$$\mathcal{C} \to \mathsf{Vec} : X \mapsto \mathsf{Hom}_{\mathcal{C}}(\mathbf{1}, X \otimes A_F).$$

For a fiber functor $F : \mathcal{C} \to \mathsf{Vec}$, let G_F be the group of tensor automorphisms of F. A fiber functor $F : \mathcal{C} \to \mathsf{Vec}$ has an obvious lifting

$$\tilde{F} : \mathcal{C} \to \mathsf{Rep}(G_F). \tag{9.17}$$

EXERCISE 9.9.21. Prove that the group G_F is isomorphic to the group $\mathsf{Aut}(A_F)$ of automorphisms of the algebra A_F.

THEOREM 9.9.22. ([**De1, De2**]) *Let $\mathcal{C}$ be a positive symmetric fusion category. Then*

(i) *there exists a fiber functor $\mathcal{C} \to \mathsf{Vec}$;*
(ii) *all fiber functors $\mathcal{C} \to \mathsf{Vec}$ are isomorphic to each other;*
(iii) *for any fiber functor $F : \mathcal{C} \to \mathsf{Vec}$, the functor $\tilde{F} : \mathcal{C} \to \mathsf{Rep}(G_F)$ constructed above is an equivalence.*

REMARK 9.9.23. In the setting of C^*-algebras, a similar theorem was proved by Doplicher and Roberts [**DoR**]. This theorem gives a categorical characterization of the category of finite dimensional complex representations of a compact topological group.

PROOF. We will prove (i) in Section 9.10. Let us prove (ii). Let F_1 and F_2 be two fiber functors and let A_1 and A_2 be the corresponding regular algebras. Let us consider the (A_1, A_2)-bimodule $A_1 \otimes A_2$. For any (A_1, A_2)-bimodule M we have $\mathsf{Hom}_{(A_1,A_2)}(A_1 \otimes A_2, M) = \mathsf{Hom}(\mathbf{1}, M)$. In particular, the algebra

$$\mathsf{End}_{(A_1,A_2)}(A_1 \otimes A_2) = \mathsf{Hom}_{\mathcal{C}}(\mathbf{1}, A_1 \otimes A_2)$$

is commutative (indeed, this algebra identifies with $F_1(A_2)$). Thus the algebra $\mathsf{End}_{(A_1,A_2)}(A_1 \otimes A_2)$ has a one dimensional representation. Equivalently, there exists an (A_1, A_2)-bimodule M such that $\mathsf{Hom}_{\mathcal{C}}(\mathbf{1}, M)$ is one dimensional. But then M is isomorphic to A_1 as a left A_1-module and isomorphic to A_2 as a right A_2-module. Combining these isomorphisms, we get an isomorphism of algebras $A_1 \simeq A_2$. This implies (ii) by Exercise 9.9.20.

Let us prove (iii). Let $A = A_F$ be the regular algebra. Observe that $A \otimes A$ is a generator of the category of (A, A)-bimodules since $\mathsf{Hom}_{(A,A)}(A \otimes A, M) = \mathsf{Hom}_{\mathcal{C}}(\mathbf{1}, M)$ and $\mathsf{Hom}(\mathbf{1}, M) \neq 0$ for any right (or left) A-module M. On the other hand, as above, we see that $\mathsf{End}_{(A,A)}(A \otimes A)$ is commutative, so for any simple (A, A)-bimodule M we have that $\mathsf{Hom}(\mathbf{1}, M)$ is one dimensional. In other words, any simple (A, A)-bimodule is isomorphic to A with right action twisted by an automorphism of A. Thus the category of simple (A, A)-bimodules is closed under tensor product and is equivalent to the category $\mathcal{C}_{G_F}$ from Example 2.3.6 (recall that $\mathsf{Aut}(A) = G_F$). Consider $\mathsf{Mod}_{\mathcal{C}}(A) = \mathsf{Vec}$ as a module category over $\mathcal{C}$, so the category of (A, A)-bimodules identifies with $\mathcal{C}^*_{\mathsf{Vec}}$. We can identify the category $\mathsf{Rep}(G_F)$ with functors from $\mathsf{Mod}_{\mathcal{C}}(A) = \mathsf{Vec}$ to itself commuting with the action of the category $\mathcal{C}_{G_F} \subset \mathcal{C}^*_{\mathsf{Vec}}$ above; in this language $\tilde{F}(X)$ (see (9.17)) is a functor $\mathsf{Mod}_{\mathcal{C}}(A) \ni L \mapsto X \otimes L$ with obvious commutation with $L \mapsto L \otimes_A M$ for any simple (A, A)-bimodule M. Since every (A, A)-bimodule is a direct sum of simple ones, the category $\mathsf{Rep}(G_F)$ above identifies with $(\mathcal{C}^*_{\mathsf{Vec}})^*_{\mathsf{Vec}}$, and the functor $\tilde{F}$ identifies with the canonical functor $\mathcal{C} \to (\mathcal{C}^*_{\mathsf{Vec}})^*_{\mathsf{Vec}}$. Thus, (iii) follows from Theorem 7.12.11. □

COROLLARY 9.9.24. *Let $\mathcal{C}$ be a pointed braided fusion category such that the associated quadratic form (see Section 8.4) equals 1 identically. Then there exists a braided tensor functor $\mathcal{C} \to \mathsf{Vec}$.*

PROOF. By Example 9.9.12 the category $\mathcal{C}$ is a positive symmetric category. Thus Theorem 9.9.22 (i) implies the result (see also Exercise 9.10.14). □

COROLLARY 9.9.25. *Any symmetric fusion category $\mathcal{C}$ is braided equivalent to a category of the form $\mathsf{Rep}(G, z)$. Moreover, the pair (G, z) is uniquely determined, up to isomorphism, by the category $\mathcal{C}$.*

PROOF. We know that $\mathcal{C}$ has a unique $\mathbb{Z}/2\mathbb{Z}$-grading such that the modified category $\mathcal{C}^{mod}$ is positive. Thus by Theorem 9.9.22, we have a braided equivalence $\mathcal{C}^{mod} \to \mathsf{Rep}(G)$ for a unique group G. The $\mathbb{Z}/2\mathbb{Z}$-gradings of the category $\mathsf{Rep}(G)$ are in bijective correspondence with central elements $z \in G$ such that $z^2 = 1$. Let us fix such a grading. It is immediate that $\mathsf{Rep}(G)^{mod} = \mathsf{Rep}(G, z)$. Thus the result follows from Exercise 9.9.15 (2). □

This result can be described more intrinsically in the following way. For a symmetric fusion category $\mathcal{C}$, a *super fiber functor* is a braided tensor functor $F : \mathcal{C} \to \mathsf{sVec}$. For a super fiber functor $F : \mathcal{C} \to \mathsf{sVec}$, let G_F be the group of tensor automorphisms of F. Let $z_F \in G_F$ denote the parity automorphism of F (i.e., for each $c \in \mathcal{C}$ the automorphism of $F(c) \in \mathsf{sVec}$ corresponding to c is the parity automorphism). It is clear that z_F is central and $z_F^2 = 1$. A super fiber functor $F : \mathcal{C} \to \mathsf{sVec}$ has an obvious lifting $\tilde{F} : \mathcal{C} \to \mathsf{Rep}(G_F, z_F)$.

THEOREM 9.9.26. (**[De1, De2]**) *Let $\mathcal{C}$ be a symmetric fusion category. Then*
- (i) *there exist super fiber functors $\mathcal{C} \to \mathsf{sVec}$;*
- (ii) *all super fiber functors $\mathcal{C} \to \mathsf{sVec}$ are isomorphic to each other;*
- (iii) *for any super fiber functor $F : \mathcal{C} \to \mathsf{sVec}$, the functor $\tilde{F} : \mathcal{C} \to \mathsf{Rep}(G_F, z_F)$ constructed above is an equivalence.*

EXERCISE 9.9.27. Deduce Theorem 9.9.26 from Theorem 9.9.22.

Corollary 9.9.25 implies the following classification of triangular semisimple Hopf algebras over a field $\Bbbk$ of characteristic zero. Let G be a finite group, and z a central element in G of order 2. Recall that we have a triangular structure R_z on $\Bbbk G$ given by the formula

$$R_z = \frac{1}{2}(1 \otimes 1 + 1 \otimes z + z \otimes 1 - z \otimes z);$$

one has $\mathsf{Rep}(\Bbbk G, R_z) = \mathsf{Rep}(G, z)$ as symmetric categories. Also, if J is a twist for $\Bbbk G$, then we can define the triangular Hopf algebra $H = \Bbbk G^J$ with the triangular structure $R_{z,J} = J_{21}^{-1} R_z J$.

PROPOSITION 9.9.28. (**[EtG3]**) *Any triangular semisimple Hopf algebra (H, R) over $\Bbbk$ is of the form $(\Bbbk G^J, R_{z,J})$ for some G, z, J.*

PROOF. The category $\mathsf{Rep}(H, R)$ is a symmetric fusion category, so by Corollary 9.9.25 it is of the form $\mathsf{Rep}(G, z)$ for some G, z. Moreover, (H, R) corresponds to a (non-symmetric) fiber functor on this category, which corresponds to some twist J. □

REMARK 9.9.29. Recall that the possible twists J for $\Bbbk G$ are explicitly classified in Corollary 7.12.24.

EXERCISE 9.9.30. The following conditions are equivalent for a symmetric fusion category $\mathcal{C}$:
- (i) $\mathcal{C}$ is positive;
- (ii) $\mathcal{C}$ is Tannakian;
- (iii) the essential image of any super fiber functor $F : \mathcal{C} \to \mathsf{sVec}$ is contained in $\mathsf{Vec} \subset \mathsf{sVec}$;
- (iv) there exists a finite group G such that $\mathcal{C}$ is equivalent to $\mathsf{Rep}(G)$ as a braided fusion category.

LEMMA 9.9.31. *If $\mathcal{C}_1, \mathcal{C}_2$ are symmetric categories and $F : \mathcal{C}_1 \to \mathcal{C}_2$ is a surjective braided tensor functor, then $\mathcal{C}_1$ is Tannakian if and only if $\mathcal{C}_2$ is.*

PROOF. If $\mathcal{C}_2$ satisfies property (ii) from Exercise 9.9.30 then so does $\mathcal{C}_1$. By surjectivity, if $\mathcal{C}_1$ satisfies (iii) then so does $\mathcal{C}_2$. □

Recall that a symmetric fusion category $\mathcal{C}$ has a canonical $\mathbb{Z}/2\mathbb{Z}$-grading, see Corollary 9.9.14. It is clear from Theorem 9.9.22 that $\mathcal{C}_1 \subset \mathcal{C}$ is a unique maximal Tannakian subcategory of $\mathcal{C}$. The following result is immediate:

COROLLARY 9.9.32.
- (i) *If $\mathsf{FPdim}(\mathcal{C})$ is odd then $\mathcal{C}_1 = \mathcal{C}$. In other words, a symmetric fusion category of odd Frobenius-Perron dimension is Tannakian.*
- (ii) *In general, either $\mathcal{C}_1 = \mathcal{C}$ or $\mathsf{FPdim}(\mathcal{C}_1) = \frac{1}{2}\mathsf{FPdim}(\mathcal{C})$.* □

EXERCISE 9.9.33. (1) Let $\mathcal{C}$ be a Tannakian fusion category. By Theorem 9.9.22 (iii), a choice of a fiber functor $F : \mathcal{E} \to \mathsf{Vec}$ defines a finite group $G_F := \mathsf{Aut}\, F$ and a braided equivalence $\tilde{F} : \mathcal{C} \simeq \mathsf{Rep}(G_F)$. Let $\mathsf{Fun}(G_F)$ be the algebra of functions $G_F \to \Bbbk$; the group G_F acts on $\mathsf{Fun}(G_F)$ via left translations, so $\mathsf{Fun}(G_F)$ is a commutative algebra in the category $\mathsf{Rep}(G_F)$. Therefore $A_{\mathcal{C},F} := \tilde{F}^{-1}(\mathsf{Fun}(G_F))$ is a commutative algebra in $\mathcal{C}$. Prove that the algebra $A_{\mathcal{C},F}$ is canonically isomorphic to the regular algebra A_F.

(2) Prove that the regular algebra A_F is *separable*, that is, the multiplication morphism $A_F \otimes A_F \to A_F$ considered as a morphism of A_F-bimodules, admits a unique splitting.

In practice, we will often denote A_F by $A_{\mathcal{C}}$ and call it the regular algebra in $\mathcal{C}$. To some extent, this abuse of language and notation is justified because the isomorphism class of $A_{\mathcal{C},F}$ does not depend on F (recall that by Theorem 9.9.22(ii), all fiber functors are isomorphic).

REMARK 9.9.34. The results of this section fail dramatically over fields of positive characteristic. It was shown in [**GelfK**], [**GeoM**] (see also [**Ande**]) that over fields of positive characteristic there exist symmetric fusion categories which are non-integral and hence do not admit a fiber functor. For example, in characteristic 5, there exists a symmetric fusion category with the Grothendieck ring isomorphic to the Yang-Lee fusion ring (see Example 3.1.9). The question of classification of symmetric fusion categories in positive characteristic is currently open. However, see [**Os8**] for a generalization of Theorem 9.9.26 to characteristic p.

9.10. Existence of a fiber functor

This section is devoted to the proof of Theorem 9.9.22 (i). For this we need to generalize some notions that we introduced previously. Let $\mathcal{C}$ be a symmetric fusion category. In this section an algebra $A \in \mathcal{C}$ is a nonzero ind-object of $\mathcal{C}$ with a structure of a commutative algebra with unit.

EXAMPLE 9.10.1. (1) In the case $\mathcal{C} = \mathsf{Vec}$ an algebra in the above sense is just a commutative algebra with unit; while an algebra in the sense of Section 7.8 is a finite dimensional commutative algebra with unit.

(2) For an object $X \in \mathcal{C}$ its symmetric algebra $S^\bullet(X) = \oplus_i S^i(X)$ is an example of algebra in the above sense.

For an algebra $A \in \mathcal{C}$ we consider the category $\mathcal{C}_A$ of right A-modules; thus the objects of $\mathcal{C}_A$ are ind-objects of $\mathcal{C}$ with a structure of a right A-module. The category $\mathcal{C}_A$ endowed with the bifunctor $\otimes_A$ has an obvious structure of a symmetric tensor category. In particular, for any $M \in \mathcal{C}_A$ we have its symmetric and exterior powers $S^i(M) \in \mathcal{C}_A$ and $\wedge^i(M) \in \mathcal{C}_A$, $i \geq 0$.

EXERCISE 9.10.2. Show that for any $M, N \in \mathcal{C}_A$ we have

$$S^p(M \oplus N) = \bigoplus_{i+j=p} S^i(M) \otimes_A S^j(N),\ p \geq 0,$$

and

$$\wedge^p(M \oplus N) = \bigoplus_{i+j=p} \wedge^i(M) \otimes_A \wedge^j(N),\ p \geq 0.$$

Note that objects of $\mathcal{C}_A$ need not be rigid, in general.

EXERCISE 9.10.3. Let $\mathcal{C} = \mathsf{Vec}$. Show that $M \in \mathcal{C}_A$ is rigid if and only if it is a finitely generated projective module over A.

EXERCISE 9.10.4. Show that a direct sum and tensor product of rigid objects in $\mathcal{C}_A$ is rigid. Show that a direct summand of a rigid object is rigid.

For any rigid object $M \in \mathcal{C}_A$ we define its dimension $\dim(M) \in \mathsf{Hom}_A(A, A) = \mathsf{Hom}_{\mathcal{C}}(\mathbf{1}, A)$ as the composition

$$A \to M \otimes_A M^* \cong M^* \otimes_A M \to A.$$

EXERCISE 9.10.5. (1) Show that the dimension is additive and multiplicative.
(2) Show that for any rigid $M \in \mathcal{C}_A$ the objects $S^i(M)$ and $\wedge^i(M)$ are rigid.

We have an obvious braided tensor functor $\mathcal{C} \to \mathcal{C}_A$ sending $X \in \mathcal{C}$ to $X_A := X \otimes A \in \mathcal{C}_A$. It is immediate that X_A is rigid for any $X \in \mathcal{C}$ and $\dim(X_A) = \dim(X)$. Furthermore, $S^i(X_A) = S^i(X)_A$ and $\wedge^i(X_A) = \wedge^i(X)_A$.

We will say that a commutative unital algebra $B \in \mathcal{C}_A$ is an A-algebra. Equivalently, an A-algebra is an algebra $B \in \mathcal{C}$ together with a homomorphism $1 : A \to B$.

EXAMPLE 9.10.6. Let $M \in \mathcal{C}_A$. Then $S^\bullet(M)$ is an example of an A-algebra.

EXERCISE 9.10.7. Show that the algebra $S^\bullet(M)$ has the following universal property: for any A-algebra B and a morphism of A-modules $u : M \to B$ there is a unique A-algebra homomorphism $u_{alg} : S^\bullet(M) \to B$ such that the map $M = S^1(M) \subset S^\bullet(M) \xrightarrow{u_{alg}} B$ coincides with u.

For an A-algebra B and any $M \in \mathcal{C}_A$ we define its *extension of scalars* to be $M_B := M \otimes_A B \in \mathcal{C}_B$. Clearly, extension of scalars is a braided tensor functor. It is obvious that extension of scalars of a rigid module is again rigid module and $\dim(M_B)$ is the image of $\dim(M)$ under the obvious map $\mathsf{Hom}_{\mathcal{C}}(\mathbf{1}, A) \to \mathsf{Hom}_{\mathcal{C}}(\mathbf{1}, B)$. Also, $S^i(M_B) = S^i(M)_B$ and $\wedge^i(M_B) = \wedge^i(M)_B$.

The following result is crucial.

LEMMA 9.10.8. *Let M be a rigid A-module. The existence of an A-algebra B such that M_B has $\mathbf{1}_B = B$ as a direct summand is equivalent to the condition $S^n(M) \neq 0$ for all $n \in \mathbb{Z}_{\geq 0}$.*

PROOF. One implication is trivial since the natural map $\mathbf{1}_B = S^n(\mathbf{1}_B) \to S^n(M_B) = S^n(M)_B$ induced by the embedding $\mathbf{1}_B \to M_B$ is injective, see Exercise 9.10.2.

Let us prove the other implication. Consider an A-algebra B with multiplication map $m : B \otimes_A B \to B$. Assume that we have maps of A-modules $u : M^* \to B$ and $v : M \to B$ such that the map δ defined as a composition

$$\mathbf{1}_A \xrightarrow{coev} M \otimes_A M^* \xrightarrow{v \otimes u} B \otimes_A B \xrightarrow{m} B$$

equals the unit map $1 : A \to B$. Then we have the maps of B-modules $\beta : M_B \to \mathbf{1}_B$ defined as $M_B = M \otimes_A B \xrightarrow{v \otimes \mathsf{id}} B \otimes_A B \xrightarrow{m} B = \mathbf{1}_B$ and $\alpha : \mathbf{1}_B \to M_B$ defined as $B \xrightarrow{coev \otimes \mathsf{id}} M \otimes_A M^* \otimes_A B \xrightarrow{\mathsf{id} \otimes u \otimes \mathsf{id}} M \otimes_A B \otimes_A B \xrightarrow{\mathsf{id} \otimes m} M_B$. The condition $\delta = 1$ implies that $\beta\alpha = \mathsf{id}_B$, that is $\mathbf{1}_B$ is a direct summand of M_B.

It remains to construct an algebra B with the properties above. For this we consider an algebra $S := S^\bullet(M \oplus M^*)$. Let $\tilde{v} : M = S^1(M) \subset S$ and

$\tilde{u}: M^* = S^1(M^*) \subset S$ be the obvious inclusions. We define the map $\tilde{\delta}: A \to S$ as a composition

$$\mathbf{1}_A \xrightarrow{coev} M \otimes_A M^* \xrightarrow{\tilde{v}\otimes\tilde{u}} S \otimes_A S \xrightarrow{m} S.$$

Finally, we set $B = S^\bullet(M \oplus M^*)/I$, where I is the ideal generated by the image of the map $1 - \tilde{\delta}: A \to S$. In other words, the algebra B is defined via the generators $M \oplus M^*$ and the relation $1 = \tilde{\delta}$, see Exercise 9.10.7. It is clear that the algebra B endowed with the maps $u: M^* \xrightarrow{\tilde{u}} S \to B$ and $v: M \xrightarrow{\tilde{v}} S \to B$ satisfies the condition $1 = \delta$. However, we still have to show that the algebra B is nonzero.

Thus, we need to show that $I \neq S$ or, equivalently, that the map

$$p: S = S \otimes_A \mathbf{1}_A \xrightarrow{\mathrm{id}_S \otimes (1-\tilde{\delta})} S \otimes_A S \xrightarrow{m} S$$

is not surjective. Assume the contrary. Then the A-module map $1: A \to S$ factors through p since A is a projective object of $\mathcal{C}_A$. In other words, there exists a morphism $x: \mathbf{1}_A \to S$ such that $1 = (1-\delta)x$. Recall that the algebra $S = S^\bullet(M) \otimes_A S^\bullet(M^*)$ is naturally graded by $\mathbb{Z}_{\geq 0} \times \mathbb{Z}_{\geq 0}$; clearly 1 lands in the $(0,0)$-graded component and $\tilde{\delta}$ lands in the $(1,1)$-graded component. Decompose x in the sum of its graded components: $x = x_{0,0} + x_{1,0} + x_{0,1} + x_{1,1} + \ldots$. Clearly, we can assume that $x_{p,q} = 0$ for $p \neq q$ (since if x is a solution of $1 = (1-\tilde{\delta})x$ then $x' = x_{0,0} + x_{1,1} + \ldots$ is a solution too). Now, the equation $1 = (1-\tilde{\delta})x$ is equivalent to the following graded equations:

$$x_{0,0} = 1;\ x_{1,1} - \tilde{\delta}x_{0,0} = 0;\ x_{2,2} - \tilde{\delta}x_{1,1} = 0; \ldots$$

This means that $x_{p,p} = \tilde{\delta}^p$ and $\tilde{\delta}^n = 0$ for large enough n since the sum $x = x_{0,0} + x_{1,1} + \ldots$ is finite. Conversely, if $\tilde{\delta}^n = 0$ then $1 = (1-\tilde{\delta})(1 + \tilde{\delta} + \ldots + \tilde{\delta}^{n-1})$.

So the universal algebra B is nontrivial if and only if $\tilde{\delta}^n \neq 0$ for all n. Now $\tilde{\delta}^n: \mathbf{1}_A \to S^n(M) \otimes_A S^n(M^*)$ equals to the coevaluation map $\mathbf{1}_A \to S^n(M) \otimes_A S^n(M)^*$ and is zero if and only if $S^n(M) = 0$. The lemma is proved. □

COROLLARY 9.10.9. *Let $X \in \mathcal{C}$ and $\dim(X) = d \in \mathbb{Z}_{>0}$. Then for any $0 \leq i \leq d$ there exist an algebra $A_i \in \mathcal{C}$ and an A_i-module N_i such that $X_{A_i} \simeq \mathbf{1}^i_{A_i} \oplus N_i$.*

PROOF. We use induction in i. In the case $i = 0$ we set $A_0 = \mathbf{1}$ and $N_0 = X$. Now assume that $i < d$ and A_i and N_i are already constructed. Observe that $\dim(N_i) = \dim(X_{A_i}) - \dim(\mathbf{1}^i_{A_i}) = d - i > 0$. Thus $\dim S^n(N_i) = \binom{d-i+n-1}{n} \neq 0$ for any $n \in \mathbb{Z}_{\geq 0}$. Hence by Lemma 9.10.8 we can construct the A_i-algebra A_{i+1} such that $(N_i)_{A_{i+1}} = \mathbf{1}_{A_{i+1}} \oplus N_{i+1}$ for some $N_{i+1} \in \mathcal{C}_{A_{i+1}}$. The result follows. □

LEMMA 9.10.10. *Assume that $\mathcal{C}$ is a positive symmetric fusion category. Then in the setup of Corollary 9.10.9 we have $N_d = 0$, so $X_{A_d} \simeq \mathbf{1}^d_{A_d}$.*

PROOF. We have $\dim \wedge^{d+1}(X) = 0$. Since the category $\mathcal{C}$ is positive, this implies that $\wedge^{d+1}(X) = 0$. On the other hand, by Exercise 9.10.2

$$\wedge^{d+1}(X)_{A_d} = \wedge^{d+1}(X_{A_d}) = \wedge^{d+1}(\mathbf{1}^d_{A_d} \oplus N_d) = \bigoplus_{i+j=d+1} \wedge^i(\mathbf{1}^d_{A_d}) \otimes_A \wedge^j(N_d)$$

contains $\wedge^d(\mathbf{1}^d_{A_d}) \otimes_A \wedge^1(N_d) = N_d$ as a direct summand. The result follows. □

COROLLARY 9.10.11. *Let $\mathcal{C}$ be a positive symmetric fusion category. There exists an algebra $A \in \mathcal{C}$ such that for any object $X \in \mathcal{C}$ with $\dim(X) = d$ we have $X_A \simeq \mathbf{1}^d_A$.*

PROOF. By Lemma 9.10.10 for each simple $X \in \mathcal{C}$ there is an algebra $A(X)$ such that $X_{A(X)} \simeq \mathbf{1}_{A(X)}^{\dim(X)}$. Now take $A = \bigotimes_X A(X)$. □

Let A be an algebra as in Corollary 9.10.11. Then the tensor functor $X \mapsto X_A$ lands into subcategory of $\mathcal{C}_A$ consisting of finite direct sums of A.

EXERCISE 9.10.12. Let $R = \mathsf{Hom}_{\mathcal{C}}(\mathbf{1}, A)$. Then the subcategory of $\mathcal{C}_A$ consisting of finite direct sums of copies of A is braided tensor equivalent to the category of free R-modules of finite rank via the functor $M \mapsto \mathsf{Hom}_{\mathcal{C}}(\mathbf{1}, M)$.

COROLLARY 9.10.13. *For any positive symmetric fusion category $\mathcal{C}$ there exists a commutative $\Bbbk$-algebra R and a braided tensor functor from $\mathcal{C}$ to free R-modules of finite rank.* □

Proof of Theorem 9.9.22 (i). In order to specify a fiber functor $F : \mathcal{C} \to \mathsf{Vec}$ one needs to assign a vector space $F(X)$ of dimension $\dim(X)$ for any simple object $X \in \mathcal{C}$ and to define isomorphisms $F(X \otimes Y) \simeq F(X) \otimes F(Y)$ satisfying the axioms of a braided tensor functor. If we choose a basis in each vector space $F(X)$ and $\mathsf{Hom}_{\mathcal{C}}(X \otimes Y, Z)$, the isomorphisms above become matrices and the axioms become some polynomial relations on entries of these matrices. Thus the existence of the fiber functor is equivalent to the existence of solutions to some polynomial system of equations. We know from Corollary 9.10.13 that this system of equations has a solution with values in some commutative $\Bbbk$-algebra R. Thus by Hilbert Nullstellensatz there is a solution with values in $\Bbbk$. □

EXERCISE 9.10.14. (1) Let $X \in \mathcal{C}$ be an invertible object such that $c_{X,X} = \mathsf{id}_{X \otimes X}$. Prove that $S^\bullet(X) = T^\bullet(X)$. Deduce that $R = \mathsf{Hom}_{\mathcal{C}}(\mathbf{1}, S^\bullet(X))$ is a polynomial algebra in one variable, $R = \Bbbk[y]$.

(2) Prove that the algebra $A = S^\bullet(X)/(y-1)$ is commutative and semisimple.

(3) Give an alternative proof of Corollary 9.9.24.

9.11. Deligne's theorem for infinite categories

In the previous sections we described Deligne's classification of symmetric fusion categories. In fact Deligne's theory developed in [**De1, De2**] is significantly more general and his results extend to tensor categories which are not necessarily finite or semisimple. Let us review this generalization, without detailed proofs (we refer the reader to [**De2**] for details). This material is not used in subsequent sections.

At first sight, one might expect that Deligne's theorem should extend to this setting in a straightforward way, just replacing finite groups by affine proalgebraic groups. However, this turns out not to be the case, because in a general tensor category, there may be no twist of the symmetric structure making all dimensions non-negative, and there may be simple objects of dimension 0.

To give an example, let V be a supervector space of dimension $(m|n)$, where $m, n > 0$, i.e., $V = V_+ \oplus V_-$, where V_+ is the even part and V_- is the odd part, so that $\dim V_+ = m$ and $\dim V_- = n$. Consider the Lie algebra $\mathfrak{g} = \underline{\mathsf{End}}(V) = V \otimes V^* \in \mathsf{sVec}$ (as usual, the Lie bracket is obtained from the associative product by taking the commutator). Let $\mathcal{C}$ be the tensor category of representations of $\mathfrak{g}$ in sVec (see Exercise 9.9.7). Then $\wedge^r V = \oplus_{i+j=r} \wedge^i V_+ \otimes S^j V_-$, and $S^r V = \oplus_{i+j=r} S^i V_+ \otimes \wedge^j V_-$. It is not difficult to show that these objects are simple (in particular, nonzero) for all r, while if $m - n \geq 0$ then $\dim \wedge^r V = 0$ for $r > m - n$, and if $n - m \geq 0$ then $\dim S^r V = 0$ for $r > n - m$. In particular, there is clearly no twist making all dimensions positive. So, we see that in the non-fusion case, the story is more complicated, and instead of a fiber functor to Vec after a suitable twist of the symmetric braiding, which does not always exist, the best we can hope for is a fiber functor to sVec.

Deligne showed that this hope is indeed realized, under a certain additional assumption. Namely, there is the following necessary condition for the existence of such a fiber functor.

DEFINITION 9.11.1. We say that a tensor category $\mathcal{C}$ is of *subexponential growth* if for each $X \in \mathcal{C}$, there exists a positive number C_X such that for any $n \geq 0$, $\mathrm{length}(X^{\otimes n}) \leq C_X^n$.

EXAMPLE 9.11.2. The category Vec has subexponential growth, with $C_X = \dim(X)$. This is clearly the best constant, so the subexponential growth here is honestly exponential. So we see a stark contrast with other fields of mathematics (noncommutative algebra, theory of computation), where exponential growth is viewed as fast. In the theory of tensor categories, this means the slowest possible growth, and we will see that things can get worse.

LEMMA 9.11.3. *Let $\mathcal{C}$ be a tensor category, and $\mathcal{D}$ a finite tensor category. If there is a quasi-tensor functor $F : \mathcal{C} \to \mathcal{D}$ then $\mathcal{C}$ has subexponential growth; namely, one can take $C_X = \mathsf{FPdim}(F(X))$.*

PROOF. This follows immediately, since $\mathsf{FPdim}(Y) \geq 1$ for any $Y \in \mathcal{D}$, so

$$\mathrm{length}(X^{\otimes n}) \leq \mathrm{length}(F(X)^{\otimes n}) \leq \mathsf{FPdim}(F(X)^{\otimes n}) = C_X^n, \ n \geq 0.$$

□

Lemma 9.11.3 implies that subexponential growth is necessary for having a fiber functor to sVec, and we will see that this condition is not always satisfied. However, as Deligne showed, this condition is also sufficient.

THEOREM 9.11.4. ([**De2**]) *Let $\mathcal{C}$ be a symmetric tensor category of subexponential growth over a field $\Bbbk$ of characteristic zero (for example, a finite category). Then $\mathcal{C}$ admits a fiber functor $F : \mathcal{C} \to \mathsf{sVec}$, and this functor is unique up to an isomorphism.*

More precisely, Deligne proves the following theorem, which implies Theorem 9.11.4. For each partition λ of n let π_λ be the irreducible representation of S_n corresponding to λ, and let $S^\lambda : \mathcal{C} \to \mathcal{C}$ be the corresponding Schur functor, $S^\lambda(X) = (\pi_\lambda \otimes X^{\otimes n})^{S_n}$.

THEOREM 9.11.5. *Let $\mathcal{C}$ be a symmetric tensor category over a field $\Bbbk$ of characteristic zero, such that for every object $X \in \mathcal{C}$ there exists a partition λ such that $S^\lambda(X) = 0$. Then $\mathcal{C}$ admits a fiber functor $F : \mathcal{C} \to \mathsf{sVec}$, and this functor is unique up to an isomorphism.*

For the proof of Theorem 9.11.5 (which is nontrivial), see [**De2**]; the method is similar to the proof of Deligne's theorem for fusion categories. Note that the result for fusion categories is a special case of Theorem 9.11.5, since for positive fusion categories we have $\wedge^n X = 0$ for $n = \dim(X) + 1$, i.e., we can take $\lambda = (1^n)$, $n = \dim(X) + 1$.

PROPOSITION 9.11.6. *Any symmetric tensor category of subexponential growth satisfies the condition of Theorem 9.11.5; so Theorem 9.11.5 implies Theorem 9.11.4.*

PROOF. Suppose that $S^\lambda(X) \neq 0$ for any λ. We have $X^{\otimes n} = \oplus_{\lambda:|\lambda|=n}\pi_\lambda^* \otimes S^\lambda(X)$. Thus,

$$\text{length}((X \otimes X^*)^{\otimes n}) \geq \dim_{\Bbbk} \mathsf{Hom}_{\mathcal{C}}(\mathbf{1}, (X \otimes X^*)^{\otimes n}) =$$

$$\dim_{\Bbbk} \mathsf{End}_{\mathcal{C}}(X^{\otimes n}) \geq \sum_{\lambda:|\lambda|=n} \dim_{\Bbbk}(\pi_\lambda)^2 = n!,$$

which contradicts subexponential growth. □

REMARK 9.11.7. Theorem 9.11.4 allows one to obtain an explicit classification of finite dimensional triangular Hopf algebras in characteristic zero, which generalizes Proposition 9.9.28 to the non-semisimple case. See [**AEG, EtG5**] for details.

EXERCISE 9.11.8. Let $V \in \mathsf{sVec}$ be a supervector space of dimension $(m|n)$ (i.e., the even part has dimension m and the odd part has dimension n). Show that $S^\lambda(V) = 0$ if and only if $\lambda_{m+1} \geq n+1$, i. e., the Young diagram of λ contains the square (n, m) (where both coordinates run from 0 to ∞).

Let us now explain how Theorem 9.11.4 is applied. Suppose $F : \mathcal{C} \to \mathsf{sVec}$ is a fiber functor. Let $F_1 : \mathcal{C} \to \mathsf{Vec}$ be the composition of F and of the forgetful tensor functor Forg : $\mathsf{sVec} \to \mathsf{Vec}$. Let $H = \mathsf{Coend}(F)$. By the reconstruction theory (see Theorem 5.4.1) H is a Hopf algebra and the functor F_1 induces a tensor equivalence $\mathcal{C} \simeq \mathsf{Rep}(H)$. If the functor F_1 is braided (e.g. when the functor F factors through $\mathsf{Vec} \subset \mathsf{sVec}$) the multiplication in H is commutative and thus $H = O(G)$ for a suitable affine group scheme G. However in general the functor F_1 is not braided (since Forg is not) and the multiplication in H is not commutative. In order to deal with this possibility observe that the tensor functor Forg has a canonical *parity automorphism* which acts by 1 on even vector spaces and by (-1) on odd vector spaces. We will denote by z the induced tensor automorphism of the functor $F_1 = \text{Forg} \circ F$; clearly $z^2 = 1$. The automorphism z gives rise to a linear form $\chi_z : H \to \Bbbk$; since z is a tensor automorphism, the form χ_z is a homomorphism of algebras. The algebra H carries an "adjoint action" of z given by

$$H \xrightarrow{(\Delta\otimes\mathsf{id})\circ\Delta} H \otimes H \otimes H \xrightarrow{\chi_z\otimes\mathsf{id}\otimes\chi_z} H$$

Thus $H = H_0 \oplus H_1$ where z acts by 1 on H_0 and by (-1) on H_1. One shows that H is supercommutative with respect to this grading. Moreover, modifying suitably (see Theorem 3.1.1 in [**AEG**]) the coproduct in H, one gets a commutative Hopf algebra $\mathcal{H}$ in sVec, so according to Exercise 9.9.7, $\mathcal{H} = O(G)$ for some affine group scheme G in sVec. By definition, for any $X \in \mathcal{C}$, $F(X)$ has a natural structure of a G-module (i.e., an $O(G)$-comodule). Moreover, the action of $z \in G$ coincides with the parity automorphism of $F(X)$. Let $\mathsf{Rep}(G, z)$ denote the full subcategory of $\mathsf{Rep}(G)$ consisting of all representations such that the action of $z \in G$ coincides with parity automorphism (the tensor category $\mathsf{Rep}(G, z)$ is defined for any $z \in G$ such that $z^2 = 1$ and the adjoint action of z on G coincides with the parity automorphism). The remarks above imply

THEOREM 9.11.9. *The functor $F : \mathcal{C} \to \mathsf{Rep}(G, z)$ is an equivalence of categories.*

REMARK 9.11.10. (i) It is instructive to apply Theorem 9.11.9 to the symmetric category $\mathsf{Rep}(G)$ where G is a an affine group scheme in sVec with the obvious fiber functor. One gets $\mathsf{Rep}(G) \simeq \mathsf{Rep}(\tilde{G}, z)$ where $\tilde{G}$ is a semi-direct product of G and $\mathbb{Z}/2\mathbb{Z}$ where the nontrivial element $z \in \mathbb{Z}/2\mathbb{Z}$ acts on G by the parity automorphism, see [**De2**].

(ii) One shows that the category $\mathsf{Rep}(G, z)$ is fusion if and only if G is an (even) finite group, see Exercise 9.11.13 below. Thus z should act trivially on G, i.e., z should be central. Thus Theorem 9.11.9 can be considered as a direct generalization of Theorem 9.9.26 (iii).

Theorem 9.11.9 reduces the study of symmetric tensor categories with a fiber functor $\mathcal{C} \to \mathsf{sVec}$ to the study of group schemes G in sVec. Let us describe such group schemes in more explicit terms. For this purpose, let us apply the forgetful functor Forg : $\mathsf{sVec} \to \mathsf{Vec}$. Namely, let $A = \mathrm{Forg}(O(G))$. Then, since Forg is not a braided functor, A is not, in general, a commutative algebra, but rather it is supercommutative, i.e., $A = A_+ \oplus A_-$ (a $\mathbb{Z}/2$-grading), and elements of A_+ commute with everything, while elements of A_- anticommute with each other. Also, A has a coassociative coproduct, counit, and antipode, and the coproduct $\Delta : A \to A \otimes A$ is an algebra map, if $A \otimes A$ is understood in the supersense. Such a structure is called a (supercommutative) *Hopf superalgebra*.

We may think of A geometrically using the notion of an *affine proalgebraic supergroup*, the geometric object attached to any supercommutative Hopf superalgebra A. This notion can be defined in terms of the functor of points: for every supercommutative local Artinian $\Bbbk$-algebra R we have the group of points $G(R) = \mathrm{Hom}(A, R)$ (the group of even homomorphisms).

DEFINITION 9.11.11. The functor $G : R \mapsto G(R)$ is called the *proalgebraic supergroup* corresponding to A; one writes $G = \mathrm{Spec}(A)$.

Note that an affine algebraic supergroup is just an affine algebraic supermanifold with a group structure.

Let $I \subset A$ be the ideal generated by A_-. It is easy to see that I is a Hopf ideal which consists of nilpotent elements (as $\Delta(A_-) \subset A_- \otimes A_+ \oplus A_+ \otimes A_-$ and $a^2 = 0$ for all $a \in A_-$). So $A/I = A_0$ is an ordinary commutative Hopf algebra. Since we are in characteristic zero, by Corollary 5.10.5, $A_0 = O(G_0)$, where G_0 is an ordinary affine proalgebraic group (called the *even part* of G). Thus, I is the Jacobson radical of A, and $G_0 = G(\Bbbk)$.

If we restrict ourselves to the case when $\mathcal{C}$ is tensor-generated by a single object X (i.e., any object of $\mathcal{C}$ is a subquotient of a direct sum of tensor products of X's and X^*'s; note that any tensor category is a union of such tensor subcategories), then A is finitely generated, and G is an affine algebraic supergroup.

EXERCISE 9.11.12. (i) Let $G = GL_{m|n}$ be the algebraic supergroup attached to the affine group scheme $GL(V)$ in sVec, $V = V_+ \oplus V_-$, $\dim V_+ = m, \dim V_- = n$ (see Exercise 9.9.7). Describe explicitly the corresponding function algebra $A = O(G)$. Compute $GL_{m|n}(R)$, where $R = \wedge(\varepsilon_1, ..., \varepsilon_N)$. Show that the even part of $GL_{m|n}$ is $GL_m \times GL_n$.

(ii) Construct a character $GL_{m|n} \to GL_1$ which generates a group of such characters (it is called the Berezinian, in honor of F. Berezin who invented it). Compute this character explicitly on $GL_{m|n}(R)$ for a supercommutative local Artinian algebra R, and show that its value on $(a, b) \in GL_m \times GL_n$ is $\det(a) \det(b)^{-1}$.

(iii) Suppose that V is equipped with a symmetric or skew-symmetric isomorphism to V^*. Define the corresponding orthogonal/symplectic supergroup $OSp(m|n)$, respectively $OSp(n|m)$, and do the analog of (i) for this supergroup.

(iv) For any algebraic supergroup G, let $\mathfrak{g}$ be the space of left-invariant derivations of $A = O(G)$. That is, $\mathfrak{g} = \mathfrak{g}_0 \oplus \mathfrak{g}_1$, where $\mathfrak{g}_0$ is the space of even derivations of A, and $\mathfrak{g}_1$ is the space of odd derivations of A, i.e., odd linear maps $D : A \to A$ such that $D(ab) = D(a)b + \mathrm{sgn}(a)aD(b)$. Show that $\mathfrak{g}$ is a Lie superalgebra (see Exercise 9.9.7). Compute this Lie superalgebra in (i) and (iii).

(v) Let G be an algebraic supergroup. Show that G-modules is the same thing as G_0-modules with a compatible action of $\mathfrak{g}$ (i.e., "Harish-Chandra modules" for the pair $(G_0, \mathfrak{g})$).

EXERCISE 9.11.13. Let G be a finite supergroup, i.e., the algebra $A = O(G)$ is finite dimensional.

(i) Show that G is a semidirect product $G_0 \ltimes V$ of a finite group G_0 with a finite dimensional vector space V, i.e., $A^* = \Bbbk G_0 \ltimes \wedge V$.

(ii) Deduce that any finite symmetric tensor category over $\Bbbk$ is of the form $\mathsf{Rep}(G_0 \ltimes V, z)$, where $z \in G_0$ is a central element which acts by -1 on V. In particular, such categories have Chevalley property (see Section 4.12), and do not have simple modules of zero dimension.

(iii) Classify symmetric finite tensor categories with two simple objects which are invertible.

(iv) Is (i) satisfied for $G = GL_{m|n}$?

9.12. The Deligne categories $\mathsf{Rep}(S_t)$, $\mathsf{Rep}(GL_t)$, $\mathsf{Rep}(O_t)$, $\mathsf{Rep}(Sp_{2t})$

In this section we want to describe examples of tensor categories of faster-than-exponential growth which were found by Deligne and Deligne-Milne ([**De3, DelM**]). This material is not used in subsequent sections.

9.12.1. The category $\mathsf{Rep}(S_t)$. We start with the category $\mathsf{Rep}(S_t)$, $t \in \Bbbk$, which interpolates the categories $\mathsf{Rep}(S_n)$ of representations of the symmetric group S_n to non-integer values of n.

To construct such an interpolation, we should take a closer look at the category $\mathsf{Rep}(S_n)$, and try to describe it in such a way that n occurs "analytically", i.e., as a parameter which can be replaced by a non-integer number. To this end, note that $\mathsf{Rep}(S_n)$ contains the permutation representation $V = \mathsf{Fun}(X, \Bbbk)$, where $X = \{1, ..., n\}$, and every irreducible representation of S_n occurs in $V^{\otimes m}$ for a sufficiently large m. Therefore, we can start with the category $\mathcal{C}_n$ whose objects are $V^{\otimes m}$, $m \geq 0$, and $\mathrm{Mor}(V^{\otimes m}, V^{\otimes k}) = \mathsf{Hom}_{S_n}(V^{\otimes m}, V^{\otimes k})$, and then define $\mathsf{Rep}(S_n)$ to be the *Karoubian envelope* of the additive completion of $\mathcal{C}_n$, obtained by adding to $\mathcal{C}_n$ the operation of direct sum and images of all the idempotents. It is easy to see that this recovers the usual representation category of S_n.

To interpolate to non-integer t, we need to describe the space $\mathrm{Mor}(V^{\otimes m}, V^{\otimes k})$ without mentioning S_n. To do so, note that $V^* \cong V$, so $\mathsf{Hom}_{S_n}(V^{\otimes m}, V^{\otimes k}) \cong \mathsf{Fun}(X^{m+k}, \Bbbk)^{S_n}$. The space $\mathsf{Fun}(X^r, \Bbbk)^{S_n}$ has a basis δ_O of characteristic functions of orbits of S_n on $X^r = \{1, ..., n\}^r$. These orbits are labeled by equality patterns: if $x = (x_1, ..., x_r) \in X^r$, the orbit of x is determined by which x_i are equal to each other. Such an equality pattern is encoded by a set partition of $\{1, ..., r\}$; namely, the indices i for which x_i are equal to a particular element $b \in X$ are included in the same subset. E.g., if $x = (x_1, x_2, x_3, x_4, x_5) \in X^5$, and $x_1 = x_3, x_2 = x_5$, and no other equalities hold, then the orbit of x is labeled by the set partition $([1,3],[2,5],[4])$.

Clearly, the set partitions that arise from orbits in this way are exactly those which have the number of subsets $\leq n = |X|$. So, if $n \geq r$, all set partitions arise. Thus, for $n \geq m+k$, the space $\mathsf{Hom}_{S_n}(V^{\otimes m}, V^{\otimes k})$ is independent of n (it has a basis δ_O, where the orbits $O = O(P)$ are labeled by set partitions P of $\{1, ..., m+k\}$).

Now let us consider the composition of morphisms. For this purpose, it is convenient to use a somewhat different basis of $\mathsf{Hom}_{S_n}(V^{\otimes m}, V^{\otimes k})$. Namely, let us say that a set partition P is a refinement of P' ($P \geq P'$) if P' is obtained by taking unions of some of the subsets of P. E.g., $([1,3],[2,5],[4])$ is a refinement of $([1,3],[2,4,5])$. Now for a set partition P let

$$e_P = \sum_{P' \geq P} \delta_{O(P')}.$$

That is, e_P is the characteristic function of all the configurations $x = (x_1, ..., x_r)$ such that $x_i = x_j$ if i, j belong to the same subset in P, but there may also be other equalities. Clearly, $\{e_P\}$ is a basis of $\mathsf{Hom}_{S_n}(V^{\otimes m}, V^{\otimes k})$, since it differs from $\{\delta_{O(P)}\}$ by a triangular matrix with 1's on the diagonal.

Now consider the composition of morphisms in the basis $\{e_P\}$. Let $e_P : V^{\otimes m} \to V^{\otimes k}$, $e_Q : V^{\otimes r} \to V^{\otimes m}$ (so that P is a set partition of $\{1, ..., m+k\}$ and Q is a set partition of $\{1, ..., r+m\}$). Introduce variables $x_1, ..., x_k, y_1, ..., y_m, z_1, ..., z_r \in X$; so P is a certain equality pattern involving x_i and y_j, and Q is an equality pattern involving y_j and z_ℓ. Let $P * Q$ be the least restrictive equality pattern between x_i and z_ℓ which is implied by P and Q. Also, let $N(P,Q)$ be the number of equality classes (under the pattern $P * Q$) of the variables y_j that are not required to equal any of the x_i or z_ℓ by the equality pattern $P * Q$. Then we have

LEMMA 9.12.1.

$$e_P \circ e_Q = n^{N(P,Q)} e_{P*Q}.$$

PROOF. It is clear that if x_i, y_j satisfy the pattern P and y_j, z_ℓ satisfy the pattern Q then x_i, z_ℓ satisfy the pattern $P * Q$. Moreover, given x_i, z_ℓ satisfying $P * Q$, the number of choices of y_j such that x_i, y_j satisfy P and y_j, z_ℓ satisfy Q is $n^{N(P,Q)}$ (regardless of the particular values of x_i, z_ℓ). This implies the lemma. □

EXAMPLE 9.12.2. Let $k = 4, m = 5, r = 3$, $P = (y_2 = y_3, x_3 = x_4 = y_4 = y_5)$, $Q = (z_1 = z_2, y_5 = z_3)$. Then $P * Q = (x_3 = x_4 = z_3, z_1 = z_2)$, and $N(P,Q) = 2$ (the free y-variables are y_1 and $y_2 = y_3$). Thus, $e_P \circ e_Q = n^2 e_{P*Q}$.

Now for each $t \in \Bbbk$ define the category $\widehat{\mathsf{Rep}(S_t)}$ whose objects are non-negative integers $[m]$ (counterparts of $V^{\otimes m}$), and $\mathsf{Hom}([m],[k])$ is the $\Bbbk$-vector space with a basis e_P labeled by set partitions P of $\{1, ..., m+k\}$ and composition law

$$e_P \circ e_Q = t^{N(P,Q)} e_{P*Q}.$$

Note that this category has a natural strict symmetric monoidal structure. Namely, the tensor product functor is just the addition of integers for objects and taking the union of equality patterns for morphisms, with the obvious symmetric braiding. The unit object is the object $[0]$.

REMARK 9.12.3. The algebra $P_{m,t} := \mathrm{End}_{\widehat{\mathrm{Rep}(S_t)}}([m])$ is called the m-th *partition algebra*, and its structure is discussed in [**Mar1, Mar2, Jon4, HaR**].

DEFINITION 9.12.4. The category $\widetilde{\mathrm{Rep}}(S_t)$ is the Karoubian envelope of the additive completion of $\widehat{\mathrm{Rep}}(S_t)$.

Clearly, $\widetilde{\mathrm{Rep}}(S_t)$ is a Karoubian category (i.e., an idempotent-closed additive category) over $\Bbbk$, which inherits the tensor structure from $\widehat{\mathrm{Rep}}(S_t)$. Moreover, it is not hard to show that this category is rigid (with all objects being self-dual). Furthermore, it is easy check that $\dim[m] = t^m$ (where $[m]$ is the object corresponding to the integer m); this is just the interpolation of the equality $\dim V^{\otimes m} = n^m$.

The following theorem summarizes the properties of $\widetilde{\mathrm{Rep}}(S_t)$.

Given a partition λ and a positive integer n, define the partition $\widetilde{\lambda}(n) := (n - |\lambda|, \lambda_1, \lambda_2, ...)$ (it is well defined if and only if $n \geq |\lambda| + \lambda_1$). Also, if μ is a partition of n, let π_μ be the corresponding irreducible representation of S_n, and let $P_\lambda(t)$ be the polynomial defined by the equality

$$\dim \pi_{\widetilde{\lambda}(n)} = P_\lambda(n), \ n >> 0.$$

THEOREM 9.12.5. ([**De3**],[**ComO**]) *(i) For any $t \in \Bbbk$, the indecomposable objects X_λ of $\widetilde{\mathrm{Rep}}(S_t)$ are labeled by all partitions λ, and for each λ, $\dim X_\lambda = P_\lambda(t)$ for all $t \in \Bbbk$ but finitely many non-negative integers t.*

(ii) If $t \notin \mathbb{Z}_+$ then $\widetilde{\mathrm{Rep}}(S_t)$ is a semisimple abelian symmetric tensor category with simple objects X_λ (also denoted by $\mathrm{Rep}(S_t)$).

(iii) If $t = n \in \mathbb{Z}_+$ then $\widetilde{\mathrm{Rep}}(S_{t=n})$ is not abelian, but admits an additive symmetric monoidal functor $F : \widetilde{\mathrm{Rep}}(S_{t=n}) \to \mathrm{Rep}(S_n)$, such that $F(X_\lambda) = \pi_{\widetilde{\lambda}(n)}$ if $\widetilde{\lambda}(n)$ is defined, and zero otherwise.

REMARK 9.12.6. Note that in Theorem 9.12.5(iii), the functor F is not a tensor functor. More precisely, it satisfies all the properties of a tensor functor except that it is not faithful: it maps some nonzero morphisms to zero.

Note that if $t \notin \mathbb{Z}_+$ then the category $\mathrm{Rep}(S_t)$ is not of subexponential growth, and does not come from a supergroup; indeed, length$([m])$ is bounded below by $\dim \mathrm{Hom}([o],[m])$, i.e., the number of set partitions of $\{1, ..., m\}$, which is the Bell number B_m, growing faster than any exponential. Also, if $t \notin \mathbb{Z}$, dimensions of objects aren't integers, which clearly rules out an equivalence with supergroup representations.

REMARK 9.12.7. The object X_λ may be informally viewed as an analog of the representation of the symmetric group corresponding to the Young diagram λ with an additional "very long" top row of "length" $t - |\lambda|$ (which may be negative and even non-integer).

9.12.2. The category $\mathrm{Rep}(GL_t)$. The category $\mathrm{Rep}(GL_t)$ is obtained similarly to the category $\mathrm{Rep}(S_t)$, by interpolating $\mathrm{Rep}(GL_n)$ to non-integer values of n. It was first defined in [**DelM**], Examples 1.26, 1.27 (see also [**De1, De2**]).

Namely, recall that in the classical category $\mathrm{Rep}(GL_n)$ we have the vector representation $V = \Bbbk^n$, and every irreducible representation of GL_n occurs in $V^{\otimes r} \otimes V^{*\otimes s}$ for some r, s. Now,

$$\mathrm{Hom}_{GL_n}(V^{\otimes r_1} \otimes V^{*\otimes s_1}, V^{\otimes r_2} \otimes V^{*\otimes s_2}) = \mathrm{Hom}_{GL_n}(V^{\otimes r_1+s_2}, V^{\otimes r_2+s_1}),$$

so it is nonzero only if $r_1 + s_2 = r_2 + s_1 = m$, and in the latter case is spanned by elements of $\Bbbk S_m$, by the Fundamental Theorem of Invariant Theory (this spanning set is a basis if $n \geq m$). The category $\mathsf{Rep}(GL_n)$ can then be defined as the Karoubian envelope of the additive completion of the subcategory with objects $[r,s] := V^{\otimes r} \otimes V^{*\otimes s}$ and morphisms as above.

Now consider composition of morphisms. To do so, note that the elements of S_m defining morphisms can be depicted as oriented planar tangles (with possibly intersecting strands) with s_1 inputs and r_1 outputs on the bottom and r_2 inputs and s_2 outputs on the top, and m arrows, each going from an input to an output. The composition of morphisms is then defined as concatenation of tangles, followed by closed loop removal, with each removed loop earning a factor of n. For example, if $A : [1,1] \to [1,1]$ is given by $A = \mathrm{coev}_V \circ \mathrm{ev}_V$, then $A^2 = nA$.

Now, given $t \in \Bbbk$, one can define the category $\widehat{\mathsf{Rep}}(GL_t)$ with objects $[r,s]$, $r, s \in \mathbb{Z}_+$, and the space of morphisms $\mathsf{Hom}([r_1,s_1],[r_2,s_2])$ being spanned by planar tangles as above, with the same composition law as above, except that every removed closed loop earns a factor of t.

REMARK 9.12.8. The endomorphism algebra $\mathsf{End}([r,s])$ is called the *walled Brauer algebra* and denoted $B_{r,s}(t)$ (see [**Tu2, Koi**]).

Note that the category $\widehat{\mathsf{Rep}}(GL_t)$ has a natural strict symmetric monoidal structure. Namely, the tensor product functor is just the addition of pairs of integers for objects and taking the union of planar tangles for morphisms, with the obvious symmetric braiding. The unit object is the object $[0,0]$.

DEFINITION 9.12.9. The category $\widetilde{\mathsf{Rep}}(GL_t)$ is the Karoubian envelope of the additive completion of $\widehat{\mathsf{Rep}}(GL_t)$.

Clearly, $\widetilde{\mathsf{Rep}}(GL_t)$ is a Karoubian category (i.e., an idempotent-closed additive category) over $\Bbbk$, which inherits the tensor structure from $\widehat{\mathsf{Rep}}(GL_t)$. Moreover, it is not hard to show that this category is rigid (with $[r,s]^* = [s,r]$). Furthermore, it is easy check that $\dim[r,s] = t^{r+s}$; this is just the interpolation of the equality $\dim(V^{\otimes r} \otimes V^{*\otimes s}) = n^{r+s}$.

THEOREM 9.12.10. *(*[**DelM, De1, De2**]*) (i) The category* $\widetilde{\mathsf{Rep}}(GL_t)$ *is a semisimple abelian symmetric tensor category if* $t \notin \mathbb{Z}$ *(in this case we also denote it by* $\mathsf{Rep}(GL_t)$*).*

(ii) The category $\widetilde{\mathsf{Rep}}(GL_t)$ *has the following universal property: if* $\mathcal{C}$ *is a rigid symmetric tensor category then isomorphism classes of (possibly non-faithful) symmetric additive monoidal functors* $\widetilde{\mathsf{Rep}}(GL_t) \to \mathcal{C}$ *are in bijection with isomorphism classes of objects* $X \in \mathcal{C}$ *of dimension* t*, via* $F \mapsto F([1,0])$*.*

(iii) If $t = n \in \mathbb{Z}$*, and if* p, q *are non-negative integers with* $p - q = n$*, then the category* $\widetilde{\mathsf{Rep}}(GL_{t=n})$ *(which is not abelian) admits a non-faithful symmetric additive monoidal functor* $\widetilde{\mathsf{Rep}}(GL_n) \to \mathsf{Rep}(GL_{p|q})$ *to the representation category of the supergroup* $GL_{p|q}$ *(introduced in Exercise* 9.11.12*), which sends* $[1,0]$ *to the supervector space* $V = \Bbbk^{p|q}$ *with* p *even and* q *odd dimensions.*

(iv) We have a natural symmetric tensor functor $\mathrm{Res}\colon \widetilde{\mathsf{Rep}}(GL_t) \to \widetilde{\mathsf{Rep}}(GL_{t-1})$.

Note that (iii) and (iv) are easy consequences of (ii).

Let us consider the case $t \notin \mathbb{Z}$. In this case, simple objects in $\mathsf{Rep}(GL_t)$ are labeled by pairs of arbitrary partitions, (λ, μ), $\lambda = (\lambda_1, ..., \lambda_r)$, $\mu = (\mu_1, ..., \mu_s)$.

Namely, letting $V = [1,0]$ be the tautological object (the interpolation of the defining representation), we have simple objects $X_{\lambda,\mu}$ which are direct summands in $S^\lambda V \otimes S^\mu V^*$, where S^λ is the Schur functor corresponding to the partition λ. More specifically, $X_{\lambda,\mu}$ is the only direct summand in $S^\lambda V \otimes S^\mu V^*$ which does not occur in $S^{\lambda'} V \otimes S^{\mu'} V^*$ with $|\lambda'| < |\lambda|$. This summand occurs with multiplicity 1. All of this is readily seen by noting that this is the case in $\mathsf{Rep}(GL_n)$ for large n, in which case $X_{\lambda,\mu}$ is the irreducible representation $V_{[\lambda,\mu]_n}$ of GL_n, with highest weight $[\lambda,\mu]_n$, where

$$[\lambda,\mu]_n = (\lambda_1, ..., \lambda_r, 0, ..., 0, -\mu_s, ..., -\mu_1)$$

(here, the string of zeros in the middle has length $n - r - s$).

Thus, we should think of $X_{\lambda,\mu}$ as the interpolation of the representation $V_{[\lambda,\mu]_n}$ to non-integer values of n; in particular, $X^*_{\lambda,\mu} = X_{\mu,\lambda}$.

EXERCISE 9.12.11. Show that the dimension of $X_{\lambda,\mu}$ is given by the interpolation of the Weyl dimension formula:

$$\dim X_{\lambda,\mu}(t) \tag{9.18}$$

$$= d_\lambda d_\mu \prod_{i=1}^{r} \frac{\binom{t+\lambda_i-i-s}{\lambda_i}}{\binom{\lambda_i+r-i}{\lambda_i}} \prod_{j=1}^{s} \frac{\binom{t+\mu_j-j-r}{\mu_j}}{\binom{\mu_j+s-j}{\mu_j}} \prod_{i=1}^{r}\prod_{j=1}^{s} \frac{t+1+\lambda_i+\mu_j-i-j}{t+1-i-j},$$

where

$$d_\lambda = \dim V_\lambda = \prod_{1\le i<j\le r} \frac{\lambda_i - \lambda_j + j - i}{j-i}$$

is the dimension of the irreducible representation of $GL_{|\lambda|}$ with highest weight λ.

(Note that since the function (9.18) takes integer values at large positive integer t, it is an integer-valued polynomial, i.e., a linear combination of binomial coefficients $\binom{t}{k}$.)

9.12.3. The categories $\mathsf{Rep}(O_t)$, $\mathsf{Rep}(Sp_{2t})$. The category $\mathsf{Rep}(O_t)$ is defined similarly to the category $\mathsf{Rep}(GL_t)$. Namely, recall that in the classical category $\mathsf{Rep}(O_n)$ we have the vector representation $V = \Bbbk^n$, and every irreducible representation of O_n occurs in $V^{\otimes r}$ for some r. Now,

$$\mathsf{Hom}_{O_n}(V^{\otimes r_1}, V^{\otimes r_2}) = (V^{\otimes r_1+r_2})^{O_n},$$

so it is nonzero only if $r_1+r_2 = 2m$, in which case it can be written as $\mathsf{End}_{O_n}(V^{\otimes m})$ and is the image of the Brauer algebra $B_m(n)$, by the Fundamental Theorem of Invariant Theory for orthogonal groups (this image is isomorphic to the Brauer algebra if $n \ge m$). The category $\mathsf{Rep}(O_n)$ can then be defined as the Karoubian envelope of the additive completion of the subcategory with objects $[r] := V^{\otimes r}$ and morphisms as above.

Now consider composition of morphisms. A basis in the Brauer algebra $B_m(n)$ is formed by matchings of $2m$ points, so we have a spanning set in $\mathsf{Hom}(V^{\otimes r_1}, V^{\otimes r_2})$ formed by unoriented planar tangles (with possibly intersecting strands) connecting r_1 points at the bottom and r_2 points at the top, which define a perfect matching. Then composition is the concatenation of tangles, followed by removal of closed loops, so that each removed loop is replaced by a factor of n.

Now, given $t \in \Bbbk$, one can define the category $\widehat{\mathsf{Rep}}(O_t)$ with objects $[r]$, $r \in \mathbb{Z}_+$, and the space of morphisms $\mathsf{Hom}([r_1],[r_2])$ being spanned by planar tangles as above, with the same composition law as above, except that every removed closed loop earns a factor of t. Thus, for instance, the endomorphism algebra $\mathrm{End}([m])$ is the Brauer algebra $B_m(t)$ (see [**Wen1**]).

Similarly to $\widehat{\mathrm{Rep}}(GL_t)$, the category $\widehat{\mathrm{Rep}}(O_t)$ has a natural strict symmetric monoidal structure. Namely, the tensor product functor is just the addition of integers for objects and taking the union of planar tangles for morphisms, with the obvious symmetric braiding. The unit object is the object $[0]$.

DEFINITION 9.12.12. The category $\widetilde{\mathrm{Rep}}(O_t)$ is the Karoubian envelope of the additive completion of $\widehat{\mathrm{Rep}}(O_t)$.

Clearly, $\widetilde{\mathrm{Rep}}(O_t)$ is a Karoubian category over $\Bbbk$, which inherits the tensor structure from $\widehat{\mathrm{Rep}}(O_t)$. Moreover, it is not hard to show that this category is rigid (with $[r]^* = [r]$). Moreover, it is easy check that $\dim[r] = t^r$.

The category $\widetilde{\mathrm{Rep}}(Sp_{2t})$ is defined in a completely parallel way, starting from the representation category of the symplectic group Sp_{2n}. It is in fact easy to see that the categories $\widetilde{\mathrm{Rep}}(O_t)$ and $\widetilde{\mathrm{Rep}}(Sp_{-t})$ are equivalent as tensor categories, and differ only by a change of the braiding. Namely, define an involutive tensor automorphism u of the identity functor of $\widetilde{\mathrm{Rep}}(O_t)$ (called the parity automorphism) by $u|_{[r]} = (-1)^r$, and define a new commutativity isomorphism on $\widetilde{\mathrm{Rep}}(O_t)$ which differs by sign from the old one if both factors are odd (i.e., $u = -1$ on them), and is the same as the old one if one of the factors is even (i.e., has $u = 1$). Then it is easy to see that $\widetilde{\mathrm{Rep}}(O_t)$ with this new commutativity is equivalent to $\widetilde{\mathrm{Rep}}(Sp_{-t})$ as a symmetric monoidal category.[1]

THEOREM 9.12.13. ([**De1, De2**]) *(i) The categories $\widetilde{\mathrm{Rep}}(O_t), \widetilde{\mathrm{Rep}}(Sp_t)$ are semisimple abelian symmetric tensor categories if $t \notin \mathbb{Z}$ (in this case they are also denoted by $\mathrm{Rep}(O_t)$, $\mathrm{Rep}(Sp_t)$).*

(ii) The category $\widetilde{\mathrm{Rep}}(O_t)$ (respectively, $\widetilde{\mathrm{Rep}}(Sp_t)$) has the following universal property: if $\mathcal{C}$ is a rigid symmetric tensor category then isomorphism classes of (possibly non-faithful) additive symmetric monoidal functors $\widetilde{\mathrm{Rep}}(O_t) \to \mathcal{C}$ (respectively $\widetilde{\mathrm{Rep}}(Sp_t) \to \mathcal{C}$) are in bijection with isomorphism classes of objects $X \in \mathcal{C}$ of dimension t with a symmetric (respectively, skew-symmetric) isomorphism $X \to X^*$, via $F \mapsto F([1])$.

(iii) If $t = n \in \mathbb{Z}$, and if p, q are non-negative integers with $p - 2q = n$, then the category $\widetilde{\mathrm{Rep}}(O_{t=n})$ (which is not abelian) admits a non-faithful additive symmetric monoidal functor $\widetilde{\mathrm{Rep}}(O_n) \to \mathrm{Rep}(OSp_{p|2q})$ to the representation category of the supergroup $OSp_{p|2q}$, which sends $[1]$ to the supervector space $V = \Bbbk^{p|2q}$.

(iv) We have a natural symmetric tensor functor Res $:\widetilde{\mathrm{Rep}}(O_t) \to \widetilde{\mathrm{Rep}}(O_{t-1})$ *and* $\widetilde{\mathrm{Rep}}(Sp_{2t}) \to \widetilde{\mathrm{Rep}}(Sp_{2t-2})$.

Again, (iii) and (iv) follow from (ii).

Now assume $t \notin \mathbb{Z}$ and let us describe the simple objects. The simple objects X_λ of $\mathrm{Rep}(O_t)$ are labelled by all partitions $\lambda = (\lambda_1, ..., \lambda_r)$; namely, X_λ is the unique direct summand in $S^\lambda V$ which does not occur in $S^{\lambda'} V$ for any λ' with $|\lambda'| < |\lambda|$ (it occurs with multiplicity 1). The object X_λ is the interpolation of the representation V_λ of O_n with highest weight $\sum \lambda_i \omega_i$, where ω_i are the fundamental weights corresponding to the representation $\wedge^i V$.

[1] There is a similar relationship between the categories $\widetilde{\mathrm{Rep}}(GL_t)$ and $\widetilde{\mathrm{Rep}}(GL_{-t})$.

EXERCISE 9.12.14. Show that the dimension of X_λ is given by the interpolation of the Weyl dimension formula:

$$\dim X_\lambda(t)$$
$$=\prod_{i=1}^{r}\frac{(\frac{t}{2}+\lambda_i-i)\binom{\lambda_i+t-r-i-1}{\lambda_i}}{(\frac{t}{2}-i)\binom{\lambda_i+r-i}{\lambda_i}}\prod_{1\le i<j\le r}\frac{(\lambda_i-\lambda_j+j-i)(\lambda_i+\lambda_j+t-i-j)}{(j-i)(t-i-j)}.$$

(Note that, as for $\mathsf{Rep}(GL_t)$, since this function takes integer values at large positive integer t, it is an integer-valued polynomial).

REMARK 9.12.15. Similarly to $\mathsf{Rep}(S_t)$, if $t\notin\mathbb{Z}$ then the categories $\mathsf{Rep}(GL_t)$, $\mathsf{Rep}(O_t)$ are not of subexponential growth, and do not come from a supergroup, since dimensions of objects are not integers.

Thus, we see that the fast-growing categories $\mathsf{Rep}(S_t)$, $\mathsf{Rep}(GL_t)$, $\mathsf{Rep}(O_t)$ are not pathological examples, but rather are glimpses of a new and largely unexplored world of combinatorics of sets of "non-integer cardinality" and linear algebra of vector spaces of "non-integer dimension".

We note, however, that by Exercise 9.9.9(ii), these symmetric tensor categories do not admit a straightforward generalization to positive characteristic (as symmetric tensor categories), since in characteristic p, dimensions of objects must lie in $\mathbb{F}_p$. The appropriate generalization is more subtle, and was recently proposed by Deligne [**De4**].

EXERCISE 9.12.16. Let $\Bbbk$ be an algebraically closed field of characteristic p. Show that the Karoubian category $\mathsf{Rep}(S_t)$ is well defined for any $t\in\Bbbk$, but it does not admit a symmetric monoidal functor (faithful or not) into a symmetric tensor category if $t\notin\mathbb{F}_p$.

9.13. Recognizing group-theoretical fusion categories

Let G be a finite group. Recall that a fusion category $\mathcal{C}$ is called a G-extension of a fusion category $\mathcal{D}$ if $\mathcal{C}$ has a faithful G-grading

$$\mathcal{C}=\bigoplus_{g\in G}\mathcal{C}_g,$$

such that $\mathcal{C}_1\cong\mathcal{D}$.

The centers of G-extensions were described in Proposition 8.23.11 in terms of de-equivariantization. In particular, the center of a G-extension contains a Tannakian subcategory $\mathsf{Rep}(G)$. Using the theory of Tannakian categories developed in Section 9.9 we obtain the following converse to this result.

PROPOSITION 9.13.1. *Let $\mathcal{C}$ be a fusion category such that $\mathcal{Z}(\mathcal{C})$ contains a Tannakian subcategory $\mathcal{E}=\mathsf{Rep}(G)$ which maps to Vec under the forgetful functor $\mathcal{Z}(\mathcal{C})\to\mathcal{C}$. Then $\mathcal{C}$ is a G-extension of some fusion category $\mathcal{D}$.*

PROOF. Let $F:\mathcal{Z}(\mathcal{C})\to\mathcal{C}$ denote the forgetful functor. Its restriction $F|_\mathcal{E}:\mathcal{E}\to\mathsf{Vec}$ is isomorphic to the fiber functor of $\mathcal{E}$ by Theorem 9.9.26. Every simple object X of $\mathcal{C}$ determines a tensor automorphism of $F|_\mathcal{E}$ as follows. Given an object Y in $\mathcal{E}$, the permutation isomorphism $\gamma_X: X\otimes F(Y)\xrightarrow{\sim}F(Y)\otimes X$ defining the central structure of Y (see (7.40)) yields an automorphism $\gamma_X\circ\delta$ of $F(Y)\otimes X$, where $\delta: F(Y)\otimes X\to X\otimes F(Y)$ is the "trivial" isomorphism, coming from the fact that $F(Y)\in\mathsf{Vec}$. Since $\mathsf{End}_\mathcal{C}(F(Y)\otimes X)=\mathsf{End}_\Bbbk F(Y)$, we obtain a linear automorphism $i_X : F(Y) \to F(Y)$. Clearly, i_X gives rise to a tensor automorphism of $F|_{\mathcal{E}}$. Let G denote the group of tensor automorphisms of $F|_{\mathcal{E}}$. We have a canonical assignment $X \mapsto i_X \in G$. It is multiplicative in X (in the sense that $i_Z = i_X i_Y$ for any simple $Z \subset X \otimes Y$), and thus defines a grading $\mathcal{C} = \bigoplus_{g\in G} \mathcal{C}_g$.

Now note that every simple object of the center $\mathcal{Z}(\mathcal{C})$ of a graded category $\mathcal{C}$ is either supported on its trivial component or is disjoint from it. By construction, the centralizer $\mathcal{E}'$ coincides with the category $\mathcal{Z}(\mathcal{C})_e$ of objects of $\mathcal{Z}(\mathcal{C})$ supported on $\mathcal{C}_e$ (indeed, X is in $\mathcal{E}'$ if and only if i_Z is the identity, where Z is a simple summand of $F(X)$). Therefore, F restricts to a surjective functor $\mathcal{E}' \to \mathcal{C}_e$. Using Lemma 6.2.4 and Theorem 7.16.6, we have

$$\frac{\mathsf{FPdim}(\mathcal{E}')}{\mathsf{FPdim}(\mathcal{C}_e)} = \frac{\mathsf{FPdim}(\mathcal{Z}(\mathcal{C}))}{\mathsf{FPdim}(\mathcal{C})} = \mathsf{FPdim}(\mathcal{C}).$$

It follows from Corollary 8.21.7 that

$$\mathsf{FPdim}(\mathcal{C}_e) = \frac{\mathsf{FPdim}(\mathcal{E}')}{\mathsf{FPdim}(\mathcal{C})} = \frac{\mathsf{FPdim}(\mathcal{C})}{|G|},$$

which means that the above grading of $\mathcal{C}$ is faithful (cf. Theorem 3.5.2). □

Let $\mathcal{C}$ be a braided fusion category. Suppose that $\mathcal{C}$ contains a Tannakian subcategory $\mathcal{E} = \mathsf{Rep}(G)$. Since $\mathcal{C}$ is embedded into $\mathcal{Z}(\mathcal{C})$, we can view $\mathcal{E}$ as a subcategory of $\mathcal{Z}(\mathcal{C})$. Let $\mathcal{C}_G$ denote the corresponding de-equivariantization of $\mathcal{C}$ (see Section 8.23). By Theorem 8.23.3 we have $\mathcal{C} \cong (\mathcal{C}_G)^G$.

PROPOSITION 9.13.2. *Let $\mathcal{C}$ be a non-degenerate braided fusion category. Suppose that $\mathcal{C}$ contains a Tannakian subcategory $\mathcal{E} = \mathsf{Rep}(G)$. Then $\mathcal{C}_G$ has a faithful G-grading.*

PROOF. The restriction of the de-equivariantization functor $F : \mathcal{C} \to \mathcal{C}_G$ to $\mathcal{E}$ is isomorphic to the fiber functor $\mathsf{Rep}(G) \to \mathsf{Vec}$. Hence for any object X in $\mathcal{C}_G$ and any object V in $\mathsf{Rep}(G)$ we have an automorphism of $F(V) \otimes X$ defined as the composition

$$F(V) \otimes X \xrightarrow{\sim} X \otimes F(V) \xrightarrow{\sim} F(V) \otimes X, \tag{9.19}$$

where the first isomorphism comes from the fact that $F(V) \in \mathsf{Vec}$ and the second comes from the central structure on F.

When X is simple, we have an isomorphism $\mathsf{Aut}_{\mathcal{C}_G}(F(V)\otimes X) \cong \mathsf{Aut}_{\mathsf{Vec}}(F(V))$, hence we obtain a tensor automorphism i_X of $F|_{\mathcal{E}}$. Since $\mathsf{Aut}_{\otimes}(F|_{\mathcal{E}}) \cong G$, we have an assignment $X \mapsto i_X \in G$. The hexagon axiom of braiding implies that this assignment is multiplicative, i.e., that $i_Z = i_X i_Y$ for any simple object Z contained in $X \otimes Y$. Thus, it defines a G-grading on $\mathcal{C}_G$:

$$\mathcal{C}_G = \bigoplus_{g\in G} (\mathcal{C}_G)_g, \quad \text{where } \mathcal{O}((\mathcal{C}_G)_g) = \{X \in \mathcal{O}(\mathcal{C}) \mid i_X = g\}. \tag{9.20}$$

By construction, the trivial component $\mathcal{C}_e$ of the above grading is $\mathcal{E}'_G$. So

$$\mathsf{FPdim}(\mathcal{C}_e) = \frac{\mathsf{FPdim}(\mathcal{E}')}{\mathsf{FPdim}(\mathcal{E})} = \frac{\mathsf{FPdim}(\mathcal{C})}{|G|^2} = \frac{\mathsf{FPdim}(\mathcal{C}_G)}{|G|}.$$

Thus, $\mathcal{C}_G$ has $|G|$ nonzero components, i.e., grading (9.20) is faithful. □

EXERCISE 9.13.3. In the notation of Proposition 9.13.2 let $g \mapsto T_g$ denote the action of G on $\mathcal{C}_G$. Prove that for every simple object X in $\mathcal{C}_G$ such that $i_X = h$ we have $i_{T_g(X)} = ghg^{-1}$.

DEFINITION 9.13.4. Let $\mathcal{C}$ be a braided fusion category. We say that a Tannakian subcategory $\mathcal{E} \subset \mathcal{C}$ is *Lagrangian* if $\mathcal{E}' = \mathcal{E}$.

The following Proposition provides a characterization of the centers of pointed fusion categories.

PROPOSITION 9.13.5. *Let $\mathcal{C}$ be a non-degenerate braided fusion category. Then $\mathcal{C}$ is equivalent to the center of a pointed fusion category if and only if $\mathcal{C}$ contains a Lagrangian subcategory.*

PROOF. The category $\mathcal{Z}(\mathsf{Vec}_G^\omega)$ contains a Tannakian subcategory $\mathcal{E} = \mathsf{Rep}(G)$ by Proposition 8.23.11. Since $\mathsf{FPdim}(\mathcal{Z}(\mathsf{Vec}_G^\omega)) = |G|^2$ we see from Corollary 8.21.7 that $\mathsf{FPdim}(\mathcal{E}') = \mathsf{FPdim}(\mathcal{E}) = |G|$. Therefore, $\mathcal{E}' = \mathcal{E}$, i.e., $\mathcal{E}$ is Lagrangian.

Conversely, if $\mathcal{C}$ contains a Lagrangian subcategory $\mathcal{E} = \mathsf{Rep}(G)$ then by Proposition 9.13.2 $\mathcal{C}_G$ is faithfully graded by G. Since

$$\mathsf{FPdim}(\mathcal{C}_G) = \frac{\mathsf{FPdim}(\mathcal{C})}{|G|} = |G|,$$

we see that each component of $\mathcal{C}_G$ has Frobenius-Perron dimension 1, i.e., $\mathcal{C}_G$ is pointed. Thus, $\mathcal{C}_G \cong \mathsf{Vec}_G^\omega$ for some group G and $\omega \in Z^3(G, \Bbbk^\times)$. Therefore, $\mathcal{C} \cong (\mathsf{Vec}_G^\omega)^G \cong \mathcal{Z}(\mathsf{Vec}_G^\omega)$. □

Recall that group-theoretical categories were introduced in Definition 9.7.1.

COROLLARY 9.13.6. *Let $\mathcal{C}$ be fusion category. Then $\mathcal{C}$ is group-theoretical if and only if $\mathcal{Z}(\mathcal{C})$ contains a Lagrangian subcategory.*

PROOF. By Theorem 8.12.3 $\mathcal{C}$ is group-theoretical if and only if there is an equivalence of braided fusion categories $\mathcal{Z}(\mathcal{C}) \cong \mathcal{Z}(\mathsf{Vec}_G^\omega)$ for some finite group G and $\omega \in Z^3(G, , \Bbbk^\times)$. So the result follows from Proposition 9.13.5. □

EXERCISE 9.13.7. Show that the words "Lagrangian subcategory" in Proposition 9.13.5 and Corollary 9.13.6 cannot be replaced by "subcategory which coincides with its centralizer" (in other words, the condition that this subcategory is Tannakian, i.e., positive, cannot be dropped).

Hint: Consider the Ising category $\mathcal{C}_2(q)$ and use Exercise 8.18.6(i).

However, it turns out that the Tannakian condition can be dropped for integral categories. Namely, the following result is a quite useful improvement of Corollary 9.13.6 for integral fusion categories.

PROPOSITION 9.13.8. *Let $\mathcal{C}$ be an integral fusion category. Then $\mathcal{C}$ is group-theoretical if and only if $\mathcal{Z}(\mathcal{C})$ contains a subcategory $\mathcal{D}$ such that $\mathcal{D} = \mathcal{D}'$ (i.e., a symmetric subcategory $\mathcal{D}$ of maximal possible dimension, $\mathsf{FPdim}(\mathcal{D}) = \mathsf{FPdim}(\mathcal{C})$).*

PROOF. The proof is given in [**DrGNO1**, Theorem 4.8]; we will not give it here. Note that this result is automatic if $\mathsf{FPdim}(\mathcal{C})$ is odd, since in this case $\mathcal{D}$ is Tannakian and, hence, Lagrangian. □

9.14. Fusion categories of prime power dimension

Let p be a prime number.

PROPOSITION 9.14.1. *Let $\mathcal{C}$ be a fusion category such that* $\mathsf{FPdim}(\mathcal{C}) = p^n$, $n \geq 1$. *Then $\mathcal{C}$ admits a faithful $\mathbb{Z}/p\mathbb{Z}$-grading.*

PROOF. By Proposition 9.6.5 $\mathcal{C}$ is pseudo-unitary. Using Theorem 9.3.7 we see that Frobenius-Perron dimensions of all simple subobjects of object $I(\mathbf{1})$ in $\mathcal{Z}(\mathcal{C})$ are divisors of p^n. Furthermore, these dimensions are integers since the forgetful image of $I(\mathbf{1})$ belongs to $\mathcal{C}_{ad}$. It follows from the class equation (9.9) that the group G of invertible subobjects of $I(\mathbf{1})$ is non-trivial. This group is precisely the group of tensor automorphisms of $\mathsf{id}_{\mathcal{C}}$, see Exercise 9.2.5. It follows from Proposition 4.14.3 that the universal grading group of $\mathcal{C}$ is a non-trivial p-group and, hence, has a quotient isomorphic to $\mathbb{Z}/p\mathbb{Z}$. □

EXERCISE 9.14.2. ([**KacG3, Mas**]) Show that a semisimple Hopf algebra of dimension p^n over $\Bbbk$ has a nontrivial central grouplike element.

COROLLARY 9.14.3. *A fusion category of prime power Frobenius Perron dimension is nilpotent.*

REMARK 9.14.4. Corollary 9.14.3 generalizes a classical result in group theory stating that p-group is nilpotent.

COROLLARY 9.14.5. *Let $\mathcal{C}$ be a fusion category such that* $\mathsf{FPdim}(\mathcal{C}) = p$. *Then* $\mathcal{C} \cong \mathsf{Vec}_{\mathbb{Z}/p\mathbb{Z}}^{\omega}$ *for some 3-cocycle* $\omega \in Z^3(\mathbb{Z}/p\mathbb{Z}, \Bbbk^\times)$.

PROOF. This is clear since by Proposition 9.14.1, $\mathcal{C}$ must be pointed (since the dimension of $\mathcal{C}_1$ for any grading of $\mathcal{C}$ must divide the dimension of $\mathcal{C}$). □

This allows us to obtain a classification of semisimple quasi-Hopf algebras of prime dimension.

COROLLARY 9.14.6. *Any semisimple quasi-Hopf algebra of prime dimension p over a field of characteristic zero is twist equivalent to the group algebra of a cyclic group $\mathbb{Z}/p\mathbb{Z}$, with associator corresponding to some* $\omega \in Z^3(\mathbb{Z}/p\mathbb{Z}, \Bbbk^\times)$.

As a corollary we also obtain the Kac-Zhu theorem on the classification of Hopf algebras of prime dimension (first proved by G. Kac [**KacG3**] for semisimple Hopf algebras and later by Zhu [**Z**] in general).

COROLLARY 9.14.7. *Any Hopf algebra of prime dimension over a field of characteristic zero is the group algebra of a cyclic group.*

PROOF. If the Hopf algebra H in question is semisimple, then the result follows from Corollary 9.14.5. So it remains to show that H must be semisimple.

The result is trivial for $p = 2$ (as in this case H is clearly commutative), so we may assume that $p > 2$. Also, if H or H^* has a nontrivial grouplike element, the result follows immediately. So it suffices to assume that there is no such grouplike elements. Then by Radford's S^4 formula (Corollary 8.9.7), $S^4 = \mathsf{id}$. Hence the eigenvalues of S^2 are ± 1. Since p is odd, these eigenvalues cannot add up to zero, so $\mathsf{Tr}(S^2) \neq 0$, hence H is semisimple by Corollary 7.18.10, as desired. □

Unfortunately, even though we proved a quasi-Hopf analog of Radford's S^4 formula in Theorem 7.19.1, it is not clear to us how to extend the proof of Theorem 9.14.7 to quasi-Hopf algebras. Thus, we have the following

CONJECTURE 9.14.8. Corollary 9.14.5 holds for any finite tensor category of prime Frobenius-Perron dimension. Equivalently, the semisimplicity assumption in Corollary 9.14.6 can be dropped.

The above method also allows us to classify fusion categories and semisimple Hopf algebras of dimension p^2.

PROPOSITION 9.14.9. *Let $\mathcal{C}$ be a fusion category of Frobenius-Perron dimension p^2. Then either $\mathcal{C}$ is pointed, or, for $p = 2$, $\mathcal{C}$ is equivalent to an Ising category $\mathcal{C}_2(q)$ where q is a primitive 8th root of unity (see Example* 8.18.5*).*

PROOF. By Proposition 9.14.1, $\mathcal{C}$ admits a faithful grading by $\mathbb{Z}/p\mathbb{Z}$: $\mathcal{C} = \mathcal{C}_0 \oplus \mathcal{C}_1 \oplus ... \oplus \mathcal{C}_{p-1}$. Since $\mathcal{C}_0$ has dimension p, by Theorem 9.14.5, it is pointed. The category $\mathcal{C}_i$ for each i is a module category over $\mathcal{C}_0$, so $\mathbb{Z}/p\mathbb{Z}$ acts transitively on its simple objects. If this action is free for all i, then $\mathcal{C}$ is pointed. If this action is trivial for some i, then $\mathcal{C}_i$ has a unique simple object of dimension $\sqrt{p}$. We know from Corollary 3.5.8 that this can only happen if $\mathsf{FPdim}(\mathcal{C})$ is even, so $p = 2$. In this case, $\mathcal{C}$ has the fusion ring of the Ising category, and one can show that $\mathcal{C}$ must be one of the two Ising categories (we leave this to the reader). □

COROLLARY 9.14.10. *Any semisimple Hopf algebra of dimension p^2 over a field of characteristic zero (for prime p) is the group algebra of $(\mathbb{Z}/p\mathbb{Z})^2$ or $\mathbb{Z}/p^2\mathbb{Z}$.*

PROOF. This follows from Proposition 9.14.9 and the fact that any group of order p^2 is isomorphic to $(\mathbb{Z}/p\mathbb{Z})^2$ or $\mathbb{Z}/p^2\mathbb{Z}$. □

REMARK 9.14.11. Note that unlike the case of dimension p, there exist non-semisimple Hopf algebras of dimension p^2, e.g., Taft algebras (Example 5.5.6). Ng showed in [**Ng**] that these are in fact the only examples; so any Hopf algebra of dimension p^2 is either the group algebra of an abelian group or a Taft algebra.

Recall that for a fusion subcategory $\mathcal{K}$ of a braided fusion category $\mathcal{C}$ the commutator subcategory $\mathcal{K}^{co}$ is introduced in Definition 4.14.10. For a pair of fusion subcategories $\mathcal{A}, \mathcal{B} \subset \mathcal{C}$ let $\mathcal{A} \vee \mathcal{B}$ denote the fusion subcategory of $\mathcal{C}$ generated by $\mathcal{A}$ and $\mathcal{B}$.

PROPOSITION 9.14.12. *Let $\mathcal{C}$ be a nilpotent braided fusion category. Then for any maximal symmetric subcategory $\mathcal{K}$ of $\mathcal{C}$ one has $(\mathcal{K}')_{ad} \subset \mathcal{K}$.*

PROOF. Let $\mathcal{K}$ be a symmetric subcategory of $\mathcal{C}$, i.e., $\mathcal{K} \subset \mathcal{K}'$. Assume that $(\mathcal{K}')_{ad} \not\subset \mathcal{K}$. Then we will construct a symmetric subcategory $\mathcal{E} \subset \mathcal{C}$ such that $\mathcal{E} \supsetneq \mathcal{K}$. Namely, set

$$\mathcal{E} := \mathcal{K} \vee (\mathcal{K}^{co} \cap (\mathcal{K}^{co})'). \tag{9.21}$$

Clearly $\mathcal{E} \supset \mathcal{K}$. It is also clear that $\mathcal{K}^{co} \cap (\mathcal{K}^{co})'$ is symmetric. Since $\mathcal{K}$ is symmetric and $\mathcal{K}^{co} \cap (\mathcal{K}^{co})' \subset (\mathcal{K}^{co})' \subset \mathcal{K}'$ we see that $\mathcal{E}$ is symmetric. It remains to show that $\mathcal{E} \neq \mathcal{K}$, or, equivalently, that $\mathcal{K}^{co} \cap (\mathcal{K}^{co})' \not\subset \mathcal{K}$. By Corollary 8.22.7, $(\mathcal{K}^{co})' \supset (\mathcal{K}')_{ad}$, so it suffices to prove that

$$\mathcal{K}^{co} \cap (\mathcal{K}')_{ad} \not\subset \mathcal{K}. \tag{9.22}$$

Consider the upper central series (see Definition 4.14.12)

$$\mathcal{K}' = (\mathcal{K}')^{(0)} \supset (\mathcal{K}')^{(1)} \supset (\mathcal{K}')^{(2)} \supset \dots, \quad (\mathcal{K}')^{(m+1)} := ((\mathcal{K}')^{(m)})_{ad}.$$

Let n be the largest integer such that $(\mathcal{K}')^{(n)} \not\subset \mathcal{K}$ (such n exists because $\mathcal{K}'$ is nilpotent). Since $(\mathcal{K}')^{(n+1)} \subset \mathcal{K}$, one has $(\mathcal{K}')^{(n)} \subset \mathcal{K}^{co}$. Since $(\mathcal{K}')_{ad} \not\subset \mathcal{K}$, one has $n > 0$, so $(\mathcal{K}')^{(n)} \subset (\mathcal{K}')_{ad}$. Thus $(\mathcal{K}')^{(n)} \subset \mathcal{K}^{co} \cap (\mathcal{K}')_{ad}$. Since $(\mathcal{K}')^{(n)} \not\subset \mathcal{K}$, we get (9.22). □

COROLLARY 9.14.13. *In the situation of Proposition* 9.14.12 *the fusion category $\mathcal{K}'$ has a faithful grading with trivial component $\mathcal{K}'$:*

$$\mathcal{K}' = \bigoplus_{g \in G} \mathcal{K}'_g, \qquad \mathcal{K}'_1 = \mathcal{K}. \tag{9.23}$$

Such a grading is unique up to equivalence. The group G is abelian.

THEOREM 9.14.14. *Let $\mathcal{C}$ be an integral braided fusion category. Suppose that $\mathcal{C}$ contains a symmetric subcategory $\mathcal{K}$ such that $(\mathcal{K}')_{ad} \subset \mathcal{K}$. Then $\mathcal{C}$ is group-theoretical.*

PROOF. We may assume that $\mathcal{K} \supset \mathcal{C}'$ (otherwise replace $\mathcal{K}$ by $\mathcal{K} \vee \mathcal{C}'$).

By Proposition 9.13.8, it suffices to construct a fusion subcategory $\mathcal{D} \subset \mathcal{Z}(\mathcal{C})$ such that $\mathcal{D} = \mathcal{D}'$. To this end, we will use the braided tensor embeddings

$$\mathcal{C} \to \mathcal{Z}(\mathcal{C}) : X \mapsto (X, c_{X,-}) \quad \text{and} \quad \mathcal{C}^{\text{op}} \to \mathcal{Z}(\mathcal{C}) : X \mapsto (X, c^{-1}_{-,X}),$$

where c denotes the braiding of $\mathcal{C}$. Let $\mathcal{C}_+, \mathcal{C}_- \subset \mathcal{Z}(\mathcal{C})$ denote the images of these embeddings. Note that $\mathcal{C}_- = (\mathcal{C}_+)'$ and $\mathcal{C}_+ \cap \mathcal{C}_- = \mathcal{C}'$.

By Corollary 9.14.13, there is a faithful grading $\mathcal{K}' = \bigoplus_{g \in G} (\mathcal{K}')_g$ with trivial component $\mathcal{K}$, and the group G is abelian. Set $(\mathcal{K}')^{\text{op}}_g := F((\mathcal{K}')_g)$, where $F : \mathcal{K}' \xrightarrow{\sim} (\mathcal{K}')^{\text{op}}$ is the canonical equivalence of $\Bbbk$-linear categories. Since G is abelian, the decomposition $(\mathcal{K}')^{\text{op}} = \bigoplus_{g \in G} (\mathcal{K}')^{\text{op}}_g$ is also a G-grading.

Let us identify $\mathcal{K}$, $\mathcal{K}'$ and $\mathcal{K}^{\text{op}}, (\mathcal{K}')^{\text{op}}$ with subcategories of $\mathcal{C}_+$ and $\mathcal{C}_-$, respectively. Define $\mathcal{D} \subset \mathcal{Z}(\mathcal{C})$ to be the fusion subcategory generated by all objects of the form $Y_1 \otimes Y_2$, $Y_1 \in (\mathcal{K}')_g$, $Y_2 \in (\mathcal{K}')^{\text{op}}_g$, $g \in G$. We will show that $\mathcal{D}' = \mathcal{D}$.

We have a non-degenerate pairing $b : G \times G \to \Bbbk^\times$ such that $c_{X_2X_1}c_{X_1X_2} = b(g_1, g_2) \cdot \mathsf{id}_{X_1 \otimes X_2}$ if $X_i \in (\mathcal{K}')_{g_i}$, $g_i \in G_i$. Since the subcategories $\mathcal{K}', (\mathcal{K}')^{\text{op}} \subset \mathcal{Z}(\mathcal{C})$ centralize each other, we see that if $Y_i = X_i \otimes \tilde{X}_i$, $X_i \in (\mathcal{K}')_{g_i}$, $\tilde{X}_i \in (\mathcal{K}')^{\text{op}}_{\tilde{g}_i}$, $g_i, \tilde{g}_i \in G$, then

$$c_{Y_2Y_1}c_{Y_1Y_2} = b(g_1, g_2)b(\tilde{g}_1, \tilde{g}_2)^{-1} \; \mathsf{id}_{Y_1 \otimes Y_2}\,. \tag{9.24}$$

This formula implies that $\mathcal{D} \subset \mathcal{D}'$ and moreover, $\mathcal{D}' \cap (\mathcal{K}' \vee (\mathcal{K}')^{\text{op}}) = \mathcal{D}$. To prove that $\mathcal{D}' = \mathcal{D}$, it remains to show that $\mathcal{D}' \subset \mathcal{K}' \vee (\mathcal{K}')^{\text{op}}$. Since $\mathcal{D} \supset \mathcal{K} \vee \mathcal{K}^{\text{op}}$, it suffices to show that $(\mathcal{K} \vee \mathcal{K}^{\text{op}})' = \mathcal{K}' \vee (\mathcal{K}')^{\text{op}}$. Clearly $(\mathcal{K} \vee \mathcal{K}^{\text{op}})' \supset \mathcal{K}' \vee (\mathcal{K}')^{\text{op}}$, so it remains to prove that

$$\mathsf{FPdim}(\mathcal{K} \vee \mathcal{K}^{\text{op}}) \cdot \mathsf{FPdim}(\mathcal{K}' \vee (\mathcal{K}')^{\text{op}}) = \mathsf{FPdim}(\mathcal{C})^2. \tag{9.25}$$

Since $\mathcal{K} \supset \mathcal{C}'$, we have $\mathsf{FPdim}(\mathcal{K}) \cdot \mathsf{FPdim}(\mathcal{K}') = \mathsf{FPdim}(\mathcal{C}) \cdot \mathsf{FPdim}(\mathcal{C}')$. Since $\mathcal{C}_+ \cap \mathcal{C}_- = \mathcal{C}'$ and $\mathcal{K} \supset \mathcal{C}'$, we get $\mathcal{K}' \cap (\mathcal{K}')^{\text{op}} = \mathcal{K} \cap \mathcal{K}^{\text{op}} = \mathcal{C}'$. Now (9.25) follows from Lemma 8.21.6. □

COROLLARY 9.14.15. *An integral nilpotent braided fusion category $\mathcal{C}$ is group-theoretical.*

PROOF. This follows from Theorem 9.14.14 and Proposition 9.14.12. □

COROLLARY 9.14.16. *Let $\mathcal{C}$ be an integral fusion category such that $\mathsf{FPdim}(\mathcal{C})$ is a prime power. Then $\mathcal{C}$ is group-theoretical. Thus, any semisimple quasi-Hopf (in particular, Hopf) algebra of prime power dimension has a group-theoretical category of representations.*

PROOF. Since $\mathsf{FPdim}(\mathcal{Z}(\mathcal{C})) = \mathsf{FPdim}(\mathcal{C})^2$, we see that $\mathcal{Z}(\mathcal{C})$ is nilpotent by Corollary 9.14.3. Thus, $\mathcal{Z}(\mathcal{C})$ is group-theoretical by Corollary 9.14.15. Hence, $\mathcal{C} \boxtimes \mathcal{C}^{\mathrm{op}}$ is group-theoretical and so is $\mathcal{C}$ by Proposition 9.7.9. □

REMARK 9.14.17. The integrality of $\mathcal{C}$ is automatic for odd primes.

9.15. Burnside's theorem for fusion categories

Let $\mathcal{C}$ be a weakly integral fusion category. We will use shortened notation d_X for the Frobenius-Perron dimension $\mathsf{FPdim}(X)$ of an object X in $\mathcal{C}$. By Corollary 9.6.6, these dimensions coincide with those coming from the canonical spherical structure of $\mathcal{C}$.

LEMMA 9.15.1. *Let X and Y be two simple objects of an integral braided category $\mathcal{C}$ with relatively prime dimensions d_X, d_Y. Let S denote the S-matrix of $\mathcal{C}$. Then one of the following two possibilities is realized:*

(i) *X and Y projectively centralize each other, i.e., the square of the braiding on $X \otimes Y$ is a scalar (see Definition 8.22.3);*
(ii) *$s_{XY} = 0$.*

PROOF. It suffices to consider the case when $\mathcal{C}$ is non-degenerate, since any braided category can be embedded into a non-degenerate one (its center). In this case, by the Verlinde formula, $\frac{s_{XY}}{d_X}$ and $\frac{s_{XY}}{d_Y}$ are algebraic integers (see Proposition 8.13.11). Since d_X and d_Y are relatively prime, $\frac{s_{XY}}{d_X d_Y}$ is also an algebraic integer. Since s_{XY} is a sum of $d_X d_Y$ roots of unity, we see that $\frac{s_{XY}}{d_X d_Y}$ is either a root of unity, in which case the square of the braiding must be a scalar (case (i)), or 0 (case (ii)). □

COROLLARY 9.15.2. *Let $\mathcal{C}$ be an integral non-degenerate braided category which contains a simple object X with dimension $d_X = p^r$, $r > 0$, where p is a prime. Then $\mathcal{C}$ contains a nontrivial symmetric subcategory.*

PROOF. We first show that $\mathcal{C}$ contains a nontrivial proper subcategory. Assume not. Take any simple $Y \neq \mathbf{1}$ with d_Y coprime to d_X. We claim that $s_{XY} = 0$. Indeed, otherwise X and Y projectively centralize each other, hence Y centralizes $X \otimes X^*$, so the centralizer of the subcategory generated by Y in $\mathcal{C}$ is nontrivial, and we get a nontrivial proper subcategory, a contradiction.

Now let us use the orthogonality of columns $(s_{X,Y})$ and (d_Y) of the S-matrix:

$$\sum_{Y \in \mathcal{O}(\mathcal{C})} \frac{s_{XY}}{d_X} d_Y = 0.$$

By Lemma 9.15.1, all the nonzero summands in this sum, except the one for $Y = \mathbf{1}$, come from objects Y of dimension divisible by p. Therefore, all the summands in this sum except for the one for $Y = \mathbf{1}$ (which equals 1) are divisible by p. This is a contradiction.

Now we prove the corollary by induction in $\mathsf{FPdim}(\mathcal{C})$. Let $\mathcal{D}$ be a nontrivial proper subcategory of $\mathcal{C}$. If $\mathcal{D}$ is degenerate, then $\mathcal{D}\cap\mathcal{D}'$ is a nontrivial proper symmetric subcategory of $\mathcal{C}$. Otherwise, $\mathcal{D}$ is non-degenerate, and by Theorem 8.21.4(i), $\mathcal{C} = \mathcal{D} \boxtimes \mathcal{D}'$. Thus $X = X_1 \otimes X_2$, where $X_1 \in \mathcal{D}$, $X_2 \in \mathcal{D}'$ are simple. Since the dimension of X_1 or X_2 is a positive power of p, we get the desired statement from the induction assumption applied to $\mathcal{D}$ or $\mathcal{D}'$ (which are non-degenerate braided categories of smaller dimension). □

DEFINITION 9.15.3. A braided fusion category $\mathcal{C}$ is called *slightly degenerate* if $\mathcal{C}' = \mathsf{sVec}$.

In other words, a degenerate braided fusion category $\mathcal{C}$ is slightly degenerate if $\mathcal{C}'$ does not contain non-trivial Tannakian subcategories.

PROPOSITION 9.15.4. (i) *Let $\mathcal{C}$ be a braided fusion category such that $\mathsf{sVec} \subset \mathcal{C}'$ (for example, a slightly degenerate category). Let χ be the invertible object generating sVec and let Y be any simple object of $\mathcal{C}$. Then $\chi \otimes Y$ is not isomorphic to Y.*

(ii) *Let $\mathcal{C}$ be slightly degenerate and pointed. Then $\mathcal{C} = \mathsf{sVec} \boxtimes \mathcal{C}_0$, where $\mathcal{C}_0$ is a non-degenerate pointed category.*

PROOF. (i) Assume the contrary, i.e., $\chi \otimes Y = Y$. Since χ centralizes Y, we have from this identity that the trace T of the isomorphism $u : Y \to Y^{**}$ from (8.30) is equal to $-T$ (as $u|_\chi = -1$). This is a contradiction, as $T \neq 0$.

(ii) Our job is to show that $\chi \neq \xi^{\otimes 2}$ for any ξ (this is the condition for the group of invertible objects of $\mathcal{C}$ to be the direct product of the $\mathbb{Z}/2\mathbb{Z}$ generated by χ with another subgroup). Assume the contrary, and let q be the quadratic form defining the braiding. Then we have $q(\xi)^4 = q(\chi) = -1$, $q(\chi \otimes \xi) = q(\xi^{\otimes 3}) = q(\xi)^9 = q(\xi)$, so the squared braiding of ξ and χ is given by the scalar

$$B(\chi, \xi) = q(\chi \otimes \xi)/q(\chi)q(\xi) = -1,$$

which contradicts to χ lying in $\mathcal{C}'$. □

COROLLARY 9.15.5. *Let $\mathcal{C}$ be a slightly degenerate integral category of dimension > 2. Then $\mathcal{C}$ contains an odd-dimensional simple object that does not belong to $\mathcal{C}'$.*

PROOF. Let χ be the invertible object generating $\mathcal{C}'$. Let X be any simple object outside of $\mathcal{C}'$. By Proposition 9.15.4(i), $\chi \otimes X \neq X$, which implies that $X \otimes X^*$ does not contain χ. Thus, either X itself is odd-dimensional, or $X \otimes X^*/\mathbf{1}$ is odd-dimensional, and is a direct sum of simple objects not contained in $\mathcal{C}'$. In this case one of the summands has to be odd-dimensional. □

PROPOSITION 9.15.6. *Let $\mathcal{C}$ be a slightly degenerate integral braided fusion category containing a simple object X of dimension p^r for some prime $p > 2$. Then $\mathcal{C}$ contains a nontrivial Tannakian subcategory.*

PROOF. The proof is by induction on the dimension of $\mathcal{C}$. Let $\mathcal{B}$ be the category spanned by the invertible objects of $\mathcal{C}$. Then $\mathcal{B} \cap \mathcal{B}'$ contains the subcategory sVec.

If $\mathcal{B}\cap\mathcal{B}'$ is bigger than sVec, then it contains a nontrivial Tannakian subcategory, and we are done. Otherwise, by Proposition 9.15.4(ii), $\mathcal{B} = \mathsf{sVec} \boxtimes \mathcal{B}_0$, where $\mathcal{B}_0$ is a pointed non-degenerate braided category. If $\mathcal{B}_0$ is nontrivial, then $\mathcal{C} = \mathcal{B}_0 \boxtimes \mathcal{B}_0'$, and $\mathcal{B}_0'$ is slightly degenerate, so we are done by the induction assumption. Thus, it suffices to consider the case $\mathcal{B} = \mathsf{sVec}$, which we do from now on.

Let $\mathbf{1}$ and χ be the simple objects of $\mathsf{sVec} \subset \mathcal{C}$ (which are the only invertible objects of $\mathcal{C}$). Let Y be a non-invertible simple object of $\mathcal{C}$ of dimension not divisible by p.

Assume that X and Y projectively centralize each other. In this case the category $\mathcal{D}$ generated by Y and χ centralizes $X \otimes X^*$, so it is a proper subcategory of $\mathcal{C}$. If it is not slightly degenerate, $\mathcal{D} \cap \mathcal{D}'$ contains more than two simple objects, hence contains a nontrivial Tannakian subcategory. So we may assume that this subcategory is slightly degenerate, in which case we are done by the induction assumption.

Thus, we may assume that X and Y do not projectively centralize each other. In this case Lemma 9.15.1 tells us that $s_{X,Y} = 0$.

Now, using Proposition 9.15.4(i), we have

$$\sum_Y \frac{s_{XY}}{d_X} d_Y = 0,$$

and all the nonzero terms in this sum correspond to either $d_Y = 1$ (there are two such terms, both equal to 1), or d_Y divisible by p, which gives terms divisible by p. So we get that 2 is divisible by p, a contradiction. $\square$

PROPOSITION 9.15.7. *Let $\mathcal{C}$ be an integral non-degenerate braided category of dimension $p^a q^b$, $a + b > 0$. Then $\mathcal{C}$ contains a nontrivial invertible object.*

PROOF. By Corollary 9.14.3 a fusion category of a prime power Frobenius-Perron dimension is nilpotent and, hence, contains a nontrivial invertible object. So we may assume that both a and b are positive. Suppose that $\mathcal{C}$ does not contain nontrivial invertible objects. By Proposition 8.14.6 the squared dimensions of simple objects of $\mathcal{C}$ divide $p^a q^b$. Therefore, $\mathcal{C}$ must contain a simple object of dimension p^r, $r > 0$. Hence by Corollary 9.15.2, it contains a nontrivial symmetric subcategory $\mathcal{D}$. By Deligne's Theorem 9.9.26, $\mathcal{D} = \mathsf{Rep}(G, z)$ for some group G. This group G is solvable, thanks to the usual Burnside theorem for finite groups (saying that a group of order $p^n q^m$ is solvable). Therefore, $\mathcal{D}$ contain nontrivial invertible objects, which is a contradiction. $\square$

LEMMA 9.15.8. *Let $\mathcal{C}$ be a slightly degenerate integral braided fusion category of dimension $2^r q^s > 2$, where $q > 2$ is a prime, and r, s are non-negative integers. Suppose that $\mathcal{C}$ contains only two invertible objects. Then $\mathcal{C}$ contains a nontrivial Tannakian subcategory.*

PROOF. It follows from Proposition 8.14.6 that there exists a non-invertible simple object Y of $\mathcal{C}$ whose dimension is a power of 2. Also, by Corollary 9.15.5, $\mathcal{C}$ contains a simple object X of dimension q^m, $m > 0$. Now the statement follows from Proposition 9.15.6. $\square$

THEOREM 9.15.9. *A fusion category of Frobenius-Perron dimension $p^r q^s$, where p, q are primes and r, s are non-negative integers, is solvable.*

PROOF. Let $\mathcal{C}$ be a fusion category such that $\mathsf{FPdim}(\mathcal{C}) = p^r q^s$. We may assume that $\mathcal{C}$ is integral (which is automatic when p, q are odd) since otherwise $\mathcal{C}$ is $\mathbb{Z}/2\mathbb{Z}$-graded with integral trivial component and we can apply Proposition 9.8.4.

It suffices to prove that $\mathcal{Z}(\mathcal{C})$ contains a Tannakian subcategory $\mathcal{T} \cong \mathsf{Rep}(G)$, where G is a cyclic group of prime order. Indeed, the restriction of forgetful functor

$F : \mathcal{Z}(\mathcal{C}) \to \mathcal{C}$ to $\mathcal{T}$ is either fully faithful or maps $\mathcal{T}$ to Vec (since $\mathcal{T}$ has no quotient categories except Vec and $\mathcal{T}$). In the former case $\mathcal{C}$ is a G-equivariantization (Theorem 8.23.3) and in the latter case $\mathcal{C}$ is a G-extension (Theorem 9.13.1). In any case, $\mathcal{C}$ is categorically Morita equivalent to a G-extension. The statement then follows from induction on the dimension of $\mathcal{C}$ and Lemma 9.8.3.

By Proposition 9.15.7 the subcategory $\mathcal{B}$ spanned by all invertible objects of $\mathcal{Z}(\mathcal{C})$ is nontrivial. If $\mathcal{B}$ is non-degenerate, then by Theorem 8.21.4(i), $\mathcal{Z}(\mathcal{C}) = \mathcal{B} \boxtimes \mathcal{B}'$, where $\mathcal{B}'$ is another non-degenerate braided category, which is nontrivial (as $\mathcal{Z}(\mathcal{C})$ is not pointed), but has no nontrivial invertible objects. Thus, by Proposition 9.15.7, this case is impossible.

Therefore, $\mathcal{B}$ is degenerate. Let $\mathcal{E} := \mathcal{B} \cap \mathcal{B}'$. It is a nontrivial pointed symmetric subcategory in $\mathcal{Z}(\mathcal{C})$. So if $\mathsf{FPdim}(\mathcal{E}) > 2$, we are done (as $\mathcal{E}$ necessarily contains a Tannakian subcategory $\mathsf{Rep}(G)$, where G is a cyclic group of prime order).

It remains to consider the case $\mathsf{FPdim}(\mathcal{E}) = 2$. In this case, we must consider the additional possibility that $\mathcal{E} = \mathsf{sVec}$ (in which case one of p, q is equal to 2). In this situation, by Proposition 9.15.4(ii), $\mathcal{B} = \mathcal{E} \boxtimes \mathcal{D}$, where $\mathcal{D}$ is non-degenerate, so if $\mathcal{D}$ is nontrivial, by Theorem 8.21.4(i) $\mathcal{Z}(\mathcal{C}) = \mathcal{D} \boxtimes \mathcal{D}'$, where $\mathcal{D}'$ is another non-degenerate braided category, whose subcategory of invertible objects is $\mathcal{E}$. Thus, it is sufficient to consider the case $\mathcal{B} = \mathcal{E} = \mathsf{sVec}$. In this case $\mathcal{E}'$ (the centralizer of $\mathcal{E}$ in $\mathcal{Z}(\mathcal{C})$) has dimension is $2^{2r-1}q^{2s} > 2$, contains only two invertible objects, and $\mathcal{E}' \cap \mathcal{E} = \mathcal{E} = \mathsf{sVec}$, i.e., $\mathcal{E}'$ is slightly degenerate. Thus, $\mathcal{Z}(\mathcal{C})$ contains a non-trivial Tannakian subcategory $\mathcal{T}$ by Lemma 9.15.8. Since $\mathcal{T}$ is the representation category of a solvable group, it contains a subcategory $\mathsf{Rep}(G)$ for a cyclic group G, thanks to the classical Burnside's theorem for finite groups. □

9.16. Lifting theory

In this Section $\Bbbk$ is any field.

The Ocneanu rigidity theory (Section 9.1) can be used to define liftings of tensor categories and related objects from characteristic p to characteristic zero, see [**ENO2**]. Let us briefly summarize the basics of lifting theory and some of its applications (we refer the reader to [**ENO2**] for more details).

Let $\mathcal{C}$ be a split fusion category over $\Bbbk$, see Section 4.16. Let $\mathcal{O}$ be a local commutative ring with maximal ideal $\mathfrak{m}$, and $\mathcal{O}/\mathfrak{m} = \Bbbk$. A *lifting* $\widetilde{\mathcal{C}}$ of $\mathcal{C}$ to $\mathcal{O}$ is a split fusion category over $\mathcal{O}$ with the same Grothendieck ring as $\mathcal{C}$, together with an equivalence $\widetilde{\mathcal{C}}/\mathfrak{m} \cong \mathcal{C}$. If we are given such a lifting, and if K is the quotient field of $\mathcal{O}$, then we can consider the split fusion category $\widehat{\mathcal{C}} := \widetilde{\mathcal{C}} \otimes_{\mathcal{O}} K$ over the field K, which is called a lifting of $\mathcal{C}$ to K.

We are interested in these notions in the "mixed characteristic" case, when $\Bbbk$ has characteristic p, and K has characteristic zero. In this case, one is faced with the question of the existence of the lifting. This question is answered by the following theorem.

THEOREM 9.16.1. *If* $\dim \mathcal{C} \neq 0$ *then* $\mathcal{C}$ *admits a lifting to* $\mathcal{O}$ *and* K*, and this lifting is unique up to equivalence.*

The proof of this theorem is based on Ocneanu rigidity (Section 9.1). Namely, one can show by standard methods of deformation theory that obstructions to lifting lie in $H^4(\mathcal{C})$, and if this obstruction vanishes, then the freedom in choosing a lifting lies in $H^3(\mathcal{C})$. But by Ocneanu rigidity, $H^i(\mathcal{C}) = 0$ for $i \geq 1$ if $\dim \mathcal{C} \neq 0$.

Similar theorems hold for braided and symmetric categories, module categories, Hopf algebras, etc., as long as the fusion categories involved have nonzero dimensions in $\Bbbk$ (see [**ENO2**], Section 9). Moreover, it is shown in [**ENO2**] that lifting is faithful, i.e., liftings of non-equivalent categories to K are not equivalent.

One early application of the lifting theory is the following theorem.

THEOREM 9.16.2. *([**EtG1**]) Let H be a semisimple and cosemisimple Hopf algebra over any field $\Bbbk$. Then $S^2 = \mathsf{id}_H$ and $\dim(H) \neq 0$ in $\Bbbk$.*

PROOF. It is shown in Corollary 8.20.17 that the categorical dimension $\mathsf{Tr}_H(S^2)$ of $\mathsf{Rep}(H)$ is nonzero. Theorem 9.16.1 then implies that $\mathsf{Rep}(H)$ admits a lifting to characteristic zero. Similarly, using Corollary 9.1.6 (Ocneanu rigidity for module categories), one shows that the fiber functor on this category associated to H admits a lifting to characteristic zero. Thus, the Hopf algebra H admits a lifting $\widetilde{H}$ to characteristic zero. By Corollary 9.6.7 one has $S^2 = \mathsf{id}$ on $\widetilde{H}$. Hence, one has $S^2 = \mathsf{id}$ on H, and hence $\dim(H) = \mathsf{Tr}_H(S^2) \neq 0$. □

9.17. Bibliographical notes

9.1. The statement of Theorem 9.1.4 is known as "Ocneanu rigidity" since it was first established by Ocneanu for unitary categories (unpublished). In the case of categories of nonzero global dimension a proof was given by Blanchard and Wassermann (unpublished). In [**ENO2**] it was pointed out that the assumption of nonzero dimension is automatically satisfied in characteristic zero. The first published proof was given in [**ENO2**]. Our exposition is an improved version of the arguments of Section 7 of [**ENO2**]: namely, unlike [**ENO2**], we do not use weak Hopf algebras, but rather use the categorical language.

9.2. Proposition 9.2.2 is Proposition 5.4 in [**ENO2**].

9.3. Theorems 9.3.2 and 9.3.7 and Propositions 9.3.4 and 9.3.9 are taken from [**ENO2**]. Corollary 9.3.5 appears as Remark 3.2 in [**ENO2**]. Theorem 9.3.11 is [**ENO2**, Theorem 8.51]. In the special case $\mathcal{C} = \mathsf{Rep}(H)$ for some Hopf algebra H Lemma 9.3.10 and Theorem 9.3.11 were proved in [**KaSZ**, 6.3].

9.4. Proposition 9.4.2 is Proposition 8.22 in [**ENO2**].

9.5. Proposition 9.5.1 is Proposition 8.23 in [**ENO2**].

9.6. The results of this Section are motivated by those of Larson and Radford, see Corollary 9.6.7. Proposition 9.6.5 is Proposition 8.24 in [**ENO2**] and Proposition 9.6.9(i) is Proposition 8.27 in [**ENO2**]. Proposition 9.6.9 (ii) is taken from [**DrGNO2**]. Proposition 9.6.11 is essentially Theorem 8.35 in [**ENO2**].

9.7. Group-theoretical fusion categories were implicitly introduced in [**Os2**] and explicitly in [**ENO2**, Section 8.8]. Proposition 9.7.9 is [**ENO2**, Proposition 8.44]. Hopf algebras obtained by an extension of a cocommutative Hopf algebra by a commutative one were first considered by Kac in [**KacG2**]. The proof of the fact that fusion categories from Exercise 9.7.5(ii) are group-theoretical is due to Natale [**Na1**]. A criterion for an equivariantization of a fusion category to be group-theoretical (cf. Remark 9.7.6) was given in [**Nik2**].

9.8. The classes of weakly group-theoretical and solvable fusion categories were introduced in [**ENO3**], and the material of Section 9.8 is taken from that paper.

9.9. Tannakian fusion categories form a very special class of Tannakian categories considered in [**Sa**]. The results of this Section is a special case of more general results in Section 9.11.

9.10. The material of this Section is taken from [**De1**].

9.11. The results of this Section are taken from [**De2**].

9.12. The category $\mathsf{Rep}(GL_t)$ was introduced by Deligne and Milne in [**DelM**] and the categories $\mathsf{Rep}(S_t)$, $\mathsf{Rep}(O_t)$, and $\mathsf{Rep}(Sp_{2t})$ were introduced by Deligne in [**De3**].

9.13. The results of this Section are taken from Section 4 in [**DrGNO2**].

9.14. Propositions 9.14.1 and 9.14.9 and Corollary 9.14.5 are from [**ENO2**]. Corollary 9.14.7 is the main result of [**Z**]. In the semisimple case it was proved in [**KacG3**]. Proposition 9.14.12 and Theorem 9.14.14 are taken from [**DrGNO1**].

9.15 The results of this Section are taken from [**ENO3**].

9.16. Theorem 9.16.1 is Theorem 9.3 from [**ENO2**] and Theorem 9.16.2 is Corollary 3.2 from [EG5].

9.18. Other results

9.18.1. Isocategorical groups. We have seen that a finite group G is completely determined by its category of (say, complex) finite dimensional representations as a symmetric tensor category. This gives rise to the question whether this is still true if we drop the symmetric structure. In other words, is a finite group G determined by its fusion category of finite dimensional complex representations?

The answer turns out to be "no". Namely, let us say that finite groups G_1 and G_2 are *isocategorical* if $\mathsf{Rep}(G_1) \cong \mathsf{Rep}(G_2)$ as fusion categories. Examples of isocategorical G_1 and G_2 were found by Davydov in [**Da4**] and later independently by Izumi and Kosaki in [**IK**] and also in [**EtG4**]. A basic example is as follows: $G_1 = \mathrm{Sp}_{2m}(\mathbb{F}_2) \ltimes \mathbb{F}_2^{2m}$ for $m \geq 3$, and G_2 is the unique nontrivial extension of $\mathrm{Sp}_{2m}(\mathbb{F}_2)$ by $\mathbb{F}_2^{2m}$ (which acts in the corresponding Weil representation). Moreover, one can show that all examples are of this type: if G_1 and G_2 are isocategorical then they are extensions of the same group G by the same G-module A over $\mathbb{F}_2$ which admits a G-invariant nondegenerate skew-symmetric isomorphism $A \to A^\vee$ of A to the character group of A (i.e., a skew-symmetric form on A). In particular, isocategorical groups always have the same order, and if this order is not divisible by 4, then they must be isomorphic.

9.18.2. Classification of weakly integral fusion categories. By Theorem 9.15.9 a weakly integral fusion category $\mathcal{C}$ such that $\mathsf{FPdim}(\mathcal{C})$ is divisible by at most two primes is weakly group-theoretical. At the moment of writing it is an open problem whether every weakly integral fusion category is weakly group-theoretical. (In particular, we do not know if the representation category of every semisimple Hopf algebra is weakly group-theoretical. Examples of non group-theoretical semisimple Hopf algebras were constructed in [**GeNN, Nik2**]). It was also shown in [**ENO3**] that a weakly group-theoretical fusion category $\mathcal{C}$ has the following Frobenius property: for every simple object X in $\mathcal{C}$ the Frobenius-Perron dimension of X divides $\mathsf{FPdim}(\mathcal{C})$ in the ring of algebraic integers. Generalization of this result to arbitrary fusion categories (in particular, to weakly integral ones) is also an open problem.

Below we list several partial results for the above-mentioned problems.

Every fusion category $\mathcal{C}$ of dimension pqr, where p, q, r are primes is weakly group-theoretical [**ENO3**]. If p, q, r are distinct odd primes then $\mathcal{C}$ is group-theoretical. Every fusion category of dimension 60 is group-theoretical [**ENO3**]. Every non-degenerate braided fusion category whose Frobenius-Perron dimension is a natural number less than 1800, or an odd natural number less than 33075, is weakly group-theoretical [**Na2**].

Proofs of results of this type are usually based on proving existence of non-trivial Tannakian subcategories of the center.

Every weakly integral fusion category of dimension less than 120 has Frobenius property [**DonNV**].

In general, classification of fusion categories of a given dimension up to equivalence is a hard problem (just like the problem of classifying finite groups or semisimple Hopf algebras of a given order is hard). Complete classification is known for fusion categories of "small" Frobenius-Perron dimension: dimension p (Corollary 9.14.5), dimension p^2 (Proposition 9.14.9), dimension pq [**EGO**], and dimension pq^2 [**JorL**].

9.18.3. Obstructions to categorification of based rings. An interesting question is whether a given based ring admits a categorification (see Definition 4.10.1). One obstruction to existence of a categorification was given in Theorem 9.3.11: if there exists a representation of a based ring such that the trace of a basis element is not cyclotomic then this ring has no categorifications. Another obstruction given in [**Os6**] is as follows: let R be the operator of left multiplication by the element $\sum_i b_i b_{i^*}$ and let $p(t)$ be the characteristic polynomial of R. Then for a categorifiable based ring a root α of $p(t)$ must be a *d-number*, that is, for any Galois automorphism σ, the number $\frac{\sigma(\alpha)}{\alpha}$ must be a unit.

These results suggest that most based rings are not categorifiable. It was conjectured in [**Os3**] that only finitely many based rings of a given rank are categorifiable. This conjecture is known to be true in rank ≤ 2 (see [**Os3**]) and in rank 3 (see [**Os7**]) under an additional assumption that the categorification admits a pivotal structure.

An interesting class of examples to test this conjecture is given by *near-group categories* introduced in [**Sie**]. Namely a fusion category is called near-group category if all its simple objects except for exactly one are invertible. It is easy to see that the Grothendieck ring of such a category is determined by the finite group G of isomorphism classes of invertible objects and one more non-negative integer n which is the multiplicity of the non-invertible object in its tensor square (thus for $n = 0$ we obtain Tambara-Yamagami fusion rings from Example 4.10.5). One can ask which pairs (G, n) admit a categorification. Some results in this direction are contained in [**EvG, I2, NikO**]. For example, it is known that if $n \geq |G|$ then G must be abelian and n must be a multiple of $|G|$. However, it is unknown if for a given group G, only finitely many values of n give rise to a categorifiable based rings.

9.18.4. C^*-tensor categories and compact quantum groups. The notion of a *C^*-tensor category* was introduced by Doplicher and Roberts in [**DoR**]. By definition, it is a tensor category $\mathcal{C}$ over complex numbers such that morphism spaces $\mathsf{Hom}_{\mathcal{C}}(U, V)$ are Banach spaces, the composition of morphisms $S \in \mathsf{Hom}_{\mathcal{C}}(U, V)$ and $T \in \mathsf{Hom}_{\mathcal{C}}(V, W)$ satisfies $\|T \circ S\| \leq \|T\|\|S\|$, and there is a contravariant conjugation functor $* : \mathcal{C} \to \mathcal{C}$ that is the identity map on objects and for any morphism S one has $S^{**} = S$ and $\|S^*S\| = \|S\|^2$ (in particular each $\mathsf{End}_{\mathcal{C}}(V)$, $V \in \mathcal{C}$, is a C^*-algebra). See the textbook [**NeT**] for an introduction to the theory of C^*-tensor categories and references. The most basic instance of a C^*-tensor category is the category Hilb of finite dimensional Hilbert spaces.

An important example of a C^*-tensor category is the category of finite dimensional unitary representations of a *compact quantum group.* The latter was defined by Woronowicz [**Wo1**] as a unital C^*-algebra A with a coassociative $*$-algebra homomorphism $\Delta : A \to A \otimes A$ satisfying the following cancellation property: the subspaces $\{(a \otimes 1)\Delta(b) \mid a, b \in A\}$ and $\{(1 \otimes a)\Delta(b) \mid a, b \in A\}$ are dense in $A \otimes A$ (note that for a finite dimensional A these conditions are equivalent to the existence of an antipode, but in general an antipode is not a part of the definition). In this setting, the role of a dual object (in the sense of Definition 2.10.1) is played by the conjugate representation.

The algebra of continuous functions on a usual compact topological group is the most general example of a compact quantum group for which the underlying C^*-algebra is commutative. Also, bimodule categories associated to finite index subfactors (cf. Section 7.25.2) are C^*-tensor categories (these categories are known under the name of Longo's Q-systems [**Lon**]).

Similarly to the reconstruction theory described in Section 5.3, compact quantum groups may be characterized as C^*-tensor categories equipped with a unitary fiber functor to Hilb (this result is known as Woronowicz's Tannaka-Krein duality) [**Wo2**]. Thus, compact quantum groups are functional analytic analogues of Hopf algebras. The theory of braided C^*-tensor categories is parallel to the one developed in the beginning of Chapter 8. A theorem by Doplicher and Roberts in [**DoR**] reconstructs a symmetric C^*-tensor category as the category of finite dimensional complex representations of a usual compact group (cf. Deligne's Theorem 9.9.22).

The Drinfeld-Jimbo deformation of a simply connected semisimple compact Lie group G (cf. Section 5.7 and Example 7.22.6) gives rise to a compact quantum group G_q (here $q > 0$, $q \neq 1$). It can be defined as the C^*-completion of the quantum function $*$-algebra $O_q(G)$ (see Definition 5.8.5) generated by matrix elements of type I representations of $U_q(\mathfrak{g})$ (where $\mathfrak{g}$ is the Lie algebra of G). The C^*-tensor category of representations of this compact quantum group is the category $\mathcal{C}_q(G)$ generated by type I representations of $U_q(\mathfrak{g})$ (see Section 5.8).

Finally, a large class of compact quantum groups appears as symmetries of non-commutative spaces. This class includes Wang's free unitary and free orthogonal quantum groups of [**WangS1**] and quantum permutation groups [**WangS2, BaBC**].

Bibliography

[AAITV] N. Andruskiewitsch, I. Angiono, A. Garcia Iglesias, B. Torrecillas, C. Vay, *From Hopf algebras to tensor categories*, Conformal field theories and toesor categories, pp. 1–32, in Mathematical Lectures from Peking University, C. Bai, J. Fuchs, Y.-Z. Huang, L. Kong, I. Runkel, C. Schweigert, editors. 2014, IX, 279 pp. arXiv:1204.5807.

[Ab] L. Abrams. Modules, comodules, and cotensor products over Frobenius algebras. *J. Algebra* **219** (1999), no. 1, 201–213.

[Ande] H. Andersen. Tensor products of quantized tilting modules. *Comm. Math. Phys.* **149** (1992), no. 1, 149–159.

[Andr] N. Andruskiewitsch. On finite-dimensional Hopf algebras. *ICM-2014 talk, arXiv:1403.7838.*

[Ang1] I. Angiono. Basic quasi-Hopf algebras over cyclic groups. *Adv. Math.* **225** (2010), 3545–3575.

[Ang2] I. Angiono. On Nichols algebras of diagonal type. *J. Reine Angew. Math.* **683** (2013), 189–251.

[AlB] D. Altschüler and A. Bruguières. Handleslide for a sovereign category. *preprint* (2001).

[AndeM] G. Anderson and G. Moore. Rationality in conformal field theory. *Comm. Math. Phys.* **117** (1988), 441–450.

[AndeP] H. Andersen and J. Paradowski. Fusion categories arising from semisimple Lie algebras. *Comm. Math. Phys.* **169** (1995), no. 3, 563–588.

[AndrM] N. Andruskiewitsch and M. Mombelli. On module categories over finite-dimensional Hopf algebras. *J. Algebra* **314** (2007), 383–418.

[AndrS] N. Andruskiewitsch and H.-J. Schneider. Pointed Hopf algebras. *MSRI series Cambridge Univ. Press* (2002), 1–68.

[AsH] M. Asaeda and U. Haagerup. Exotic subfactors of finite depth with Jones indices $(5+\sqrt{13})/2$ and $(5+\sqrt{17})/2$. *Comm. Math. Phys.* **202** (1999), no. 1, 1–63.

[AEG] N. Andruskiewitsch, P. Etingof and S. Gelaki. Triangular Hopf algebras with the Chevalley property. *Michigan Journal of Mathematics* **49** (2001), 277–298.

[Bal] B. Balsam. Turaev-Viro invariants as an extended TQFT. *Thesis (Ph.D.), State University of New York at Stony Brook* (2012).

[Ban] P. Bantay. The kernel of the modular representation and the Galois action in RCFT. *Comm. Math. Phys.* **233** (2003), no. 3, 423–438.

[Bax] R. Baxter. Partition function of the eight-vertex lattice model. *Ann. Physics* **70** (1972), 193–228.

[Ben1] J. Bénabou. Catégories avec multiplication. *C. R. Acad. Sci. Paris* **256** (1963), 1887–1890.

[Ben2] J. Bénabou. Algebre elementaire dans les categories avec multiplication. *C. R. Acad. Sci. Paris* **258** (1964), 771–774.

[Ber] J. Bernstein. Sackler lectures. *q-alg/9501032.*

[Bi] J. Bichon. The representation category of the quantum group of a non-degenerate bilinear form. *Comm. Algebra* **31** (2003), no. 10, 4831–4851.

[Bo] R. Borcherds. Vertex algebras, Kac-Moody algebras, and the Monster. *Proc. Natl. Acad. Sci. USA.* **83** (1986), 3068–3071.

[Br] K. Brown. Cohomology of groups. *Graduate Texts in Mathematics* **87** (1994).

[Bro] R. Brown. From groups to groupoids: A brief survey. *Bull. London Math. Soc.* **19** (1987), 113–134.

[Bru] A. Bruguiéres. Catégories prémodulaires, modularization et invariants des variétés de dimension 3. *Mathematische Annalen* **316** (2000), no. 2, 215–236.

[Bu] D. Buchsbaum. Exact categories and duality. *Trans. Amer. Math. Soc.* **80** (1955), 1–34.

[BakK] B. Bakalov and A. Kirillov Jr. *Lectures on Tensor categories and modular functors. University Series Lectures*, vol. 21, AMS (2001).

[BalK] B. Balsam and A. Kirillov Jr. Turaev-Viro invariants as an extended TQFT. *arXiv:1004.1533* (2010).

[BarM] J. Barrett and M. Mackaay. Categorical representations of categorical groups. *Theory and Applications of Categories* **16** (2006), no. 20, 529–557.

[BarW] J. Barrett and B. Westbury. Spherical categories. *Adv. Math.* **143** (1999), 357–375.

[BeD] A. Beilinson and V. Drinfeld. Chiral algebras. *American Mathematical Society Colloquium Publications* **51** (2004).

[BiL] L. Biedenharn and J. Louck. The Racah-Wigner Algebra in Quantum Theory. *Encyclopedia of Mathematics and its Applications* **9** (1981).

[BocE] J. Böckenhauer and D. Evans. Modular invariants and subfactors. *Fields Inst. Commun.* **30** (2001), 11–37.

[BonN] C. Bontea and D. Nikshych. On the Brauer-Picard group of a finite symmetric tensor category. *arXiv:1408.6445* (2014).

[BoyD] M. Boyarchenko and V. Drinfeld. Character sheaves on unipotent groups in positive characteristic: foundations. *Selecta Math. (N.S.)* **20** (2014), no. 1, 125–235.

[BuT] D. Bulacu and B. Torrecillas. Factorizable quasi-Hopf algebras – applications. *J. Pure Appl. Algebra* **194** (2004), no. 1-2, 39–84.

[BaBC] T. Banica, J. Bichon and B. Collins. Quantum permutation groups: a survey. In *Noncommutative harmonic analysis with applications to probability*, Banach Center Publ., **78**, Polish Acad. Sci. Inst. Math., Warsaw (2007), 13–34.

[BeiFM] A. Beilinson, B. Feigin and B. Mazur. Notes on conformal field theory. *http://www.math.sunysb.edu/~kirillov/manuscripts/bfmn.pdf.*

[BezFO] R. Bezrukavnikov, M. Finkelberg and V. Ostrik. Character D-modules via Drinfeld center of Harish-Chandra bimodules. *Invent. Math.* **188** (2012), no. 3, 589–620.

[BezFS] R. Bezrukavnikov, M. Finkelberg and V. Schechtman. Factorizable sheaves and quantum groups. *Lecture Notes in Mathematics* **1691** (1998).

[BocEK1] J. Böckenhauer, D. Evans and Y. Kawahigashi. On α-induction, chiral generators and modular invariants for subfactors, *Commun. Math. Phys.* **208** (1999), 429-487.

[BocEK2] J.Böckenhauer, D. Evans and Y. Kawahigashi. Chiral structure of modular Invariants for subfactors. *Commun. Math. Phys.* **210** (2000), 733–784.

[BocEK3] J.Böckenhauer, D. Evans and Y. Kawahigashi. Longo-Rehren subfactors arising from α-induction. *Publ. Res. Inst. Math. Sci.* **37** (2001), no. 1, 1–35.

[BohNS] G. Böhm, F. Nill and K. Szlachányi. Weak Hopf algebras I. Integral theory and C^*-structure. *J. Algebra* **221** (1999), 385–438.

[BehPPZ] R. Behrend, P. Pearce, V. Petkova and J-B. Zuber. Boundary conditions in Rational Conformal Field Theories. *Nuclear Phys. B* **579** (2000), no. 3, 707–773.

[BesKLT] Y. Bespalov, T. Kerler, V. Lyubashenko and V. Turaev. Integrals for braided Hopf algebras. *J. Pure Appl. Algebra* **148** (2000), no. 2, 113–164.

[BiMPS] S. Bigelow, S. Morrison, E. Peters and N. Snyder. Constructing the extended Haagerup planar algebra. *Acta Math.* **209** (2012), no. 1, 29–82.

[BrNRW] P. Bruillard, S.-H. Ng, E. Rowell and Z. Wang. On modular categories. *arXiv:1310.7050* (2014).

[Car1] P. Cartier. Dualité de Tannaka des groupes et algèbres de Lie. *C.R. Acad. Sci. Paris* **242** (1956), 322–325.

[Car2] P. Cartier. A primer of Hopf algebras. *Frontiers in number theory, physics, and geometry* **II** (2007), 537–615.

[Cas] J. Cassels. Local fields. *London Math. Soc. Student Texts* **3** (1986).

[Che] C. Chevalley. Théorie des groupes de Lie, III. (1954), Paris, Hermann.

[Chi] W. Chin. A brief introduction to coalgebra representation theory. In: "Hopf Algebras" *Lecture Notes in Pure and Appl. Math.* **237** (2004), 109–131.

[ChP] V. Chari and A. Pressley. A guide to quantum groups. *Cambridge University Press* (1995).

[ComO] J. Comes and V. Ostrik. On blocks of Deligne's category $\mathsf{Rep}(S_t)$. *Advances in Mathematics* **226** (2011), 1331–1377.

[CosG] A. Coste and T. Gannon. Remarks on Galois symmetry in rational conformal field theories. *Phys. Lett. B* **323** (1994), no. 3–4, 316–321.

[CraF] L. Crane and I. Frenkel. Four-dimensional topological quantum field theory, Hopf categories, and the canonical bases. *J. Math. Phys.* **35** (1994), no. 10, 5136–5154.

[CraY] L. Crane and D. Yetter. Deformations of (bi)tensor categories. *Cahiers Topologie Géom. Différentielle Catég.* **39** (1998), no. 3, 163–180.

[CreR] T. Creutzig and D. Ridout. Logarithmic Conformal field theory: Beyond an Introduction. *J. Phys. A: Math. Theor.* **46** (2013), no. 49.

[CapIZ] A. Cappelli, C. Itzykson and J. B. Zuber. The ADE classification of minimal and $A_1^{(1)}$ conformal invariant theories. *Commun. Math. Phys.* **113** (1987), 1–26.

[CarFS] J. Carter, D. Flath and M. Saito. The Classical and Quantum $6j$-Symbols. *Princeton University Press* (1995).

[Da1] A. Davydov. Twisting of monoidal structures. *arXiv:q-alg/970300, preprint MPI 95-123* (1997).

[Da2] A. Davydov. On some Hochschild cohomology classes of fusion algebras. The International Conference on Secondary Calculus and Cohomological Physics (Moscow, 1997), Diffiety Inst. Russ. Acad. Nat. Sci., Pereslavl' Zalesskiy, (1997).

[Da3] A. Davydov. Finite groups with the same character tables, Drinfeld algebras and Galois algebras. Algebra (Moscow, 1998), 99–111, de Gruyter, Berlin, (2000).

[Da4] A. Davydov. Galois algebras and monoidal functors between categories of representations of finite groups. *J. Algebra* **244** (2001), no. 1, 273–301.

[Da5] A. Davydov. Modular invariants for group-theoretical modular data. I. *J. Algebra* **323** (2010), no. 5, 1321–1348.

[Da6] A. Davydov. Bogomolov multiplier, double class-preserving automorphisms, and modular invariants for orbifolds. *J. Math. Phys.* **55**, 092305 (2014).

[De1] P. Deligne. Catégories tannakiennes. In: "The Grothendieck Festschrift, Vol. II" *Progr. Math.* **87** (1990), 111–195.

[De2] P. Deligne. Catégories tensorielles. *Moscow Math. J.* **2** (2002), no. 2, 227–248.

[De3] P. Deligne. La categorie des représentations du groupe symétrique S_t, lorsque t n'est pas un entier naturel. In: "Algebraic groups and homogeneous spaces" *Tata Inst. Fund. Res. Stud. Math.* (2007), 209–273.

[De4] P. Deligne. A letter to V. Ostrik, March 4, 2015.

[Di] J. Dieudonne. Introduction to the theory of formal groups. *Pure and Applied Mathematics* **20** (1973).

[Dr1] V. Drinfeld. Constant quasi-classical solutions of the Yang-Baxter quantum equation. *Dokl. Akad. Nauk SSSR* **273** (1983), no. 3, 531–535.

[Dr2] V. Drinfeld. Hopf algebras and the quantum Yang-Baxter equation. *Sov. Math. Dokl.* **32** (1985), 254–258. *Dokl. Akad. Nauk Ser. Fiz.* **283** (1985), 1060–1064.

[Dr3] V. Drinfeld. Quantum Groups. *Proceedings of the International Congress of Mathematicians, Berkeley* (1987), 798–820.

[Dr4] V. Drinfeld. Quasi-Hopf algebras. *Leningrad Math. J.* **1** (1989), 1419–1457.

[Dr5] V. Drinfeld. On almost cocommutative Hopf algebras. *Leningrad Math. J.* **1** (1990), 321–342.

[Dr6] V. Drinfeld. On quasi-triangular quasi-Hopf algebras and a group closely related with $Gal(\overline{\mathbb{Q}}/\mathbb{Q})$. *Leningrad Math. J.* **2** (1991), 829–860.

[DaN] A. Davydov and D. Nikshych. The Picard crossed module of a braided tensor category. *Algebra and Number Theory* **7** (2013), no. 6, 1365–1403.

[deBG] J. de Boere and J. Goeree. Markov traces and II_1 factors in conformal field theory. *Comm. Math. Phys.* **139** (1991), no. 2, 267–304.

[DelM] P. Deligne and J. Milne. Tannakian categories. *Lecture Notes in Mathematics* **900** (1982). http://www.jmilne.org/math/xnotes/tc.pdf

[DemG] M. Demazure and A. Grothendieck. Schemas en groupes, I,II,III, SGA3. *Lecture Notes in Mathematics* **151–153** (1970).

[DoR] S. Doplicher and J. Roberts. A new duality theory for compact groups. *Inventiones Mathematicae* **98** (1989), 157–218.

[DuL] M. Dubois-Violette and G. Launer. The quantum group of a non-degenerate bilinear form. *Phys. Lett. B* **245** (1990), no. 2, 175–177.

[DaNO] A. Davydov, D. Nikshych and V. Ostrik. On the structure of the Witt group of non-degenerate braided fusion categories. *Selecta Mathematica* **19** (2013), no. 1, 237–269.

[DiMS] P. Di Francesco, P. Mathieu and D. Senechal. Conformal Field Theory. *Graduate Texts in Contemp. Physics* (1997).

[DonNV] J. Dong, S. Natale and L. Vendramin. Frobenius property for fusion categories of small integral dimension. *arXiv:1209.1726.*

[DouSS] C. Douglas, C. Schommer-Pries and N. Snyder. Dualizable tensor categories. *arXiv:1312.7188.*

[DaMNO] A. Davydov, M. Müger, D. Nikshych and V. Ostrik. The Witt group of non-degenerate braided fusion categories. *Journal für die reine und angewandte Mathematik* **677** (2013), 135–177.

[DiVVV] R. Dijkgraaf, C. Vafa, E. Verlinde and H. Verlinde. The operator algebra of orbifold models. *Commun. Math. Phys.* **123** (1989), 485–526.

[DrGNO1] V. Drinfeld, S. Gelaki, D. Nikshych and V. Ostrik. Group-theoretical properties of nilpotent modular categories. e-print *arXiv:0704.0195* (2007).

[DrGNO2] V. Drinfeld, S. Gelaki, D. Nikshych and V. Ostrik. On braided fusion categories I. *Selecta Math.* **16** (2010), no. 1, 1–119.

[E] P. Etingof. On Vafa's theorem for tensor categories. *Math. Res. Lett.* **9** (2002), no. 5, 651–658.

[EnH] B. Enriquez and G. Halbout. Quantization of quasi-Lie bialgebras. *J. Amer. Math. Soc.* **23** (2010), 611–653.

[EtG1] P. Etingof and S. Gelaki. On finite dimensional semisimple and cosemisimple Hopf algebras in positive characteristic. *Internat. Math. Res. Notices* **16** (1998), 851–864.

[EtG2] P. Etingof and S. Gelaki. Some properties of finite-dimensional semisimple Hopf algebras. *Math. Res. Lett.* **5** (1998), no. 1-2, 191–197.

[EtG3] P. Etingof and S. Gelaki. The classification of triangular semisimple and cosemisimple Hopf algebras over an algebraically closed field. *Internat. Math. Res. Notices* **5** (2000), 223–234.

[EtG4] P. Etingof and S. Gelaki. Isocategorical groups. *Internat. Math. Res. Notices* **2** (2001), 59–76.

[EtG5] P. Etingof and S. Gelaki. The classification of finite-dimensional triangular Hopf algebras over an algebraically closed field of characteristic 0. *Moscow Mathematical Journal* **3** (2003), no. 1, 37–43.

[EtG6] P. Etingof and S. Gelaki. Finite-dimensional quasi-Hopf algebras with radical of codimension 2. *Math. Res. Lett.* **11** (2004), 685–696.

[EtG7] P. Etingof and S. Gelaki. On radically graded finite dimensional quasi-Hopf algebras. *Moscow Mathematical Journal* **5** (2005), 371–378.

[EtG8] P. Etingof and S. Gelaki. Liftings of graded quasi-Hopf algebras with radical of prime codimension. *Journal of Pure and Applied Algebra* **205** (2006), 310–322.

[EtG9] P. Etingof and S. Gelaki. Quasisymmetric and unipotent categories. *Math. Res. Lett.* **15** (2008), no. 5, 857–866.

[EtG10] P. Etingof and S. Gelaki. Descent and forms of tensor categories. *Int. Math. Res. Notices* **13** (2012), 3040–3063.

[EtKa1] P. Etingof and D. Kazhdan. Quantization of Lie bialgebras, I. *Selecta Math. (N.S.)* **2** (1996), no. 1, 1–41.

[EtKa2] P. Etingof and D. Kazhdan. Quantization of Lie bialgebras, II. *Selecta Math. (N.S.)* **4** (1998), 213–231.

[EtKa3] P. Etingof and D. Kazhdan. Quantization of Lie bialgebras, VI: Quantization of generalized Kac-Moody algebras. *Transf. Groups* **13** (2008), nos. 3–4, 527–539.

[EtKh] P. Etingof and M. Khovanov. Representations of tensor categories and Dynkin diagrams. *Internat. Math. Res. Notices* **5** (1995), 235–247.

[EtO1] P. Etingof and V. Ostrik. Finite tensor categories. *Moscow Math. Journal* **4** (2004), 627–654.

[EtO2] P. Etingof and V. Ostrik. Module categories over representations of $SL_q(2)$ and graphs. *Math. Res. Lett.* **11** (2004), no. 1, 103–114.

[EtS] P. Etingof and O. Schiffmann. Lectures on quantum groups. *International Press* (1998).

[EvG] D. Evans and T. Gannon. Near-group fusion categories and their doubles. *arXiv:1208.1500* (2012).

[EGO] P. Etingof, S. Gelaki and V. Ostrik. Classification of fusion categories of dimension *pq*. *Int. Math. Res. Notices* **57** (2004), 3041–3056.

[ENO1] P. Etingof, D. Nikshych and V. Ostrik. An analog of Radford's S^4 formula for finite tensor categories. *Int. Math. Res. Not.* **54** (2004), 2915–2933.

[ENO2] P. Etingof, D. Nikshych and V. Ostrik. On fusion categories. *Annals of Mathematics* **162** (2005), 581–642.

[ENO3] P. Etingof, D. Nikshych and V. Ostrik. Weakly group-theoretical and solvable fusion categories. *Advances in Mathematics* **226** (2011), no. 1, 176–205.

[ENO4] P. Etingof, D. Nikshych and V. Ostrik. Fusion categories and homotopy theory. *Quantum Topology* **1** (2010), no. 3, 209–273.

[ERW] P. Etingof, E. Rowell and S. Witherspoon. Braid representations from twisted quantum doubles of finite groups. *Pacific J. Math.* **234** (2008), no. 1, 33–42.

[EHKR] P. Etingof, A. Henriquez, J. Kamnitzer and E. Rains. The cohomology ring of the real locus of the moduli space of stable curves of genus 0 with marked points. *Annals of Mathematics* **171** (2010), no. 2, 731–777.

[Fi1] M. Finkelberg. Fusion categories. *Ph.D. thesis, Harvard University* (1993).

[Fi2] M. Finkelberg. An equivalence of fusion categories. *Geom. Funct. Anal.* **6** (1996), no. 2, 249–267. See also Erratum: *Geom. Funct. Anal.* bf 23 (2013), no. 2, 810–811.

[Fr] P. Freyd. Abelian categories. An introduction to the theory of functors. *Harper & Row, New York, Evanston, and London* (1964).

[FreB] E. Frenkel and D. Ben-Zvi. Vertex algebras and algebraic curves. *Mathematical Surveys and Monographs* (2001).

[FroK] J. Fröhlich and T. Kerler. Quantum groups, quantum categories and quantum field theory. *Lecture Notes in Math.* **1542** (1993).

[FuS] J. Fuchs and C. Schweigert. Category theory for conformal boundary conditions. *Fields Inst. Commun.* **39** (2003), 25–70.

[FaRT] L. Faddeev, N. Reshetikhin and L. Takhtajan. Quantization of Lie groups and Lie algebras. *Leningrad Math. J.* **1** (1990), no. 1, 193–225 [*Alg. i Anal.* **1** (1989), no. 1, 178–206].

[FrLM] I. Frenkel, J. Lepowsky and A. Meurman. Vertex operator algebras and the Monster. *Pure and Applied Mathematics* **134** (1988).

[FuRS1] J. Fuchs, I. Runkel and C. Schweigert. TFT construction of RCFT correlations I: Partition functions. *Nuclear Phys. B* **646** (2002), 353–497.

[FuRS2] J. Fuchs, I. Runkel and C. Schweigert. Conformal correlation functions, Frobenius algebras and triangulations. *Nuclear Phys. B* **624** (2002), 452–468.

[Gabe] M. Gaberdiel. An algebraic approach to logarithmic conformal field theory. *Internat. J. Modern Phys. A* **18** (2003), no. 25, 4593–4638.

[Gabr] P. Gabriel. Des catégories abéliennes. *Bulletin de la S. M. F.* **90** (1962), 323–448.

[Gal] C. Galindo. On braided and ribbon unitary fusion categories. Canad. Math. Bull. **57** (2014), no. 3, 506–510.

[Gann1] T. Gannon. The classification of affine $SU(3)$ modular invariant partition functions. *Comm. Math. Phys.* **161** (1994), no. 2, 233–263.

[Gann2] T. Gannon. Boundary conformal field theory and fusion ring representations. *Nucl. Phys. B* **627** (2002), 506–564.

[Gann3] T. Gannon. Modular data: the algebraic combinatorics of conformal field theory. *J. Algebraic Combin.* **22** (2005), no. 2, 211–250.

[Gant] F. Gantmacher. The theory of matrices. *AMS Chelsea Publishing, Providence* (1998).

[Gee] N. Geer. The Etingof-Kazhdan quantization of Lie superbialgebras. *Advances in Mathematics* **207** (2006), no. 1, 1–38.

[Gel1] S. Gelaki. Basic quasi-Hopf algebras of dimension n^3. *Journal of Pure and Applied Algebra* **198** (2005), 165–174.

[Gel2] S. Gelaki. Module categories over affine group schemes. *Quantum Topology*, **6** (2015), 1–37, doi: 10.4171/QT/59. *arXiv:1209.1155* (2013)

[Gel3] S. Gelaki. Twisting of affine algebraic groups, I. *International Mathematics Research Notices*, doi: 10.1093/imrn/rnu172. *arXiv:1406.2987* (2014).

[Ger] M. Gerstenhaber. On the deformation of rings and algebras. *Ann. of Math.* **79** (1964), 59–103.

[Gh] S. Ghosh. Planar algebras: a category theoretic point of view. *J. Algebra* **339** (2011), 27–54.

[Gr] A. Grothendieck. Sur quelques points dalgébre homologique. *Tóhoku Math. J. (2)* **9** (1957), 119–221.

[GabR] P. Gabriel and A. Roiter. Representations of finite dimensional algebras. *Springer* (1997).

[GalM] C. Galindo and M. Mombelli. Module categories over finite pointed tensor categories. *Selecta Math. (N.S.)* **18** (2012), no. 2, 357–389.

[GarM] A. Garcia Iglesias and M. Mombelli. Representations of the category of modules over pointed Hopf algebras over S_3 and S_4. *Pacific J. Math.* **252** (2011), no. 2, 343–378.

[GelaN] S. Gelaki and D. Nikshych. Nilpotent fusion categories. *Advances in Mathematics* **217** (2008), no. 3, 1053–1071.

[GelfK] S. Gelfand and D. Kazhdan. Examples of tensor categories. *Invent. Math.* **109** (1992), no. 3, 595–617.

[GelfM] S. Gelfand and Yu. Manin. Methods of homological algebra. *Springer* (2003).

[GeoM] G. Georgiev and O. Mathieu. Fusion rings for modular representations of Chevalley groups. *Contemp. Math.* **175** (1994), 89–100.

[GeNN] S. Gelaki, D. Naidu and D. Nikshych. Centers of graded fusion categories. *Algebra and Number Theory* **3** (2009), no. 8, 959–990.

[GodKO] P. Goddard, A. Kent and D. Olive. Unitary representations of the Virasoro algebra and super-Virasoro algebras. *Comm. Math. Phys.* **103** (1986), 105–119.

[GooHJ] F. Goodman, P. de la Harpe and V. Jones. Coxeter graphs and towers of algebras. *M.S.R.I. Publ.* **14** (1989).

[He] I. Heckenberger. Classification of arithmetic root systems. *Adv. Math.* **220** (2009), 59–124.

[Hi] D. Higman. Indecomposable representations at characteristic *p*. *Duke Math. J.* **21** (1954), 377–381.

[Ho] H. Hopf. Über die Topologie der Gruppen-Mannigfaltigkeiten und ihrer Verallgemeinerungen. *Ann. of Math.* **42** (1941), 22–52. Reprinted in Selecta Heinz Hopf, 119–151, Springer, Berlin (1964).

[HaN] F. Hausser and F. Nill. Integral theory for quasi-Hopf algebras. *math.QA/9904164* (1999).

[HaR] T. Halverson and A. Ram. Partition algebras. *European J. Combin.* **26** (2005), 869–921.

[HecV] I. Heckenberger and L. Vendramin. A classification of Nichols algebras of semi-simple Yetter-Drinfeld modules over non-abelian groups. *arXiv:1412.0857.*

[HenK] A. Henriques and J. Kamnitzer. Crystals and coboundary categories. *Duke Math. J.* **132** (2006), no. 2, 191–216.

[HiS] P. Hilton and U. Stammbach. A course in homological algebra (second edition). *Graduate Texts in Mathematics* **4** (1997).

[HoI] R. Howlett and I. Isaacs. On groups of central type. *Math. Z.* **179** (1982), 555–569.

[HuL] Y.-Z. Huang and J. Lepowsky. Tensor categories and the mathematics of rational and logarithmic conformal field theory. *J. Phys. A* **46** (2013), no. 49.

[I1] M. Izumi. The structure of sectors associated with Longo-Rehren inclusions, II. Examples. *Rev. Math. Phys.* **13** (2001), 603–674.

[I2] M. Izumi. Notes on near-group categories. *preprint.*

[IK] M. Izumi and H. Kosaki. On a subfactor analog of the second cohomology. *Rev. Math. Phys.* **14** (2002), 733–737.

[Ja] J. Jantzen. Lectures on quantum groups. *Graduate Studies in Mathematics* **6** (1995).

[Ji] M. Jimbo. A q-difference analog of $U(\mathfrak{g})$ and the Yang-Baxter equation. *Lett. Math. Phys.* **10** (1985), 63–69.

[Jon1] V. Jones. Index for subfactors. *Invent. Math.* **72** (1983), 1–25.

[Jon2] V. Jones. A polynomial invariant for knots via von Neumann algebras. *Bull. Amer. Math. Soc. (N.S.)* **12** (1985), no. 1, 103–111.

[Jon3] V. Jones. The annular structure of subfactors. *Essays on geometry and related topics* **1**, **2**, 401–463, Monogr. Enseign. Math., **38** (2001).

[Jon4] V. Jones, The Potts model and the symmetric group, in: Subfactors: Proceedings of the Taniguchi Symposium on Operator Algebras (Kyuzeso, 1993), River Edge, NJ, World Sci. Publishing, 1994, pp. 259–267.

[Jos] A. Joseph. Quantum groups and their primitive ideals. *Ergebnisse der Mathematik* **29** (1995).

[JorL] D. Jordan and E. Larson. On the classification of certain fusion categories. *Journal of Noncommutative Geometry* **3**, no. 3, (2009), 481–499.

[JoyS1] A. Joyal and R. Street. Braided monoidal categories. *Macquarie Mathematics Reports* **860081** (1986).

[JoyS2] A. Joyal and R. Street. Tortile Yang-Baxter operators in tensor categories. *Journal of Pure and Applied Algebra* **71** (1991), no.1, 43–51.

[JoyS3] A. Joyal and R. Street. The geometry of tensor calculus, I. *Adv. Math.* **88** (1991), 55–112.

[JoyS4] A. Joyal and R. Street. An introduction to Tannaka duality and quantum groups. *Lectures Notes in Mathematics* **1488** (1991), 411–492.

[JoyS5] A. Joyal and R. Street. Braided tensor categories. *Adv. Math.* **102** (1993), 20–78.

[JMS] V. Jones, S. Morrison and N. Snyder. The classification of subfactors of index at most 5. *Bull. Amer. Math. Soc. (N.S.)* **51** (2014), no. 2, 277–327.

[KacG1] G. Kac. Ring groups and the principle of duality, I, II. *Trans. Moscow Math. Soc.* **12** (1963), 291–339; **13** (1965), 94–126.

[KacG2] G. Kac. Extensions of groups to ring groups. *Math. USSR Sbornik* **5** (1968), 451–474.

[KacG3] G. Kac. Arithmetic properties of ring groups. *Funct. Anal. Appl.* **6** (1972), 158–160.

[KacV] V. Kac. Vertex algebras for beginners. *University Lecture Series* **10** (1998), 2nd ed.

[Kas] C. Kassel. Quantum groups. *Graduate Texts in Mathematics* **155** (1995).

[Kat] A. Kato. Classification of modular invariant partition functions in two dimensions. *Modern Phys. Lett. A* **2** (1987), 585–600.

[Ke] G. Kelly. On Mac Lane's conditions for coherence of natural associativities, commutativities, etc. *J. Algebra* **1** (1964), 397–402.

[Kir1] A. Kirillov, Jr. Modular categories and orbifold models II. *arXiv:0110221.*

[Kir2] A. Kirillov, Jr. On Inner products in modular tensor categories, I. *J. Amer. Math. Soc.* **9** (1996), 1135–1169.

[Kit1] A. Kitaev. Fault-tolerant quantum computation by anyons. *Ann. Physics* **303** (2003), no. 1, 2–30.

[Kit2] A. Kitaev. Anyons in an exactly solved model and beyond. *Ann. Physics* **321** (2006), no. 1, 2–111.

[Koi] K. Koike. On the decomposition of tensor products of the representations of classical groups: by means of universal characters. *Adv. Math.* **74** (1989), 57–86.

[Kon] L. Kong. Anyon condensation and tensor categories. *Nuclear Physics B* **886** (2014), 436–482.

[Kos] B. Kostant. Graded manifolds, graded Lie theory and prequantization. In: "Diff. Geom. Methods in Math. Phys.". *Lecture Notes in Math.* **570** (1977), 177–306.

[Kr] M. Krein. A principle of duality for a bicompact group and square block algebra. *Dokl. Akad. Nauk. SSSR* **69** (1949), 725–728.

[Ku] G. Kuperberg. Finite, connected semisimple, rigid tensor categories are linear. *Math. Res. Lett.* **10** (2003), no. 4, 411–421.

[KacGP] G. Kac and V. Paljutkin. Finite ring groups. *Trans. Moscow Math. Soc.* (1966), 251–294.

[KacVW] V. Kac and M. Wakimoto. Modular and conformal invariance constraints in representation theory of affine algebras. *Adv. Math.* **70** (1988), no. 2, 156–236.

[KashS] M. Kashiwara and P. Schapira. Categories and sheaves. *Springer* (2006).

[KassT] C. Kassel and V. Turaev. Braid groups. *Graduate Texts in Mathematics* **247** (2008).

[KazL1] D. Kazhdan and G. Lusztig. Affine Lie algebras and quantum groups. *Duke Math. J., IMRN* **2** (1991), 21–29.

[KazL2] D. Kazhdan and G. Lusztig. Tensor structures arising from affine Lie algebras, I. *J. Amer. Math. Soc.* **6** (1993), 905-947.

[KazL3] D. Kazhdan and G. Lusztig. Tensor structures arising from affine Lie algebras, II. *J. Amer. Math. Soc.* **6** (1993), 949–1011.

[KazL4] D. Kazhdan and G. Lusztig. Tensor structures arising from affine Lie algebras, III. *J. Amer. Math. Soc.* **7** (1994), 335–381.

[KazL5] D. Kazhdan and G. Lusztig. Tensor structures arising from affine Lie algebras, IV. *J. Amer. Math. Soc.* **7** (1994), no. 2, 383–453.

[KazW] D. Kazhdan and H. Wenzl. Reconstructing monoidal categories. *I. M. Gelfand Seminar, Adv. Soviet Math.* **16**, Part 2, (1993), 111–136.

[KirO] A. Kirillov Jr. and V. Ostrik. On q-analog of McKay correspondence and ADE classification of $\hat{\mathfrak{sl}}_2$ conformal field theories. *Adv. Math.* **171** (2002), no. 2, 183–227.

[KitK] A. Kitaev and L. Kong. Models for gapped boundaries and domain walls. *Comm. Math. Phys.* **313** (2012), no. 2, 351–373.

[KoS] L. Korogodsky and Y. Soibelman. Algebras of functions on quantum groups. *Amer. Math. Soc.* (1997).

[KaSZ] Y. Kashina, Y. Sommerhaeuser and Y. Zhu. On higher Frobenius-Schur indicators. *Mem. Amer. Math. Soc.* **181** (2006), no. 855, viii+65 pp.

[KiSV] A. Kitaev, A. Shen and M. Vyalyi. Classical and quantum computation. *Graduate Studies in Mathematics* **47** (2002).

[KoKR] R. Koenig, G. Kuperberg and B. Reichardt. Quantum computation with Turaev-Viro codes. *Ann. Physics* **325** (2010), no. 12, 2707–2749.

[Le] T. Leinster. Higher operads, higher categories. *Cambridge University Press, Lecture Notes Series* **298** (2004).

[Lon] R. Longo. A duality for Hopf algebras and for subfactors I. *Comm. Math. Phys.*, **159** (1994), no. 1, 133–150.

[Loo] E. Looijenga. From WZW models to modular functors. *Handbook of moduli. Vol. II, Adv. Lect. Math.* **25** (2013), 427–466,

[Lor] M. Lorenz. Representations of finite dimensional Hopf algebras. *J. Algebra* **188** (1997), no. 2, 476–505.

[Lur] J. Lurie. On the classification of topological field theories. *Current developments in mathematics* (2008), 129–280.

[Lus1] G. Lusztig. Character sheaves, I-V. *Advances in Mathematics* **56** (1985), no. 3, 193–237; **57** (1985), no. 3, 226–315; **59** (1986), no. 1, 1–63; **61** (1986), no. 2, 103–155.

[Lus2] G. Lusztig. Leading coefficients of character values of Hecke algebras. *Proc. Symp. in Pure Math.* **47** (1987), 235–262.

[Lus3] G. Lusztig. On quantum groups at roots of 1. *Geom. Dedicata* **35** (1990), 89–113.

[Lus4] G. Lusztig. Exotic Fourier transform. *Duke Math. J.* **73** (1994), no. 1, 227–241.

[Lus5] G. Lusztig. Introduction to Quantum groups. *Birkhauser* (2010).

[Lus6] G. Lusztig. Truncated convolution of character sheaves. *Bull. Inst. Math. Acad. Sinica (N.S.)*, to appear.

[Ly1] V. Lyubashenko. Hopf algebras and vector symmetries. *Uspekhi Mat. Nauk* **41** (1986), no. 5, 185–186.

[Ly2] V. Lyubashenko. Modular transformations for tensor categories. *J. Pure Appl. Algebra* **98** (1995), no. 3, 279–327.

[LaR1] R. Larson and D. Radford. Semisimple cosemisimple Hopf algebras. *Amer. J. Math.* **109** (1987), 187–195.

[LaR2] R. Larson and D. Radford. Finite dimensional cosemisimple Hopf algebras in characteristic 0 are semisimple. *J. Algebra* **117** (1988), no. 2, 267–289.

[LaS] R. Larson and M. Sweedler. An associative orthogonal bilinear form for Hopf algebras. *American J. of Math.* **91** (1969), 75–94.

[LeW] M. Levin and X.-G. Wen. String-net condensation: A physical mechanism for topological phases. *Phys. Rev. B* **71**, no. 4, (2005).

[Mac1] S. Mac Lane. Natural associativity and commutativity. *Rice Univ. Studies* **49** (1963), no. 4, 28–46.

[Mac2] S. Mac Lane. Categories for the working mathematician (2nd edition). *Graduate texts in mathematics* **5** (1998).

[Maj1] S. Majid. Representations, duals and quantum doubles of monoidal categories. *Rend. Circ. Math. Palermo (2)* Suppl., **26** (1991), 197–206.

[Maj2] S. Majid. Foundations of quantum group theory. *Cambridge University Press* (2000).

[Maj3] S. Majid. Cross Products by Braided Groups and Bosonization, *J. Algebra*, **163** (1994), no. 1, 165–190.

[Maj4] S. Majid, Braided groups and algebraic quantum field theories, Lett. Math. Phys., 22 (1991), pp. 167–176.

[Mal] G. Malle. An exotic Fourier transform for H_4. *Duke J. Math.* **73** (1994), 243–248.

[Mar1] P. Martin, Potts models and related problems in statistical mechanics, Series on Advances in Statistical Mechanics, vol. 5, World Scientific Publishing Co. Inc., Teaneck NJ, 1991.

[Mar2] P. Martin, Temperley-Lieb algebras for nonplanar statistical mechanics the partition algebra construction, J. Knot Theory Ramifications 3 (1994), pp. 51–82.

[Mas] A. Masuoka. The p^n theorem for semisimple Hopf algebras. *Proc. Amer. Math. Soc.* **124** (1996), 187–195.

[Mit] B. Mitchell. The full imbedding theorem. *Amer. J. Math.* **86** (1964), 619–637.

[Miy1] M. Miyamoto. A theory of tensor products for vertex operator algebra satisfying C_2-cofiniteness. *arXiv:math/0309350.*

[Miy2] M. Miyamoto. Flatness and semi-rigidity of vertex operator algebras. *arXiv:1104.4675.*

[Mom1] M. Mombelli. Module categories over pointed Hopf algebras. *Math. Z.* **266** (2010), no. 2, 319–344.

[Mom2] M. Mombelli. Representations of tensor categories coming from quantum linear spaces. *J. Lond. Math. Soc. (2)* **83** (2011), no. 1, 19–35.

[Mon] S. Montgomery. Hopf algebras and their actions on rings. *CBMS Regional Conference Series in Mathematics* **82** (1993).

[Mov] M. Movshev. Twisting in group algebras of finite groups. *Funct. Anal. Appl.* **27** (1994), 240–244.

[Mu1] M. Müger. Galois theory for braided tensor categories and the modular closure. *Adv. Math.* **150** (2000), no. 2, 151–201.

[Mu2] M. Müger. From subfactors to categories and topology I. Frobenius algebras in and Morita equivalence of tensor categories. *J. Pure Appl. Algebra* **180** (2003), 81–157.

[Mu3] M. Müger. On the structure of modular categories. *Proc. Lond. Math. Soc.* **87** (2003), 291–308.

[Mu4] M. Müger. From subfactors to categories and topology II. The quantum double of tensor categories and subfactors. *J. Pure Appl. Algebra* **180** (2003), no. 1-2, 159–219.

[Mu5] M. Müger. Galois extensions of braided tensor categories and braided crossed G-categories. *J. Algebra* **277** (2004), no. 1, 256–281.

[Mu6] M. Müger. Tensor categories: A selective guided tour. *Rev. Un. Mat. Argentina* **51** (2010), no. 1, 95–163.

[MiM] J. Milnor and J. Moore. On the structure of Hopf algebras. *Ann. of Math.* **81** (1965), 211–264.

[MooS1] G. Moore and N. Seiberg. Classical and quantum conformal field theory. *Commun. Math. Phys.* **123** (1989), 177–254.

[MooS2] G. Moore and N. Seiberg. Lectures on RCFT. *Physics, Geometry, and Topology* (1990), 263–361.

[MorS] S. Morrison and N. Snyder. Non-cyclotomic fusion categories. *Trans. Amer. Math. Soc.* **364** (2012), no. 9, 4713–4733.

[Na1] S. Natale. On group theoretical Hopf algebras and exact factorizations of finite groups. *J. Algebra* **270** (2003), 199–211.

[Na2] S. Natale. On weakly group-theoretical non-degenerate braided fusion categories. *arXiv:1301.6078* (2013).

[Ng] S.-H. Ng. Non-semisimple Hopf algebras of dimension p^2. *J. of Algebra* **255** (2002), no. 1, 182–197.

[Nic] W. Nichols. Bialgebras of type one. *Comm. Algebra* **6** (1978), no. 15, 1521–1552.

[Nik1] D. Nikshych. On the structure of weak Hopf algebras. *Adv. Math.* **170** (2002), 257–286.

[Nik2] D. Nikshych. Non group-theoretical semisimple Hopf algebras from group actions on fusion categories. *Selecta Math.* **14** (2008), 145–161.

[NaR] D. Naidu and E. Rowell. A finiteness property for braided fusion categories. *Algebras and Representation Theory* **14** (2011), no. 5, 837–855.

[NeT] S. Neshveyev and L. Tuset. Compact quantum groups and their representation categories. *Cours Spécialisés* **20**. Société Mathématique de France, Paris (2013).

[NgS1] S.-H. Ng and P. Schauenburg. Frobenius-Schur indicators and exponents of spherical categories. *Adv. Math.* **211** (2007), no. 1, 34–71.

[NgS2] S.-H. Ng and P. Schauenburg. Congruence subgroups and generalized Frobenius-Schur indicators. *Comm. Math. Phys.* **300** (2010), no. 1, 1–46.

[NicZ] W. Nichols and M. Zoeller. A Hopf algebra freeness theorem. *Amer. J. Math.* **111** (1989), 381–385.

[NikO] D. Nikshych and V. Ostrik. On the structure of near-group categories. *preprint.*

[NikR] D. Nikshych and B. Riepel. Categorical Lagrangian Grassmannians and Brauer-Picard groups of pointed fusion categories. *J. Algebra* **411** (2014), 191–214.

[NikV1] D. Nikshych and L. Vainerman. Finite quantum groupoids and their applications. *MSRI Publications* **43** (2002), 211–262.

[NikV2] D. Nikshych and L. Vainerman. A characterization of depth 2 subfactors of II_1 factors. *J. Funct. Anal.* **171** (2000), no. 2, 278–307.

[NikV3] D. Nikshych and L. Vainerman. A Galois correspondence for II_1 factors and quantum groupoids. *J. Funct. Anal.* **178** (2000), no. 1, 113–142.

[NSSFS] C. Nayak, S. H. Simon, A. Stern, M. Freedman, and D. S. Sarma, *Non-Abelian anyons and topological quantum computation.* Review of Modern Physics (80), p.1083, 2008.

[NTV] D. Nikshych, V. Turaev and L. Vainerman. Invariants of knots and 3-manifolds from quantum groupoids. *Topology and its Applications* **127** (2003), 91–123.

[Oc1] A. Ocneanu. Quantized groups, string algebras, and Galois theory for algebras. *London Math. Soc. Lecture Notes Series* **136** (1989), 119–172.

[Oc2] A. Ocneanu. Paths on Coxeter diagrams: From Platonic solids and singularities to minimal models and subfactors. *Lectures on operator theory, the Fields Institute Monographs, Amer. Math. Soc.* (2000), 243–323.

[Oc3] A. Ocneanu. The classification of subgroups of quantum SU(N). *Quantum symmetries in theoretical physics and mathematics (Bariloche, 2000)*, 133–159, Contemp. Math., **294**, Amer. Math. Soc., Providence, RI, 2002.

[Oh1] C. Ohn. Quantum $SL(3,\mathbb{C})$'s with classical representation theory. *J. Algebra* **213** (1999), no. 2, 721–756.

[Oh2] C. Ohn. Quantum $SL(3,\mathbb{C})$'s: the missing case. *Lecture Notes in Pure and Appl. Math.* **239** (2005), 245–255.

[Os1] V. Ostrik. Module categories, weak Hopf algebras and modular invariants. *Transform. Groups* **8** (2003), 177–206.

[Os2] V. Ostrik. Module categories over the Drinfeld double of a finite group. *Int. Math. Res. Not.* (2003), 1507–1520.

[Os3] V. Ostrik. Fusion categories of rank 2. *Math. Res. Lett.* **10** (2003), 177–183.

[Os4] V. Ostrik. Tensor categories (after P. Deligne). *arXiv:math/0401347.*

[Os5] V. Ostrik. Module categories over representations of $SL_q(2)$ in the non-semisimple case. *Geom. Funct. Anal.* **17** (2008), no. 6, 2005–2017.

[Os6] V. Ostrik. On formal codegrees of fusion categories. *Math. Res. Lett.* **16** (2009), no. 5, 899–905.

[Os7] V. Ostrik. Pivotal fusion categories of rank 3. *arXiv:1309.4822* (2013).

[Os8] V. Ostrik. On symmetric fusion categories in positive characteristic, *arXiv:1503.01492.*

[Pa] B. Pareigis. On braiding and dyslexia. *J. Algebra* **171** (1995), no. 2, 413–425.

[Po1] S. Popa. Symmetric enveloping algebras, amenability and AFD properties for subfactors. *Math. Res. Lett.* **1** (1994), no. 4, 409–425.

[Po2] S. Popa. An axiomatization of the lattice of higher relative commutants of a subfactor. *Invent. Math.* **120** (1995), no. 3, 427–445.

[PZ] V. Petkova and J. Zuber. The many faces of Ocneanu cells. *Nuclear Physics B* **603** (2001), no. 3, 449–496.

[Ra1] D. Radford. On the coradical of a finite dimensional Hopf algebra. *Proc. of the AMS* **53** (1975), no. 1, 9–15.

[Ra2] D. Radford. The order of the antipode of a finite dimensional Hopf algebra is finite. *Amer. J. Math.* **98** (1976), 333–355.

[Ra3] D. Radford. Minimal quasi-triangular Hopf algebras. *J. Algebra* **157** (1993), 285–315.

[Ra4] D. Radford. On Kauffman's knot invariants arising from finite-dimensional Hopf algebras. *Advances in Hopf algebras, Lecture Notes in Pure and Appl. Math.* **158** (1994), 205–266.

[Ra5] D. Radford. Hopf algebras. *World Scientific* (2011).

[Ra6] D. Radford. The structure of Hopf algebras with a projection. *J. Algebra* **92** (1985), 322–347.

[Ren] J. Renault. A groupoid approach to C^*-algebras. *Lecture Notes in Math.* **793** (1980).

[Res] N. Reshetikhin. Quasitriangular Hopf algebras and invariants of links. (Russian) Algebra i Analiz 1 (1989), no. 2, 169–188; translation in Leningrad Math. J. 1 (1990), no. 2, 491–513.

[RS] N. Reshetikhin and M. Semenov-Tian-Shansky. Quantum R-matrices and factorization problems. *J. Geom. Phys.* **5** (1988), 533–550.

[RT1] N. Reshetikhin and V. Turaev. Ribbon graphs and their invariants derived from quantum groups. *Comm. Math. Phys.* **127** (1990), 1–26.

[RT2] N. Reshetikhin and V. Turaev. Invariants of 3-manifolds via link polynomials and quantum groups. *Invent. Math.* **103** (1991), 547–597.

[Sa] N. Saavedra Rivano. Catégoties Tannakiennes. *Lecture Notes in Math.* **265** (1972).

[Schar] W. Scharlau. Quadratic and Hermitian Forms. *Grundlehren der mathematischen Wissenschaften* **270** (1985).

[Schau1] P. Schauenburg. Tannaka duality for arbitrary Hopf algebras. *Algebra Berichte* [Algebra Reports] **66** (1992), 57 pp.

[Schau2] P. Schauenburg. Bialgebras over noncommutative rings and a structure theorem for Hopf bimodules. *Applied Categorical Structures* **6** (1998), 193-222.

[Schau3] P. Schauenburg. A quasi-Hopf algebra freeness theorem. *Proc. Amer. Math. Soc.* **132** (2004), no. 4, 965–972.

[Schau4] P. Schauenburg. Quotients of finite quasi-Hopf algebras. In: "Hopf algebras in noncommutative geometry and physics". *Lecture Notes in Pure and Appl. Math.* **239** (2005), 281–290.

[Se] J.-P. Serre. Local Fields. *Graduate Texts in Mathematics* **67** (1979).

[Sh] M. Shum. Tortile tensor categories. *Ph.D. Thesis, Macquarie University* (1989).

[Sie] J. Siehler. Near-group categories. *Alg. Geom. Topol.* **3** (2003), 719–775.

[Sin] H. Sinh. Gr-categories. *Université Paris VII doctoral thesis* (1975).

[Stef] D. Stefan. The set of types of n-dimensional semisimple and cosemisimple Hopf algebras is finite. *J. Algebra* **193** (1997), no. 2, 571–580.

[Ster] A. Stern. Anyons and the quantum Hall effect - A pedagogical review. *Annals of Physics* **323** (2008), 204–249.

[Sw] M. Sweedler. Hopf algebras. *Mathematics Lecture Note Series* (1969), W. A. Benjamin, Inc., New York.

[Sz] K. Szlachányi. Finite quantum groupoids and inclusions of finite type. *Fields Inst. Commun.* **30** (2001), 393–407.

[SaS] Š. Sakáloš and P. Ševera. On quantization of quasi-Lie bialgebras. *arXiv:1304.6382* (2013).

[ScheW] A. Schellekens and N. Warner. Conformal subalgebras of Kac-Moody algebras. *Phys. Rev. D* **34** (1986), no. 10, 3092–3096.

[ScheY] A. Schellekens and S. Yankielowicz. Field identification fixed points in the coset construction. *Nucl. Phys. B* **334** (1990), 67–102.

[SchwF] C. Schweigert and J. Fuchs. Hopf algebras and Frobenius algebras in finite tensor categories. *Progr. Math.* **295** (2012), 189–203.

[SoZ] Y. Sommerhauser and Y. Zhu. Hopf algebras and congruence subgroups. *Mem. Amer. Math. Soc.* **219** (2012), no. 1028, vi+134 pp.

[Taf] E. Taft. The order of the antipode of finite-dimensional Hopf algebra. *Proc. Nat. Acad. Sci. U.S.A.* **68** (1971), 2631–2633.

[Tak1] M. Takeuchi. Free Hopf algebras generated by coalgebras. *J. Math. Soc. Jap.* **23** (1971), 561–582.

[Tak2] M. Takeuchi. Morita Theorems for categories of comodules. *J. Fac. Sci. Univ. Tokyo* **24** (1977), 629–644.

[Tan] T. Tannaka. Über den Dualitätssatz der nichtkommutativen topologischen Gruppen. *Tohoku Math. J.* **45** (1939), 1–12.

[Tu1] V. Turaev. The Yang-Baxter equation and invariants of links. *Invent. Math.* **92** (1988), 527–553.

[Tu2] V. Turaev. Operator invariants of tangles, and R-matrices (Russian). *Izv. Akad. Nauk SSSR Ser. Mat.* **53** (1989), no. 5, 1073–1107, 1135; translation in *Math. USSR-Izv.* **35** (1990), no. 2, 411–444.

[Tu3] V. Turaev. Modular categories and 3-manifold invariants. *Int. J. Modern Phys. B* **6** (1992), 1807–1824.

[Tu4] V. Turaev. Quantum invariants of knots and 3-manifolds. *W. de Gruyter* (1994).

[Tu5] V. Turaev. Homotopy field theory in dimension 3 and crossed group-categories. *arXiv:math/0005291.*

[Tu6] V. Turaev. Crossed group-categories. *Arabian Journal for Science and Engineering* **33** (2008), no. 2C, 484–503.

[TaY] D. Tambara and S. Yamagami. Tensor categories with fusion rules of self-duality for finite abelian groups. *J. Algebra* **209** (1998), no. 2, 692–707.

[TubW] I. Tuba and H. Wenzl. On braided tensor categories of type BCD. *J. Reine Angew. Math.* **581** (2005), 31–69.

[TurVire] V. Turaev and A. Virelizier. On two approaches to 3-dimensional TQFTs. *arXiv:1006.3501* (2010).

[TurViro] V. Turaev and O. Viro. State sum invariants of 3-manifolds and quantum $6j$-symbols. *Topology* **31** (1992), no. 4, 865–902.

[TUY] A. Tsuchiya, K. Ueno and Y. Yamada. Conformal field theory on universal family of stable curves with gauge symmetries. *Adv. Stud. Pure Math.* **19** (1989), 459–566.

[Vaf] C. Vafa. Toward classification of conformal theories. *Physics Letters B* **206** (1988), no. 3, 421–426.

[Var] A. Varchenko. Multidimensional hypergeometric functions and representation theory of Lie algebras and quantum groups. *Advances in Math. Phys.* **21** (1995).

[Verl] E. Verlinde. Fusion rules and modular transformations in 2D conformal field theory. *Nucl. Phys. B* **300** (1988), 360–376.

[Vers] D. Verstegen. New exceptional modular invariant partition functions for simple Kac-Moody algebras. *Nucl. Phys. B* **346** (1990), 349–386.

[WangS1] S. Wang. Free products of compact quantum groups. *Comm. Math. Phys.* **167** (1995), 671–692.

[WangS2] S. Wang. Quantum symmetry groups of finite spaces. *Comm. Math. Phys.* **195** (1998), 195–211.

[WangZ] Z. Wang. Topological quantum computation. *CBMS Regional Conference Series in Mathematics* **112** (2010), xiv+115 pp.

[Weib] C. Weibel. An introduction to homological algebra. *Cambridge University Press* (1994).

[Wein] A. Weinstein. Groupoids: unifying internal and external symmetry – a tour through some examples. *Notices of the AMS* **43**, no. 7.

[Wen1] H. Wenzl. On the structure of Brauer's centralizer algebras. *Ann. of Math.* (2) **128** (1988), no. 1, 173–193.

[Wen2] H. Wenzl. C^*- tensor categories from quantum groups. *J. Amer. Math. Soc.* **11** (1998), 261–282.

[Wi] F. Wilczek. Fractional statistics and anyon superconductivity. *World Scientific Publishing Co., Inc., Teaneck, NJ* (1990), x+447 pp.

[Wo1] S. Woronowicz. Compact matrix pseudogroups. *Comm. Math. Phys.* **111** (1987), 613–665.

[Wo2] S. Woronowicz. Tannaka-Krein duality for compact matrix pseudogroups. Twisted SU(N) groups. *Invent. Math.* **93** (1988), 35–76.

[X1] F. Xu. Standard λ-lattices from quantum groups. *Invent. Math.* **134** (1998), no. 3, 455–487.

[X2] F. Xu. Some computations in the cyclic permutations of completely rational nets. *Comm. Math. Phys.* **267** (2006), no. 3, 757–782.

[Ya] C. Yang. Some exact results for the many-body problem in one dimension with repulsive delta-function interaction. *Phys. Rev. Lett.* **19** (1967), 1312–1315.

[Ye1] D. Yetter. Quantum groups and representations of monoidal categories. *Math. Proc. Cambridge Philos. Soc.* **108** (1990), no. 2, 261–290.

[Ye2] D. Yetter. Markov algebras. In: "Braids" (Santa Cruz, CA, 1986), *Contemp. Math.* **78** (1988), 705–730.

[Ye3] D. Yetter. Braided deformations of monoidal categories and Vassiliev invariants. In: "Higher category theory". *Contemp. Math. AMS* **230** (1998), 117–134.

[Ye4] D. Yetter. Abelian categories of modules over a (Lax) monoidal functor. *Adv. Math.* **174** (2003), no. 2, 266–309.

[Z] Y. Zhu. Hopf algebras of prime dimension. *Int. Math. Res. Not.* **1** (1994), 158–160.

Index